Modern Control Technology

COMPONENTS AND SYSTEMS

by

Christopher T. Kilian

Anne Arundel Community College

West Publishing Company

Minneapolis/St. Paul • New York • Los Angeles • San Francisco

PRODUCTION CREDITS

Production management: Michael Bass & Associates

Copyediting: Betty Duncan

Text design: R. Kharibian & Associates

Compositor: Publication Services of Boston

WEST'S COMMITMENT TO THE ENVIRONMENT

In 1906, West Publishing Company began recycling materials left over from the production of books. This began a tradition of efficient and responsible use of resources. Today, 100% of our legal bound volumes are printed on acid-free, recycled paper consisting of 50% new paper pulp and 50% paper that has undergone a de-inking process. We also use vegetable-based inks to print all of our books. West recycles nearly 27,700,000 pounds of scrap paper annually—the equivalent of 229,300 trees. Since the 1960s, West has devised ways to capture and recycle waste inks, solvents, oils, and vapors created in the printing process. We also recycle plastics of all kinds, wood, glass, corrugated cardboard, and batteries, and have eliminated the use of polystyrene book packaging. We at West are proud of the longevity and the scope of our commitment to the environment.

West pocket parts and advance sheets are printed on recyclable paper and can be collected and re-cycled with newspapers. Staples do not have to be removed. Bound volumes can be recycled after removing the cover.

Production, Prepress, Printing and Binding by West Publishing Company.

 TEXT IS PRINTED ON 10% POST CONSUMER RECYCLED PAPER  Printed with **Printwise** Environmentally Advanced Water Washable Ink

British Library Cataloguing-in-Publication Data. A catalogue record for this book is available from the British Library.

Copyright © 1996   By WEST PUBLISHING COMPANY
                   610 Opperman Drive
                   P.O. Box 64526
                   St. Paul, MN 55164-0526

All rights reserved

Printed in the United States of America

03 02 01 00 99 98 97 96   8 7 6 5 4 3 2 1 0

LIBRARY OF CONGRESS CATALOGING-IN-PUBLICATION DATA

Kilian, Christopher T.
    Modern control technology components and systems / Christopher T. Kilian.
      p.   cm.
    Includes index.
    ISBN 0-314-06631-4 (hard : alk. paper)
    1. Automatic control.   I. Title.
TJ213.K516   1996
629.8—dc20
                                            95-49816
                                              CIP

I dedicate this book to my wife, Teresa W. Kilian,
and my children
Nancy, Laura, and Geoffrey.

CONTENTS

Preface   **xv**

CHAPTER 1   **Introduction to Control Systems**   1

   1.1   Control Systems   2

         Introduction and Background   2
         Open-Loop Control Systems   3
         Closed-Loop Control Systems   5
         Transfer Functions   6

   1.2   Analog and Digital Control Systems   9

   1.3   Classifications of Control Systems   10

         Process Control   10
             Sequentially Controlled Systems   13
         Servomechanisms   14
             Numerical Control   15
             Robotics   16

CHAPTER 2   **Introduction to Microprocessor-Based Control**   21

   2.1   Introduction to Microprocessor System Hardware   23
   2.2   Introduction to Microprocessor Operation   25
   2.3   Interfacing to a Microprocessor Controller   27

         The Parallel Interface   27
             Digital-to-Analog Conversion   27
             Analog-to-Digital Conversion   29
             A Control System Using Parallel Ports   32
         The Serial Interface   34

   2.4   Introduction to Controller Programming   35
   2.5   Microprocessor-Based Controllers   37

         Single-Chip Microcomputers (Microcontrollers)   37
         Single-Board Computers   39
         Programmable Logic Controllers   40
         Personal Computers Used in Control Systems   41

CHAPTER 3   Operational Amplifiers and Signal Conditioning   46

3.1   Operational Amplifiers   47

Introduction   47
Voltage Follower   53
Inverting Amplifier   54
Noninverting Amplifer   56
Summing Amplifier   58
Differential and Instrumentation Amplifiers   60
Integrators and Differentiators   64
Active Filters   66
Comparator   68

3.2   Special Interface Circuits   70

The Current Loop (Voltage–Current Conversion)   71
Analog Switch Circuit   73
Sample and Hold Circuit   75

3.3   Signal Transmission   77

Earth Ground and Ground Loops   77
Isolation Circuits   78
Shielding   80
Shield-Grounding Considerations   82

CHAPTER 4   Switches, Relays, and Power-Control Semiconductors   87

4.1   Switches   88

Toggle Switches   88
Push-Button Switches   90
Other Switch Types   90

4.2   Relays   93

Electromechanical Relays   93
Solid-State Relays   95

4.3   Power Transistors   98

Bipolar Junction Transistors   98
Field Effect Transistors (FET)   104

4.4   Silicon-Controlled Rectifiers   109

4.5   Triacs   115

Calculation of Delay and Conduction Periods   118

4.6   Trigger Devices   119

Unijunction Transistors   119
Programmable Unijunction Transistors   120
Diac   120

CHAPTER 5   Mechanical Systems   **126**

5.1   Behavior of Mechanical Components   127

        Overview   127
        Friction   127
        Springs   129
        Mass and Inertia   132
        Basic Equations of Motion for Linear Systems   134
        Basic Equations of Motion for Rotational Systems   138

5.2   Energy   141

        Energy Conversion   141
        Heat Transfer   144

5.3   Response of the Whole Mechanical System   147

        Underdamped, Critically Damped, and Overdamped
            Mechanical Systems   147
        Mechanical Resonance   149

5.4   Gears   153

        Spur Gears   154
        Using Gears to Change Speed   156
        Using Gears to Transfer Power   160
        Long Gear Trains   162
        Worm Gears   163
        Harmonic Drive   163

5.5   Other Power-Transmitting Techniques   164

        Belts   164
        Roller Chain   168

CHAPTER 6   Sensors   **173**

6.1   Position Sensors   174

        Potentiometers   174
        Optical Rotary Encoders   182
            Absolute Optical Encoders   183
            Incremental Optical Encoders   184
                Decoding V_1 and V_2
        Linear Variable Differential Transformers   190

6.2   Velocity Sensors   191

        Velocity from Position Sensors   191
        Tachometers   193
            Optical Tachometers   193
            Direct Current Tachometers   193

6.3   Proximity Sensors   196

Limit Switches   196
Optical Proximity Sensors   197
Hall-Effect Proximity Sensors   198

6.4   Load Sensors   201

Bonded-Wire Strain Gauges   202
Semiconductor Force Sensors   206
Low-Force Sensors   207

6.5   Pressure Sensors   208

Bourdon Tubes   209
Bellows   209
Semiconductor Pressure Sensors   210

6.6   Temperature Sensors   210

Bimetallic Temperature Sensors   212
Thermocouples   212
Resistance Temperature Detectors   215
Thermistors   216
Integrated Circuit Temperature Sensors   217

6.7   Flow Sensors   220

Pressure-Based Flow Sensors   220
Turbine Flow Sensors   222
Magnetic Flowmeters   223

6.8   Liquid-Level Sensors   224

Discrete-Level Detectors   224
Continuous-Level Detectors   225

CHAPTER 7   **Direct Current Motors**   233

7.1   Theory of Operation   234

7.2   Wound-Field DC Motors   238

Series-Wound Motors   238
Shut-Wound Motors   240
Compound Motors   243

7.3   Permanent Magnet Motors   243

Torque and Speed Relationship   244
Circuit Model of the PM Motor (Optional)   247

7.4   DC Motor–Control Circuits   251

DC Motor Control Using an Analog Drive   252
Reversing the PM Motor   255
DC Motor Control Using Pulse-Width Modulation   256
PWM Control Circuits   259
DC Motor Control for Larger Motors   261

7.5   A Comprehensive Application Using a Small DC Motor   264

7.6   Brushless DC Motors   267

CHAPTER 8   Stepper Motors   **272**

8.1   Permanent Magnet Stepper Motors   273

Effect of Load on Stepper Motors   274

Modes of Operation   276

Excitation Modes for PM Stepper Motors   279

    Two-Phase (Bipolar) Stepper Motors   279

    Four-Phase (Unipolar) Stepper Motors   281

    Available PM Stepper Motors   283

8.2   Variable Reluctance Stepper Motors   283

8.3   Hybrid Stepper Motors   286

8.4   Stepper Motor Control Circuits   287

Controlling the Two-Phase Stepper Motor   289

Controlling the Four-Phase Stepper Motor   290

Microstepping   293

Improving Torque at Higher Stepping Rates   294

8.5   Stepper Motor Application: Positioning a Disk Drive Head   297

CHAPTER 9   Alternating Current Motors   **302**

9.1   AC Power   303

Background   303

Single-Phase AC   304

Ground-Fault Interrupters   305

Three-Phase AC   306

9.2   Induction Motors   308

Theory of Operation   308

Three-Phase Motors   314

Single-Phase Motors   314

Split-Phase Control Motors (Two-Phase Motors)   318

AC Servomotors   319

9.3   Synchronous Motors   320

Theory of Operation   320

Power-Factor Correction and Synchronous Motors   321

Small Synchronous Motors   321

9.4   Universal Motors   323

9.5   AC Motor Control   323

Start–Stop Control   323

Jogging   325
Reduced-Voltage Starting   326
Variable-Speed Control of AC Motors   328

CHAPTER 10   Actuators: Electric, Hydraulic, and Pneumatic   334

10.1   Electric Linear Actuators   335

Leadscrew Linear Actuators   335
Solenoids   338

10.2   Hydraulic Systems   340

Basic Principles of Hydraulics   341
Hydraulic Pumps   343
Hydraulic Actuators   345
Pressure-Control Valves   347
Accumulators   348
Flow-Control Valves   349

10.3   Pneumatic Systems   351

Compressors, Dryers, and Tanks   352
Pressure Regulators   353
Pneumatic Control Valves   355
Pneumatic Actuators   355

10.4   Flow Valves   357

CHAPTER 11   Feedback Control Principles   361

11.1   Performance Criteria   363

11.2   On–Off Controllers   364

Two-Point Control   364
Three-Position Control   365

11.3   Proportional Control   366

The Steady-State-Error Problem   369
The Gravity Problem   371
Bias   374
Analog Proportional Controllers   374

11.4   Integral Control   376

11.5   Derivative Control   379

11.6   Proportional + Integral + Derivative Control   381

Analog PID Controllers   382
Digital PID Controllers   383
Stability   386
Turning the PID Controller   389
Sampling Rate   394
Autotuning   395

11.7 PIP Controllers   396

11.8 Fuzzy Logic Controllers   397

Introduction   397
Example of a One-Input System   400
Example of a Two-Input System   402
Closing Thoughts   405

CHAPTER 12   Relay Logic and Programmable Logic Controllers   411

12.1 Relay Logic Control   412

Relay Logic   412
Ladder Diagrams   414
Timers, Counters, and Sequencers   417

12.2 Programmable Logic Controllers   420

Introduction   420
PLC Hardware   420
    Power Supply   421
    Processor   421
    Program Memory   421
    Data Memory   422
    Programming Port   422
    Input and Output Modules   422
    PLC Bus   425
PLC Setup Procedure   425
PLC Operation   425

12.3 Programming the PLC   428

Ladder Diagram Programming   428
Bit Instructions   429
Timers   431
Counters   434
Sequencers   434
Advanced Instructions   438
Using a PLC as a Two-Point Controller   439
Other PLC Programming Languages   442

**APPENDIX A   *The Getting Started Guide for APS* published by Allen–Bradley
(tutorial for setting up and using the Allen–Bradley SLC 500
Programmable Logic Controller)**   449

APPENDIX B   Glossary   511

APPENDIX C   Answers to Odd-Numbered Exercises   521

Index   545

INTENDED USE AND LEVEL

This text is intended for use in two- or four-year technology programs that include such courses as industrial controls, industrial electronics, control systems, and electromechanical systems.

The math level required is algebra, and no calculus or Laplace transforms are used. Minimum prerequisites are basic DC/AC, digital, and basic solid-state circuits. Useful prerequisites (or corequisites) are op-amps, microprocessors, and physics (although these topics are covered in this book).

This book is about midrange in terms of difficulty. In most cases, the subject matter is taken beyond the description level to include practical examples that demonstrate the fundamental concepts involved.

SUBJECT AND APPROACH

Most technology students take only one course that deals with controls and the electromechanical interface. This one course has the responsibility to cover (1) electromechanical components such as sensors, motors, and driver circuits; (2) interface circuits between the mechanics and the electronics; (3) some understanding of the mechanical realm such as gears, springs, friction, and inertia; and (4) some basic feedback control theory that ties together all the concepts. This text covers these topics in an orderly way, with each chapter illuminating one aspect of the control system:

❑ Chapter 1 introduces the basic concepts of control systems.

❑ Chapters 2, 3, and 4 lay some groundwork of industrial control systems by presenting the basics of microprocessors, op-amp circuits and interfacing, switches, relays, and power-control devices (all of which might be a review for some students).

❑ Chapter 5 introduces some basic mechanical concepts such as springs, inertia, energy, and gears.

❑ Chapter 6 introduces all major types of sensors—for example, temperature, pressure, and position.

❑ Chapters 7, 8, and 9 cover motors—DC, stepper, and AC.

❑ Chapter 10 covers other types of actuators such as electric linear actuators, hydraulic actuators, and pneumatic actuators.

❑ Chapter 11 puts all the foregoing concepts together and presents control strategies, from simple two-point on–off systems, to proportional control, to PID systems. Also discussed are fuzzy logic controllers.

❑ Chapter 12 continues the subject of control by introducing relay ladder logic and programmable logic controllers (PLCs).

This text is designed to be a *guide* in a semester's study of modern industrial control systems, not an encyclopedia on the subject. This concept makes it the author's responsibility to provide a balanced treatment of the important subjects so that the student will emerge from the course with a sound understanding of all germane topics. The text contains somewhat more material than could reasonably be covered in a single semester or quarter. This allows instructors to select topics that accommodate local needs or lets them cover the material in two semesters or quarters. If the second term covers PLCs, Chapter 12 should be saved until then, supplemented by the Allen–Bradley tutorial in the appendix.

FEATURES

This text contains a number of important and unique features:

❑ A conversational writing style with an emphasis on explaining principles

❑ Focus on the current digital technology while including analog concepts

❑ A reasonably complete discussion of related mechanical concepts (friction, springs, gears, etc.)

❑ Basic physics concepts such as motion and inertia, heat transfer, and pressure

❑ Use of both traditional (English) and SI units in examples (sometimes repeating the same example with different units)

SUPPLEMENTAL PACKAGE

In addition to this text, the following aids are offered:

❑ A laboratory manual entitled *Modern Control Experiments* by Gerald E. Williams that contains 24 labs, 2 geared to each chapter of the text

❑ An instructors' solutions manual that gives solutions to all text problems (only odd problems are answered in the back of the text)

❑ A set of transparency masters for many of the illustrations in this book

All these may be ordered from the publisher.

How I came to write this book:

As along as I can remember, I have been interested in robots. At age 15, I built a robot "brain" consisting of over 50 relays, which could step through a series of instructions. Later, I majored in engineering at UCLA, taking courses in electronics, mechanics, and control theory. This text was started when I was teaching a course on digital control in the technology department of Arizona State University. The original idea was to develop a "thin volume" on digital control. Over the next several years (and my return to Anne Arundel Community College in Maryland), the project widened in scope and became a mainstream comprehensive text on industrial electronics and controls, which I was able to "field test" in my classrooms.

ACKNOWLEDGMENTS

I would like to thank Tom Tucker and Chris Conty, the two editors at West Publishing who have worked with me on this project. Tom encouraged me to start writing this book and helped shape it into its present form; Chris (my current editor), with his encouragement and insight, has helped bring this project to fruition. I also thank the many reviewers for their valuable suggestions, in particular:

Bob Baldacci
Northeast Community College (NE)

Don Barrett
DeVry Institute of Technology–Irving (TX)

John Bart
Erie Community College–North (NY)

David Beyer
Middlesex County College (NJ)

Lowell W. Blakley
Vincennes University (IN)

Doug Buckley
Springfield Technical Community
College (MA)

Richard Castellucis
Southern College of Technology (GA)

Gorden Chen
Niagara College (Ontario, Canada)

Christopher J. Conant
Broome Community College (NY)

Tom Fonger
CS Mott Community College (MI)

Rahmatollah Gloshan
Wayne County Community College (MI)

Richard A. Honeycutt
Davidson County Community College (NC)

Fred Kerr
DeVry Institute of Technology–Decatur (GA)

Len Klochek
Seneca College of Applied Arts &
Technology (Ontario, Canada)

John D. Meese
DeVry Institute of Technology (OH)

Abraham Musalem
Hudson Valley Community College (NY)

Mike Pelletier
Northern Essex Community College (MA)

Greg Rasmussan
St. Paul Technical College (MN)

Donald R. Remington
Wentworth Institute of Technology (MA)

John Sands
Moraine Valley Community College (IL)

Terry Schulz
Mt. Hood Community College (OR)

Saeed Shaikh
Miami–Dade Community College (FL)

David D. Smith
California State University–Chico (CA)

Ronald Warner
Mohawk Valley Community College (NY)

William Welch
Western Wisconsin Technical College (WI)

Modern Control Technology

COMPONENTS AND SYSTEMS

by

Christopher T. Kilian

Anne Arundel Community College

West Publishing Company

Minneapolis/St. Paul • New York • Los Angeles • San Francisco

1 Introduction to Control Systems

After studying this chapter, you should be able to:

- Distinguish between open-loop and closed-loop control systems.
- Understand control system block diagrams.
- Explain transfer functions.

- Differentiate between analog and digital control systems.
- Know how process control systems work.
- Know how servomechanisms work.

INTRODUCTION

A control system is a collection of components working together under the direction of some machine intelligence. In most cases, electronic circuits provide the intelligence, and electromechanical components such as sensors and motors provide the interface to the physical world. A good example is the modern automobile. Various sensors supply the on-board computer with information about the engine's condition. The computer then calculates the precise amount of fuel to be injected into the engine and adjusts the ignition timing. The mechanical parts of the system include the engine, transmission, wheels, and so on. To design, diagnose, or repair these sophisticated systems, you must understand the electronics, the mechanics, and control system principles.

In days past, so-called automatic machines or processes were controlled either by analog electronic circuits, or circuits using switches, relays, and timers. Since the advent of the inexpensive microprocessor, more and more devices and systems are being redesigned to incorporate a microprocessor controller. Examples include copying machines, soft-drink machines, robots, and industrial process controllers. Many of these machines are taking advantage of the increased processing power that comes with the microprocessor and, as a consequence, are becoming more sophisticated and are including new features. Taking again the modern automobile as an example, the original motivation for the on-board computer was to replace the mechanical and vacuum-driven subsystems used in the distributor and carburetor. Once a computer was in the design, however, making the system more sophisticated was relatively easy—for example, self-adjusting fuel/air ratio for changes in altitude. Also, features such as computer-assisted engine diagnostics could be had without much additional cost. This trend toward computerized control will no doubt continue in the future.

1.1   CONTROL SYSTEMS

Introduction and Background

In a modern **control system**, electronic intelligence controls some physical process. Control systems are the "automatic" in such things as automatic pilot and automatic washer. Because the machine itself is making the routine decisions, the human operator is freed to do other things. In many cases, machine intelligence is better than direct human control because it can react faster or slower (keep track of long-term slow changes), respond more precisely, and maintain an accurate log of the system's performance.

Control systems can be classified in several ways. A **regulator system** automatically maintains a parameter at (or near) a specified value. An example of this is a home-heating system maintaining a set temperature despite changing outside conditions. A **follow-up system** causes an output to follow a set path or procedure that has been specified in advance. An example is an industrial robot moving parts from place to place. An **event control system** controls a sequential series of events. An example is a washing machine cycling through a series of programmed steps.

Natural control systems have existed since the beginning of life. Consider how the human body regulates temperature. If the body needs to heat itself, food calories are converted to produce heat; on the other hand, evaporation causes cooling. Because evaporation is less effective (especially in humid climates), it is not surprising that our body temperature (98.6°F) was set near the high end of Earth's temperature spectrum (to reduce demand on the cooling system). If temperature sensors in the body notice a drop in temperature, they signal the body to burn more fuel. If the sensors indicate too high a temperature, they signal the body to sweat.

Man-made control systems have existed in some form since the time of the ancient Greeks. One interesting device described in the literature is a pool of water that could never be emptied. The pool had a concealed float-ball and valve arrangement similar to a toilet tank mechanism. When the water level lowered, the float dropped and opened a valve that admitted more water.

Electrical control systems are a product of the twentieth century. Electromechanical relays were developed and used for remote control of motors and devices. Relays and switches were also used as simple logic gates to implement some intelligence. Using vacuum-tube technology, significant development in control systems was made during World War II. Dynamic position control systems (servomechanisms) were developed for aircraft applications, gun turrets, and torpedoes. Today, position control systems are used in machine tools, industrial processes, robots, cars, and office machines, to name a few.

Meanwhile, other developments in electronics were having an impact on control system design. Solid-state devices started to replace the power relays in motor control circuits. Transistors and integrated circuit operational amplifiers (IC op-amps) became available for analog controllers. Digital integrated circuits replaced bulky relay logic. Finally, and perhaps most significantly, the microprocessor allowed for the creation of digital controllers that are inexpensive, reliable, and easily programmed, yet able to control complex processes.

The subject of control systems is really many subjects: electronics (both analog and digital), power-control devices, sensors, motors, mechanics, and *system theory*, which ties together all these concepts. Many students find the subject of control systems to be interesting because it deals with applications of much of the theory to which they have already been exposed. In this text, we will present material in each major subject area that makes up a control system, in more or less the same order that they appear in a control system block diagram. Some readers may choose to skip over (or lightly review) chapters that may be repetitious to them.

Finally, figures in this text use *conventional current flow*, current that travels from the positive to the negative terminal. If you are familiar with *electron flow*, remember that the theory and "numbers" are the same; only the indicated direction of the current is opposite from what you are used to.

Every control system has (at least) a **controller** and an **actuator** (also called a final control element). Shown in the block diagram in Figure 1.1, the controller is the intelligence of the system and is usually electronic. The input to the controller is called the **set point**, which is a signal representing the desired system output. The actuator is an electromechanical device that takes the signal from the controller and converts it into some kind of physical action. Examples of typical actuators would be an electric motor, an electrically controlled valve, or a heating element. The last block in Figure 1.1 is labeled **process** and has an output labeled **controlled variable**. The process block represents the physical process being affected by the actuator, and the controlled variable is the measurable result of that process. For example, if the actuator is an electric heating element in a furnace, then the process is "heating the furnace," and the controlled variable is the temperature in the furnace. If the actuator is an electric motor that rotates an antenna, then the process is "rotating of the antenna," and the controlled variable is the angular position of the antenna.

Figure 1.1
A block diagram of a control system.

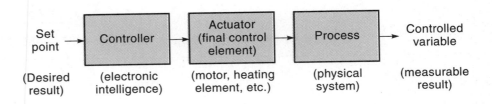

Open-Loop Control Systems

Control systems can be broadly divided into two categories: open- and closed-loop systems. In an **open-loop control system**, the controller independently calculates exact voltage or current needed by the actuator to do the job and sends it. With this approach, however, *the controller never actually knows if the actuator did what it was supposed to* because there is *no* feedback. This system absolutely depends on the controller knowing the operating characteristics of the actuator.

◆ **EXAMPLE 1.1**

Figure 1.2 shows an open-loop control system. The actuator is a motor driving a robot arm. In this case, the process is the arm moving, and the controlled variable is the angular position of the arm. Earlier tests have shown that the motor rotates the arm at 5 degrees/second (deg/s) at the rated voltage. Assume that the controller is directed to move the arm from 0° to 30°. Knowing the characteristics of the process, the controller sends a 6-second power pulse to the motor. If the motor is acting properly, it will rotate exactly 30° in the 6 seconds and stop. On particularly cold days, however, the lubricant is more viscous (thicker), causing more internal friction, and the motor rotates only 25° in the 6 seconds; the result is a 5° error. The controller has no way of knowing of the error and does nothing to correct it.

Figure 1.2
Open-loop control system

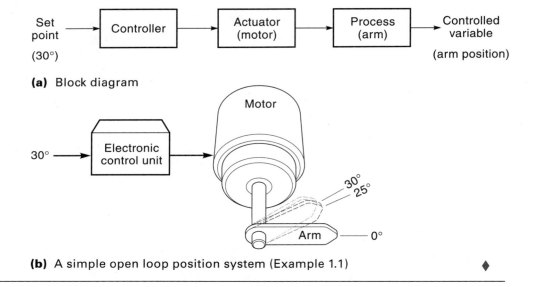

(a) Block diagram

(b) A simple open loop position system (Example 1.1)   ◆

Open-loop control systems are appropriate in applications where the actions of the actuator on the process are very repeatable and reliable. Relays and stepper motors (discussed in Chapters 4 and 8, respectively) are devices with reliable characteristics and are usually open-loop operations. Actuators such as motors or flow valves are sometimes used in open-loop operations, but they must be calibrated and adjusted at regular intervals to ensure proper system operation.

Closed-Loop Control Systems

In a **closed-loop control system**, the output of the process (controlled variable) is constantly monitored by a **sensor**, as shown in Figure 1.3(a). The sensor samples the system output and converts this measurement into an electric signal that it passes back to the controller. Because the controller knows what the system is actually doing, it can make any adjustments necessary to keep the output where it belongs. The signal from the controller to the actuator is the **forward path**, and the signal from the sensor to the controller is the **feedback**. In Figure 1.3(a), the feedback signal is subtracted from the set point at the **comparator** (just ahead of the controller). By subtracting the actual position (as reported by the sensor) from the desired position (as defined by the set point), we get the system **error**. The error signal represents the difference between "where you are" and "where you want to be." *The controller is always working to minimize this error signal.* A zero error means that the output is exactly what the set point says it should be.

Using a **control strategy**, which can be simple or complex, the controller minimizes the error. A simple control strategy would enable the controller to turn the actuator on or off—for example, a thermostat cycling a furnace on and off to

Figure 1.3

Closed-loop control system

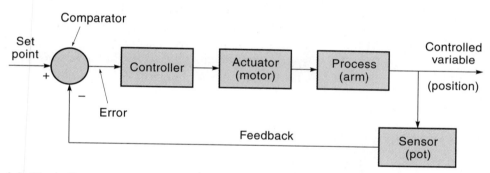

(a) Block diagram

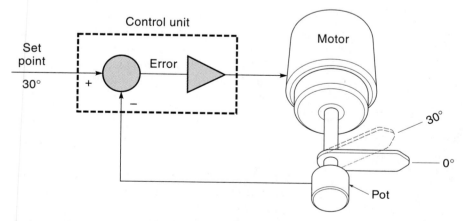

(b) A simple closed loop position system (Example 1.2)

maintain a certain temperature. A more complex control strategy would let the controller adjust the actuator force to meet the demand of the load, as described in Example 1.2.

♦ EXAMPLE 1.2

As an example of a closed-loop control system, consider again the robot arm resting at 0° [see Figure 1.3(b)]. This time a potentiometer (pot) has been connected directly to the motor shaft. As the shaft turns, the pot resistance changes. The resistance is converted to voltage and then fed back to the controller.

To command the arm to 30°, a set-point voltage corresponding to 30° is sent to the controller. Because the actual arm is still resting at 0°, the error signal "jumps up" to 30°. Immediately, the controller starts to drive the motor in a direction to reduce the error. As the arm approaches 30°, the controller slows the motor; when the arm finally reaches 30°, the motor stops. If at some later time, an external force moves the arm off the 30° mark, the error signal would reappear, and the motor would again drive the arm to the 30° position.   ♦

The self-correcting feature of closed-loop control makes it preferable over open-loop control in many applications, despite the additional hardware required. This is because closed-loop systems provide reliable, repeatable performance even when the system components themselves (in the forward path) are not absolutely repeatable or precisely known.

Transfer Functions

Physically, a control system is a collection of components and circuits connected together to perform a useful function. *Each component in the system converts energy from one form to another*; for example, we might think of a temperature sensor as converting degrees to volts or a motor as converting volts to revolutions per minute. To describe the performance of the entire control system, we must have some common language so that we can calculate the combined effects of the different components in the system. This need is behind the transfer function concept.

A **transfer function** (TF) is a mathematical relationship between the input and output of a control system component. Specifically, the transfer function is defined as the output divided by the input, expressed as

$$\text{TF} = \frac{\text{output}}{\text{input}} \tag{1.1}$$

Technically, the transfer function must describe both the time-dependent and the steady-state characteristics of a component. For example, a motor may have an initial surge of current that levels off at a lower steady-state value. The mathematics necessary to account for the time-dependent performance is beyond the scope of this text. In this text, we will consider only *steady-state values* for the transfer function, which is sometimes called simply the **gain**, expressed as

$$\text{TF}_{\text{steady state}} = \text{gain} = \frac{\text{steady-state output}}{\text{steady-state input}} \qquad (1.2)$$

♦ EXAMPLE 1.3

A potentiometer is used as a position sensor [see Figure 1.3(b)]. The pot is configured in such a way that 0° of rotation yields 0 V and 300° yields 10 V. Find the transfer function of the pot.

Solution

The transfer function is output divided by input. In this case, the input to the pot is "position in degrees," and output is volts:

$$\text{TF} = \frac{\text{output}}{\text{input}} = \frac{10 \text{ V}}{300°} = 0.0333 \text{ V/deg}$$

The transfer function of a component is an extremely useful number. It allows you to calculate the output of a component if you know the input. The procedure is simply to multiply the transfer function by the input, as shown in Example 1.4.

♦ EXAMPLE 1.4

For a temperature-measuring sensor, the input is temperature, and the output is voltage. The sensor transfer function is given as 0.01 V/deg. Find the sensor-output voltage if the temperature is 600°F.

Solution

If $\text{TF} = \dfrac{\text{output}}{\text{input}}$, then

$$\text{Output} = \text{input} \times \text{TF}$$

$$= \frac{600° \times 0.01 \text{ V}}{\text{deg}} = 6 \text{ V}$$

♦

As mentioned previously, transfer functions can be used to analyze an entire system of components. One common situation involves a series of components where the output of one component becomes the input to the next and each component has its own transfer function. Figure 1.4(a) shows the block diagram for this situation. This

Figure 1.4 A series of transfer functions reduced to a single transfer function.

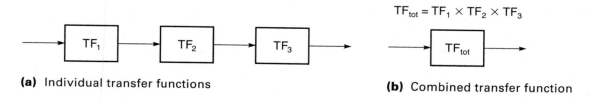

$$\text{TF}_{\text{tot}} = \text{TF}_1 \times \text{TF}_2 \times \text{TF}_3$$

(a) Individual transfer functions

(b) Combined transfer function

diagram can be reduced into a single block that has a TF_{tot}, which is the product of all the individual transfer functions. This concept is illustrated in Figure 1.4(b) and stated in Equation 1.3:*

$$\text{TF}_{tot} = \text{system gain} = \text{TF}_1 \times \text{TF}_2 \times \text{TF}_3 \times \ldots \qquad (1.3)$$

where

TF_{tot} = total steady-state transfer function for the entire (open-loop) system

$\text{TF}_1, \text{TF}_2, \ldots$ = individual transfer functions

These concepts are explained in Example 1.5.

◆ **EXAMPLE 1.5**

Consider the system shown in Figure 1.5. It consists of an electric motor driving a gear train, which is driving a winch. Each component has its own characteristics: The motor (under these conditions) turns at 100 rpm$_m$ for each volt (V_m) supplied; the output shaft of the gear train rotates at one-half of the motor speed; the winch (with a 3-inch shaft circumference) converts the rotary motion (rpm$_w$) to linear speed. The individual transfer functions are given as follows:

$$\text{Motor:} \qquad \text{TF}_m = \frac{\text{output}}{\text{input}} = \frac{100 \text{ rpm}_m}{1 \text{ V}_m} = 100 \text{ rpm}_m / \text{V}_m$$

$$\text{Gear train:} \quad \text{TF}_g = \frac{\text{output}}{\text{input}} = \frac{1 \text{ rpm}_w}{2 \text{ rpm}_m} = 0.5 \text{ rpm}_w / \text{rpm}_m$$

$$\text{Winch:} \qquad \text{TF}_w = \frac{\text{output}}{\text{input}} = \frac{3 \text{ in./min}}{1 \text{ rpm}_w} = 3 \text{ in./min/rpm}_w$$

Using Equation 1.3, we can calculate the system transfer function. If everything is correct, all units will cancel except for the desired set:

$$\text{TF}_{tot} = \text{TF}_m \times \text{TF}_g \times \text{TF}_w$$

$$= \frac{100 \text{ rpm}_m}{1 \text{ V}_m} \times \frac{0.5 \text{ rpm}_m}{1 \text{ rpm}_m} \times \frac{3 \text{ in./min}}{1 \text{ rpm}_m}$$

$$= 150 \text{ in./min/V}_m$$

We have shown that the transfer function of the complete system is 150 in./min/V_m. *Knowing this value, we can calculate the system output for any system input.* For example,

---

*Equation 1.3 is for open-loop systems only. If there is a feedback path (as shown in the accompanying diagram), then the overall system gain can be calculated as follows: $\text{TF}_{tot} = G/(1 + GH)$, where G is the total gain of the forward path and H is the total gain of the feedback path.

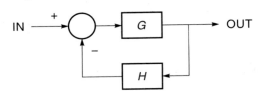

if the input to the this system is 12 V (to the motor), the output speed of the winch is calculated as follows:

$$\text{Output} = \text{input} \times \text{TF} = \frac{12\ \text{V} \times 150\ \text{in./min}}{1\ \text{V}_m} = 1800\ \text{in./min}$$

Figure 1.5
A system with three transfer functions (Example 1.5).

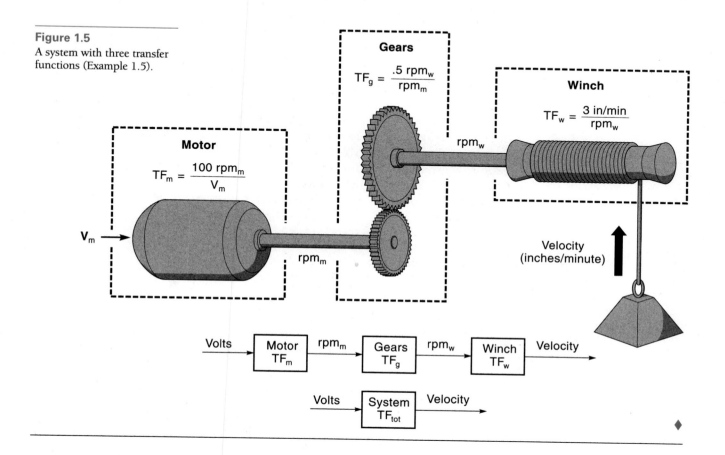

1.2   ANALOG AND DIGITAL CONTROL SYSTEMS

In an **analog control system**, the controller consists of traditional analog devices and circuits, that is, linear amplifiers. The first control systems were analog because it was the only available technology. In the analog control system, any change in either set point or feedback is sensed immediately, and the amplifiers adjust their output (to the actuator) accordingly.

In a **digital control system**, the controller uses a digital circuit. In most cases, this circuit is actually a computer, either mainframe or microprocessor-based. The computer executes a program that repeats over-and-over (each repetition is called an **iteration** or **scan**). The program instructs the computer to read the set point and sensor data and then use these numbers to calculate the output (which is sent to the actuator). The program then loops back to the beginning and starts over again. The total time for one pass through the program may be less than 1 millisecond (ms). The digital system only "looks" at the inputs at a certain time in the scan and gives the updated output later. If an input changes just after the computer looked at it, that change will remain undetected until the next time through the scan. This is fundamentally different than the analog system, which is continuous and responds immediately to any changes. However, for most digital control systems, the scan time is so short compared with the response time of the process being controlled that, for all practical purposes, the controller response is instantaneous.

The physical world is basically an "analog place." Natural events take time to happen, and they usually move in a continuous fashion from one position to the next. Therefore, most control systems are controlling analog processes. This means that, in many cases, the digital control system must first convert real-world analog input data into digital form before it can be used. Similarly, the output from the digital controller must be converted from digital form back into analog form. Figure 1.6 shows a block diagram of a digital closed-loop control system. Notice the two additional blocks: the digital-to-analog converter (DAC) and the analog-to-digital converter (ADC). (These devices, which convert data between the digital and analog formats, are discussed in Chapter 2.) Also note that the feedback line is shown going directly into the controller. This emphasizes the fact that the computer, not a separate subtraction circuit, makes the comparison between the set point and the feedback signal.

Figure 1.6 Block diagram of a digital closed-loop control system. (*Note*: A digital actuator, such as a stepper motor, would not need a DAC; similarly, a digital sensor, such as an optical shaft encoder, would not need an ADC.)

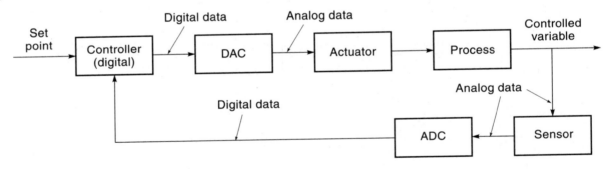

1.3 CLASSIFICATIONS OF CONTROL SYSTEMS

So far we have discussed control systems as being either open or closed loop, analog or digital. Yet we can classify control systems in other ways, which have to do with applications. Some of the most common applications are discussed next.

Process Control

Process control refers to a control system that oversees some industrial process so that a uniform, correct output is maintained. It does this by monitoring and adjusting the control parameters (such as temperature or flow rate) to ensure that the output product remains as it should.

The classic example of process control is a closed-loop system maintaining a specified temperature in an electric oven, as illustrated in Figure 1.7. In this case, the actuator is the heating element, the controlled variable is the temperature, and the sensor is a thermocouple (a device that converts temperature into voltage). The controller regulates power to the heating element in such a way as to keep the temperature (as reported by the thermocouple) at the value specified by the set point.

Figure 1.7

A closed-loop oven-heating system.

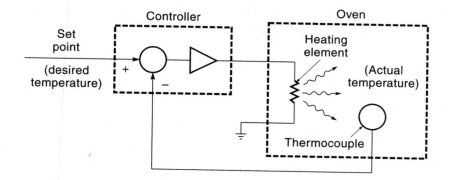

Another example of process control is a paint factory in which two colors, blue and yellow, are mixed to produce green (Figure 1.8). To keep the output color constant, the exact proportions of blue and yellow must be maintained. The setup illustrated in Figure 1.8(a) accomplishes this with flow valves 1 and 2, which are manually adjusted until the desired hue of green is achieved. The problem is that, as the level of paint in the vats changes, the flow will change and the mixture will not remain constant.

To maintain an even flow from the vats, we could add two flow valves (and their controls) as shown in Figure 1.8(b). Each valve would maintain a specified flow of paint into the mixer, regardless of the upstream pressure. Theoretically, if the blue and yellow flows are independently maintained, the green should stay constant. In practice, however, other factors such as temperature or humidity may affect the mixing chemistry and therefore the output color.

A better approach might be the system shown in Figure 1.8(c); a single sensor monitors the output color. If the green darkens, the controller increases the flow of yellow. If the green gets too light, the flow of yellow is decreased. This system is desirable because it monitors the actual parameter that needs to be maintained. In real life, such a straightforward system may not be possible because sensors that can measure the output directly may not exist and/or the process may involve many variables.

In a large plant such as a refinery, many processes are occurring simultaneously and must be coordinated because the output of one process is the input of another. In the early days of process control, separate independent controllers were used for each process, as shown in Figure 1.9(a). The problem with this approach was that, to change the overall flow of the product, each controller had to be readjusted manually.

In the 1960s, a new system was developed in which all independent controllers were replaced by a single large computer. Illustrated in Figure 1.9(b), this system is called **direct digital control** (DDC). The advantage of this approach is that all local processes

Figure 1.8

A paint-mixing example of process control.

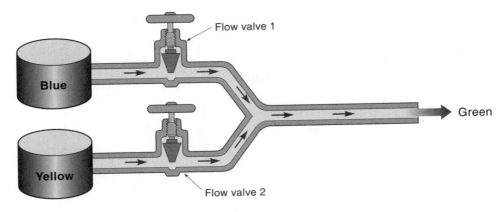

Flow valve 1

Blue

Green

Yellow

Flow valve 2

(a) Manual control

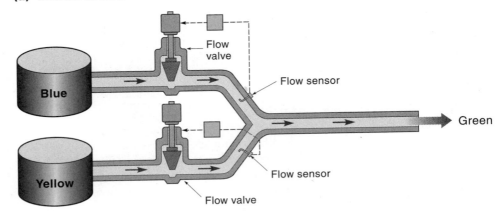

Flow valve

Flow sensor

Blue

Green

Yellow

Flow sensor

Flow valve

(b) Automatic flow control

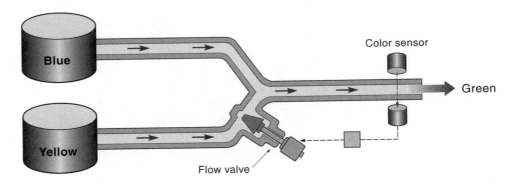

Color sensor

Blue

Green

Yellow

Flow valve

(c) Automatic color control

can be implemented, monitored, and adjusted from the same place. Also, because the computer can "see" the whole system, it is in a position to make adjustments to enhance total system performance. The drawback is that the whole plant is dependent on that one

Figure 1.9

Approaches of multiprocess control.

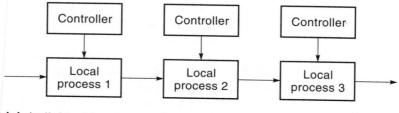

(a) Individual local controllers

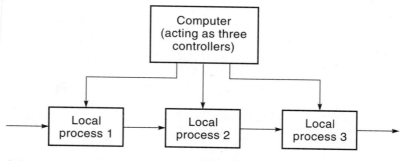

(b) Direct computer control of three processes

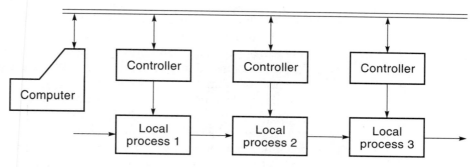

(c) Distributed computer control using local controllers

computer. If the computer goes off line to fix a problem in one process, the whole plant shuts down.

The advent of small microprocessor-based controllers has led to a new approach called **distributed computer control** (DCC), illustrated in Figure 1.9(c). In this system, each process has its own separate controller located at the site. These local controllers are interconnected via a local area network so that all controllers on the network can be monitored or reprogrammed from a single computer. Once programmed, each process is essentially operating independently.

Sequentially Controlled Systems

A **sequentially controlled system** controls a process that is defined as a series of tasks to be performed—that is, a sequence of operations, one after the other. Each operation in the sequence is performed either for a certain amount of time, in which case it is **time-driven**, or until the task is finished (as indicated by, say, a limit switch), in which case it is **event-driven**.

The classic example of a sequentially controlled system is the automatic washing machine. The first event in the wash cycle is to fill the tub. This is an event-driven task because the water is admitted until it gets to the proper level as indicated by a float and limit switch (closed loop). The next two tasks, wash and spin-drain, are each done for a specified period of time and are time-driven events (open loop). A timing diagram for a washing machine is shown in Figure 1.10.

Figure 1.10

Timing diagram for an automatic washing machine.

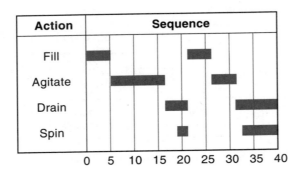

Another example of a sequentially controlled system is a traffic signal. The basic sequence may be time-driven: 45 seconds for green, 3 seconds for yellow, and 45 seconds for red. The presence or absence of traffic, as indicated by sensors in the roadbed, however, may alter the basic sequence, which is an event-driven control.

Many automated industrial processes could be classified as sequentially controlled systems. An example is a process where parts are loaded into trays, inserted into a furnace for 10 minutes, then removed and cooled for 10 minutes, and loaded into boxes in groups of six. In the past, most sequentially controlled systems used switches, relays, and electromechanical timers to implement the control logic. These tasks are now performed more and more by small computers known as **programmable logic controllers** (PLCs), which are less expensive, more reliable, and easily reprogrammed to meet changing needs—for example, to put eight items in a box instead of six. (PLCs are discussed in Chapter 12.)

Servomechanisms

A **servomechanism** is an electromechanical control system that directs the movement of a physical object such as a radar antenna or robot arm. Typically, either the output position or the output velocity (or both) is controlled. An example of a servomechanism is the positioning system for a radar antenna, as shown in Figure 1.11. In this case, the controlled variable is the antenna position. The antenna is rotated with an electric motor connected to the controller located some distance away. The user selects a direction, and the controller directs the antenna to rotate to a specific position.

Numerical control and robotics are two special categories of servomechanism systems that deserve separate treatment because of their widespread use. These are described next.

Figure 1.11

A servomechanism: a remote antenna-position system.

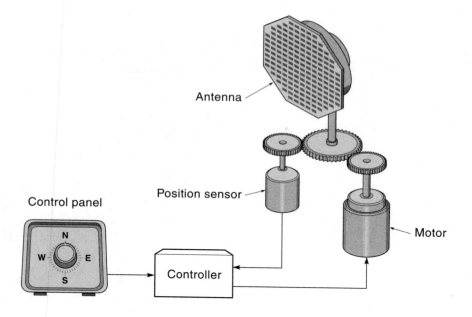

Numerical Control

Numerical control (NC) is the type of digital control used on machine tools such as lathes and milling machines. These machines can automatically cut and shape the workpiece without a human operator. Each machine has its own set of axes or parameters that must be controlled; as an example, consider the milling machine shown in Figure 1.12. The workpiece that is being formed is fastened to a movable table. The table can be moved (with electric motors) in three directions: X, Y, and Z. The cutting-tool speed is automatically controlled as well. To make a part, the table moves the workpiece past the cutting tool at a specified velocity and cutting depth. In this example, four parameters (X, Y, Z, and rpm) are continuously and independently controlled by the controller. The controller takes as its input a series of numbers that completely describe how the part is to be made. These numbers include the physical dimensions and such details as cutting speeds and feeds.

NC machines have been used since the 1960s, and certain standards that are unique to this application have evolved. Traditionally, data from the part drawing were entered manually into a computer program. This program converted the input data into a series of numbers and instructions that the NC controller could understand, and either stored them on a floppy disk or tape, or sent the data directly to the machine tool. These data were read by the machine-tool controller as the part was being made. With the advent of **computer-aided design** (CAD), the job of manually programming the manufacturing instructions has been eliminated. Now it is possible for a special computer program (called a *postprocessor*) to read the CAD-generated drawing and then produce the necessary instructions for the NC machine to make the part. This whole process—from CAD to finished part—is called **computer-aided manufacturing** (CAM).

Figure 1.12
Basics of a numerical control
milling machine.

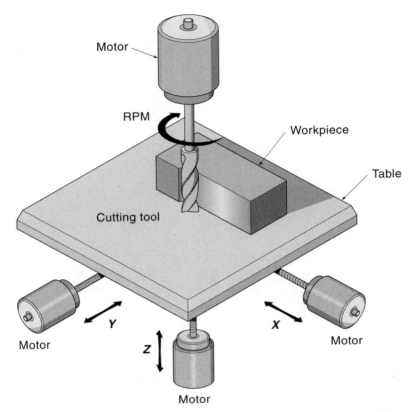

One big advantage of this process is that one machine tool can efficiently make many different parts, one after the other. This system tends to reduce the need for a large parts inventory. If the input tape (or software) is available, any needed part can be made in a short period of time. This is one example of **computer-integrated manufacturing** (CIM), a whole new way of doing things in the manufacturing industry. CIM involves using the computer in every step of the manufacturing operation—from the customer order, to ordering the raw materials, to machining the part, to routing it to its final destination.

Robotics

Industrial **robots** are classic examples of position control systems. In most cases, the robot has a single arm with shoulder, elbow, and wrist joints, as well as some kind of hand. The hand is either a gripper or other tool such as a paint spray gun. Robots are used to move parts from place to place, assemble parts, load and off-load NC machines, and perform such tasks as spray painting and welding.

Pick-and-place robots, the simplest type, pick up parts and place them somewhere else nearby. Instead of using sophisticated feedback control, they are often run open-loop using mechanical stops or limit switches (discussed in Chapter 6) to determine how far in each direction to go (sometimes called a ''bang–bang'' system). An example is shown in Figure 1.13. This robot uses pneumatic cylinders to lift, rotate, and extend the arm. It can be programmed to repeat a simple sequence of operations.

Sophisticated robots use closed-loop position systems for all joints. An example is the industrial robot shown in Figure 1.14. It has six independently controlled axes

Figure 1.13
A pick-and-place robot.

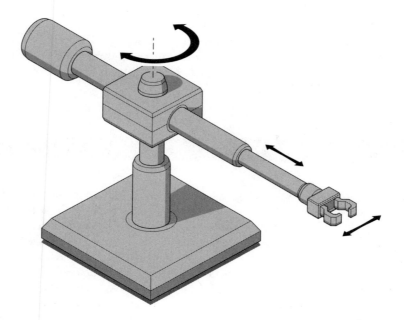

Figure 1.14 Large industrial robot.

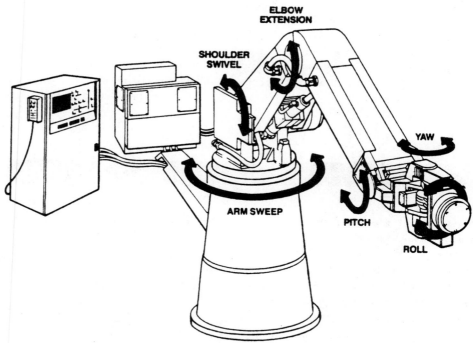

allowing it to get to difficult-to-reach places. The robot comes with and is controlled by a dedicated computer-based controller. This unit is also capable of translating human instructions into the robot program during the "teaching" phase. The arm can move from point to point at a specified velocity and arrive within a few thousandths of an inch.

SUMMARY

A control system is a system where electronic intelligence controls some physical process. This text will deal with all phases of the control system: the electronics, the power sources (such as motors), the mechanics, and system theory, which ties together all the concepts.

A control system is described in terms of a block diagram. The first block is the controller, which represents the electronic intelligence. The controller outputs a control signal to the next block, which is the actuator. The actuator is the system's first physical device to do something (for example, a motor or heating element).

There are two general categories of control systems: open loop and closed loop. In open-loop control, the controller sends a measured signal, specifying the desired action, to the actuator (the controller, however, has no way of knowing what the actuator actually does). Closed-loop control includes a sensor that feeds back a signal from the actuator to the controller, informing the controller exactly what the output is doing. This allows the controller to make correctional adjustments.

Each component in the control system can be described mathematically by a transfer function (TF), where TF = output/input. Transfer functions of individual components in a system can be mathematically combined to calculate overall system performance. A true transfer function includes time-dependent and steady-state characteristics, whereas a useful simplification (employed in this text) considers only steady-state conditions.

Control systems are classified as analog or digital. In an analog control system, the controller uses traditional analog electronic circuits such as linear amplifiers. In a digital control system, the controller uses a digital circuit, usually a computer.

Control systems are classified by application. Process control usually refers to an industrial process being electronically controlled for the purpose of maintaining a uniform correct output. A servomechanism is a control system that provides remote control motion of some object, such as a robot arm or a radar antenna. A numerical control (NC) control system directs a machine tool, such as a lathe, to machine a part automatically.

GLOSSARY

actuator   The first component in the control system which generates physical movement, typically a motor. The actuator gets its instructions directly from the controller. Another name for the actuator is the *final control element*.

analog control system   A control system where the controller is based on analog electronic circuits, that is, linear amplifiers.

CAD   *See* **computer-aided design.**

CAM   *See* **computer-aided manufacturing.**

CIM   *See* **computer-integrated manufacturing.**

closed-loop control system   A control system that uses feedback. A sensor continually monitors the output of the system and sends a signal to the controller, which makes adjustments to keep the output within specification.

comparator   Part of the control system that subtracts the feedback signal (as reported by the sensor) from the set point, to determine the error.

computer-aided design   A computer system that makes engineering drawings.

computer-aided manufacturing A computer system that allows CAD drawings to be converted for use by a numerical control (NC) machine tool.

computer-integrated manufacturing A computer system that oversees every step in the manufacturing process, from customer order to delivery of finished parts.

controlled variable The ultimate output of the process; the actual parameter of the process that is being controlled.

controller The machine intelligence of the control system.

control strategy The set of rules that the controller follows to determine its output to the actuator.

control system A system that may include electronic and mechanical components, where some type of machine intelligence controls a physical process.

DCC *See* **distributed computer control**.

DDC *See* **direct digital control**.

digital control system A control system where the controller is a digital circuit, typically a computer.

direct digital control An approach to process control where all controllers in a large process are simulated by a single computer.

distributed computer control An approach to process control where each process has its own local controller, but all individual controllers are connected to a single computer for programming and monitoring.

error In a control system, the difference between where the system is supposed to be (set point) and where it really is.

event control system A control system that cycles through a predetermined series of steps.

event-driven operation In a sequentially controlled system, an action that is allowed to start or continue based on some parameter changes. This is an example of closed-loop control.

feedback The signal from the sensor, which is fed back to the controller.

follow-up system A control system where the output follows a specified path.

forward path The signal-flow direction of the controller to the actuator.

gain The steady-state relationship between input and output of a component. (In this text, *gain* and *transfer function* are used interchangeably, although this is a simplification.)

iteration *See* **scan**.

NC *See* **numerical control**.

numerical control A digital control system that directs machine tools, such as a lathe, to automatically machine a part.

open-loop control system A control system that does not use feedback. The controller sends a measured signal to the actuator, which specifies the desired action. This type of system is *not* self-correcting.

pick-and-place robot A simple robot that does a repetitive task of picking up and placing an object somewhere else.

PLC *See* **programmable logic controller**.

programmable logic controller A small, self-contained microprocessor-based controller used primarily to replace relay logic controllers.

process The physical process that is being controlled.

process control A control system that maintains a uniform, correct output for some industrial process.

regulator system A control system that maintains an output at a constant value.

robot A servomechanism control system in the form of a machine with a movable arm.

scan One cycle through the program loop of a computer-based controller.

sensor Part of the control system that monitors the system output, the sensor converts the output movement of the system into an electric signal, which is fed back to the controller.

sequentially controlled system A control system that performs a series of actions in sequence, an example being a washing machine.

servomechanism An electromechanical feedback control system where the output is linear or rotational movement of a mechanical part.

set point The input signal to the control system, specifying the desired system output.

time-driven operation In a sequentially controlled system, an action that is allowed to happen for a specified period of time. This is an example of open-loop control.

TF *See* **transfer function**.

transfer function A mathematical relationship between the input and output of a control system component: TF = output/input. (In this text, *transfer function* and *gain* are used interchangeably, although this is a simplification.)

EXERCISES

Section 1.1

1. a. Draw a block diagram of an open-loop control system.
    b. Use the block diagram to describe how the system works.
    c. What basic requirements must the components meet for this system to work?
    d. What is the advantage of this system over a closed-loop system?

2. a. Draw a block diagram of a closed-loop control system.
    b. Use the block diagram to describe how the system works.
    c. What is the advantage of this system over an open-loop system?

3. The controlled variable in a closed-loop system is a robot arm. Initially, it is at 45°; then it is commanded to go to 30°. Describe what happens in terms of set point, feedback signal, error signal, and arm position.

4. Identify the following as open- or closed-loop control.
    a. Controlling the water height in a toilet tank
    b. Actuation of street lights at 6 PM
    c. Stopping a clothes dryer when the clothes are dry
    d. Actuation of an ice maker when the supply of cubes is low

5. A potentiometer has a transfer function of 0.1 V/deg. Find the pot's output if the input is 45°.

6. A potentiometer has a transfer function of 0.05 V/deg. Find the pot's output if the input is 89°.

7. A motor was measured to rotate (unloaded) at 500 rpm with a 6-V input and 1000 rpm with a 12-V input. What is the transfer function (steady state) for the unloaded motor?

8. In a certain system, an electric heating element was found to increase the temperature of a piece of metal 10° for each ampere of current. The metal expands 0.001 in./deg and pushes on a load sensor which outputs 1 V/0.005 in. of compression.
    a. Find the transfer functions of the three components and draw the block diagram.
    b. Calculate the overall transfer function of this system.

Section 1.2

9. Describe the differences between an *analog* and a *digital control system*.

10. The iteration time of a digital controller is 1 s. Would this controller be appropriate for the following?
    a. A robot that paint sprays cars
    b. A solar panel control system that tracks the sun across the sky

Section 1.3

11. What is the difference between a *process control system* and a *servomechanism*?

12. What is the difference between *direct digital control* and *distributed computer control*?

13. Give an example (other than in this book) of the following:
    a. A time-driven control system
    b. An event-driven control system
    c. A combined time- and event-driven control system

14. Give an example (other than in this book) of a servomechanism.

2 Introduction to Microprocessor-Based Control

After studying this chapter, you should be able to:

❑ Understand what a microprocessor is, what it does, and how it works.

❑ Understand the concepts of RAM and ROM computer memory and how memory is accessed via the address and data buses.

❑ Understand how parallel and serial data interfaces work.

❑ Perform relevant calculations pertaining to analog-to-digital converters and digital-to-analog converters.

❑ Understand the principles of digital controller software.

❑ Recognize and describe the characteristics of the various types of available digital controllers, that is, microcontrollers, single-board computers, programmable logic controllers, and personal computers.

INTRODUCTION

The digital integrated circuit (IC) called a **microprocessor** [Figure 2.1(a)], has ushered in a whole new era for control systems electronics. This revolution has occurred because the microprocessor brings the flexibility of program control and the computational power of a computer to bear on any problem. Automatic control applications are particularly well suited to take advantage of this technology, and microprocessor-based control systems are rapidly replacing many older control systems based on analog circuits or electromechanical relays. A microprocessor by itself is not a computer; additional components such as memory and input/output circuits are required to make it operational. However, the **microcontroller** [Figure 2.1(b)], which is a close relative of the microprocessor, *does* contain all the computer functions on a single IC. Microcontrollers lack some of the power and speed of the newer microprocessors, but their compactness is ideal for many control applications; most so-called microprocessor-controlled devices, such as vending machines, are really using microcontrollers. Some

Figure 2.1
Microprocessor and
microcontroller.

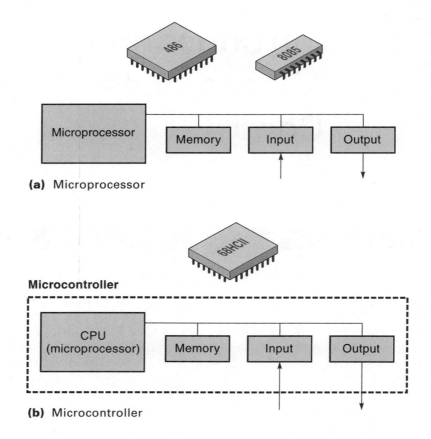

(a) Microprocessor

Microcontroller

(b) Microcontroller

specific reasons for using a digital, microprocessor design in control systems are the
following:

❑ Low-level signals from sensors, once converted to digital, can be transmitted long
distances virtually error-free.

❑ A microprocessor can easily handle complex calculations and control strategies.

❑ Long-term memory is available to keep track of parameters in slow-moving
systems.

❑ Changing the control strategy is easy by loading in a new program; no hardware
changes are required.

❑ Microprocessor-based controllers are more easily connected to the computer
network within an organization. This allows designers to enter program changes
and read current system status from their desk terminals.

In this chapter, we will present the basic concepts of a microprocessor- and micro-
controller-based system with particular emphasis on control system applications. It is by no
means an in-depth treatment, but enough to make the rest of the text more meaningful.

In the first sections of this chapter the basic concepts of microprocessor hardware
and operation are introduced (these concepts also apply to microcontrollers). I have
included this material because the student of modern control systems should have at
least a general knowledge of how the microprocessor performs its job. Readers who are

familiar with microprocessors from previous course work or other sources may choose to skip this section.

2.1   INTRODUCTION TO MICROPROCESSOR SYSTEM HARDWARE

A computer is made up of three basic functional units: the central processing unit (CPU), memory, and input/output (I/O). The **central processing unit** does the actual computing and is composed of two subparts: the arithmetic logic unit and control sections (Figure 2.2). The arithmetic logic unit (ALU) performs the actual numerical and logic calculations such as addition, subtraction, AND, OR, and so on. The control section of the CPU manages the data flow, such as reading and executing the program instructions. If data require calculations, the control section hands it over to the ALU for processing. In a microprocessor-based computer, the microprocessor is the CPU.

Figure 2.2   A block diagram of a microprocessor-based computer.

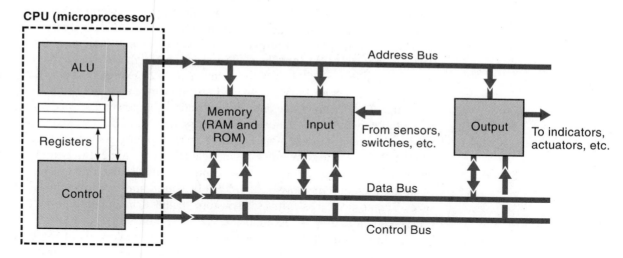

Digital data is in the form of **bits**, where each bit has a value of either 1 or 0. Digital circuits usually use 5 Vdc to represent logic 1 and 0 Vdc to represent logic 0. Eight bits together is called a **byte**. A microprocessor handles digital data in **words**, where a word may be 8, 16, or 32 bits wide. For example, an 8-bit microprocessor has a byte-sized word, with a maximum decimal value of 255. (Computers represent numbers in the *binary number system*; for example, 11111111 binary = 255 decimal.) The rightmost bit in a binary number has the least value (usually 1) and is called the **least significant bit** (LSB). The leftmost bit represents the highest value and is called the **most significant bit** (MSB). The conversion between binary and decimal can be performed directly with most scientific calculators or manually using a technique described in any textbook on the subject. To express values larger than 255, two or more words are put together. In this text, we will assume 8-bit microprocessors are used unless otherwise stated.

The **memory** section of the computer is a place where digital data in binary form (1s and 0s) are stored. Memory consists of cells organized in 8-bit groups. Each byte is given a unique numeric **address**, which represents its location just as a street address represents the location of a house. Data are written into memory and read out of memory, based solely on their address. In a particular memory circuit, the addresses might start at 1000 and run consecutively to 2000. Figure 2.3 diagrams a section of memory. Note that the first byte of data has a decimal value of 2 (00000010 = 2 decimal) and an address of 1000.

Figure 2.3

A section of memory.

Address	Data
1000	00000010
1001	01100100
1002	00000000
1003	01011100
1004	10011010

Computers usually have two kinds of addressable memory. The first is **random-access memory** (RAM), which allows the computer to read and write data at any of its addresses (it is also called read/write memory, or RWM). All data in this type of memory are lost when the power is turned off and is called **volatile memory** (an exception is designs where RAM is kept "alive" with a small battery). The second type of memory is **read-only memory** (ROM), which is similar to RAM except that new data cannot be written in; all data in ROM are loaded at the factory and cannot be changed by the computer. This memory does not lose its data when power is turned off and is called **nonvolatile memory**. Most microprocessor systems have both RAM and ROM. RAM is used for temporary program storage and as a temporary scratch-pad memory for the CPU. ROM is used to store programs and data that need to be always available. Actually, many computers use EPROM (erasable programmable read-only memory) instead of ROM for long-term memory. EPROM can be erased and reprogrammed, but only with special equipment. Disk drives also store digital data but in a form that must be processed before they are accessible to the microprocessor.

The **input/output** (I/O) section of the computer allows it to interface with the outside world. The input section is the conduit through which new programs and data are entered into the computer, and the output section allows the computer to communicate its results. An I/O interface is called a **port**. An input port is a circuit that connects input devices to the computer; examples of input devices are keyboards, sensors, and switches. An output port is a circuit that connects the computer to output devices. Examples of output devices are indicator lamps, actuators, and monitors. Input/output is discussed in more detail in the next section.

Referring again to Figure 2.2, we see that the blocks are connected by three lines labeled address bus, data bus, and control bus. The **address bus** is a group of wires that carries an address (in binary form) from the CPU to the memory and I/O circuits. The need for memory to receive addresses has already been discussed, but you may wonder why I/O ports need addresses. It turns out that all I/O ports are assigned addresses and are treated like memory locations by the CPU. The CPU outputs data to the outside world by sending them to a port address. When the circuitry of the designated output port detects its own address, it opens and allows data to pass from the data bus to whatever is connected to the port. There are two ways that I/O addressing is done. Some microprocessors use what is called **memory-mapped input/output**, where an I/O address is treated just like another memory address. Other microprocessors treat I/O addresses completely separate from memory addresses.

The **data bus** is a group of eight wires that carries the actual numerical data from place to place within the computer. Figure 2.2 shows how the data bus interconnects all blocks. Data flow in both directions on the data bus. For example, input data enter through the input port and proceed through the data bus to the CPU. If the CPU needs to store these data, it will send them back through the data bus to memory. Data to be outputted are sent (by the CPU) through the data bus to the output port. If the data bus connects to all blocks, how do the data know which block to go to? The answer is the address system. For example, when the CPU sends data to memory, it does it in two steps: First, it puts the destination memory address on the address bus; second, it puts the data on the data bus. When the designated memory detects its own address, it "wakes up" and takes the data from the data bus. The other blocks connected to the buses will ignore the whole sequence because they were not addressed. A good analogy here is the phone system, where the phone number is analogous to the memory address. Even though thousands of phones may be connected to the system, when you dial a number, only the designated phone rings. The beauty of the bus system is that it is expandable. Memory or addressable I/O units can be added to the system by simply connecting them to the buses.

The **control bus** (see Figure 2.2) consists of timing and event-control signals from the CPU. These signals are used to control the data flow on the data bus. For example, one of the control signals is the **read/write (R/W) line**. This signal informs the memory if the CPU wishes to read existing data out of memory or write new data into memory. Non-memory-mapped machines have a *memory–I/O control line*. This signal informs the system if the current data exchange involves memory or an I/O port. In general, the control bus is not as standardized as are the address and data buses.

2.2   INTRODUCTION TO MICROPROCESSOR OPERATION

The microprocessor works by executing a program of instructions. Creating the program is similar in concept to programming in BASIC, C, or any other high-level computer language. Each type of microprocessor has its own **instruction set**, which is the set of commands that it was designed to recognize and obey. Microprocessor instructions are very elemental and specific, and it usually takes more than one to accomplish what a single, high-level language instruction would. Many microprocessor instructions simply move data from one place to another within the computer; others perform mathematical or logic operations. Still another group of instructions control program flow, such as jumping forward or backward in the program. Each instruction in the instruction set is assigned its own unique 8-bit **operation code** (referred to as the **op-code**). The CPU uses this 8-bit number to identify the instruction.

All microprocessors have at least one **accumulator** [Figure 2.4(a)], which is an 8-bit data-holding register in the CPU. The accumulation acts as a "staging area" for data. All data coming to the CPU first go to the accumulator and from there to the next destination. Similarly, all data leaving the CPU exit from the accumulator. Mathematical operations usually store the result in the accumulator. Many of the instructions involve the accumulator in one way or another.

A **machine language** program is a list of instructions (in op-code form) for the microprocessor to follow. Before the program can be executed, it must first be loaded sequentially into memory. The op-code for the first instruction is loaded at the first address location, the op-code for the second instruction is loaded next in line, and so on.

Figure 2.4(b) shows a section of memory with a short program loaded in. The program listing includes the address, op-code, mnemonic, and a brief explanation. (A **mnemonic** is an English abbreviation of an instruction. A program listing using only mnemonics is called **assembly language**.) The program in Figure 2.4(b) directs the CPU to get 1 byte of data from input port 01, add 1 to it, and send the result to output port 02. Before execution can start, the address of the first instruction must be loaded into the program counter. The **program counter** is a special address-storage register that the CPU uses to keep track of where it is in the program, much like a bookmark. The program counter always holds the address of the next instruction to be executed. Once the microprocessor is activated, execution of the program is completely automatic. The execution process is a series of **fetch–execute cycles**, whereby the microprocessor first fetches the instruction from memory and then executes it. The following are the specific steps the microprocessor would go through to execute the program of Figure 2.4(b):

1. The microprocessor fetches the first instruction from memory. It knows where to find the instruction because its address is in the program counter.

2. Once in the CPU, the op-code is decoded to see which instruction it is, then the proper hardware is activated to execute this instruction. In the example program

Figure 2.4 The CPU uses an accumulator and a program counter to execute a simple program.

Adr	Instruction		
	Op Code	Mnemonic	Explanation
00	DB	IN 01	Get data from input port 01.
01	01		
02	3C	INR A	Increment accumulator.
03	D3	OUT 02	Send data to output port 02.
04	02		
05	76	HLT	Halt.
06			

Op codes are for Intel 8085 in hexidecimal. In this case, hexidecimal is used as a shorthand form of binary.

(a) Microprocessor registers

(b) Example microprocessor program

of Figure 2.4(b), the first instruction (IN 01) is 2 bytes long and causes data from input port 01 to travel along the data bus to the accumulator. Also, the program counter advances to 02 (the address of the next instruction). Execution of the first instruction is now complete.

3. The next fetch–execute cycle starts, this time fetching the instruction from address 02. The new instruction (INR A) is "increment the accumulator," so the accumulator is sent to the ALU to be incremented (add 1) and the result put back in the accumulator. The program counter advances to 03, which is the address of the next instruction.

4. The next fetch–execute cycle starts, this time fetching the instruction from address 03. The instruction (OUT 02) is executed, causing the accumulator data to be sent to output port 02.

5. The final instruction is fetched. It is a "halt," which causes the microprocessor to cease operating and go into a wait mode.

2.3 INTERFACING TO A MICROPROCESSOR CONTROLLER

An important part of any control system is the link between the controller and the real world. For a digital controller, data enter and exit through a parallel interface or through a serial interface. Both data formats are discussed next.

The Parallel Interface

The **parallel interface** transfers data 8 bits (or more) at the same time, using eight separate wires. It is essentially an extension of the data bus into the outside world. The parallel interface is ideal for inputting or outputting data from devices that are either on or off. For example, a single limit switch uses only one input bit, and an on–off signal to a motor requires only one output bit. These 1-bit signals are called **logic variables**, and eight such signals can be provided from a single (8-bit) port. This concept will be expanded on later in this section.

In other applications, the controller may use a parallel interface to connect to an analog device—for example, driving a variable-speed DC motor. In such a case, the binary output of the controller must first be converted into an analog voltage before it can drive the motor. This operation is performed by a special circuit called a digital-to-analog converter.

Digital-to-Analog Conversion
The **digital-to-analog converter** (DAC) is a circuit that converts a digital word into an analog voltage. It is not within the scope of this text to describe the internal workings of the DAC, but a general understanding of the operating parameters is appropriate.

Figure 2.5 shows the block diagram of a typical 8-bit DAC. The input is an 8-bit digital word. The output is a current that is proportional to the binary input value and must be converted to a voltage with an op-amp. A stable reference voltage (V_{ref}) must be

Figure 2.5

A digital-to-analog converter
(DAC) block diagram.

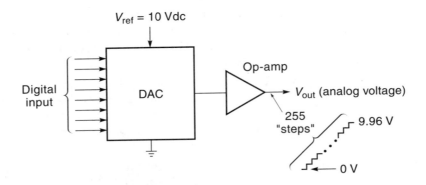

supplied to the DAC. This voltage defines the maximum analog voltage—that is, for a digital input of 11111111, V_{out} is essentially V_{ref}. If the input is 00000000, the V_{out} will be 0 Vdc. For all values in between, the output voltage is a linear percentage of V_{ref}. Specifically, the output voltage for any digital input (for the 8-bit DAC) is

$$V_{out} = \frac{input \times V_{ref}}{256} \tag{2.1}$$

where

$$V_{out} = \text{DAC output analog voltage}$$
$$input = \text{decimal value of the binary input}$$
$$V_{ref} = \text{reference voltage to the DAC}$$

◆ **EXAMPLE 2.1**

An 8-bit DAC has a V_{ref} of 10 V. The binary input is 10011011. Find the analog output voltage.

Solution

The binary input of 10011011 has a decimal value of 155. Applying Equation 2.1, we can calculate the analog output voltage:

$$V_{out} = \frac{input \times V_{ref}}{256} = \frac{155 \times 10\ V}{256} = 6.05\ V$$

Therefore, 6.05 V is the voltage we would expect on the analog output pin. [It is interesting to note that if the input were all 1s (which is a decimal value of 255), the output would be (255/256) × 10 V = 9.96 V, not 10 V as you might expect. This is a characteristic of the DAC.] ◆

An important consideration of digital-to-analog conversion is **resolution**. The resolution of a DAC is the worst case error that is introduced when converting between digital and analog. This loss in accuracy occurs because digital words can only represent discrete values, as indicated by the stair-step diagram in Figure 2.5. For example, the maximum value of an 8-bit number is 255 decimal, which means there are 255 possible "steps" of the output voltage. The difference between steps is the value of the least significant bit (LSB). Because the smallest increment is one step, the resolution (for 8-bit data) is 1 part in 255, or 0.39%. This resolution is adequate for many applications, but if more is needed, two (or more) 8-bit ports can be used together. Two ports provide 16 bits of data. The maximum decimal value of 16 bits is 65,535. Being able to divide an analog number into 65,535 parts means that each part will be much smaller, so we can more precisely represent that number.

♦ **EXAMPLE 2.2**

An antenna can rotate 180°. The controller needs to know the antenna position to within 1°. Can an 8-bit port be used?

Solution

The resolution required is 1 part in 180. Because 8 bits provide a resolution of 1 part in 255, an 8-bit port is certainly adequate. In fact, we have a choice: We could have the LSB = 1°, in which case the input values would range from 0 to 180, or we could equate 180° with 255, which makes the LSB = 0.706°. The latter makes maximum use of the 8 bits to give a better resolution, but if the system really doesn't need it, the clear, simple relationship of LSB = 1° is desirable.   ♦

Figure 2.6 shows a data sheet for an 8-bit DAC (DAC0808). This device comes as a 16-pin DIP (dual in-line package) and uses an external op-amp (such as the LF 351), two resistors, and a capacitor to complete the circuit. It requires plus and minus power-supply voltages. The time to complete a conversion is a fast 150 ns (nanoseconds). (The circuit shown in Figure 2.6 has a V_{ref} of 10 Vdc.)

Analog-to-Digital Conversion

The modern **analog-to-digital converter** (ADC) consists of a single IC with a few support components and is capable of converting an analog voltage into a digital word. This is a more complicated process (than for the DAC), and the hardware requires some conversion time, which is typically in the microsecond range. The conversion time required depends on the type of ADC, the applied clock frequency, and the number of bits being converted. Figure 2.7 shows a block diagram for an 8-bit ADC. The input V_{in} can be any voltage between 0 V and V_{ref}. When V_{in} is 0 Vdc, the output is 00000000; when V_{in} is V_{ref}, the output is 11111111 (255 decimal). For input voltages between 0 and V_{ref}, the output increases linearly with V_{in}; therefore, we can develop a simple ratio for the ADC:

$$\frac{\text{output}}{V_{in}} = \frac{255}{V_{ref}} \qquad \text{(for 8 bits)}$$

Solving for output gives the following:

$$\text{Output} = \frac{V_{in} \times 255}{V_{ref}} \qquad\qquad (2.2)$$

where

output = decimal output value of an 8-bit ADC
V_{in} = analog input voltage to the ADC
V_{ref} = ADC reference voltage

Figure 2.6

The data sheet for the DAC0808, an 8-bit digital-to-analog converter. (Courtesy of National Semiconductor Corp.)

Digital-to-Analog Converters

DAC0808, DAC0807, DAC0806 8-Bit D/A Converters

General Description

The DAC0808 series is an 8-bit monolithic digital-to-analog converter (DAC) featuring a full scale output current settling time of 150 ns while dissipating only 33 mW with ±5V supplies. No reference current (I_{REF}) trimming is required for most applications since the full scale output current is typically ±1 LSB of 255 I_{REF}/256. Relative accuracies of better than ±0.19% assure 8-bit monotonicity and linearity while zero level output current of less than 4 μA provides 8-bit zero accuracy for $I_{REF} \geq 2$ mA. The power supply currents of the DAC0808 series are independent of bit codes, and exhibits essentially constant device characteristics over the entire supply voltage range.

The DAC0808 will interface directly with popular TTL, DTL or CMOS logic levels, and is a direct replacement for the MC1508/MC1408. For higher speed applications, see DAC0800 data sheet.

Features

- Relative accuracy: ±0.19% error maximum (DAC0808)
- Full scale current match: ±1 LSB typ
- 7 and 6-bit accuracy available (DAC0807, DAC0806)
- Fast settling time: 150 ns typ
- Noninverting digital inputs are TTL and CMOS compatible
- High speed multiplying input slew rate: 8 mA/μs
- Power supply voltage range: ±4.5V to ±18V
- Low power consumption: 33 mW @ ±5V

Block and Connection Diagrams

Typical Application

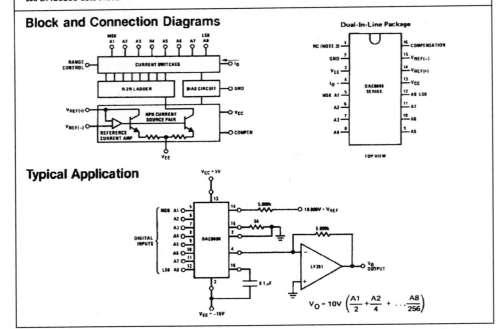

$$V_O = 10V \left(\frac{A1}{2} + \frac{A2}{4} + \cdots \frac{A8}{256} \right)$$

To start the conversion process, a start-conversion pulse is sent to the ADC. The ADC then samples the analog input and converts it to binary. When completed, the ADC activates the data-ready output. This signal can be used to alert the computer to read in the binary data.

Figure 2.7

An analog-to-digital converter (ADC) block diagram.

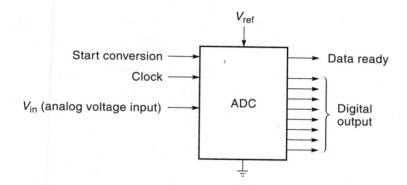

♦ **EXAMPLE 2.3**

An 8-bit ADC has a V_{ref} of 7 Vdc; the analog input is 2.5 Vdc. What is the binary output of the ADC?

Solution

The output is an 8-bit word that has a maximum decimal value of 255 (decimal) when $V_{in} = V_{ref}$. Therefore, an analog input voltage (V_{in}) of 7 Vdc would be converted to 255 decimal. Using this set of I/O data, we can develop a ratio and then use that to find the output for the specific input of 2.5 Vdc:

$$\frac{\text{output}}{V_{in}} = \frac{255}{7\text{ Vdc}}$$

Solving for output gives the following:

$$\text{Output} = \frac{2.5\text{ Vdc} \times 255}{7\text{ Vdc}} = 91$$

The result is 91 decimal = 01011011 binary. This is the output that would appear on the eight output lines of the ADC.   ♦

Figure 2.8 shows a data sheet for an 8-bit ADC (ADC0804). Packaged as a 20-pin DIP, this device can operate on a single 5-Vdc power supply and requires an external resistor and capacitor to complete the ADC circuit. The start-conversion pulse is applied to pin 3 ($\overline{\text{WR}}$), and the data-ready signal comes from pin 5 ($\overline{\text{INTR}}$). This particular ADC can be connected in a free-running mode where it performs one conversion after the other as fast as it can. Notice also that the pin labeled V_{ref} (pin 9) must be set at half of the actual V_{ref}. For example, if the requirements call for an analog voltage range of 0–5 Vdc, then pin 9 would be set to 2.5 Vdc. The time to complete a conversion is approximately 100 μs (micro-seconds), making it almost 700 times slower than the DAC0808 discussed earlier.

Figure 2.8

The data sheet for the ADC0804, an 8-bit analog-to-digital converter. (Courtesy of National Semiconductor Corp.)

 **National Semiconductor**

Analog-to-Digital Converters

ADC0801, ADC0802, ADC0803, ADC0804 8-Bit µP Compatible A/D Converters

General Description

The ADC0801, ADC0802, ADC0803, ADC0804 are CMOS 8-bit, successive approximation A/D converters which use a modified potentiometric ladder—similar to the 256R products. They are designed to meet the NSC MICROBUS™ standard to allow operation with the 8080A control bus, and TRI-STATE® output latches directly drive the data bus. These A/Ds appear like memory locations or I/O ports to the microprocessor and no interfacing logic is needed.

A new differential analog voltage input allows increasing the common-mode rejection and offsetting the analog zero input voltage value. In addition, the voltage reference input can be adjusted to allow encoding any smaller analog voltage span to the full 8 bits of resolution.

Features

- MICROBUS (8080A) compatible—no interfacing logic needed
- Easy interface to all microprocessors, or operates "stand alone"

- Differential analog voltage inputs
- Logic inputs and outputs meet T^2L voltage level specifications
- Works with 2.5V (LM336) voltage reference
- On-chip clock generator
- 0V to 5V analog input voltage range with single 5V supply
- No zero adjust required
- 0.3" standard width 20-pin DIP package

Key Specifications

- Resolution   8 bits
- Total error   ±1/4 LSB, ±1/2 LSB and ±1 LSB
- Conversion time   100 µs
- Access time   135 ns
- Single supply   5 V_{DC}
- Operates ratiometrically or with 5 V_{DC}, 2.5 V_{DC}, or analog span adjusted voltage reference

Typical Applications

Connection Diagrams

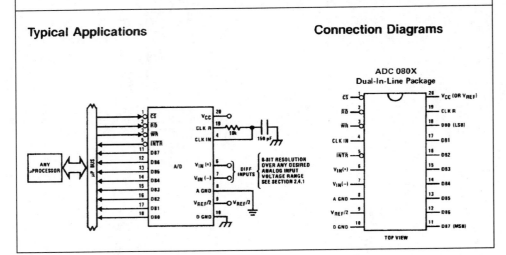

A Control System Using Parallel Ports

Figure 2.9 shows a position control system using a microprocessor-based controller with parallel ports. This particular system has one output port and three input ports (each port has its own address). The output port is partitioned: Six bits are converted in a DAC to provide the analog motor-drive signal, the seventh bit specifies motor direction (1 = clockwise, 0 = counterclockwise), and the eighth bit turns on an audio alarm if some emergency situation is detected. The first input port inputs the set-point data, the

Figure 2.9

A control system using parallel interface.

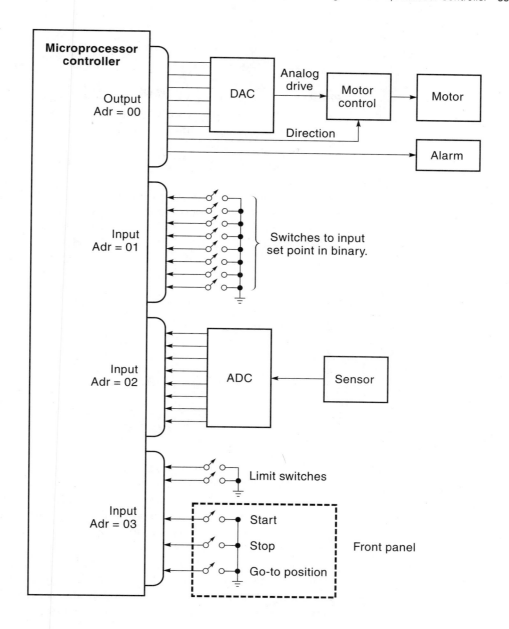

second inputs the ADC data from the sensor, and the third inputs various 1-bit logical variables. In this case, the system has two mechanical limit switches, as well as three front-panel switches.

Operation of the system proceeds as follows: The controller inputs the data from port 03 to determine if the start button has been pressed. If so, then the set point is read in from port 01 and the digitized sensor data is read in from port 02. Based on its control strategy, the controller outputs to port 00 a binary word representing the motor-control voltage. This digital data is converted to an analog voltage with the DAC. This entire sequence is repeated over and over until the stop button is pushed.

The Serial Interface

In a **serial interface**, the data are sent 1 bit after the other on a single wire. There are a number of good reasons for doing this. First, the cabling is simpler because only two wires are needed (at a minimum), those being "data" and "return." Second, shielding a small group of wires, which is often necessary in an electrically noisy industrial environment, is easier. Third, serial data can make use of existing single-channel data lines such as the telephone system. For these reasons, serial data transfer is usually recommended for distances greater than 10–30 ft.

Because data always exist in a parallel form inside the computer, it must be converted to serial data before coming out the serial port. This is accomplished with a special parallel-to-serial converter IC called a **universal asynchronous receiver transmitter** (UART). On the other end of the line, a receiver must convert the serial data back into parallel data, which is done with another UART. Figure 2.10 shows the basic serial data circuit.

Figure 2.10

Components in a serial interface circuit.

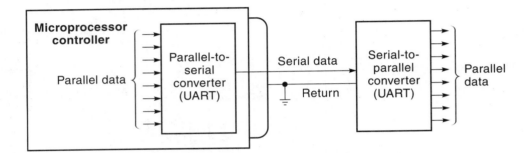

Serial data are classified as being either synchronous or asynchronous. *Synchronous data* require that the data bytes be sent as a group in a "package." It is used in sophisticated communication systems that move a lot of data and will not be further discussed here. *Asynchronous data* transfer is the more common (but slower) type of serial transfer and allows for individual bytes to be sent when needed.

Figure 2.11 shows the standard format for asynchronous serial data. First, a start bit is sent, then the data (LSB first), then a parity-error checking bit, and finally the stop bit(s). Some variation is allowed to this format, but both transmitter and receiver must use the same format. The other important parameter in serial transmission is the number of bits sent per second (frequently called the **baud** rate, although the term is technically incorrect in most cases). Standard bit rates are 300 bps (bits per second), 1200 bps, 2400 bps, 9600 bps, 14,400 bps and highway. Serial data transmission is much slower than parallel transmission. At 300 bps, it takes almost 37 ms to transmit 1 byte of data, compared to a few microseconds for parallel—this is over 1000 times slower. Still, for many applications, particularly process control, the longer data-transfer times are not a problem.

In many cases, serial data are transmitted over telephone lines, which requires using a modem. A **modem** is a device that converts digital data into audio tones. The interface between the microcomputer and the modem uses the **RS-232 standard**. Officially, this standard specifies the serial data interface between **data terminal equipment** (DTE) and

Figure 2.11

Serial data format for the binary word 10110010 (least significant bit).

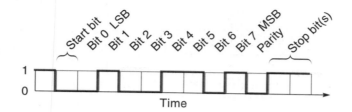

data communication equipment (DCE), as illustrated in Figure 2.12. The DTE is usually the computer, and the DCE is usually the modem. The RS-232 standard specifies connector types, signal names, pin numbers, voltages, and timing. In practice, the RS-232 standard can be applied to any serial interface as long as one unit acts like a DTE and the other acts like a DCE. If two DTE units need to interface with each other, a special cable called a **null modem** is required.

Serial data transfer is somewhat more complicated than parallel data transfer, but it offers advantages such as two-wire communications and a true standard interface. The hardware to handle serial data is standardized, readily available, and reliable.

Figure 2.12

The RS-232 interface between data terminal equipment (DTE) and data communication equipment (DCE). Serial data are transferred on pins 2 and 3; the other signals control the flow of data.

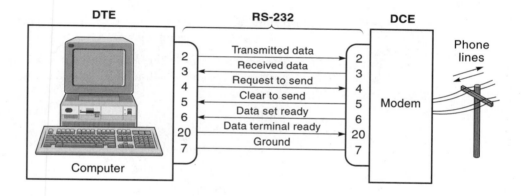

2.4   INTRODUCTION TO CONTROLLER PROGRAMMING

It is beyond the scope of this text to present a detailed discussion of how to program a microprocessor in machine language. Still, it is useful to investigate in a general way what the software must do. A digital controller is a computer operating in **real time**. This means that *the program is running all the time—repeatedly taking in the newest sensor data and then calculating a new output for the actuator.*

The basic structure of a controller program is a *loop*. In a loop structure, the same sequence of instructions is executed over and over again, and each pass through the loop

Figure 2.13
A generalized controller
program.

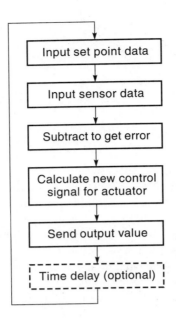

is called an **iteration**, or **scan**. Figure 2.13 shows a generalized controller program, and an explanation of the program follows:

1. The program reads in the set-point data (recall that the set point is the desired position of the controlled variable). This data could be read in from an input port or from memory.

2. The program directs the computer to read (from a sensor) the actual value of the controlled variable.

3. The actual data are subtracted from the set point to get the error.

4. Based on the error data, the computer calculates a new actuator control signal.

5. The new output is sent to the actuator.

6. The programs loops back to step 1 and starts over again.

The time it takes for the computer to execute one pass through the loop determines the time interval between input readings (known as the **sampling rate**). If this interval is too long, the computer may not get an accurate picture of what the controlled variable is really doing (see Chapter 11 for a discussion of aliasing). Execution of the loop can be accelerated by specifying a faster computer or streamlining the program. In other situations, the computer must pause and wait. For example, a pause might be inserted to give an operator time to make some adjustment or to allow time for a motor to "spin down." This is done by inserting time-delay loops in the program. A **time-delay loop** is simply a do-nothing, "wheel-spinning" loop where the computer is instructed to count up to some large number. Using this technique, we can make the program pause for any length of time—from a few microseconds to hours. If a time-delay loop is inserted in the main program loop (as shown in Figure 2.13), the effect is to slow the cycle time for the main loop. This is sometimes done to force matching of the sample rate to some predetermined value.

At one time, people thought that the best and most efficient microprocessor programs were those written directly in assembly language—that is, the programmer would directly select the machine language instructions. Today, sophisticated programs (called *compilers*) can convert a program written in a high-level language, primarily C, into very efficient machine language. High-level languages use English-sounding words and a set of powerful commands to specify simple and complicated programming operations with a minimum of instructions. Using a high-level language to write programs for a microprocessor offers big advantages, such as more compact program listing, ease of writing equations, and more comprehensible documentation. Also, programs written in a high-level language can be compiled to run on any model of microprocessor.

2.5 MICROPROCESSOR-BASED CONTROLLERS

Single-Chip Microcomputers (Microcontrollers)

A microprocessor by itself is not a computer. To be functional, the microprocessor must be connected to other integrated circuits that provide the memory and I/O capability. A microcontroller is a computer on a single IC, designed specifically for control applications. It consists of a microprocessor, memory (both RAM and ROM), I/O ports, and possibly other features such as timers and ADCs/DACs. Having the complete controller on a single chip allows the hardware design to be simple and very inexpensive. Microcontrollers are showing up increasingly in products as varied as industrial applications, home appliances, and toys. In such uses as these, they are called **embedded controllers** because the controller is located physically in the equipment being controlled.

The main difference between microprocessors and microcontrollers is that microprocessors are being designed for use in microcomputers where greater speed and larger word size are the driving requirements, whereas microcontrollers are evolving toward reduced chip count by integrating more hardware functions on the chip. Most control applications do not need the 32-bit word size and 50-MHz (megahertz) speed of the newer microprocessors. Eight bits and 1 MHz will work just fine in many applications, *and* the single-chip microcontroller costs much less.

Microcontrollers have a small, fixed-memory size (although external memory can be added) and a fixed number of I/O ports. For example, the Motorola 68HC11, a popular 8-bit microcontroller, has 256 bytes of RAM, 8K of ROM, and 512K bytes of EPROM. It also has five 8-bit ports with built-in serial data transfer and ADC capability. Another popular 8-bit microcontroller is the Intel 8051, which has 128 bytes of RAM and 4K bytes of ROM, four parallel data ports, and a serial port. For control applications, these hardware arrangements usually are adequate: ROM is used to store the control program, and RAM is used as data registers and a "scratch pad." The I/O signal lines can usually be connected directly to the microcontroller without additional port circuitry. Figure 2.14(a) shows a block diagram for the 68HC11 microcontroller, and Figure 2.14(b) shows the block diagram of the 8051 microcontroller.

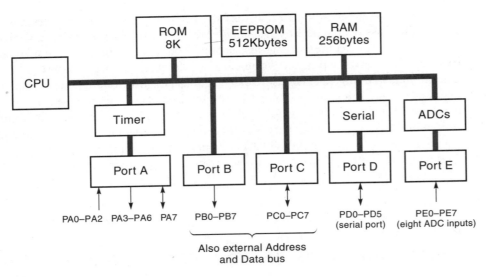

(a) Motorola 68HC11 microcontroller block diagram

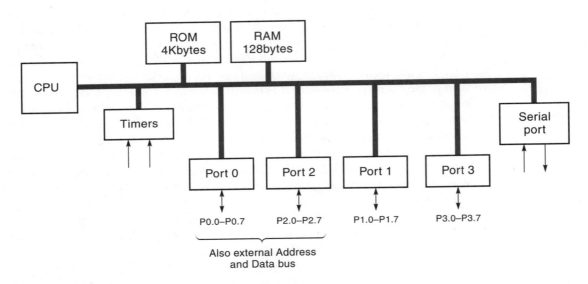

(b) Intel 8051 microcontroller block diagram

Figure 2.14 Block diagrams of microcontrollers.

Another difference between microprocessors and microcontrollers concerns the instruction set. The microprocessor tends to be rich in instructions dealing with moving data in and out of memory but has relatively few instructions that manipulate individual bits within a byte. The microcontroller tends to be just the opposite: fewer memory-move instructions and more bit-handling instructions. The reason for the lack of memory-move instructions is obvious; the microcontroller has little or no external memory. The additional bit-handling instructions were included because they are so useful in control system applications. For example, in a control system, each separate bit of a parallel output word might control a different device such as a motor or indicator light. The bit-handling instructions allow the software to easily turn on or off one device without affecting the others.

A wide variety of microcontrollers are available. At the low end are the 4-bit models, which are more than adequate for appliances and toys. These tend to be large-volume, low-cost applications. A typical 4-bit microcontroller is the National COP420. It comes in a 28-pin package and has 64 bytes of RAM and 1K byte of ROM. Eight-bit microcontrollers (such as the 68HC11 and 8051 mentioned earlier) are very popular because 8 bits turn out to be a convenient size for both numeric and character data. At the high end, 16- and even 32-bit microcontrollers are now available for control systems requiring sophisticated, high-speed calculating power for such applications as complicated servomechanisms, avionics, or image processing.

Single-Board Computers

Single-board computers are off-the-shelf microprocessor-based computers built on a single printed-circuit card (Figure 2.15). They come in many configurations, but in general they use a standard microprocessor such as the Zilog Z80, the Intel 8085 or

Figure 2.15   A single-board computer. (Courtesy of Vesta Technology, Inc.)

8088 family, or the Motorola 68000. They also include memory ICs (both RAM and ROM), I/O capability, and perhaps special interface circuits such as ADCs or DACs. Single-board computers are manufactured by major microprocessor producers such as Intel and Motorola as well as many other smaller companies. The obvious advantage of using a ready-made microprocessor board is that it eliminates design- and board-testing time. This is particularly important in small-volume production or one-of-kind systems.

Programmable Logic Controllers

A **programmable logic controller** (PLC) is a self-contained microprocessor-based unit, designed specifically to be a controller. The PLC includes an I/O section that can interface directly to such system components as switches, relays, small motors, and lights. Developed in the late 1960s to replace relay logic controllers, PLCs come in various sizes and capabilities; Figure 2.16 shows a selection of PLCs. The big difference between PLCs and the other devices discussed in this section is that the PLC comes packaged and ready to go. Installation is very easy because in many cases the sensors and actuators can be connected directly to the PLC. Once installed, the microprocessor program is **downloaded** into the PLC from some source such as a personal computer. The PLC manufacturer usually supplies software to facilitate the programming operation. This software can convert a relay logic-wiring diagram (ladder diagram) directly into

Figure 2.16 Programmable logic controllers. (Courtesy of Allen-Bradley Company)

a PLC program. Multiple PLCs in a plant can be networked so the individual units can be monitored and programmed from a single station. This is a form of distributed computer control (DCC) discussed in Chapter 1. PLCs are discussed in detail in Chapter 12.

Personal Computers Used in Control Systems

The availability of relatively low-cost, off-the-shelf **personal computers** (PCs) has made them an attractive alternative for small, one-of-kind control applications. Control system software packages are commercially available for the PC. These programs are adaptable and allow the user to tailor the software to fit the control application. Most of these packages are menu-driven and use interactive graphics to link animation with changing process values. Some programs have provisions to mathematically simulate the process being controlled to help optimize the controller coefficients.

A standard PC comes with four to eight identical **expansion slots**, which are circuit-card connectors emanating from the *motherboard* (main board) of the computer. **Expansion cards** plug into these slots and form a bridge between the computer and the outside world. Many different types of interface cards are available, such as I/O serial and parallel data ports, ADCs, DACs, and computer-controlled output relays, to name a few. Figure 2.17 shows an example of an interface expansion card.

Figure 2.17 A 24-bit parallel I/O interface expansion card for a personal computer. (Courtesy of Omega Engineering, Inc.)

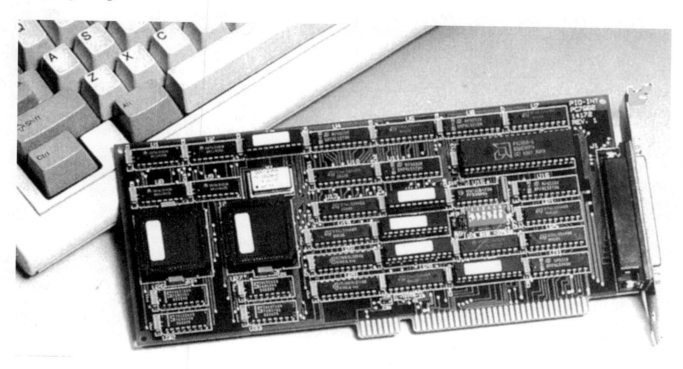

Historically, data-acquisition and control functions were kept separate. Controllers ran the process, and other instruments measured and recorded the result. The concept of having a single PC perform both tasks seems logical; after all, the PC can use its computing ability first as the controller and then tabulate system performance data. These data can be stored on disk and/or displayed on the monitor.

A potential problem may arise because the controller must operate in real time. If a computer is to control a process and monitor it at the same time, the data-reduction process must not take so long as to interfere with the control duties; a control response can't wait. One way to overcome this problem is to divide the control and data-acquisition tasks among multiple processors. Using the PC as the master computer, a separate microprocessor on an expansion card can perform data collection uninterrupted. One type of I/O controller card has slots for three smaller boards. These smaller boards have various combinations of analog and digital I/O ports and counter-timers. Some boards are available with solid-state relays, which can be used to directly control AC and DC motors.

A PC with I/O expansion cards often costs less than a stand-alone computerized control system. The cards do not need a separate enclosure and use the PC's power supply, keyboard for input, and monitor for display. Also, using a standard PC means that programs can be developed on another compatible computer, eliminating process downtime.

Numerous manufacturers are selling rugged PCs that can survive in harsh industrial environments. These computers typically use a membrane-type keyboard (the keyboard appears as one continuous sheet of flexible plastic) and have sealed cases and filters covering the air vents. Some models of these computers are rack-mountable and contain their own battery-backup power supply.

SUMMARY

A microprocessor is a digital integrated circuit that performs the basic operations of a computer. Microprocessors are used extensively as the basis of a digital controller. Digital control systems are advantageous because digital data can be transferred and stored virtually error-free, and the control strategy can be changed by simply reprogramming.

A computer consists of three basic functional units: (1) the CPU (microprocessor), which executes the programmed instructions and performs the calculations; (2) the memory, which stores the program and data; and (3) input/output, which interfaces the computer to the outside world. A microprocessor-based computer interconnects these units with three groups of signals called buses. The address bus carries the address of the data to be processed. The data bus carries the data, and the control bus carries timing and control signals. Computers handle data as groups of binary bits. Many microprocessor-based controllers handle data in 8-bit groups called a byte.

A microprocessor has a set of instructions that it can execute (called the instruction set). Each instruction is identified by a digital code called the operation code (op-code). A program consists of a list of these op-codes stored in memory. The microprocessor automatically fetches the instructions from memory and executes them, one by one.

A digital controller may have two kinds of data interfaces: parallel and serial. The parallel interface is the most straightforward system, where all 8 bits are sent at the same time on eight separate wires. In the serial interface, data is sent 1 bit after the other on a single wire. Serial data transfer is better for longer distances.

Many control systems use components that require an analog signal interface; therefore, the signals to or from the digital controller must be converted with an ADC (analog-to-digital converter) or a DAC (digital-to-analog converter). Both circuits are available in IC form.

The digital controller program has a standard format. First, it reads the set point and sensor values. Then it subtracts these values to determine the system error. Based on the error value, it next calculates the appropriate actuator response signal and sends it out. Then it loops back to the beginning of the program and executes the same set of instructions over and over.

Microprocessor-based controllers come in a number of standard forms. A microcontroller includes a microprocessor, memory, and input/output all on a single IC. A single-board computer is an off-the-shelf microprocessor-based computer, assembled onto a single printed circuit board. A programmable logic controller (PLC) is a self-contained unit specifically designed to be a controller. A personal computer (PC) is a general-purpose, self-contained computer; however, with the addition of interface expansion cards, a PC becomes a very adaptable and cost-effective controller.

GLOSSARY

accumulator A temporary, digital data-storage register in the microprocessor used in many math, logic, and data-moving operations.

ADC *See* **analog-to-digital converter.**

address A number that represents the location of 1 byte of data in memory or a specific input/output port.

address bus A group of signals coming from the microprocessor to memory and I/O ports, specifying the address.

ALU *See* **arithmetic logic unit.**

arithmetic logic unit (ALU) The part of the CPU that performs arithmetic and logical operations.

analog-to-digital converter (ADC) A device (usually an IC) that can convert an analog voltage into its digital binary equivalent.

assembly language A computer program written in mnemonics, which are English-like abbreviations for machine-code instructions.

baud The rate at which the signal states are changing; frequently used to mean "bits per second."

bit The smallest unit of digital data, which has a value of 1 or 0.

byte An 8-bit digital word.

central processing unit (CPU) The central part of a computer, the CPU performs all calculations and handles the control functions of the computer.

control bus A group of timing and control signals coming from the microprocessor to memory and I/O ports.

CPU *See* **central processing unit.**

DAC *See* **digital-to-analog converter.**

data bus A group of signals going to and from the microprocessor, memory, and I/O ports. The data bus carries the actual data that are being processed.

data communication equipment (DCE) One of two units specified by the RS-232 standard (for serial data transfer); the DCE is usually a modem.

data terminal equipment (DTE) One of two units specified by the RS-232 standard (for serial data transfer). The DTE is usually the computer.

DCE *See* **data communication equipment.**

digital-to-analog converter (DAC) A circuit that translates digital data into an analog voltage.

download To transfer a computer program or data into a computer (from another computer).

DTE *See* **data terminal equipment.**

embedded controller A small microprocessor-based controller that is permanently installed within the machine it is controlling.

expansion card/slot An expansion card is a printed circuit card that plugs into an expansion slot on the motherboard of a personal computer (PC). The expansion card usually interfaces the PC to the outside world.

fetch–execute cycle A computer cycle where the CPU fetches an instruction and then executes it.

input/output (I/O) Data from the real world moving in and out of a computer.

I/O *See* **input/output.**

instruction set The set of program commands that a particular microprocessor is designed to recognize and execute.

iteration   One pass through the computer program being executed by the digital controller; each iteration "reads" the set-point and sensor data and calculates the output to the actuator.

least significant bit (LSB)   The rightmost bit of a binary number. Can also mean the smallest increment of change.

logical variable   A single data bit in those cases where a single bit is used to represent an on–off switch, motor on–off control, and so on.

LSB   *See* **least significant bit**.

machine language   The set of operation codes that a CPU can execute.

memory   The part of the computer that stores digital data. Memory data is stored as bytes, where each byte is given an address.

memory-mapped input/output   A system where I/O ports are treated exactly like memory locations.

microcontroller   An integrated circuit that includes a microprocessor, memory, and input/output; in essence, a "computer on a chip."

microprocessor   A digital integrated circuit that performs the basic operations of a computer but requires some support integrated circuits to be functional.

most significant bit (MSB)   The leftmost bit in a binary number.

MSB   *See* **most significant bit**.

mnemonic   An English-like abbreviation of an operation code.

modem   A circuit that converts serial data from digital form into tones that can be sent through the telephone system.

nonvolatile memory   Computer memory such as ROM that will not lose its data when the power is turned off.

null modem   A cable that allows two DTE units to communicate with each other (*see* **RS-232**).

operation code (op-code)   A digital code word used by the microprocessor to identify a particular instruction.

parallel interface   A type of data interface where 8 bits enter or leave a unit at the same time on eight wires.

PC   *See* **personal computer**.

personal computer (PC)   A microprocessor-based, self-contained, general-purpose computer (usually refers to an IBM or compatible computer).

PLC   *See* **programmable logic controller**.

port   The part of a computer where I/O data lines are connected; each port has an address.

program counter   A special address-holding register in a computer that holds the address of the next instruction to be executed.

programmable logic controller (PLC)   A rugged, self-contained microprocessor-based controller designed specifically to be used in an industrial environment.

RAM   *See* **random-access memory**.

random-access memory (RAM)   Sometimes called read/write memory, a memory arrangement using addresses where data can be written in or read out; RAM loses its contents when the power is turned off.

read-only memory (ROM)   Similar to RAM in that it is addressable memory, but it comes preprogrammed and cannot be written into; also, it does not lose its data when the power is turned off.

read/write (R/W) line   A control signal that goes from the microprocessor to memory.

real time   Refers to a computer that is processing data *at the same time* that the data are generated by the system.

resolution   In digital-to-analog conversion, the error that occurs because digital data can only have certain discrete values.

ROM   *See* **read-only memory**.

RS-232 standard   A serial data transmission standard that specifies voltage levels and signal protocol between a DTE (computer) and a DCE (modem).

R/W   *See* **read/write line**.

sampling rate   The times per second a digital controller reads the sensor data.

scan   *See* **iteration**.

serial interface   A type of interface where data are transferred 1 bit after the other on a single wire.

single-board computer   A premade microprocessor-based computer assembled onto a single printed-circuit card.

time-delay loop   A programming technique where the computer is given a "do-nothing" job such as counting to some large number for the purpose of delaying time.

UART   *See* **universal asynchronous receiver transmitter**.

universal asynchronous receiver transmitter (UART)   A special purpose integrated circuit that converts data from parallel to serial format and vice versa.

volatile memory   Computer memory such as RAM that will lose its data when the power is turned off.

word   A unit of digital data that a particular computer uses; common word sizes are 4, 8, 16, and 32 bits.

Section 2.1

1. Briefly describe the functions of the ALU, control unit, CPU memory, and input/output.

2. What steps does the microprocessor take to read data at address 1020? (Specify the actions of the address bus and data bus in your answer.)

3. Briefly define *address bus*, *data bus* and *control bus*.

Section 2.2

4. What is a microprocessor *instruction set*, and how is it different from a high-level language such as BASIC?

5. A certain microprocessor has a simple instruction set shown below.

```
INSTRUCTION SET
Op-code        Explanation
76             Halt the microprocessor.
C6*            Add next byte to accumulator.
D6*            Subract next byte from accumulator.
3C             Increment the accumulator.
3D             Decrement the accumulator.
3E*            Move the next byte into the
               accumulator.
*These instructions use two bytes.
```

What number would be in the accumulator after the program shown below was run?

```
PROGRAM
Address        Op-code
001            3E
002            05
003            D6
004            02
005            3C
006            76
```

Section 2.3

6. Temperature values from $-20°F$ to $120°F$ are input data for a microprocessor computer. Are 8 bits sufficient? If so, what is the resolution?

7. Explain the function of the following: *parallel data port* and *serial data port*.

8. Serial data are sent at 1200 bps using the format of Figure 2.11, with one stop bit. How long would it take to send 1000 bytes of data?

9. An 8-bit DAC has a reference voltage of 9 V. The binary input is 11001100. Find the analog output voltage.

10. The binary data from the computer in a certain application are expected to go from 00000000 to only 00111111. These data are the input of a DAC. The analog output should go 0–5 V. Find the DAC reference voltage necessary to make this happen.

11. An 8-bit ADC has a reference voltage of 12 V and an analog input of 3.7 V. Find the binary output.

12. The binary output of an ADC should have the range 00000000–11111111 corresponding to an input of 0–6 V. Find the necessary reference voltage.

Section 2.4

13. What is *real-time computing*, and is it necessary for control systems?

14. Describe the basic steps in a control program scan (loop).

15. At some point in the program it is desired to have the computer wait 5 s for an operator response. How would this delay be accomplished in software?

16. A program contains 150 instructions, and the average execution time per instruction is 2 μs. Find the sample rate of this program.

Section 2.5

17. What is a *microcontroller*, and what are some differences between a microcontroller and a microprocessor?

18. What is a *programmable logic controller*?

19. You want to use a personal computer to control a simple robot arm. The arm has two joints, an elbow and a wrist. Each joint has a DC motor and a position sensor that outputs a DC voltage. You already have a "plain vanilla" PC; make a list of what you would need to acquire to make this system work.

20. Compare and contrast the following: a *microprocessor*, a *microcontroller*, a *programmable logic controller*, and a *personal computer*.

3 Operational Amplifiers and Signal Conditioning

OBJECTIVES

After studying this chapter, you should be able to:

❑ Recognize the characteristics of an operational amplifier and describe how they can be used as the basis for different types of useful amplifiers.

❑ Design the following types of op-amp circuits to meet specific requirements: voltage follower, inverting amplifier, noninverting amplifier, summing amplifier, differential amplifier, and comparator.

❑ Understand the operation of the following types of circuits: integrators and differentiators, active filters, current-loop signal transmission, analog switches and multiplexers, and sample and hold.

❑ Understand the concepts of the earth ground and ground loops, magnetic and electrostatic shielding, and the importance of a single-point ground.

INTRODUCTION

One of the necessary conditions of any real system is the successful **interfacing**, or connecting together, of the various components. In a block diagram, an interface is represented by a line between two blocks, indicating that some sort of signal passes between the blocks. If it were only this easy! The fact is, interfacing is sometimes the most difficult task in getting a system operational. There are several different categories of interfacing requirements.

One type of interfacing is between analog and digital circuits. Most controllers are digital, whereas many sensors and actuators use analog signals. This means analog-to-digital or digital-to-analog converters may be required (as discussed in Chapter 2). When designing this type of interface, you must consider such things as resolution (number of bits), analog voltage level, and conversion speed required.

Another interface problem is matching voltage levels between components. A sensor may put out a voltage range of 0–0.5 V, whereas the receiving component may need a signal in the voltage range of 0–10 V. Or a sensor may put out a high-impedance signal (easily loaded down) and needs to be converted to a stronger low-impedance signal.

Sometimes several sensors must share the same input port of a controller. This requires an electronic switching circuit capable of connecting different analog channels to the same destination. Another requirement may be to add or subtract analog signals—for example, when the feedback signal is subtracted from the set-point in an analog controller.

Some situations require that there be little or no signal loss between components, even if they are some distance apart. This might seem to be impossible because all wires have resistance; however, the *current-loop* technique virtually eliminates signal attenuation.

Another set of interface problems deals with handling electrical noise from the outside world. Although some types of interference can be filtered out, it is usually best to try to prevent the noise from entering the system. This is done with proper shielding and grounding.

We will deal with these topics in this chapter. It will not be an exhaustive treatment because whole books are available on each subject, but it will at least introduce accepted solutions to various problems.

3.1   OPERATIONAL AMPLIFIERS

Introduction

An **operational amplifier (op-amp)** is a high-gain linear amplifier. Op-amps are usually packaged in IC form (one to four op-amps per IC) and are relatively inexpensive. The op-amp approaches the ideal amplifier of the analog designer's dreams because it has such ideal characteristics:

1. Very high open-loop gain: $A = 200,000+$, but unpredictable
2. Very high input resistance: $R_{in} > 1\ M\Omega$
3. Low output resistance: $R_{out} = 50–75$ ohm

These characteristics make designing with op-amps relatively easy. As we will see, the high open-loop gain makes it possible to create an amplifier with a very predictable stable gain of anywhere from 1 to 1000 or more. The significance of the very high input resistance (R_{in}) is that the op-amp draws very little input current. This means it will not load down whatever circuit or sensor is driving it. The op-amp's low output resistance (R_{out}) means it can drive a load without being loaded down itself. However, an op-amp is a signal amplifier, not a power amplifier. It is not designed to output large currents and so is not usually used to drive loads such as loudspeakers or motors directly.

Figure 3.1 shows the symbol for a typical op-amp. It has two inputs (V_1 and V_2) and one output (V_{out}). Also shown are the two power-supply inputs, which are typically $+12$ V and -12 V. The output voltage can swing to within about 80% of the supply voltages. Notice there is no ground connection at all. Most op-amps are actually differential amplifiers, which means they amplify the difference between V_1 and V_2. This is shown in Equation 3.1:

Figure 3.1

The op-amp symbol.

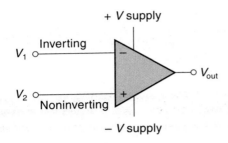

$$V_{out} = A(V_2 - V_1) \tag{3.1}$$

where

V_{out} = output voltage
A = open-loop gain
V_1 = inverting input
V_2 = noninverting input

The **open-loop gain** (A) is the raw unmodified gain of the op-amp; it is high, typically 200,000 or more. V_2 is called the **noninverting input**. As the name implies, *the output is in phase with the noninverting input* (when V_2 goes positive, V_{out} goes positive; when V_2 goes negative, V_{out} goes negative). The noninverting input is identified by the + sign in the symbol of Figure 3.1.

The other input to the op-amp is called the **inverting input**. *The output will be out of phase with the signal at the inverting input* (when V_1 goes more positive, the output will go more negative, and vice versa). The inverting input is identified by the − sign in the symbol.

Even though the op-amp has two separate inputs, there is just one input voltage, which is the difference between V_2 and V_1. This is illustrated in Example 3.1.

♦ **EXAMPLE 3.1**

Figure 3.2 shows an op-amp with an open-loop gain of 100,000. Find the output for the following conditions:

a. V_1 and V_2 are both 4 μV.
b. V_1 is 2 μV, and V_2 is 4 μV.
c. V_1 is 6 μV, and V_2 is 3 μV.

Figure 3.2   Various input voltage combinations (A = 100,000) (Example 3.1).

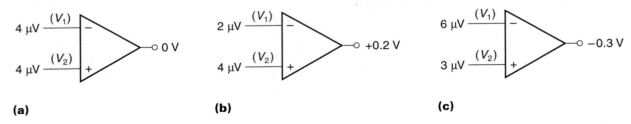

(a)   (b)   (c)

Solution   We will use Equation 3.1 to solve this problem.

a. Both V_1 and V_2 are 4 μV:

$$V_{out} = 100,000 \times (4 \text{ μV} - 4 \text{ μV})$$

$$= 100,000 \times (0 \text{ μV})$$

$$= 0 \text{ V}$$

This shows that the output of the op-amp is zero if the inputs are the same voltage, regardless of their actual value.

b. The noninverting input V_2 is 4 μV, and the inverting input V_1 is 2 μV:

$$V_{out} = 100,000 \times (4 \text{ μV} - 2 \text{ μV})$$

$$= 100,000 \times (2 \text{ μV})$$

$$= 0.2 \text{ V}$$

This result shows that the output is positive if the $(V_2 - V_1)$ quantity has a net positive value.

c. The inverting input is 6 μV, and the noninverting input is 3 μV:

$$V_{out} = 100,000 \times (3 \text{ μV} - 6 \text{ μV})$$

$$= 100,000 \times (-3 \text{ μV})$$

$$= -0.3 \text{ V}$$

This result is negative because the $(V_2 - V_1)$ quantity has a negative net value.   ◆

Example 3.1 begins to illustrate one aspect of op-amps that may seem strange at first—the sign of the output. Consider the three op-amps in Figure 3.3. In Figure 3.3(a), both inputs are positive, yet the output is negative. Why? The − input has the larger magnitude, so the quantity $(V_2 - V_1)$ is negative (2 μV − 3 μV = −1 μV). From Equation 3.1 (the op-amp equation),

Figure 3.3   Various input voltage combinations (A = 100,000).

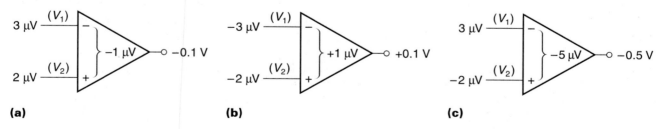

(a)   (b)   (c)

$$V_{out} = A(V_2 - V_1)$$

you can see that if $(V_2 - V_1)$ is negative, V_{out} will be negative.[*]

Now consider the circuit of Figure 3.3(b). The inputs are both negative, yet the output is positive. To understand this, again examine the $(V_2 - V_1)$ quantity, paying attention to the signs. In this case, $[-2 \ \mu V - (-3 \ \mu V)] = +1 \ \mu V$, which is positive. The circuit in Figure 3.3(c) is more straightforward. In this case, $(V_2 - V_1) = (-2 \ \mu V - 3 \ \mu V) = -5 \ \mu V$, which is clearly negative.

Now consider the case where only a single input is required. There are two possibilities: The output will be either in phase or out of phase with the input. To make a noninverting amplifier (where the output is in phase with the input), the $-$ input is grounded, and the input signal is connected to the noninverting input ($+$), as shown in Figure 3.4(a). If we want an inverting amplifier, where the output is out of phase with the input, we connect the signal into the inverting input ($-$) and ground the $+$ input, as shown in Figure 3.4(b). All the amplifier circuits discussed so far are called *open loop* because they operate at maximum gain. As we will see, this is not the typical application of an op-amp—we are doing it here because it simplifies the discussion of how the differential inputs work.

Figure 3.4

Single-input, open-loop amplifiers: (a) noninverting and (b) inverting.

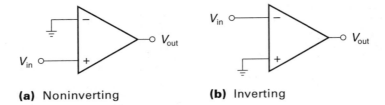

(a) Noninverting   **(b)** Inverting

Most op-amp circuits incorporate **negative feedback**. This means that a portion of the output signal is fed back and subtracted from the input. Negative feedback results in a very stable and predictable operation at the expense of lowered gain (which we can easily afford because the open-loop gain is so high to begin with). Analyzing op-amp circuits is actually easier than analyzing traditional discrete transistor amplifiers because the op-amp's impressive parameters allow us to make three circuit-simplifying assumptions:

- ❏ Assumption 1: $V_1 = V_2$.   *Explanation*: How can we possibly assume that the inputs V_1 and V_2 are always the same? Could we not force V_1 and V_2 to be anything we like? Yes, but the argument goes like this: The output voltage is equal to $A(V_2 - V_1)$ where A, being the open-loop gain, is a very high number. Thus, even a small difference between V_1 and V_2 will cause a very large output. However, the output has a practical upper limit established by the power supply; therefore, to keep the output from exceeding its limits, the difference between V_1 and V_2 must be very small. This is illustrated in Figure 3.5. The power supply is $+15$ V and -15 V, which limits the output voltage swing to about $+12$ V and

---

[*]Another way to determine the output polarity is to use the following rule: The output will assume the polarity that matches the symbol of the most positive input. In the case of Figure 3.3(a), the $-$ input has the largest positive value, so the output is negative.

Figure 3.5

Op-amp inputs are always virtually the same voltage.

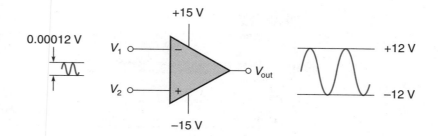

−12 V (being 80% of the supply). If the open-loop gain is 100,000, then the difference between V_1 and V_2 that would cause an output of 12 V is computed using Equation 3.1: $V_{out} = A(V_2 - V_1)$. Rearranging gives us

$$(V_2 - V_1) = \frac{V_{out}}{A} = \frac{12\ \text{V}}{100{,}000} = 0.00012\ \text{V}$$

So we see that, to keep the amplifier operating linearly with the output within its bounds, the difference between V_2 and V_1 must be less than 0.00012 V, which is a very small voltage. Hence, we say that V_1 *is virtually the same as* V_2.

❏ Assumption 2: Input current is zero.   *Explanation*: The input resistance of an op-amp is very high, typically 1 MΩ or more. It is so high that we can model the inputs as being open circuits as shown in Figure 3.6; of course, no current can flow into an open circuit.

❏ Assumption 3: Output resistance is zero.   *Explanation*: A low-output resistance means that the output voltage will not be pulled down even if the load draws a lot of current. This is the weakest of the three assumptions because the output resistance is typically between 50 and 75 Ω (however, it can be much lower with negative feedback). This assumption only holds if the load being driven is considerably higher than the output resistance of the op-amp, which is the case in most applications.

Many different types of op-amps are available, with names such as *general purpose*, *wide bandwidth*, *low noise*, and *high frequency*, to name a few. For most control applications, the general-purpose types are adequate. Figure 3.7 shows a data sheet for the popular, general-purpose op-amp 741 (MC1741), which comes in four types of packages. Besides the inverting and noninverting input and output pins, this op-amp has two

Figure 3.6

Equivalent circuit model of an op-amp.

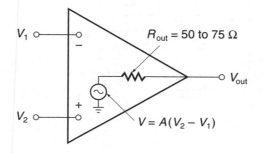

Figure 3.7
Data sheet for the 741
general-purpose op-amp.
(Courtesy of Motorola, Inc.)

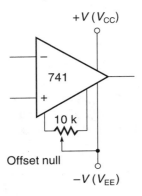

MC1741

INTERNALLY COMPENSATED, HIGH PERFORMANCE OPERATIONAL AMPLIFIERS

. . . designed for use as a summing amplifier, integrator, or amplifier with operating characteristics as a function of the external feedback components.

- No Frequency Compensation Required
- Short-Circuit Protection
- Offset Voltage Null Capability
- Wide Common-Mode and Differential Voltage Ranges
- Low-Power Consumption
- No Latch Up
- Low Noise Selections Offered — N Suffix

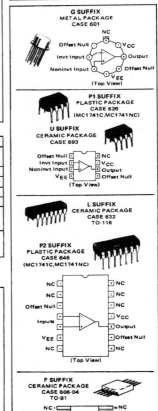

OPERATIONAL AMPLIFIER SILICON MONOLITHIC INTEGRATED CIRCUIT

MAXIMUM RATINGS (T$_A$ = +25°C unless otherwise noted)

Rating	Symbol	MC1741C	MC1741	Unit
Power Supply Voltage	V$_{CC}$	+18	+22	Vdc
	V$_{EE}$	−18	−22	Vdc
Input Differential Voltage	V$_{ID}$	±30		Volts
Input Common Mode Voltage (Note 1)	V$_{ICM}$	±15		Volts
Output Short Circuit Duration (Note 2)	t$_S$	Continuous		
Operating Ambient Temperature Range	T$_A$	0 to +70	−55 to +125	°C
Storage Temperature Range	T$_{stg}$			°C
Metal, Flat and Ceramic Packages		−65 to +150		
Plastic Packages		−55 to +125		
Junction Temperature Range	T$_J$			°C
Metal and Ceramic Packages		175		
Plastic Packages		150		

Note 1.   For supply voltages less than ± 15 V, the absolute maximum input voltage is equal to the supply voltage.
Note 2.   Supply voltage equal to or less than 15 V

EQUIVALENT CIRCUIT SCHEMATIC

ELECTRICAL CHARACTERISTICS (V$_{CC}$ = +15 V, V$_{EE}$ = −15 V, T$_A$ = 25°C unless otherwise noted).

Characteristic	Symbol	MC1741 Min	MC1741 Typ	MC1741 Max	MC1741C Min	MC1741C Typ	MC1741C Max	Unit
Input Offset Voltage (R$_S$ ≤ 10 k)	V$_{IO}$	—	1.0	5.0	—	2.0	6.0	mV
Input Offset Current	I$_{IO}$	—	20	200	—	20	200	nA
Input Bias Current	I$_{IB}$	—	80	500	—	80	500	nA
Input Resistance	r$_i$	0.3	2.0	—	0.3	2.0	—	MΩ
Input Capacitance	C$_i$	—	1.4	—	—	1.4	—	pF
Offset Voltage Adjustment Range	V$_{IOR}$	—	±15	—	—	±15	—	mV
Common Mode Input Voltage Range	V$_{ICR}$	±12	±13	—	±12	±13	—	V
Large Signal Voltage Gain (V$_O$ = ±10 V, R$_L$ ≥ 2.0 k)	A$_v$	50	200	—	20	200	—	V/mV
Output Resistance	r$_o$	—	75	—	—	75	—	Ω
Common Mode Rejection Ratio (R$_S$ ≤ 10 k)	CMRR	70	90	—	70	90	—	dB
Supply Voltage Rejection Ratio (R$_S$ ≤ 10 k)	PSRR	—	30	150	—	30	150	µV/V
Output Voltage Swing	V$_O$							V
(R$_L$ ≥ 10 k)		±12	±14	—	±12	±14	—	
(R$_L$ ≥ 2 k)		±10	±13	—	±10	±13	—	
Output Short-Circuit Current	I$_{os}$	—	20	—	—	20	—	mA
Supply Current	I$_D$	—	1.7	2.8	—	1.7	2.8	mA
Power Consumption	P$_C$	—	50	85	—	50	85	mW
Transient Response (Unity Gain — Non-Inverting)								
(V$_I$ = 20 mV, R$_L$ ≥ 2 k, C$_L$ ≤ 100 pF) Rise Time	t$_{TLH}$	—	0.3	—	—	0.3	—	µs
(V$_I$ = 20 mV, R$_L$ ≥ 2 k, C$_L$ ≤ 100 pF) Overshoot	os	—	15	—	—	15	—	%
(V$_I$ = 10 V, R$_L$ ≥ 2 k, C$_L$ ≤ 100 pF) Slew Rate	SR	—	0.5	—	—	0.5	—	V/µs

more pins called *offset null*. As indicated in the small diagram of Figure 3.7, these can be used to adjust the output voltage up or down slightly for the purpose of eliminating the **DC offset voltage**—a small DC voltage that may occur at the output, even when the inputs are exactly equal.

Looking at the electrical characteristics (Figure 3.7), the large signal voltage gain is given as 50–200 V/mV. This means that, at a minimum, the ratio is 50 V out for each millivolt in, which is the equivalent of 50,000 V out for each volt in, or a gain of 50,000 (minimum). Notice also that the input resistance is given as being typically 2 MΩ and the output resistance is typically 75 Ω.

Many useful signal-conditioning circuits can be built using op-amps. Some of the most common are presented in the pages that follow.

Voltage Follower

The voltage follower, which is very useful circuit, can boost the current of a signal without increasing the voltage. It can transform a high-impedance signal (easily loaded down) into a robust low-impedance signal. Figure 3.8 shows a voltage follower circuit. It has a voltage gain of 1, with a high R_{in} and a low R_{out}. Its operation can be explained as follows: We start with the basic op-amp equation (Equation 3.1):

$$V_{out} = A(V_2 - V_1)$$

In the circuit, V_{out} is connected to V_1; thus, $V_{out} = V_1$. Substituting in V_{out} and expanding Equation 3.1,

$$V_{out} = (AV_2) - (AV_{out})$$

Solving for V_{out}, we get

$$V_{out} = \frac{AV_2}{1+A}$$

But because A is *much* greater than 1,

$$V_{out} = \frac{AV_2}{A} \approx V_2$$

We see that the output voltage V_{out} equals the input voltage V_2, meaning the overall gain is 1. Also notice that the actual input signal goes directly into the noninverting input, so it draws essentially no current.

A more intuitive way to explain the voltage follower circuit is as follows: The input V_2 is virtually the same voltage as V_1 (from Assumption 1). V_1 is connected to V_{out}, so it's as if V_2 were connected to V_{out}, hence the gain of 1.

Figure 3.8

A voltage follower circuit.

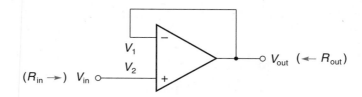

The voltage follower is a simple and very useful circuit. Consider the situation shown in Figure 3.9(a). In this case, a high-impedance sensor (10 kΩ), is connected directly to a controller with a 1 kΩ input resistance. The sensor generates 5 V internally, but this is reduced by the voltage drop across the 10 kΩ internal resistance. By redrawing the circuit [Figure 3.9(b)], we see that these two resistances form a voltage divider. The actual input voltage to the controller can be calculated as follows from the voltage-divider rule:

$$V_{in} = \frac{1\ k\Omega \times 5\ V}{1\ k\Omega + 10\ k\Omega} = 0.45\ V$$

This shows that only 0.45 V of the 5 V signal makes it to the controller. We could amplify the signal at the controller to make up for the attenuation, but that would amplify noise as well as the signal. A better solution is to insert a voltage follower near the sensor, as shown in Figure 3.9(c). Because the op-amp draws no signal current, there is no voltage drop across the 10 kΩ resistor, and the full 5 V enters the voltage follower and appears at its output. The 1 kΩ input resistance of the controller is so much higher than the output resistance of the op-amp that almost all of the 5 V will appear across the controller terminals.

Inverting Amplifier

The **inverting amplifier** is probably the most common op-amp configuration. The circuit shown in Figure 3.10 requires just two resistors, R_i and R_f. R_i is the input resistor, and R_f is the feedback resistor that feeds part of the output signal back to the input. This is

Figure 3.9
Using a voltage follower to prevent load down.

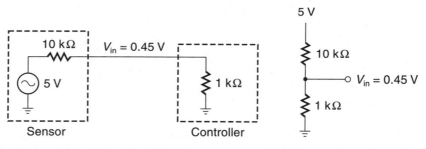

(a) Signal experiences voltage drop   **(b)** Equivalent circuit

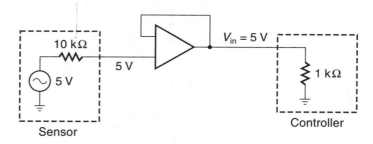

(c) No signal voltage drop

Figure 3.10

The inverting amplifier circuit.

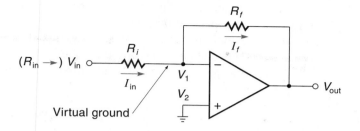

an inverting amplifier because *the input signal goes to the inverting input, which means the output is out of phase with the input.* The voltage gain is determined by the resistor values.

An explanation of how the inverting amp works is as follows: First, if the op-amp input draws no current, then all the signal current (I_{in}) must go through R_f—there is nowhere else for it to go. Therefore, $I_{in} = I_f$. By assumption, V_1 and V_2 are virtually the same voltage, and V_2 is grounded; thus, V_1 is at **virtual ground**. If V_1 is (almost) at ground, then the entire input signal voltage V_{in} is dropped across R_i. From Ohm's law,

$$I_{in} = \frac{V_{in}}{R_i}$$

As already noted, virtually all I_{in} goes through the feedback resistor R_f. The voltage across R_f is the difference between virtual ground and V_{out}. Thus, we can write Ohm's law equation for R_f :

$$I_{in} = I_f = \frac{0 - V_{out}}{R_f}$$

Combining the two previous equations,

$$\frac{V_{in}}{R_i} = \frac{0 - V_{out}}{R_f}$$

Solving for V_{out} and rearranging gives us

$$V_{out} = \frac{-V_{in}R_f}{R_i}$$

$$\frac{V_{out}}{V_{in}} = \frac{-R_f}{R_i}$$

However, V_{out}/V_{in} is the voltage gain, so

$$A_V = \frac{-R_f}{R_i} \qquad (3.2)$$

where

A_V = voltage gain of the inverting amp
R_f = value of the feedback resistor
R_i = value of the input resistor

This result (Equation 3.2) shows us that the voltage gain of the inverting amp is simply the ratio of the two resistors R_f and R_i. The minus sign reminds us that the output is inverted. The gain derived in Equation 3.2 is called the **closed-loop gain** and is always lower than the (open-loop) gain of the op-amp by itself.

Another important point is that the input impedance for the entire inverting amp is approximately R_i (*not* infinite as one might think). Figure 3.10 shows this: The right end of R_i is at virtual ground; therefore, the entire V_{in} is "dropped" across R_i.

◆ **EXAMPLE 3.2**

An inverting amp is to have a gain of 10. The signal source is a sensor with an output impedance of 1 kΩ. Draw a circuit diagram of the completed amplifier.

Solution

First, select a value for R_i. Because R_i essentially determines the amplifier's input resistance, *it should be at least ten times higher (if possible) than the signal source impedance to ensure maximum voltage transfer*. In this example, we select $R_i = 10$ kΩ. Next, rearrange Equation 3.2 to solve for R_f :

$$R_f = -AR_i$$
$$= -(-10) \times 10 \text{ k}\Omega = 100 \text{ k}\Omega$$

Figure 3.11 shows the completed circuit.

Figure 3.11

An inverting amplifier circuit (Example 3.2).

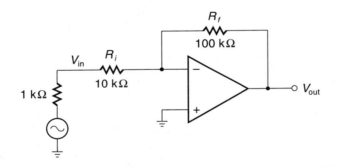

Noninverting Amplifier

Many situations call for an amplifier that does not invert the output. For example, the output of a temperature sensor might be such that as the temperature goes up, the voltage goes up. If this is the same relationship that the controller wants, we don't want the amplifier to invert it. The circuit for the **noninverting amplifier** is shown in Figure 3.12. It is similar to the inverting amp except the input signal V_{in} now goes directly to the noninverting input and R_i is grounded. Notice that the noninverting amp has an almost infinite input impedance (R_i) because V_{in} connects only to the op-amp input.

Figure 3.12

The noninverting amplifier circuit.

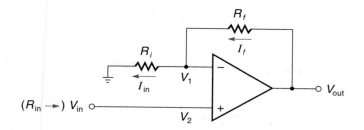

An explanation of how the circuit works is as follows: If V_1 is virtually the same as V_2, then the voltage input (V_{in}) appears across R_i. Applying Ohm's law to R_i, we can calculate I_{in}:

$$I_{in} = \frac{V_{in} - 0}{R_i}$$

The current in R_f can also be calculated using Ohm's law. We know the voltage across R_f is the difference between V_{in} and V_{out}. Therefore,

$$I_f = \frac{V_{out} - V_{in}}{R_f}$$

Because no current enters the inverting input of the op-amp, all current in R_f must go into R_i:

$$I_{in} = I_f$$

Combining these three equations gives us

$$I_{in} = I_f = \frac{V_{in} - 0}{R_i} = \frac{V_{out} - V_{in}}{R_f}$$

Solving for V_{out} and rearranging gives us

$$V_{out} - V_{in} = \frac{R_f V_{in}}{R_i}$$

$$V_{out} = \frac{R_f V_{in}}{R_i} + V_{in} = V_{in}\left(\frac{R_f}{R_i} + 1\right)$$

$$\frac{V_{out}}{V_{in}} = \frac{R_f}{R_i} + 1$$

V_{out}/V_{in} is the voltage gain, so the resulting equation for the gain of the noninverting amp is

$$A_V = \frac{R_f}{R_i} + 1 \qquad (3.3)$$

where

A_V = voltage gain for the noninverting amp
R_f = value of the feedback resistor
R_i = value of the input resistor

◆ EXAMPLE 3.3

Draw the circuit diagram of a noninverting amp with a gain of 20.

Solution   Using Equation 3.3 and putting in a gain of 20,

$$A_V = 20 = \frac{R_f}{R_i} + 1$$

Rearranging gives us

$$\frac{R_f}{R_i} = 19 \qquad \text{or} \qquad R_f = 19 \times R_i$$

Now select R_i to be an appropriate value (as explained below) and solve for R_f. If we select R_i to be 2 kΩ, then

$$R_f = 19 \times R_i = 19 \times 2\text{ k}\Omega = 38\text{ k}\Omega$$

Figure 3.13 shows the completed circuit. The basis for selecting both R_i and R_f is that the current in these external resistors should be much larger than the small current that actually enters the op-amp (recall that the op-amp equation was based on the assumption that *no* current enters the op-amp). Therefore, both R_i and R_f should be at least ten times smaller than the op-amp input resistance—in this case, no more than 100 kΩ if possible.

Figure 3.13

A noninverting amplifier circuit (Example 3.3).

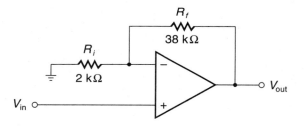

◆

Summing Amplifier

The **summing amplifier** has an output voltage that is the sum of any number of input voltages. Figure 3.14(a) depicts this situation. In this case, the amplifier would add the input voltages of 1 V, 2 V, and 4 V and give an output of 7 V. You might be tempted to think you could do this by simply connecting the wires as shown in Figure 3.14(b), but the output of that circuit would be something between 1 and 4 V depending on which source was the "strongest" (had the lowest resistance).

Figure 3.14

Connecting the wires does not sum the voltage.

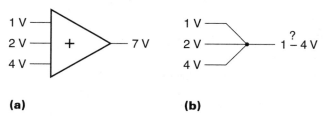

Figure 3.15

The summing amplifier circuit.

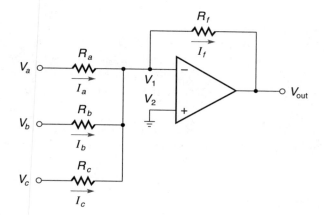

Figure 3.15 shows the summing amplifier circuit. Each input signal goes through a separate resistor that keeps it isolated from the others. The inverting input is used, so the output will be inverted. An explanation of how it works is as follows: Because the op-amp input draws no current, all individual input currents must combine and go through the feedback resistor R_f:

$$I_f = I_a + I_b + I_c \tag{3.4}$$

Note that V_2 is grounded, so V_1 is at virtual ground. Therefore, the voltage across each of the four resistors is simply V_a, V_b, V_c, and V_{out}. Applying Ohm's law to express the current through each resistor, we can rewrite Equation 3.4 as follows:

$$\frac{0 - V_{out}}{R_f} = \frac{V_a}{R_a} + \frac{V_b}{R_b} + \frac{V_c}{R_c}$$

Solving for V_{out},

$$V_{out} = -\left(\frac{R_f}{R_a} V_a + \frac{R_f}{R_b} V_b + \frac{R_f}{R_c} V_c \right)$$

In other words, the output voltage equals the sum of the products of the input voltages and their respective gains. Also, the input impedance of each input equals the value of the respective input resistor (R_a, R_b, etc.).

If $R_a = R_b = R_c = R_i$, the output of the summing amp simplifies to:

$$V_{out} = -\frac{R_f}{R_i}(V_a + V_b + V_c) \tag{3.5}$$

where

$$V_{out} = \text{output voltage of the summing amp}$$
$$R_f = \text{value of the feedback resistor}$$
$$R_i = \text{value of all input resistors}$$
$$V_a, V_b, V_c = \text{input signal voltages}$$

Equation 3.5 shows that the output voltage V_{out} is equal to the sum of the input voltages times a gain factor of R_f/R_i. The minus sign reminds us that the output is inverted.

◆ EXAMPLE 3.4

According to a comfort scale, the air conditioning in a building should come on when the sum of the temperature and humidity sensor voltages goes above 1 V. A threshold circuit in the air conditioner requires 5 V for turn-on. Design an interface circuit to connect the two sensors to the air conditioning unit.

Solution

This circuit requires a summing amplifier with two inputs and a gain of 5. By specifying both input resistors to be the same (at 1 kΩ), we can use Equation 3.5, and our only calculation concerns the gain portion of the equation:

$$A = \frac{R_f}{R_i} = 5$$

Rearranging gives us

$$R_f = 5 \times R_i$$

When $R_i = 1\ k\Omega$,

$$R_f = 5 \times 1\ k\Omega = 5\ k\Omega$$

Figure 3.16 shows the completed circuit. Notice that an inverting amp with a gain of 1 ($R_i = R_f$) was added to make the final output positive.

Figure 3.16

A summing amplifier circuit (Example 3.4).

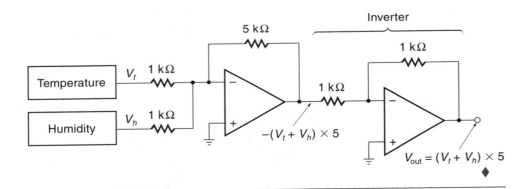

Differential and Instrumentation Amplifiers

A **differential amplifier** amplifies the difference between two input voltages. In the circuits we have examined thus far, the input voltages have been referenced to ground, but the op-amp can be the basis of a practical differential amp as well. Figure 3.17 shows such a circuit. As before, the gain is established with resistors. This circuit will amplify a **differential voltage**, which is the difference between the two voltage levels V_a and V_b, when neither is ground. The output of the amplifier (V_{out}) is a single voltage level referenced to ground, sometimes called a **single-ended voltage**. If $R_a = R_b$ and $R_f = R_g$, which is usually the case, then the equation for V_{out} is

Figure 3.17

The differential amplifier circuit.

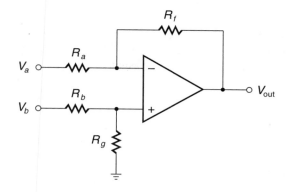

$$V_{out} = \frac{R_f}{R_a} (V_b - V_a) \qquad (3.6)$$

Rearranging gives us

$$\frac{V_{out}}{(V_b - V_a)} = \frac{R_f}{R_a}$$

$V_{out}/(V_b - V_a)$ is output/input, which is a gain, so

$$A_V = \frac{R_f}{R_a} \qquad (3.7)$$

where

A_V = voltage gain of the differential amp
R_f = value of R_f and R_g
R_a = value of input resistors, R_a and R_b

As with the basic op-amp, the polarity of the output (V_{out}) will be positive when the input V_b is more positive than input V_a. The selection of resistor values R_a and R_b is based on a compromise. If they are too high—say, over 100 kΩ—then the currents may be so small that our basic op-amp assumptions won't hold (for example, the assumption that all current in R_a goes through R_f). On the other hand, if the resistances are too low (less than ten times the source resistance), then a considerable amount of attenuation will occur before the signal even gets to the amp.

♦ **EXAMPLE 3.5**

A differential amp is needed to amplify the voltage difference between two temperature sensors. The sensors have an internal resistance of 5 kΩ, and the maximum voltage difference between the sensors will be 2 V. Design the differential amp circuit to have an output of 12 V when the difference the inputs is 2 V.

Solution   First calculate the gain required:

$$A_V = \frac{V_{out}}{V_{in}} = \frac{12\ V}{2\ V} = 6$$

By letting $R_a = R_b$ and $R_f = R_g$, we can use Equation 3.7. Noting that the sensor impedance is 5 kΩ, we would like the input resistance of the amp to be at least ten times 5 kΩ. Therefore, if we select $R_a = 50$ kΩ, then

$$A_V = \frac{R_f}{R_a} = 6$$

and

$$R_f = R_a \times 6$$
$$= 50 \text{ k}\Omega \times 6 = 300 \text{ k}\Omega$$

Figure 3.18 shows the completed design.

Figure 3.18

A differential amplifier circuit (Example 3.5).

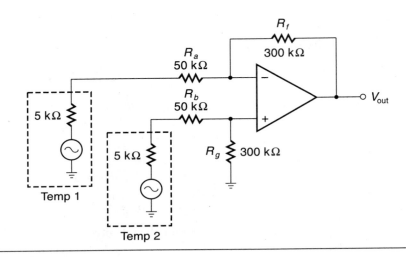

An **instrumentation amplifier** is a differential amp that has its inputs buffered with voltage followers, as shown in Figure 3.19. Voltage follower circuits on the inputs perform three desirable functions: (1) They increase the input resistance so that the source (such as a sensor) will never be loaded down, (2) they make both input resistances equal, and (3) they isolate the gain-defining resistors (R_f, R_i, etc.) from the signal source. This last quality means that instrumentation amps can be prebuilt to have a specific gain.

Figure 3.19

The instrumentation amplifier circuit.

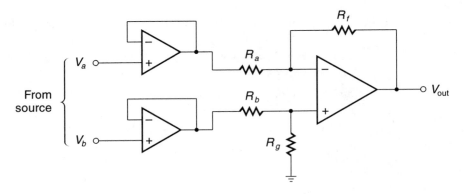

A data sheet for the LH0036 instrumentation amplifier is shown in Figure 3.20. This particular device has a very high resistance on both inputs and can provide gains of 1 to 1000 by simply connecting an external resistor (R_g) between pins 4 and 7. (Gain = 1 + 50 kΩ/R_g).

A newer type of instrumentation amplifier is known as a **programmable gain instrumentation amplifier**. These amps have gains that are selectable with digital inputs. An

Figure 3.20

Data sheet for the LH0036 instrumentation amplifier. (Courtesy of National Semiconductor Corp.)

LH0036 Instrumentation Amplifier

General Description

The LH0036C is a micro power instrumentation amplifier designed for precision differential signal processing. Extremely high accuracy can be obtained due to the 300 MΩ input impedance and excellent 100 dB common mode rejection ratio. It is packaged in a hermetic TO-8 package. Gain is programmable from 1 to 1000 with a single external resistor. Power supply operating range is between ±1V and ±18V. Input bias current and output bandwidth are both externally adjustable or can be set by internally set values. The LH0036C is specified for operation over the −25°C to +85°C temperature range.

Features

■ High input impedance   300 MΩ
■ High CMRR   100 dB
■ Single resistor gain adjust   1 to 1000
■ Low power   90 μW
■ Wide supply range   ±1V to ±18V
■ Adjustable input bias current
■ Adjustable output bandwidth
■ Guard drive output

Equivalent Circuit and Connection Diagrams

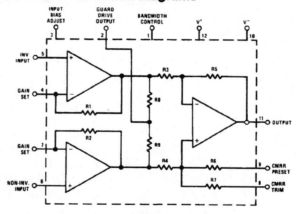

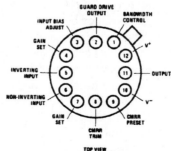

TOP VIEW
Order Number LH0036CG
See NS Package Number G12B

TL/H/5545–1

example of this type is the Burr–Brown PGA204, which is packaged as a 16-pin IC. This amp has four fixed gains of 1, 10, 100, and 1000, which are selected by two digital inputs (00, 01, 10, 11). These amplifiers are useful in applications where signals of vastly different voltages are digitized by the same ADC (analog-to-digital converter). In such a system, the digital controller could switch to the appropriate gain for each different signal level.

Integrators and Differentiators

Op-amp circuits can be designed to integrate or differentiate an incoming waveform. These special-purpose circuits are likely to be found only inside an analog controller.

Figure 3.21 shows an **integrator** circuit. Notice that the feedback element is a capacitor. The integrator gives an output voltage (V_{out}) that is proportional to the total area under a curve traced out by the input voltage waveform (the horizontal axis being time), as specified in Equation 3.8:

$$V_{out} = -\frac{1}{RC} \times (\text{area under } V_{in} \cdot \text{time curve}) \tag{3.8}$$

where

V_{out} = output voltage of the integrator
R, C = values of components in Figure 3.21

Figure 3.21

An integrator circuit.

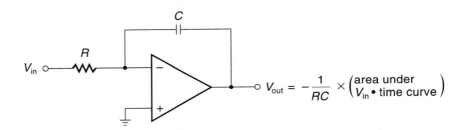

The integrator concept can best be explained by the sample waveforms shown in Figure 3.22 (which assumes $RC = 1$). Notice that the integrator input voltage (V_{in}) rises from 0 to 1 V in the first 10 s. The triangular area under that portion of the curve (a–b) is 5 V · s, so the output (V_{out}) of the integrator rises from 0 to − 5 V during the same time. From time b to c, V_{in} remains at 1 V, so the new area added is 10 V · s. Consequently, V_{out} increases by 10 to become − 15 V at time c. Then, V_{in} returns to 0 V. Because no new area is added between c and d, V_{out} remains at − 15 V. Integrators can be useful because they keep a record of what has gone on before.

Like all delicate analog circuits, the integrator must be maintained in proper adjustment to work well. If there is even a small unwanted DC offset voltage at V_{out}, it will cause the integrator output to build up over time and eventually saturate at the power-supply voltage.

Figure 3.23 shows a **differentiator** circuit. The differentiator gives an output voltage that is proportional to the rate of change (slope) of the input voltage, as specified in Equation 3.9:

$$V_{out} = -RC \times \frac{\Delta V_{in}}{\Delta t} \tag{3.9}$$

Figure 3.22
The voltage waveform of an integrator circuit. (*V_{out} = $-V \cdot s$ in this case because $RC = 1$.)

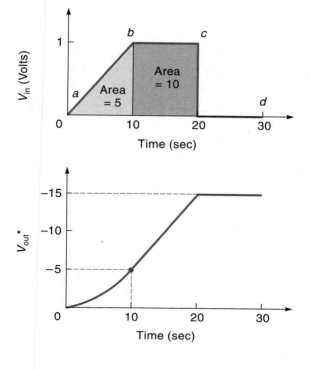

where

$$V_{out} = \text{output voltage of the differentiator}$$
$$R, C = \text{values of components in Figure 3.23}$$
$$\Delta V_{in}/\Delta t = \text{rate of change, or slope, of } V_{in}$$

The differentiator concept is illustrated in Figure 3.24 (which assumes $RC = 1$). From time a to b, the input voltage (V_{in}) is 0 V, and because it is not changing, the output voltage (V_{out}) is 0 V. During the time period b–c, V_{in} increases at a constant rate of 1 V/s, so the V_{out} curve reflects this by staying at a constant -1 V. From time c to d, the slope of V_{in} increases to 2 V/s, so V_{out} jumps to -2 V. After time d, V_{in} stays at 3 V, and because it is not changing, V_{out} is 0. Differentiators can tell us how fast a variable is changing. In practice, however, they suffer from the problem that even a small amount of noise in the input will be accentuated, giving a very "noisy" output.

Figure 3.23
A differentiator circuit.

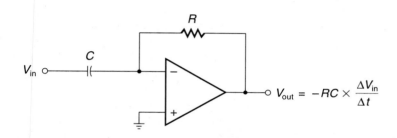

$$V_{out} = -RC \times \frac{\Delta V_{in}}{\Delta t}$$

Figure 3.24

Voltage waveforms of a differentiator. ($^*V_{out} = -V$ in this case because $RC = 1$.)

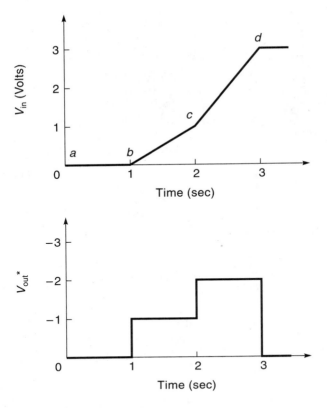

Active Filters

Filter circuits either pass or stop signals, depending on frequency. Figure 3.25 shows the responses of some basic types of filters. The **low pass filter** [Figure 3.25(a)] allows only frequencies below the **cutoff frequency** (f_c) to pass. Frequencies above f_c are attenuated. The steepness of the attenuation depends on the type of filter. A common single-stage R-C filter (described below) decreases the signal by a factor of 10 each time the frequency goes up by 10. The **high-pass filter** [Figure 3.25(b)] tends to reject signals with frequencies below the cutoff frequency and pass those above. The **band-pass filter** [Figure 3.25(c)] passes signals with a range of frequencies between f_{c1} and f_{c2} and rejects all others. The **notch filter** [Figure 3.25(d)] rejects only a narrow range of frequencies and passes all others.

Some of these filters are particularly useful for signal conditioning in control systems. For example, sensors reporting such things as temperature or flow rate have relatively slow-changing signals. A low-pass filter would allow the sensor signal to pass while rejecting higher-frequency electrical noise, as from motors or relays. The notch filter is also particularly useful. It can be used to attenuate a particular noise frequency, such as 60 Hz, and pass everything else.

Theoretically, filters can be constructed entirely from passive components (R, C, L). In practice, such filters tend to change characteristics when inserted in a circuit because the impedances interact. The op-amp can solve this problem by providing impedance

Figure 3.25
Basic filter responses.

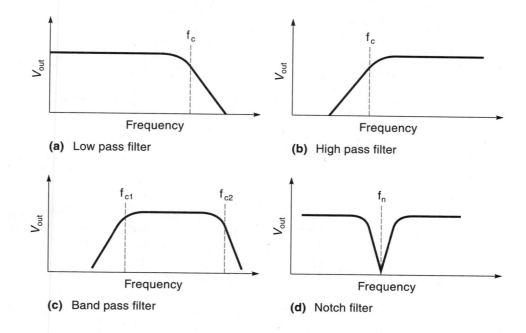

(a) Low pass filter

(b) High pass filter

(c) Band pass filter

(d) Notch filter

isolation between the filter and the circuit it's driving, and provide some gain as well. A filter using an op-amp is called an **active filter**. Figure 3.26 shows a single-stage low-pass filter incorporating an op-amp. Notice that this is basically a noninverting amp with an R-C filter connected to the input. The performance of this filter can be described with Equations 3.3 and 3.10:

$$\text{Gain} = A_V = \frac{R_f}{R_i} + 1 \qquad (3.3)$$

$$\text{Cutoff frequency (low- and high-pass filters)} = f_c = \frac{1}{2\pi RC} \qquad (3.10)$$

Figure 3.26
A low-pass filter circuit.

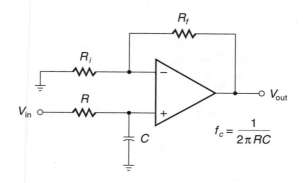

$$f_c = \frac{1}{2\pi RC}$$

Figure 3.27

A high-pass filter circuit.

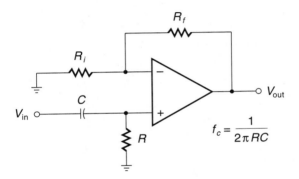

$$f_c = \frac{1}{2\pi RC}$$

Figure 3.27 shows a single-stage high-pass filter. It is similar to the low-pass filter but the positions of the R and C are reversed. The equations for the gain and cutoff frequency are exactly the same as for the low-pass filter (but the meaning of f_c is different—see Figure 3.25).

A band-pass filter can be built by cascading a low-pass filter and a high-pass filter together. The cutoff frequency of the low-pass filter (f_{c2}) must be higher than the cutoff frequency of the high-pass filter (f_{c1}), as indicated in Figure 3.25(c).

Figure 3.28 shows an op-amp notch filter. The filter itself is a Wein Bridge type. The gain of the signal is 1 at all frequencies except near the notch. The frequency (f_n), which is suppressed or "notched out," can be calculated from Equation 3.11:

$$\text{Notched frequency (Wein Bridge notch filter)} = f_n = \frac{1}{2\pi RC} \qquad (3.11)$$

Comparator

A typical situation in a control system is a slow-moving analog signal from a sensor being used to trigger some event. Such an interface requires a threshold detector circuit that will switch from off to on when a specified input voltage level is reached. A **comparator** is such a circuit (Figure 3.29). It is really an op-amp that is specially designed for this application (regular op-amps usually aren't stable enough). Comparators are operated open-loop so that if V_2 is even slightly more positive than V_1, the tremendous gain will

Figure 3.28

A Wein Bridge notch filter circuit.

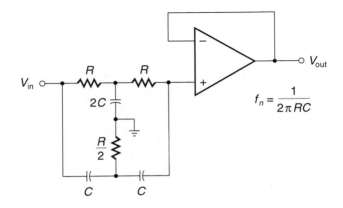

$$f_n = \frac{1}{2\pi RC}$$

Figure 3.29
A comparator circuit.

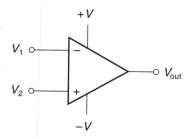

amplify the small difference and drive the output into positive saturation (close to $+V$). On the other hand, if V_1 is slightly more positive than V_2, the output will go to negative saturation ($-V$). The output is essentially digital in nature—either on or off depending on a very small change in the inputs. This concept is best demonstrated in an example.

♦ **EXAMPLE 3.6**

The blower on a hot-air solar panel should come on when the temperature reaches 100°F. An analog temperature sensor in the solar panel needs to be interfaced to a digital controller such that the controller receives a 5 V switch-on signal when the sensor voltage reaches 2.7 V. Design the interface circuit.

Solution

Figure 3.30(a) shows the interface circuit. The signal from the sensor is connected to the noninverting input of the comparator. The inverting input comes from a voltage divider that yields a precise reference voltage of 2.7 V. Notice also that the supply voltages of the comparator are 5 V and ground.

As long as the sensor voltage is below 2.7 V, the reference voltage at the inverting input predominates, and the output will try to go negative. In this case, the output will go to about 0 V because that is what the negative supply voltage is. When the sensor voltage goes only slightly above 2.7 V, the noninverting input becomes positive compared with the inverting input, and the output saturates positive, which is about 5 V. The switch-on point can easily be adjusted by changing the reference voltage resistors.   ♦

One practical problem with comparators is known as **chatter**, the condition that occurs when the output (V_{out}) oscillates back-and-forth when V_{in} is near the threshold. Chatter is caused by noise on the V_{in} signal or some sort of undesirable feedback, say, through the power supply. Practical circuits overcome chatter by using a **window comparator** [Figure 3.30(b)], a comparator with built-in hysteresis. In this discussion, *hysteresis* means that the switch-on voltage will be greater than the switch-off voltage. For example, if hysteresis were added to the system of Example 3.6, the switch-on voltage might be 2.8 V, and the switch-off voltage might be 2.6 V. Illustrating this situation, Figure 3.31(a) shows the ideal single-threshold system with a noiseless signal. Figure 3.31(b) shows how chatter results if the signal is noisy. Finally, Figure 3.31(c) shows how a system with a switch-on and switch-off threshold can eliminate chatter.

Figure 3.30
Comparator circuits.

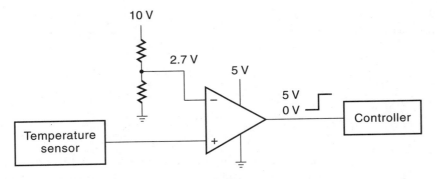

(a) Comparator circuit for Example 3.6

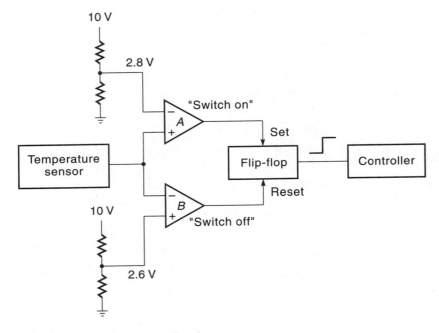

(b) Window comparator circuit

Figure 3.30(b) shows a window comparator built from two op-amps and an *R-S* flip-flop. When the temperature-sensor voltage rises above 2.8 V, op-amp A switches on and sets the flip-flop. Once set, the flip-flop will stay set until the temperature voltage goes below 2.6 V, in which case op-amp B will switch on and reset the flip-flop.

3.2 SPECIAL INTERFACE CIRCUITS

A number of special-purpose interface circuits are useful in certain applications. Some common ones are discussed in this section.

Figure 3.31

How a window comparator eliminates chatter.

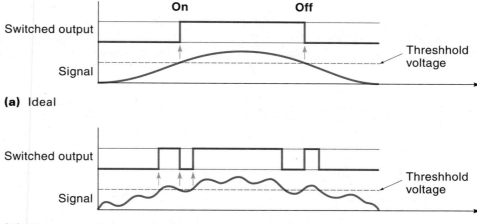

(a) Ideal

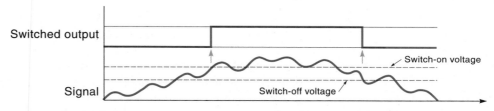

(b) Noisy signal causes chatter on output

(c) Window comparator has two thresholds to eliminate chatter

The Current Loop (Voltage–Current Conversion)

Most signals are voltage signals, which means that the information being conveyed is proportional to the voltage. Two potential problems with this can arise: Susceptibility to electrical noise increases (which is usually in the form of voltage spikes), and any resistance in the signal wire causes a voltage drop. For short distances, wire can usually be effectively shielded from noise, and voltage drops are not a problem. For longer cable runs with numerous connectors, such as might be found in a process control system in a large factory, noise and total cable resistance may become significant. Figure 3.32 illustrates this situation. Notice that the signal from the sensor has been greatly diminished by the cable resistance.

Figure 3.32

Signal voltage is reduced from wire resistance.

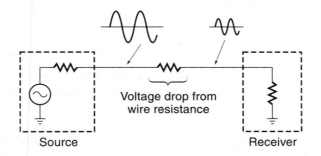

One solution to the problems of noise and signal attenuation is to use current instead of voltage to convey the information. This is effective because current, unlike voltage, is not as susceptible to noise and does not drop when it goes through a resistance. As Kirchhoff's current law tells us, whatever current goes into a branch, comes out. This is illustrated in Figure 3.33. Notice that the signal emerges from the source, travels to the receiver, and then loops back again to the source. The receiver does not siphon off any current but merely senses it; therefore, we have a **current loop**, which goes from source to receiver and back to source. Any resistance in the wires cannot alter the fact that the current remains the same throughout the entire loop.

Figure 3.33

A current-loop circuit.

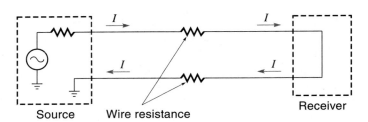

Source   Wire resistance   Receiver

Op-amps can be used to implement both the transmitter and the receiver in a current-loop system. Figure 3.34 shows a transmitter circuit. This circuit converts a voltage signal (V_{in}) into an output current (I_{out}), which is proportional to V_{in}. The op-amp circuit causes the output current to be independent of any resistance in the line. It accomplishes this by automatically increasing or decreasing its output voltage (V_{out}) in response to any increasing or decreasing line resistance.

Figure 3.34

The op-amp as a current-loop transmitter.

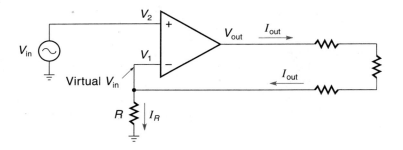

The explanation for how the circuit of Figure 3.34 works is as follows: Recall that for an op-amp we can assume V_1 is virtually the same as V_2; therefore, the voltage across R must be V_{in}. Applying Ohm's law to R, we get

$$I_R = \frac{V_{in}}{R} \tag{3.12}$$

where I_R is the value of the current in resistor R. The source of I_R cannot be the − input of the op-amp, so it must be coming through the loop. The loop current (as specified in Equation 3.12) is dependent only on the input signal voltage (V_{in}) and the resistor R, not by any line or load resistance.

Once the current signal gets to the receiver, it needs to be reconverted back into a voltage. In other words the receiver is a **current-to-voltage converter**. A resistor is the

Figure 3.35 The complete current-loop system.

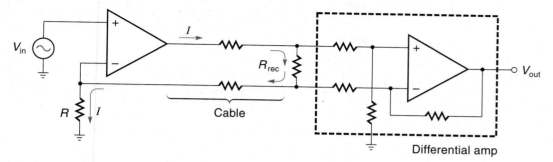

most direct way to convert a current into a voltage, but we have a special problem here. *We have to detect the voltage across the load resistor, without knowing where this voltage is with respect to ground.* Consider the complete current-loop circuit in Figure 3.35. The receiver resistor (R_{rec}) is "floating," that is, neither end is grounded. (If we ground the bottom of R_{rec}, current can "escape" from the loop, and the whole concept will fall apart.) The problem is solved by using a differential amplifier as the receiver, as illustrated in Figure 3.35. The differential amplifier samples and amplifies only the voltage difference across the receiver resistor. The differential amp must be designed to have a high input resistance to prevent any significant amount of current from leaving the loop at the receiver. *The output of the differential amp is a single-ended voltage that can be referenced to ground.* The magnitude of the output voltage is computed as follows:

$$\text{Voltage across the receiver resistor:} \quad V_{rec} = IR_{rec}$$

where

$$I = \text{loop current}$$
$$R_{rec} = \text{receiver resistance}$$

$$\text{Output voltage of the receiver:} \quad V_{out} = A_V V_{rec} = A_V IR_{rec}$$

where

$$A_V = \text{differential amp voltage gain}$$

The current-loop technique is one of the industry standard methods of connecting controllers to sensors and/or actuators. The standard system specifies a current range of 4–20 mA, where 4 mA corresponds to the minimum signal and 20 mA corresponds to the maximum signal. For example, if a sensor puts out a signal in the range of 0–5 V, then 4 mA in the current loop would correspond to 0 V from the sensor, and 20 mA would correspond to 5 V. Transmitters and receivers in the range 4 to 20 mA are available in IC form; for example, the Burr–Brown RCV420 receiver converts 4–20 mA to 0–5 V.

Analog Switch Circuit

An **analog switch** (also known as a *transmission gate*) is a solid-state device that allows analog signals to pass through or not. It performs the same function as a mechanical switch but has some definite advantages: It is smaller and more reliable, it uses less

Figure 3.36
The analog switch symbol.

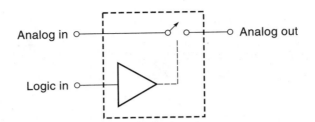

power but is faster acting, and it doesn't have mechanical parts to wear out. The analog switch is usually activated by a digital control signal.

Figure 3.36 shows the symbol of an analog switch. Each switch has three terminals: analog in, analog out, and logic in. Some models allow signals to pass in either direction like a mechanical switch, whereas others are unidirectional. The logic-in input is the control signal that closes or opens the switch. Typically, the circuits are packaged in an IC. Figure 3.37 shows the LF 11331 IC which contains four separate analog switches.

The switch itself is an FET (field effect transistor) (FETs are discussed in Chapter 4). Being a semiconductor, the FET has some resistance even when the switch is on (closed). This on resistance is typically 50–300 Ω. To keep this resistance from dropping signal voltage, the load resistance should be much greater than 300 Ω. The off resistance of the analog switch is very high, typically 1 MΩ or more.

The analog switch is frequently used to connect two or more sensors to a single ADC. Figure 3.38 shows a system where three sensors are serviced by one ADC. The controller can activate each analog switch, thus connecting itself with the sensor it wants to read. The operation of switching through one input at a time is called **multiplexing**. Analog multiplexing circuits are available in IC form. Figure 3.39 shows the LF 11508, an eight-channel analog multiplexer. A binary code placed on inputs A0, A1, and A3 will cause one of the eight input signals to be switched through to the output.

The ADC0809 multiplexer (not shown) is an IC that includes an 8-bit multiplexer and an ADC in one package. It is a very useful chip and found in many designs.

Figure 3.37 The LF1131 quad analog switch circuits and schematic diagram. (Courtesy of National Semiconductor Corp.)

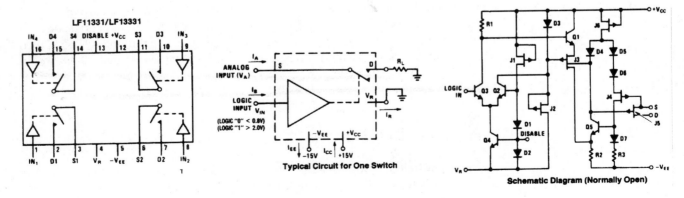

Figure 3.38

Three sensors switched into a single analog-to-digital converter (ADC).

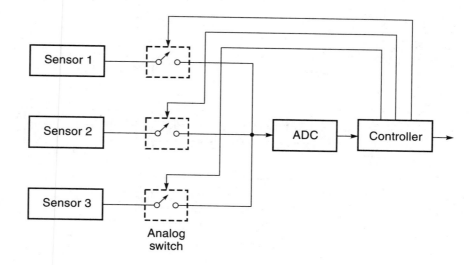

Sample-and-Hold Circuit

A **sample-and-hold circuit** can read in, or sample, a voltage and then remember, or hold, it for a period of time. The function of this circuit can best be described in an example. Consider the system discussed in the previous section, where three analog signals were interfaced to a digital controller. Now add an additional constraint that all three sensors be read at exactly the same time. Assuming that just one ADC is available, the only way to meet this requirement is to include three sample-and-hold circuits as shown in Figure 3.40. By command from the controller, all three sample-and-hold circuits will take voltage readings and store the results. Then at the convenience of the system, these values can be read one-at-a-time by the controller.

There is another reason why a sample-and-hold circuit might be used in conjunction with an ADC. For the ADC to give an accurate output, the analog input should be

Figure 3.39   The eight-channel analog LF11508 multiplexer circuit. (Courtesy of National Semiconductor Corp.)

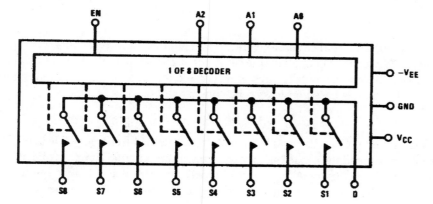

EN	A2	A1	A0	SWITCH ON
H	L	L	L	S1
H	L	L	H	S2
H	L	H	L	S3
H	L	H	H	S4
H	H	L	L	S5
H	H	L	H	S6
H	H	H	L	S7
H	H	H	H	S8
L	X	X	X	NONE

Figure 3.40

An example of using
sample-and-hold circuits.

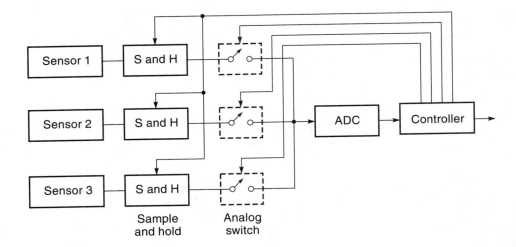

constant during the read-in time. If the analog signal is changing too fast, the sample-and-hold circuit can be used to "freeze" the input voltage during the conversion.

Figure 3.41 shows the schematic of a sample-and-hold circuit, which consists of an analog switch, a capacitor, and a voltage follower amplifier. To take a sample, the analog switch is closed for a period of time long enough for the capacitor to charge up to V_{in}; then the switch is opened. The signal voltage is trapped on the capacitor because (theoretically) it can't discharge through either the op-amp or analog switch. The voltage can be read anytime via the output of the voltage follower. Recall that the gain of a voltage follower is 1.

Figure 3.41

A sample-and-hold circuit.

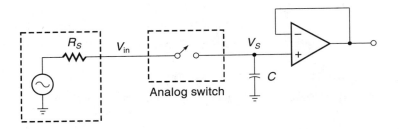

For the sample-and-hold circuit to work, certain conditions must be met. First, the switch must be closed long enough for the capacitor to charge up to the full value of V_{in}. The time required depends on the sizes of R_s (source resistance) and C, according to Equation 3.13 (which is based on 5 RC time constants):

$$t = 5R_sC \qquad (3.13)$$

where t is the time for C to charge to 99% of V_{in}. Eventually the charge will leak off, so the controller must read the capacitor voltage within a certain time. Herein is one of the trade-offs in the circuit design. A larger capacitor will hold the charge longer, but it will increase the read-in time because it takes longer to be charged up. As always, some compromise value of C is selected.

3.3   SIGNAL TRANSMISSION

A control system that performs just fine on a laboratory test bench may not work at all in the real world. In the lab, the cable runs are short, and the electrical noise is minimum. The real installation may require cable runs of perhaps hundreds of feet, in a dirty, electrically noisy environment. In this section, we examine some of the problems and their solutions in the area of signal transmission.

Earth Ground and Ground Loops

Earth ground refers to an electrical connection to the surface of the earth, usually through a metal rod driven into the ground. Because the earth is not a particularly good conductor, the ground voltage is different from place to place. Often confused with the earth ground is the **signal common** (or **signal return**), which is a conductor in the circuit and serves as a common reference point for the other circuit voltages. The signal common of a circuit is often referred to as the *ground*, and it may or may not be connected to the earth ground. For a DC circuit, the signal common is usually connected to the negative terminal of the power supply or battery. An automobile chassis ground is an example of this. AC signals also have a designated signal return.

With the exception of antenna systems, circuits do not have to be attached to the earth to work, but most are.* This is done primarily for safety reasons so that high voltages cannot build up between the chassis and the local ground. Consider the situation illustrated in Figure 3.42(a). A 200-Vac signal is generated in one unit and sent by two wires to a second unit. The signal common is connected to the chassis, but the chassis is not connected to the ground. This circuit would work fine electrically, but there is a potential safety hazard. Due to inductive and capacitive coupling, it is likely that the 200 V signal voltage would align itself as shown, + 100 V for the signal and − 100 V for the signal common (and chassis). This is a dangerous situation because anyone standing on the ground and touching the chassis would get a -100 V shock. In Figure 3.42(b), the chassis has been connected to an earth ground. Now the voltage of the chassis is 0 V with respect to the earth and is safe to touch, as is the signal common wire.

Sometimes both the source and the receiver are earth-grounded, as shown in Figure 3.43(a). One might be tempted to use the earth as the return path, but this won't work because the earth is not a good conductor and probably the two ground points are not at the same voltage. Assume that the cable in Figure 3.43(a) is 200 ft long and the source voltage ranges from 0 to 2 V. The difference in earth voltage for that distance could easily be 3 V (due to local conditions). It's a case where the "noise" (3 V) is larger than the signal (2 V). Connecting a separate signal return wire, as shown in Figure 3.43(b), is not a good idea because we now have a **ground loop**. A ground loop occurs when large currents flow in the return wire because of the difference in ground voltages. In the circuit of Figure 3.43(b), the wire resistance is 0.3 Ω (for wire size AWG 12), and if there really is a 3 V difference between the two ground points, the ground-loop current would be $3\,V/0.3\,\Omega = 10\,A$. Clearly, it is not desirable to have large, unpredictable currents running

---

*Mobile and other nongrounded transmitters must make do with an *artificial ground plane*, such as the hood of a car, or they inductively and capacitively couple to real ground.

Figure 3.42
Connecting the signal
common to an earth ground
for safety.

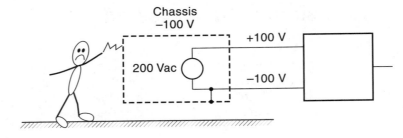

(a) Common connected to chassis only

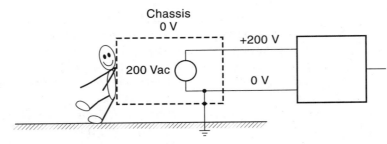

(b) Common connected to chassis and earth

around in your cabling! The solution to the ground-loop problem is indicated in Figure 3.43(c). This circuit does have a return wire for the signal, but it is grounded only at one end.

Isolation Circuits

The ground-loop problem can be avoided by simply not connecting both ends of the signal return to an earth ground. However, sometimes the ground connections have already been made inside the equipment, in which case the only solution is to use **isolation circuits** between the source and receiver. An isolation circuit can transfer a signal voltage without any metal-to-metal contact. This allows for the signal common of the source to be at a different voltage than the signal common of the receiver.

Figure 3.43 Grounding signal returns can cause ground loops.

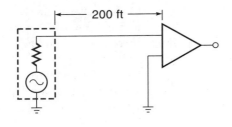
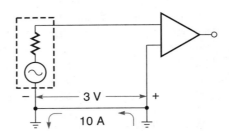
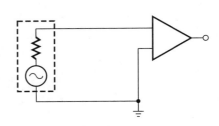

(a) Using earth as common   **(b)** Connecting grounds causes a "ground loop"   **(c)** Grounded at one end (no "ground loop")

Figure 3.44 shows a transformer-coupled isolation circuit. The input voltage is used to amplitude modulate an internally generated AC voltage. The modulated AC voltage is transformer-coupled to a demodulator, which reconstructs the original input voltage. The modulation is necessary because most sensor outputs are slow-changing DC, not AC. Because there is no electrical connection across the transformer coils, the signal commons can exist at different voltages, and the ground-loop path is broken.

Figure 3.44

A transformer-coupled isolation circuit.

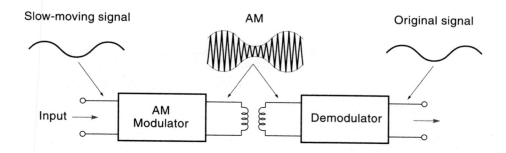

Figure 3.45 shows an optical linear signal isolator. The input signal is amplified to drive a light-emitting diode (LED). A phototransistor receives the light signal and converts it back to a voltage. Isolation is achieved because the input and output are two separate circuits with no electrical interconnection. To ensure that the output is an exact copy of the input, a feedback system using a second phototransistor is incorporated.

Figure 3.45

A linear signal optical isolator.

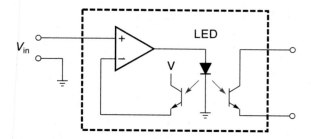

A simpler optical device can be used if the signal to be isolated is digital. Digital signals do not require a linear relationship between input and output (just on and off). Figure 3.46 shows the TIL-112 digital **optical coupler**. A standard TTL digital device can drive the input. When the input (V_{in}) is driven low, a voltage is developed across the LED causing it to light. This light turns on the phototransistor, which pulls its output (V_{out}) to 0 V. The output of the photocoupler can be connected directly to a TTL gate with a pull-up resistor as shown.

Optical couplers (sometimes called opto-isolators) are very useful in providing noise isolation because signals can only travel in one direction through the coupler. Consider the case shown in Figure 3.47 where a microprocessor controller is required to drive a relay and a motor. Both the motor and relay can put large voltage spikes on the line,

Figure 3.46

The TIL-112 optical coupler (for digital signals).

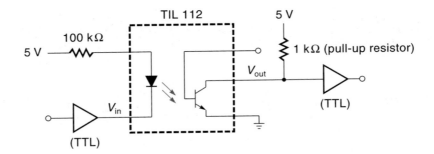

which could cause the microprocessor to act erratically. The optical coupler (OC) will not let this noise from the outside world go back upstream into the controller because the LED is designed to generate light, not receive it.

Figure 3.47

Interfacing components with optical isolators reduce noise spikes.

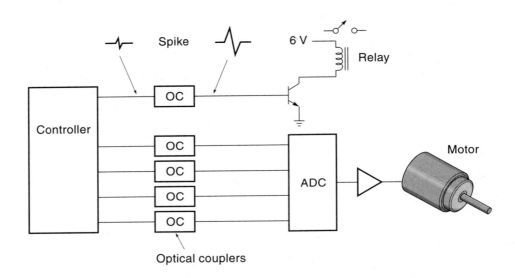

Another use of the optical coupler is high-voltage protection. Because there is no electrical connection between input and output, high voltage cannot pass either way. This protects the relatively delicate digital controller circuitry from high voltage spikes in the actuator power lines. The TIL-112 can resist up to 2.5 kV.

Shielding

A current through a wire generates a magnetic field around that wire. If another wire happens to be in the vicinity, the magnetic field from the first wire will induce a current in the second wire, proportional to the rate of change of the magnetic field. Therefore, DC will not induce current in a nearby wire, but AC will. This undesired induced current is electrical noise. Good design practice dictates that wires with large AC currents, such as 60-Hz power wires, be kept apart from low-level signal wires.

Shielding cannot block the **magnetic field noise**, but it can draw it away from the signal wire. This is shown in Figure 3.48 in a cross-section view. A shielding material is used that has a low resistance to the magnetic field. For this application, steel is better than copper or aluminum. The magnetic lines of force are more inclined to go through the shield than the air, so the field within the shield is sharply reduced. For this to be effective, the shield must be relatively thick-walled, and in fact, even a heavy conductor placed near the signal wire will act as a magnetic shield.

Figure 3.48

Shielding a conductor from a magnetic field.

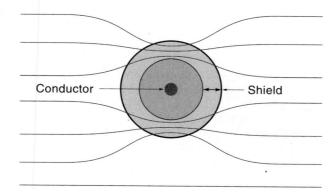

Another source of noise is from an **electric field**. An isolated wire placed in an electric field will assume a voltage proportional to that field. Figure 3.49 shows a wire in the middle of a 100-V field. The wire will assume a voltage of 50 V. If the field voltage is AC, then an AC voltage will be induced on the wire. Large time-varying electric fields are common in an industrial environment.

Figure 3.49

A wire in an electric field.

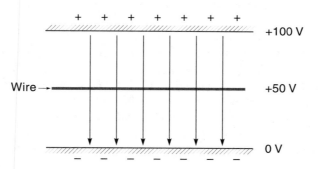

Merely putting a conductive shield around a wire will not keep the field out. As Figure 3.50 shows, the shield itself would acquire a voltage from the field and then reradiate this voltage to the wire. However, if the shield is grounded (forced to be 0 V), then there can be no electric field within the shield, and the wire is protected (Figure 3.51). For best results, the shield should not be used as the signal return path because the shield will probably have noise on it (from the shielding action). The ideal setup is shown in Figure 3.52, where both the signal and the return are contained within a shield and the shield is grounded at one end.

Figure 3.50

A wire in an ungrounded shield.

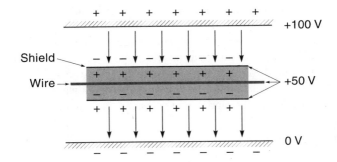

Figure 3.51

An electrostatic shield grounded at one end.

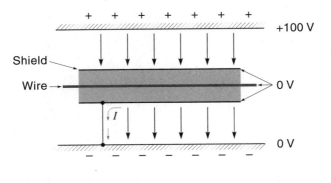

Figure 3.52

A signal and return contained within a shield.

Shield-Grounding Considerations

The time to think about grounding is when a system is being designed. A grounding strategy built in from the beginning is better than trying to patch things after the system is built. We have discussed the fact that shields must be grounded to be effective and that they should be grounded at only one end. Because the controller is the hub of the control system, it makes sense to tie all shields together at the controller and ground them there. This is called a **single-point ground** and is illustrated in Figure 3.53. It is important that the single-point ground have a solid, low-impedance connection to a real earth ground; otherwise, noise coming in on one shield will reradiate out on all the other shields.

Figure 3.53

All shields are grounded at one place.

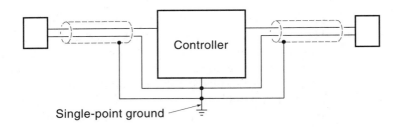

Figure 3.54

Using an isolation circuit to avoid ground loops.

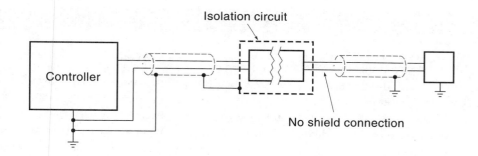

Sometimes the shield cannot be isolated from ground at the component end. In such cases, the best approach is to insert an isolation circuit and break the shield at that point (Figure 3.54). This approach has the advantage of allowing individual components to be grounded for safety reasons, yet preventing ground loops.

SUMMARY

The operational amplifier (op-amp) is a linear amplifier packaged as an integrated circuit, which has the following desirable properties: (1) very high open-loop gain (typically, 200,000), (2) very high input resistance (1 + MΩ), and (3) low output resistance (50–75 Ω). With the addition of a few passive components, many different useful interface circuits can be built from op-amps, such as the following:

❏ A voltage follower has a voltage gain of 1 but can boost the current and isolate different stages in a circuit.

❏ Inverting and noninverting amplifiers are stable linear amplifiers where the gain can be set with two resistors.

❏ A summing amplifier gives an output voltage that is the sum of multiple input voltages.

❏ An instrumentation amplifier is a practical differential amplifier usually packaged as an IC. It amplifies the difference between two voltages and includes high-resistance buffered inputs and an easily selectable gain.

❏ Integrators and differentiators give an output voltage that is proportional to the change of input over time.

❏ An active filter is a circuit using an amplifier that passes certain frequencies and rejects others.

❏ A comparator compares two analog voltages and identifies which one is larger (with a digital output).

A current loop is a special type of signal interface that uses current instead of voltage to convey the informa-tion. This system is immune to wire resistance in the connecting cables because the current is the same everywhere in a closed loop, regardless of the resistance. The current-loop system requires special transmitter and receiver circuits, both of which can be made with op-amps.

The analog switch is a solid-state device that can switch analog signals. A common use of analog switches is for analog multiplexing. An example of multiplexing is where a number of analog sensor signals are connected, one at a time, to an ADC. Another interface device is a sample-and-hold circuit, which can temporarily hold an analog voltage on a charged capacitor until it can be processed.

Earth ground refers to the voltage on the physical earth and is *not* the same as signal return or signal common in a circuit. Equipment is grounded to the earth primarily for safety reasons. However, if the same conductor is grounded at two locations, different local ground voltages can cause large ground-loop currents in the conductor, which is un-desirable. To avoid this situation, conductors should be grounded at only one end; if that is not possible, isolation circuits such as optical isolators can be used to break the electrical continuity.

Shielding protects a signal from electrical noise. Electri-cal noise comes in two forms, magnetic fields and electrical fields, and shielding considerations are different for each type.

GLOSSARY

active filter   A circuit that incorporates an op-amp.

analog switch   A solid-state device that performs the same function as a low-power mechanical switch.

band-pass filter   A circuit that allows only a specified range of frequencies to pass and attenuates all others above and below the pass band.

chatter   the condition that occurs when the output of a comparator oscillates when the input voltage is near the threshold voltage.

closed-loop gain   The gain of an amplifier when feedback is being used. The value of closed-loop gain is less than open-loop gain (for negative feedback).

comparator   A type of op-amp that is used open-loop to determine if one voltage is higher or lower than another voltage.

current loop   In a signal transmission system, a single loop of wire that goes from the transmitter to the receiver and back to the transmitter. The signal intelligence is conveyed by the current level instead of the voltage. This system is immune to voltage drops caused by wire resistance.

current-to-voltage converter   An op-amp based circuit used as a transmitter for a current-loop system.

cutoff frequency   The frequency at which a filter circuit starts to attenuate.

DC offset voltage   The small voltage that may occur on the output of an op-amp, even when the inputs are equal; DC offset can be eliminated with a resistor adjustment.

differential amplifier   A circuit that produces an output voltage that is proportional to the instantaneous voltage difference between two input signals. Op-amps by themselves are difference amplifiers; however, a practical differential amplifier incorporates additional components.

differential voltage   A signal voltage carried on two wires, where neither wire is at ground potential.

differentiator   An op-amp circuit that has an output voltage proportional to the instantaneous rate of change of the input voltage.

earth ground   The voltage at (or connection to) the surface of the earth at some particular place.

electric field   A condition that exists in the space between two objects that are at a different voltage potential. A wire in this space will assume a ''noise'' voltage proportional to the field strength.

high-pass filter   A circuit that allows higher frequencies to pass but attenuates lower-frequency signals.

instrumentation amplifier   A practical differential amplifier, usually packaged in an IC, with features such as high input resistance, low output resistance, and selectable gain.

integrator   An op-amp based circuit that has an output voltage proportional to the area under the curve traced out by the input voltage.

interfacing   The interconnection between system components.

inverting amplifier   A simple op-amp voltage amplifier circuit with one input, where the output is out of phase with the input; one of the most common op-amp circuits.

inverting input   The minus ($-$) input of an op-amp; the output will be out of phase with this input.

isolation circuit   A circuit that can transfer a signal voltage without a physical electrical connection.

low-pass filter   A circuit that allows lower-frequency signals to pass but attenuates higher-frequency signals.

magnetic field noise   An unwanted current induced in a wire because the wire is in a time-varying magnetic field.

multiplexing   The concept of switching input signals (one at a time) through to an output; is typically used so that multiple sensors can use a single ADC (analog-to-digital converter).

negative feedback   A circuit design where a portion of the output signal is fed back and subtracted from the input signal. This results in a lower but predictable gain and other desirable properties.

noninverting amplifier   A simple op-amp voltage amplifier circuit with one input, where the output is in phase with the input.

noninverting input   The positive ($+$) input of an op-amp; the output will be in phase with this output.

notch filter   A circuit that attenuates a very narrow range of frequencies.

operational amplifier (op-amp)   A high-gain linear amplifier packaged in an integrated circuit; the basis of many special-purpose amplifier designs.

open-loop gain   The gain of an amplifier when no feedback is being used; it is usually the maximum gain possible.

optical coupler   An isolation circuit that uses a light-emitting diode (LED) and a photocell to transfer the signal.

programmable gain instrumentation amplifier   An instrumentation amplifier with fixed gains that can be selected with digital inputs.

sample-and-hold circuit   A circuit that can temporarily store or remember an analog voltage level.

signal common   The common voltage reference point in a circuit, usually the negative terminal of the power supply (or battery).

signal return   *See* **signal common.**

single-ended voltage   A signal voltage that is referenced to the ground.

single-point ground   A single connection point in a system, where all signal commons are connected together and then connected to an earth ground.

summing amplifier   An op-amp circuit that has multiple inputs and one output. The value of the output voltage is the sum of the individual input voltages.

virtual ground   A point in a circuit that will always be practically at ground voltage but that is not physically connected to the ground.

voltage follower   A very simple but useful op-amp circuit with a voltage gain of 1, used to isolate circuit stages and for current gain.

window comparator   A comparator with built-in hysteresis, that is, two thresholds: an upper switch-in point and a lower switch-in point.

EXERCISES

Section 3.1

1. Find V_{out} for each amplifier in Figure 3.55. Assume the gain is 10 in all cases.

2. An op-amp has an open-loop gain of 200,000 and is being supplied with $+12$ V and -12 V. What is the maximum voltage difference that can exist between V_1 and V_2 before saturation? (Is this why V_1 and V_2 are considered "virtually" the same?)

3. Use the basic op-amp assumptions to describe how a voltage follower works.

4. Draw the circuit of an inverting amplifier with a gain of 35.

5. Draw the circuit of an inverting amplifier with a gain of 50 and show the pin numbers of the 8 pin 741 general-purpose op-amp. Power supply voltage is ± 15V.

6. Draw the circuit of a noninverting amplifier with a gain of 24.

7. Draw the circuit of a noninverting amplifier with a gain of 30 and draw the pin numbers of the 8 pin 741 general-purpose op-amp. Power supply voltage is ± 15V.

Figure 3.55

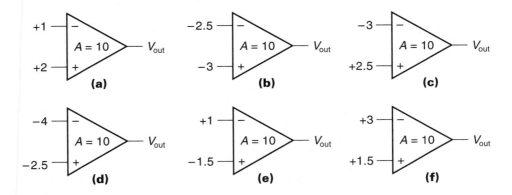

Figure 3.56

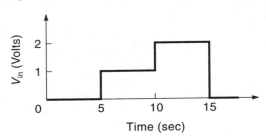

Time (sec)

8. Draw the circuit of a summing amplifier with four inputs. The gain for each input should be 15. Assume the impedance of each source connected to the summing amp is 1 kΩ.

9. Draw the circuit of a difference amplifier with a gain of 20. Assume the impedance of each source is 6 kΩ.

10. Sketch the output of an integrator circuit that has an input waveform shown in Figure 3.56 (assume $RC = 1$):

11. Sketch the output of an integrator circuit that has $R = 2$ MΩ and $C = 10$ μF. The input waveform is shown in Figure 3.56.

12. Sketch the output of a differentiator circuit that has an input waveform as shown in Figure 3.57 (assume $RC = 1$).

13. Sketch the output of a differentiator circuit that has R $= 2$ MΩ and $C = 10$ μF. The input waveform is shown in Figure 3.57.

Figure 3.57

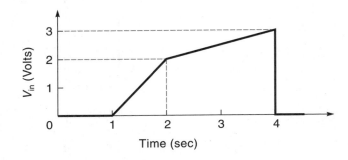

Time (sec)

14. Draw the circuit of a low-pass filter with $f_c = 1$ kHz and a gain of 10. Use $R = 1$ kΩ.

15. Draw the circuit of a low-pass filter with $f_c = 5$ kHz and a gain of 20. Use $R = 1$ kΩ.

16. Draw the circuit of a high-pass filter with $f_c = 10$ kHz and a gain of 10. Use $R = 1$ kΩ.

17. Draw the circuit of a high-pass filter with $f_c = 8$ kHz and a gain of 15. Use $R = 1$ kΩ.

18. Draw the circuit of a notch filter with $f_N = 1.6$ kHz. Use $R = 2$ kΩ.

19. Draw the circuit of a comparator that will switch from 0 to 5 V when the analog input goes above 2.9 V.

Section 3.2

20. Explain why the signal is never attenuated in a current-loop system.

21. Draw the circuit of a current-loop system. When V_{in} is 1 V, the loop current should be 5 mA. At all times, V_{out} should equal V_{in}.

22. What is an *analog switch*, and how is it different than a mechanical switch?

23. The sample-and-hold circuit of Figure 3.41 is being used to hold a signal for 0.2 s. If $R_s = 1$ kΩ, what size capacitor is needed?

Section 3.3

24. Explain the difference between *earth ground* and *signal return*.

25. What is a *ground loop*? How is it possible to have large currents in the return line when there is no apparent power supply?

26. Find the current in a ground loop that is 200 ft long, has a wire resistance of 0.05 Ω/ft, and has a difference of 4 V between grounds.

27. Describe how the TIL-112 optical isolator in Figure 3.46 works.

28. What are the reasons for using isolation circuits in the interconnection of a control system?

29. Explain the principle behind magnetic shielding.

30. Why is it necessary to ground an electrostatic shield?

31. Explain the concept of single-point ground and why it is important.

4 Switches, Relays, and Power-Control Semiconductors

OBJECTIVES

After studying this chapter, you should be able to:

- Understand mechanical switch types, configurations, and terms.
- Describe the types and operation of electromechanical relays.
- Understand the characteristics of solid-state relays and the advantages and disadvantages of solid-state relays compared with electromechanical relays.
- Comprehend the operation of power transistors and power amplifiers, calculate transistor circuit currents, and select a replacement power transistor.
- Understand basic JFET and power MOSFET characteristics and applications.

- Understand the characteristics and operation of silicon-controlled rectifiers and basic silicon-controlled rectifier circuits.
- Understand how a triac operates and some triac applications.
- Explain the use of some trigger circuits for the silicon-controlled rectifiers and triac, in particular those that use the unijunction transistor, programmable unijunction transistor, or diac.

INTRODUCTION

An important ability of almost all control systems is to regulate the flow of electrical power because electricity is usually the link between the controller and the actuator (which is typically an electric motor). The problem is that the small-signal output of the controller is usually not large enough to drive the load directly and must be amplified in some way.

Three classifications of components are used to control electrical power: (1) electromechanical devices such as switches and relays, (2) power transistors and field effect transistors, and (3) a group of semiconductor devices called thyristors, which includes silicon-controlled rectifiers and triacs. All these power control devices are the subject of this chapter.

4.1 SWITCHES

A *mechanical switch* is a device that can open or close, thereby allowing a current to flow or not. As you have no doubt observed, switches come in many different shapes, sizes, and configurations.

Toggle Switches

Probably the most common switch type is the **toggle switch**, which is available in various contact configurations. Each switch consists of one or more *poles*, where each pole is actually a separate switch. The contact arrangement for the **single-pole/single-throw switch** (SPST), the simplest switch, is illustrated in Figure 4.1(a). Notice that it has a single set of contacts that can either open or close. Next, up the line of complication, is the **single-pole/double-throw switch** (SPDT) illustrated in Figure 4.1(b). Notice that the movable "arm" called the *common* (C), or *wiper*, can connect with either contact A or B. Figure 4.1(c) illustrates a **double-pole/double-throw switch** (DPDT), which consists of two electrically separate SPDT switches in one housing that operate together. In Figure 4.1(d), notice how the terminals are arranged on the back of the DPDT switch housing, with the three terminals for each pole running the "long way" on the switch body. Although not as common, switches with up to six poles are available.

Figure 4.1

Toggle switch contact arrangements.

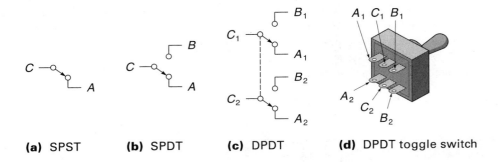

(a) SPST **(b)** SPDT **(c)** DPDT **(d)** DPDT toggle switch

So far the discussion has centered on the **two-position switch**, which comes in two configurations: the simple *on–off* [Figure 4.1(a)] and the *on–on* [Figure 4.1(b)]. A **three-position switch**, called *on–off–on*, has a stable position in the middle, where the C contact is not connected to either A or B, as illustrated in Figure 4.2. Such a switch could be used to control a motor, where the three positions might correspond to "forward," "off," and "reverse."

Another variation of the toggle switch is to have one or more of the positions be spring-actuated, meaning that physical pressure must be maintained on the handle to keep it in position. These are called **momentary-contact switches**, an example of which is a car's ignition switch where you must maintain continuous pressure on the key to engage the starter motor. Table 4.1 shows some of the variations possible for a particular

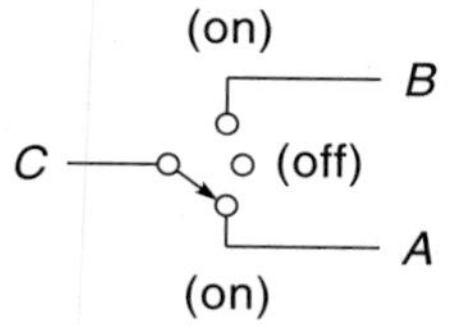

Figure 4.2

A three-position switch: on–off–on.

style of toggle switch. For example, the third switch (1SFX191), is a three-position switch with one of the on positions being momentary.

Toggle switch contacts have maximum voltage and current ratings for AC and DC operation. A typical toggle switch can handle less DC voltage than AC voltage at the rated current; for example, a switch might be rated at 5 A for 125 Vac or 5 A for 28 Vdc. The reason for the difference is that when the contacts are just opening, the current will continue (briefly) to arc across, which tends to burn and pit the contact surface. AC will arc less than DC for the same voltage because AC is going to 0 V twice each cycle.

Closely related to the toggle switch is the **slide switch** illustrated in Figure 4.3. Although mechanically different internally, the slide switch performs the same functions as the toggle switch and is available in the same configurations. Slide switches tend to be less expensive but are not available with the high current rating of toggle switches.

TABLE 4.1 Typical Toggle Switches[*]

Type	Number of poles	Circuit
1SBX191	1	On–Off–On
1SCY191	1	On On
1SFX191	1	On–Off–On[†]
1SGX191	1	On On[†]
1SHX191	1	On[†]–Off–On[†]
2SBX191	2	On–Off–On
2SCY191	2	On On
2SFX191	2	On–Off–On[†]
2SGX191	2	On On[†]
2SHX191	2	On[†]–Off–On[†]
3SBX191	3	On–Off–On
3SCY191	3	On On
3SFX191	3	On–Off–On[†]
3SCX191	3	On On[†]
3SHX191	3	On[†]–Off–On[†]
4SGX191	4	On–Off–On
4SCY191	4	On On
4SFX191	4	On–Off–On[†]
4SGX191	4	On On[†]
4SHX191	4	On[†]–Off–On[†]

[*]Rated 5 A at 125 Vac; 5 A at 28 Vdc.
[†]Momentary contact.

Figure 4.3
A slide switch.

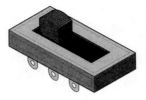

Push-Button Switches

Push-button switches [Figure 4.4(a)] are almost always the momentary type—pressure must be maintained to keep the switch activated. Figure 4.4(b–d) shows the symbol for the push-button switch. Notice there are two configurations possible: **normally open** (NO) and **normally closed** (NC). For the NO switch [Figure 4.4(b)], the contacts are open until the button is pushed; and for the NC switch [Figure 4.4(c)], the contacts are closed when the switch is "at rest" and open when the button is pushed.

Figure 4.4
Push-button switches.

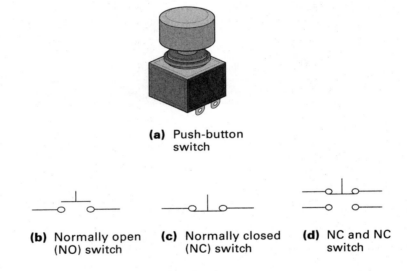

(a) Push-button
switch

(b) Normally open
(NO) switch

(c) Normally closed
(NC) switch

(d) NC and NC
switch

Other Switch Types

A **limit switch** is a push-button switch mounted in such a position that it is activated by physical contact with some movable object. An example is a car door switch, which senses whether or not the door is closed. Limit switches are available with different kinds of actuator hardware, such as the "paddle" or roller. Often these are mounted on a small standard-sized switch body called a **microswitch** (developed by the MicroSwitch Company). A microswitch requires a very small throw of a few thousandths of an inch. Figure 4.5 illustrates some examples of limit switches. (Limit switches and other proximity sensors are discussed in Chapter 6.)

Figure 4.5

Limit switches (using a microswitch type).

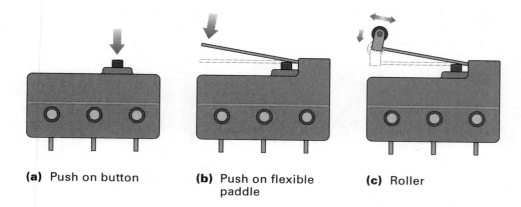

(a) Push on button

(b) Push on flexible paddle

(c) Roller

A **DIP switch** is a set of small SPST switches built into a unit shaped like an integrated circuit (IC) (DIP stands for dual in-line package). The DIP switch can be plugged into an IC socket or soldered into a circuit board. As shown in Figure 4.6, each individual switch uses two pins directly across from each other; that is, switch 1 uses pins 1 and 14, switch 2 uses pins 2 and 13, and so on. (Do not use a pencil to set the switches; the graphite can work its way into the switch and short it out.)

Figure 4.6

A DIP switch.

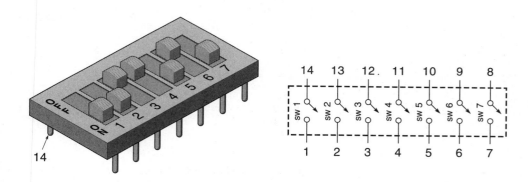

A **rotary switch**, considerably different from the switches discussed so far, can perform multiple, complicated switching operations. As shown in Figure 4.7, the rotary switch is constructed of **switch wafers** mounted along a single shaft. The inner part of each wafer rotates in steps with the shaft, and the outer part remains stationary. To understand how the switch works, consider Figure 4.7(a), which shows the switch in the off position. In this position, the C terminal is making contact with the pad, but the pad is not touching either terminals 1 or 2. In Figure 4.7(b), the shaft has been rotated clockwise to position *X*. Notice that the C terminal is still connected to the pad, but now the little extension of the pad is making contact with terminal 1. In Figure 4.7(c), the shaft has been rotated to position *Y*, and now the pad is in contact with terminal 2. Notice that there could be three or four such switches on the same wafer, as well as multiple wafers.

Figure 4.7
Rotary switch operation.

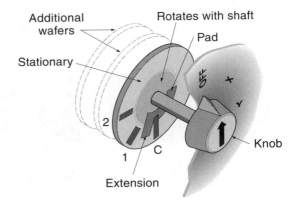

(a) Off position

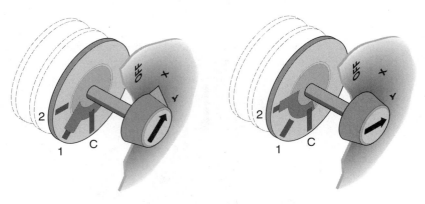

(b) Contact C connects
to contact 1

(c) Contact C connects
to contact 2

A **thumbwheel switch**, a special type of rotary switch, is used to input numeric data. The operator selects a number by rotating the numbered wheel (Figure 4.8), and each number corresponds to a different switch setting. The switch schematic in Figure 4.8(b) shows that one of ten separate terminals is connected with the C terminal. Also available are thumbwheel switches that output 4-bit BCD (binary coded decimal) and other codes.

Another type of data-input device is the **membrane switch** keypad. This type of keypad consists of push-button switches built up from a number of layers, as shown in Figure 4.9(a). The bottom layer is usually a printed-circuit board with two nonconnecting pads for each switch. Lying over the circuit board is a spacer layer with a hole over each switch position. The third layer is flexible and has a conductive pad over each switch. Overlying the whole thing is a flexible waterproof membrane, which has the key lettering. When a key is pressed, the conductive pad is forced into contact with the circuit board and provides a path for the current, as illustrated in Figure 4.9(b). Membrane switches are especially appropriate in dirty industrial environments because the membrane keeps dirt from getting into the switch assembly.

Figure 4.8
A thumbwheel switch
(ten-decimal position).

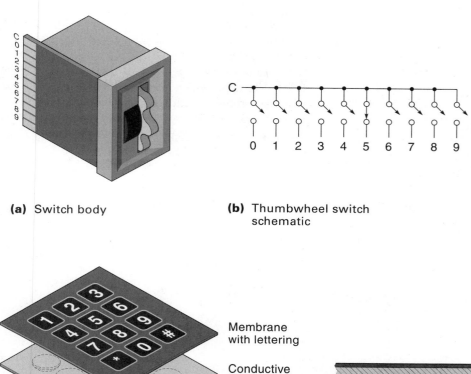

(a) Switch body

(b) Thumbwheel switch
schematic

Figure 4.9
A membrane switch.

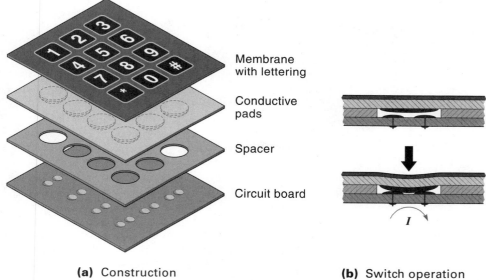

Membrane
with lettering

Conductive
pads

Spacer

Circuit board

(a) Construction

(b) Switch operation

4.2   RELAYS

Electromechanical Relays

The **electromechanical relay** (EMR) is a device that uses an electromagnet to provide the force to close (or open) switch contacts, in other words, an electrically powered switch. A diagram of a simple relay is shown in Figure 4.10(a). When the electromagnet, called the *coil*, is energized, it pulls down on the spring-loaded *armature*. Relay contacts are described as being one of two kinds: normally open contacts (NO), which are open in the unenergized state, and normally closed contacts (NC), which are closed in the unenergized state. Figure 4.10(b) shows one common schematic symbol for the relay. By convention, the symbol always depicts the relay in the unenergized state, so you can easily determine which are the NC and NO contacts from the schematic. Another schematic representation for the relay is shown in Figure 4.10(c). This symbol is used in a type of drawing called a *ladder diagram*. In a ladder diagram, the relay coil and its contacts are separated on the drawing. (Ladder diagrams are discussed in Chapters 9 and 12.)

Figure 4.10
An electromagnetic relay.

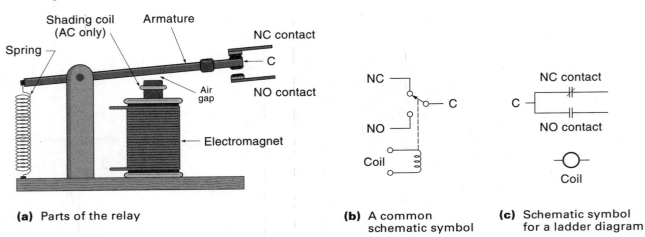

(a) Parts of the relay

(b) A common schematic symbol

(c) Schematic symbol for a ladder diagram

The electrical specifications for the contacts are different than for the coil. For the contacts, the maximum current and voltage for DC and AC operation is specified. For the coil, the intended voltage and coil resistance are usually specified. You can see these numbers on the specification (spec) sheet shown in Table 4.2. The coil voltage and resistance can be used to calculate the steady-state coil current. Actually, it takes more voltage and current to pull in the relay contacts than it does to hold them there because the armature must be pulled in across an air gap. Hence, these quantities are called, respectively, **pull-in voltage** and **pull-in current**. For example, the contacts of a particular 6-V relay actually close at 2.1 V and stay closed until the voltage is decreased to 1 V. The values of voltage and current needed to keep the relay energized are called the **minimum holding voltage** and **sealed current**. Notice that the actual pull-in voltage is much less than the rated coil voltage. This is to guarantee that the relay will pull-in quickly and reliably when operated at the rated voltage. Coil voltages are specified to be AC or DC. The difference is that AC coils are constructed with **shaded poles** to prevent "buzzing" with 60-cycle power. A shaded-pole relay has a metal ring around the pole face of the electromagnet (see Figure 4.10). Magnetic flux induced into this ring keeps the relay closed when the AC cycles through 0 V.

TABLE 4.2 Typical General-Purpose Relays*

| Type | Coils | | Action |
	Input	Ohm	
Y1-SS1.OK	6 DC	1,000	SPDT
Y1-SS220	3 DC	220	SPDT
Y2-V52	6 DC	52	2PDT
Y2-V185	12 DC	185	2PDT
Y2-V700	24 DC	700	2PDT
Y2-V2.5K	48 DC	2,500	2PDT
Y2-15K	115 DC	15,000	2PDT
Y4-V52	6 DC	52	4PDT
Y4-V185	12 DC	185	4PDT
Y4-V700	24 DC	700	4PDT
Y4-V2.5K	48 DC	2,500	4PDT
Y4-V15K	115 DC	15,000	4PDT
Y6-V25	6 DC	25	6PDT
Y6-V90	12 DC	90	6PDT
Y6-V430	24 DC	430	6PDT
Y6-V1.5K	48 DC	1,500	6PDT
Y6-V9.0K	115 DC	9,000	6PDT

*Contacts: 2 A typically, 3 A maximum 125 Vac or 28 Vdc.

Relays are available in a variety of sizes, contact configurations, and power-handling capabilities. Some miniature relays can plug into an IC socket and be powered directly from a digital logic gate. On the other end of the spectrum, a power relay, often called a **contactor**, is used to switch the current directly to larger machines and may handle 50 A. Figure 4.11 shows a selection of different relays.

The **reed relay** is unique because the small reedlike contacts are encapsulated in a small sealed glass tube that is evacuated or filled with an inert gas like dry nitrogen. The contacts are activated by an external magnetic field, as shown in Figure 4.12. Contacts are either dry or mercury-wetted. Mercury-wetted contacts have a thin coating of mercury that fills in surface irregularities, making a larger conduction area, and reduces pitting. Generally, reed relays have a long life and low coil voltage (frequently TTL compatible), and are immune to dirty environments; however, they are generally low power (contacts rated at 2 A or less) and vibration sensitive.

Solid-State Relays

A **solid-state relay** (SSR) is a purely solid-state device that has replaced the EMR in many applications, particularly for turning on and off AC loads such as motors. Physically, the SSR is packaged in a box (about the same size as an EMR), with four electrical terminals, as shown in Figure 4.13(a). The two input terminals are analogous to the coil of an EMR, and the two output terminals are analogous to the contacts of the EMR (usually SPST, normally open).

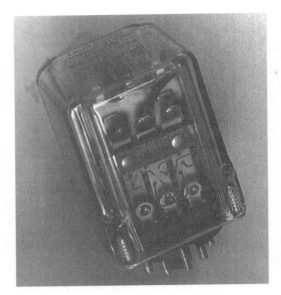

(a) General purpose relay

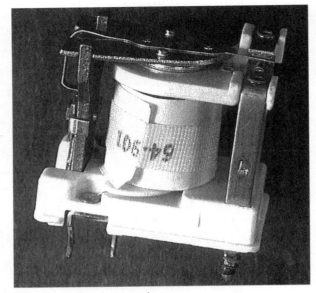

(b) General purpose relay

(c) High current relay

(d) Miniature relay

Figure 4.11

A selection of relays.

The input or *control voltage* of the SSR is typically 5 Vdc, 24 Vdc, or 120 Vac. The 5-Vdc models are designed to be driven directly from TTL digital logic circuits. Turning our attention to the output side of the SSR, we see that the load is placed in series with the 120-Vac or 240-Vac power. The output current can be as high as 50 A in some models.

Figure 4.12
A Reed relay.

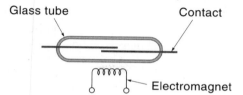

Glass tube   Contact

Electromagnet

Figure 4.13
A solid-state relay.

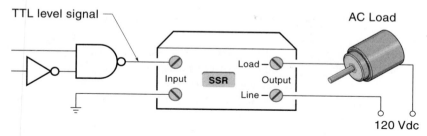

TTL level signal

Input   **SSR**   Output

Load

Line

AC Load

120 Vdc

(a) Using an SSR to drive a motor

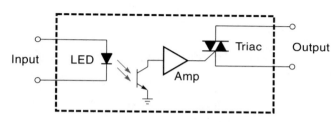

Input   LED   Amp   Triac   Output

(b) SSR circuit

Many SSRs incorporate a feature called **zero-voltage switching**: The line current is switched on at the precise time that the AC voltage is crossing 0 V. This eliminates sharp output voltage–rise times and so minimizes electromagnetic interference noise (EMI).

Figure 4.13(b) shows a block diagram of the interior of an SSR. The input voltage drives an LED, and the light from the LED turns on a photo transistor, which in turn turns on the triac (a solid-state switching device discussed later in this chapter). The LED electrically isolates the input and output sections of the SSR. This is important for two reasons: First, it allows the control electronics to have a separate ground from the power lines; second, it prevents high-voltage spikes in the power circuit from working their way back upstream to the more delicate control electronics.

Solid-state relays have a number of important advantages over electromechanical relays. Having no moving parts means that (theoretically) they will never wear out and makes them practically immune to shock and vibration. Also, because of the built-in electronics, they can be driven with a low-voltage source (such as TTL) regardless of the output-current capability. The main disadvantages of the SSR are the following: (1) They can be "false triggered" by electrical noise; (2) even when on, the output resistance is not exactly 0 ohm, so there is some small voltage drop and consequent power loss within the relay, and when off they may have lethal levels of leakage current; (3) although they are long-lasting, unlike an EMR they do not fail predictably; (4) contact arrangements are limited, so they may not work for all relay applications.

The **hybrid solid-state relay** is similar to the SSR but uses a low-voltage, fast-acting reed relay instead of an LED to turn on the output triac. Using the reed relay provides good electrical isolation and may work better than the SSR in some situations.

4.3   POWER TRANSISTORS

Power transistors are used extensively in control circuits as both switches and power amplifiers. Power transistors are fundamentally the same as small-signal transistors but are designed to carry more current. Many readers of this text will have already studied transistor operation, but for those who haven't (and perhaps as a review for those who have) the following basic material on transistors is presented. Also, the reader is reminded that this text uses *conventional current flow*, which means that the current is assumed to flow from the positive battery terminal to the negative. If you are familiar with *electron flow*, remember that the theory and "numbers" are the same, only the *direction* of the current is opposite from what you are used to.

Bipolar Junction Transistors

The **bipolar junction transistor** (BJT), referred to as a *transistor*, is a three-terminal solid-state device that operates on electric current much like a valve does on water in a pipe. This concept is illustrated in Figure 4.14. Figure 4.14(a) shows that the transistor has three terminals; the **base**, **emitter**, and **collector** and that it is connected into a simple series circuit with the (conventional) current entering the collector (C), and leaving through the emitter (E). Functionally, this is analogous to the system illustrated in Figure 4.14(b) where water is being pumped through a partially open valve. The flow of water can be regulated by opening or closing the valve. In the transistor circuit, the flow of current (I_C) can be regulated by adjusting a small control current in the base (I_B); the more I_B, the more I_C. In fact, I_C could be 100 times (or more) larger than I_B.

Figure 4.14

Basic transistor action (conventional current).

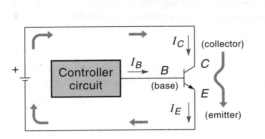

(a) Simple transistor circuit

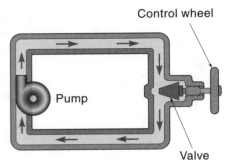

(b) Water-in-pipe analogy

Figure 4.15

NPN and PNP transistors (conventional current).

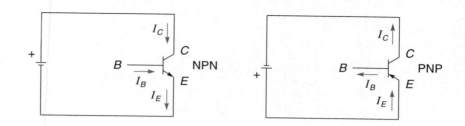

There are two basic types of transistors: **NPN** and **PNP**, both of which are made from three layers of semiconductor material. As illustrated in Figure 4.15, the only functional difference between the two types is the direction of current flow. The arrowhead of the emitter indicates the (conventional) current direction. The NPN type is more common and will be the type used in the following discussions.

Basic transistor operation can be summarized with the following statements:

1. Under the proper operating conditions, I_C will be some multiple of I_B; in other words, *the transistor is a current amplifier*. The **forward current gain** (h_{FE} or B) (which varies according to transistor type) is expressed as

$$h_{FE} = \frac{I_C}{I_B} \tag{4.1}$$

where

h_{FE} = forward current gain
I_C = collector current
I_B = base current

2. Inside the transistor, the small base current joins with the collector current to become the emitter current (or the emitter current divides to become the base and collector currents). This is expressed as

$$I_E = I_C + I_B \tag{4.2}$$

where

I_E = emitter current
I_C = collector current
I_B = base current

However, because the collector current is usually so much larger than the base current, a useful approximation is to consider the collector current to be equal to the emitter current:

$$I_E = I_C \tag{4.3}$$

3. The transistor dissipates power anytime there is a current flowing through it *and* a voltage is across it, expressed as

$$P_D = I_C V_{CE} \tag{4.4}$$

where

P_D = power dissipated by the transistor
I_C = collector current
V_{CE} = voltage between the collector and emitter

◆ EXAMPLE 4.1

A power transistor with an h_{FE} of 50 is operating with a load current (I_C) of 3 A. What is the base current?

Solution   Recall that a transistor is basically a current amplifier where the I_C is h_{FE} times larger than the base current (I_B). By rearranging Equation 4.1, we can solve for I_B:

$$I_B = \frac{I_C}{h_{FE}} = \frac{3\ A}{50} = 60\ mA$$ ◆

We have established that the transistor will "amplify" its base current, but how do you know what the base current is? The base–emitter junction acts like a forward-biased diode, meaning it must first be given a **forward-bias voltage** (about 0.7 V for silicon transistors, 0.3 V for germanium) before any appreciable base current flows at all. Once the base has been elevated to the bias voltage, any further increase in base voltage will start the base current flowing. This is shown in graph form in Figure 4.16. The **biasing**

Figure 4.16

A silicon transistor input (base) curve.

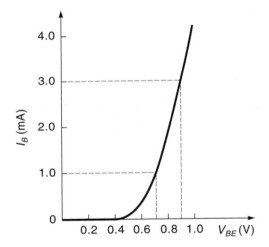

circuit of a transistor amplifier establishes a background "ballpark" DC voltage at the base. The signal to be amplified is then superimposed on this bias voltage. One common method of creating the bias voltage is with a resistor-type voltage-divider circuit as shown in Figure 4.17. Notice that resistors R_1 and R_2 create a base bias voltage of 0.8 V. Consulting the input curve of Figure 4.16, we see that 0.8 V would cause a base current I_B of 2 mA. The h_{FE} of the transistor is given as 100, so rearranging Equation 4.1 we can calculate the collector current I_C:

$$I_C = h_{FE}I_B = 100 \times 2\ mA = 200\ mA$$

Notice that the load is in series with the collector so that all 200 mA is going through the transistor and the load.

Figure 4.17

A voltage-divider bias circuit.

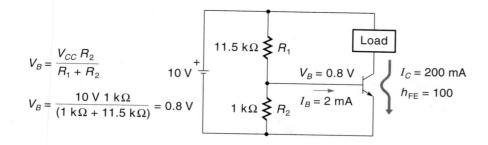

$$V_B = \frac{V_{CC}\, R_2}{R_1 + R_2}$$

$$V_B = \frac{10\text{ V } 1\text{ k}\Omega}{(1\text{ k}\Omega + 11.5\text{ k}\Omega)} = 0.8\text{ V}$$

We are now ready to discuss the operation of the simple transistor power amplifier shown in Figure 4.18. The DC bias conditions have already been discussed in the preceding paragraph. Now consider an input signal voltage of ±0.1 V, which is super-imposed onto the base through a capacitor. In other words, the input is added to the 0.8-Vdc bias voltage to form an AC base voltage that cycles from 0.7 to 0.9 V, as shown in Figure 4.18. Referring again to the base curve (Figure 4.16), we see that a base voltage of 0.7–0.9 V translates into a base current of 1–3 mA, which the transistor multiplies into a collector current of 100–300 mA, and all of this current goes through the load. Notice that the output voltage V_C is out of phase with the output current I_C. This occurs because the transistor lowers its resistance to increase the current, and a lower resistance creates a lower voltage drop.

Figure 4.18

A simple transistor amplifier.

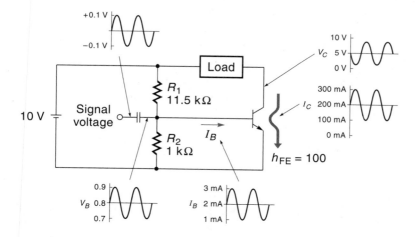

The circuit in Figure 4.18 is a class A amplifier. Specifically, **class A operation** is when the bias point is set somewhere in the middle of the operating range, and the actual output current is above and below this value. In other words, the transistor is biased to be "half-on" so that the input signal can turn it "more on" or "less on" from that middle point. Class A operation allows for both positive and negative input voltages to be amplified but has the drawback that power is being dissipated at all times, even when no input signal is present (because I_C is always flowing).

◆ **EXAMPLE 4.2**

Find the load current in the transistor circuit shown in Figure 4.19. Use the input curves in Figure 4.16 and assume that the h_{FE} of the transistor is 70.

Solution

We want to find the output current, and because we know the current gain is 70, the problem comes down to finding out what the base current is. In most cases, such as this one, it is the base voltage that we can calculate (or measure), so we will use the input curves to find the corresponding base current. First, calculate the base bias voltage from R_1 and R_2 using the voltage-divider rule:

$$V_{BE} = \frac{V_C R_2}{(R_1 + R_2)} = \frac{10\ \text{V} \times 1\ \text{k}\Omega}{(12\ \text{k}\Omega + 1\ \text{k}\Omega)} = 0.77\ \text{V}$$

Now, knowing that the base voltage is 0.77 V, we use the input curve of Figure 4.16 to determine that I_B is about 1.7 mA. The load current I_C can now be calculated by rearranging Equation 4.1:

$$I_C = h_{FE} I_B = 70 \times 1.7\ \text{mA} = 119\ \text{mA}$$

Note: The purpose of this example was to give you insight on how the class A transistor actually works. Unfortunately, in most cases, the input curves are not presented in spec sheets, so transistor amplifier designers make use of feedback and various empirical relationships, which are presented in any good transistor design text.

Figure 4.19

A transistor circuit (Example 4.2).

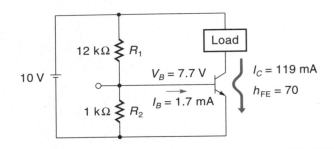

There are two other biasing possibilities besides class A. In **class B operation**, the base bias voltage is set just under 0.7 V (for silicon), which is the point where the transistor is just at the brink of turning on. The advantage of this system is that no power is dissipated when there is no input signal (because the transistor is off). The disadvantage is that it can only amplify positive voltages.

The third biasing possibility is **class C operation** where the transistor is operated like a switch, either full-on or full-off. The base voltage will be either 0 V, which will turn it completely off, or high enough (say, 1 V) to turn it on so much that the transistor acts like a short circuit. The advantage of this system is that it is very power efficient (for those applications that can use the on–off type of power). Applications of class C amplifiers are discussed in Chapter 7.

Power transistors are physically bigger than signal transistors and are designed to carry large currents. In control systems, they are used to provide the drive current for motors and other electromechanical devices. When a transistor has a large current and voltage at the same time, the resulting power ($V_C I_C$) must be dissipated in the form of heat. A typical power transistor is designed to operate up to 200°C (360°F) above ambient temperature. However, its power capacity is derated proportionally for temperatures above 25°C (as denoted by the spec sheet). The power transistor case has a flat metal surface to provide a thermal escape path for the heat. Therefore, to operate at anywhere near the rated power, the transistor must be mounted firmly to the chassis or a metal **heat sink**—a piece of metal with cooling fins to dissipate the heat into the air. Figure 4.20 illustrates two standard types of power transistors and how they are to be mounted. Many times the case itself is the collector terminal. If the collector must be kept electrically insulated from the mounting chassis, then a special mica insulator is used, together with a thermally conducting white grease.

Figure 4.20

Styles and mounting of power transistors.

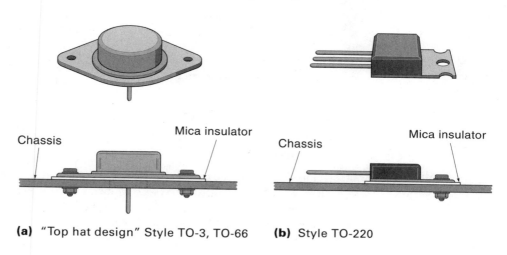

Chassis   Mica insulator   Chassis   Mica insulator

(a) "Top hat design" Style TO-3, TO-66   **(b)** Style TO-220

♦ **EXAMPLE 4.3**

A 100-watt (W) transistor fails, apparently from overheating. The transistor was replaced, and measurements showed that the load current (I_C) was 2 A, V_{CE} was 35 V, and the case temperature rose to 125°C. What is the problem?

Solution   First, calculate the power being dissipated by the transistor (using Equation 4.4):

$$P_D = I_C V_{CE} = 2\,\text{A} \times 35\,\text{V} = 70\,\text{W}$$

Seventy watts is certainly less than the manufacturer's designation of 100 W. The actual power capacity, however, will be less in this situation because the case is 100°C over ambient temperature (ambient is taken to be 25°C). Consulting a spec sheet, we find that the power must be derated 0.57 W/°C (above 25°). We now calculate the actual reduced power capacity:

$$P_D = 100° - (0.57\,\text{W/°C} \times 100°\text{C}) = 43\,\text{W}$$

Clearly, the problem is that 70 W exceeds the derated power capacity of 43 W. Apparently, the transistor is not "heat sinked" well enough; that is, the heat cannot conduct itself away from the transistor fast enough.   ♦

Figure 4.21 presents a sample selection sheet for some power transistors. For example, consider the 2N4912 (the first NPN type on the list). This device can have a continuous load current (I_C) of 1 A and a maximum V_{CE} of 80 V. Notice the wide range given for h_{FE} (20–100). It is common for transistors with the same part number to have widely different values of h_{FE}, so most transistor circuits are designed to accommodate this. Also, the gain of any transistor eventually decreases if the frequency gets high enough. The **gain-bandwidth product** (f_T) is a constant that is the product of the frequency and the gain at that frequency. Therefore the gain-bandwidth product represents the frequency that is high enough to cause the gain (h_{FE}) to reduce to 1.

♦ **EXAMPLE 4.4**

A transistor in a piece of equipment has gone bad and needs to be replaced with one of those listed in Figure 4.21. The bad transistor had the following specifications:

Type: NPN

Maximum load current: 5 A

Maximum voltage across the transistor: 30 V

Maximum power dissipation: 50 W

h_{FE}:   30

Maximum frequency: 5 kHz

Solution   First, consider load current of 5 A. Looking down the list of transistors, we see that the first 5-A NPN transistor (2N4231A) meets all qualifications except possibly the h_{FE} requirement. A safer choice might be the MJ3247 (8 A), which has a guaranteed minimum h_{FE} of 40 min, as well as meeting the other qualifications.   ♦

Field Effect Transistors (FET)

The **field effect transistor** (FET) is another three-terminal solid-state amplifying device. FETs perform the same job as the BJTs of the previous section, but because of some advantages are becoming more common in power applications. These advantages include high input impedance, high switching speeds, and less temperature sensitivity. There are two types of FETs, the **junction FET** (JFET) and the **metal-oxide semiconductor FET** (MOSFET). Both types have three terminals; the **drain** (D), the **source** (S), and the **gate** (G). Also, FETs are made to be either N-channel or P-channel; the type determines the direction of current through the device. The FETs discussed in this section are all N-channel.

The simple JFET circuit is shown in Figure 4.22(a). The load current I_{DS} passes through the drain (D) to the source (S). The amount of drain current is controlled by the gate voltage (as contrasted with the BJT where the collector current is controlled by the base current). The characteristics of the JFET are described by the characteristic curves shown in Figure 4.22(b). Notice that the load current (I_{DS}) is on the vertical axis, and the

POWER TRANSISTORS — BIPOLAR METAL

TO-213AA Package (Formerly TO-66)

CASE 80-02

STYLE 1:
PIN 1. BASE
2. EMITTER
CASE. COLLECTOR

I_C Cont Amps Max	$V_{CEO(sus)}$ Volts Min	NPN	PNP	h_{FE} Min/Max	@ I_C Amp	Resistive Switching			fT MHz Min	PD (Case) Watts @ 25°C
						ts μs Max	tf μs Max	@ I_C Amp		
1.0	40		2N4898	20/100	0.5	0.6 typ	0.3 typ	0.5	3.0	25
	60		2N4899	20/100	0.5	0.6 typ	0.3 typ	0.5	3.0	25
	80	2N4912	2N4900	20/100	0.5	0.6 typ	0.3 typ	0.5	3.0	25
	175	2N3583	2N6420	40/200	0.5	2.0 typ	0.23 typ	0.5	10	35
	225	2N3738	2N6424	40/200	0.1	3.0 typ	0.3 typ	0.1	10	20
	250		2N5344	25/100	0.5	0.6	0.1	0.5	60	40
	300	2N3739	2N6425	40/200	0.1	3.0 typ	0.3 typ	0.1	10	20
			2N5345	25/100	0.5	0.6	0.1	0.5	60	40
2.0	125	2N5050		25/100	0.75	3.5	1.2	0.75	10	40
	150	2N5051		25/100	0.75	3.5	1.2	0.75	10	40
	200	2N5052		25/100	0.75	3.5	1.2	0.75	10	40
	225		2N6211	10/100	1.0	2.5	0.6	1.0	20	35
	250	2N3584	2N6421	25/100	1.0	4.0	3.0	1.0	10	35
	300	2N3585	2N6212	10/100	1.0	2.5	0.6	1.0	20	35
		2N4240	2N6422	25/100	1.0	4.0	3.0	1.0	10	35
			2N6423	30/150	0.75	6.0	3.0	0.75	15	35
	350		2N6213	10/100	1.0	2.5	0.6	1.0	20	35
3.0	140	2N3441		25/100	0.5				0.2	25
4.0	60		2N3740,A	30/100	0.25	1.3 typ	0.27 typ	0.25	4.0	25
		2N3054,A	2N6049	25/100	0.5	1.0 typ	0.3 typ	0.5	3.0	75
		2N3766		40/160	0.5	0.9 typ	0.09 typ	0.5	10	20
		2N6294##	2N6296##	750/18 k	2.0	0.9 typ	0.7 typ	2.0	4.0#	50
	80		2N3741,A	30/100	0.25	1.3 typ	0.27 typ	0.25	4.0	25
		2N3767		40/160	0.5	0.9 typ	0.09 typ	0.5	10	20
		2N6295##	2N6297##	750/18 k	2.0	0.9 typ	0.7 typ	2.0	4.0	50
5.0	40	2N4231A	2N6312	25/100	1.5	0.5 typ	0.2 typ	1.5	4.0	75
	60	2N4232A	2N6313	25/100	1.5	0.5 typ	0.2 typ	1.5	4.0	75
	80	2N4233A	2N6314	25/100	1.5	0.5 typ	0.2 typ	1.5	4.0	75
	225	2N6233		25/125	1.0	3.5	0.5	1.0	20	50
	275	2N6234		25/125	1.0	3.5	0.5	1.0	20	50
	325	2N6235		25/125	1.0	3.5	0.5	1.0	20	50
7.0	60	2N6315	2N6317	20/100	2.5	1.0	0.8	2.5	4.0	90
	80	2N5427		30/120	2.0	2.0	0.2	2.0	30	40
		2N5428		60/240	2.0	2.0	0.2	2.0	30	40
		2N6316	2N6318	20/100	2.5	1.0	0.8	2.5	4.0	90
	100	2N5429		30/120	2.0	2.0	0.2	2.0	30	40
		2N5430		60/240	2.0	2.0	0.2	2.0	30	40
	250	2N6078		12/70	1.2	2.8	0.3	1.2	1.0	45
	275	2N6077		12/70	1.2	2.8	0.3	1.2	1.0	45
8.0	60	2N6300##	2N6298##	750/18 k	4.0	1.5 typ	1.5 typ	4.0	4.0#	75
	80	2N6301##	2N6299##	750/18 k	4.0	1.5 typ	1.5 typ	4.0	4.0#	75
	120	MJ3247	MJ3237	40 min	3.0	0.4 typ	0.18 typ	5.0	20	75
	150	MJ3248	MJ3238	40 min	3.0	0.4 typ	0.18 typ	5.0	20	75
10	80	2N6495		10/60	10	0.15 typ	0.05 typ	10	25	70

|h_fe| @ 1.0 MHz   ## Darlington

Figure 4.21 A selection sheet for power transistors. (Courtesy of Motorola Inc.)

Figure 4.22

A junction field effect transistor (JFET) operation.

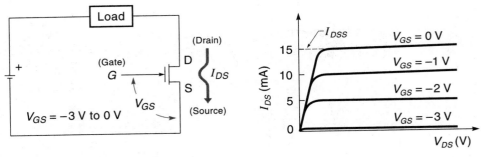

(a) Simple JFET circuit

(b) FET characteristic curves (sample)

voltage across the FET (V_{DS}) is on the horizontal axis. Each curve represents what happens for a different gate voltage. For example, when the gate voltage (V_{GS}) is -3 V, the I_{DS} will be 0 mA. This gate voltage is known as $V_{GS(off)}$. When the gate voltage is -2 V, I_{DS} will be 5 mA, and so on, until the gate voltage is 0 V, causing I_{DS} to be 15 mA. Because the gate voltage is not allowed to go positive in a JFET, 15 mA is the highest I_{DS} for this particular FET and is called I_{DSS} (drain–source current with the gate shorted).

The second type of FET is the MOSFET. MOSFETs are particularly important in this text because *power MOSFETs*, which can handle many amperes, are becoming more popular. In Figure 4.23(a), notice that the gate is not even touching the rest of the device. This means that the gate is capacitively coupled, causing a very high DC input resistance and allowing for what is called the **enhancement mode** of operation. In the enhancement mode, all gate voltages are positive. (JFETs require negative gate voltages, known as the *depletion mode* of operation). Consider the sample characteristic curves of a power-type MOSFET shown in Figure 4.23(b). When the V_{GS} is 4 V, the I_{DS} is 0 A. When the V_{GS} is 5 V, the I_{DS} is 1 A; when the V_{GS} is 6 V, the I_{DS} is 2 A (and so on, up to the maximum current allowed by the device).

Figure 4.23

A metal-oxide semiconductor field effect transistor (MOSFET).

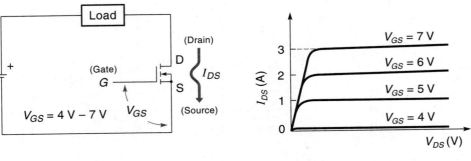

(a) Simple MOSFET circuit

(b) MOSFET characteristic curves (sample)

For both the JFET and the MOSFET, the input is the gate voltage and the output is the drain–source current. The gain of this device is called the **transconductance** (g_m) and is expressed in Equation 4.5:

$$\text{Gain} = \frac{\Delta \text{ output}}{\Delta \text{ input}} = g_m = \frac{\Delta I_{DS}}{\Delta V_{GS}} \text{ S (or mho)} \qquad (4.5)$$

where

g_m = FET gain, called the transconductance
I_{DS} = load current (from drain to source)
V_{GS} = control voltage at the gate

Notice that the unit of g_m is $1/R$, known as *siemens* (S). An older term for conductance (still in use) is *mho*.

♦ **EXAMPLE 4.5**

For the power MOSFET whose characteristic curves are shown in Figure 4.23(b), find the gain g_m.

Solution

From Equation 4.5, we see that the gain g_m is the *change in output current due to the corresponding change in input gate voltage*. Thus, we pick two convenient gate voltages—say, 5 V and 6 V (which represents a change of 1 V)—and divide that into the corresponding change in the load current I_{DS}:

$$\text{Gain} = g_m = \frac{\Delta I_{DS}}{\Delta V_{GS}} = \frac{(2\,\text{A} - 1\,\text{A})}{(6\,\text{V} - 5\,\text{V})} = \frac{1\,\text{A}}{1\,\text{V}} = 1\,\text{S (or 1 mho)} \qquad ♦$$

The specifications for the Motorola MTM6N55 power MOSFET are shown in Figure 4.24. Some important characteristics of this MOSFET are as follows:

1. Maximum load current (I_{DS}) is 6 A.
2. The gate threshold voltage ($V_{GS(th)}$), the gate voltage that just causes the FET to conduct, is between 2 and 4.5 V.
3. The minimum FET gain, called the forward transconductance (g_{FS}), is 2 mho. (*Note*: g_{FS} is the same as g_m.)
4. Total power dissipation for the FET (P_D) should be less than 150 W.

♦ **EXAMPLE 4.6**

A MTM6N55 power FET in a circuit was found to have a gate voltage of 5 V and a load current of 4 A. Is this FET operating within the specifications? (Spec sheet is given in Figure 2.24.)

Solution

The spec sheet says that the gate threshold voltage is between 2 and 4.5 V. We will calculate the expected load current for both extremes.

For the minimum gate threshold voltage of 2 V, the active gate voltage is

$$V_{GS} = 5\,\text{V} - 2\,\text{V} = 3\,\text{V}$$

Figure 4.24

Specification sheet for an MTM6N55 power MOSFET. (Courtesy of Motorola Inc.)

Designer's Data Sheet

N-CHANNEL ENHANCEMENT MODE SILICON GATE TMOS POWER FIELD EFFECT TRANSISTOR

These TMOS Power FETs are designed for high voltage, high speed power switching applications such as line operated switching regulators, converters, solenoid and relay drivers.

● Silicon Gate for Fast Switching Speeds — Switching Times Specified at 100°C
● Designer's Data — I_{DSS}, $V_{DS(on)}$ and SOA Specified at Elevated Temperature
● Rugged — SOA is Power Dissipation Limited
● Source to Drain Diode Characterized for Use With Inductive Loads
● Low Drive Requirement, $V_{G(th)}$ = 4.5 Volts (max)

6 AMPERE

N-CHANNEL TMOS POWER FET

$r_{DS(on)}$ = 1.5 OHMS
550 and 600 VOLTS

MTM6N55
MTM6N60

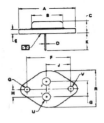

STYLE 3
PIN 1. GATE
      2. SOURCE
CASE DRAIN

CASE 1-05
TO-3 TYPE

MAXIMUM RATINGS

Rating	Symbol	MTM6N55	MTM6N60	Unit
Drain — Source Voltage	V_{DSS}	550	600	Vdc
Drain — Gate Voltage (R_{GS} = 1.0 mΩ)	V_{DGR}	550	600	Vdc
Gate — Source Voltage	V_{GS}	±20		Vdc
Drain Current Continuous Pulsed	I_D I_{DM}	6.0 30		Adc
Gate Current — Pulsed	I_{GM}	1.5		Adc
Total Power Dissipation @ T_C = 25°C Derate above 25°C	P_D	150 1.2		Watts W./°C
Operating and Storage Temperature Range	T_J, T_{stg}	−65 to 150		°C

THERMAL CHARACTERISTICS

Thermal Resistance Junction to Case	$R_{\theta JC}$	0.83	°C/W
Maximum Lead Temp. for Soldering Purposes, 1/8" from case for 5 seconds	T_L	275	°C

Designer's Data for "Worst Case" Conditions

The Designer's Data Sheet permits the design of most circuits entirely from the information presented. Limit curves—representing boundaries on device characteristics—are given to facilitate "worst case" design.

ELECTRICAL CHARACTERISTICS (T_C = 25°C unless otherwise noted)

Characteristic	Symbol	Min	Max	Unit
OFF CHARACTERISTICS				
Drain-Source Breakdown Voltage (V_{GS} = 0, I_D = 5.0 mA) MTM6N55 MTM6N60	$V_{(BR)DSS}$	550 600	— —	Vdc
Zero Gate Voltage Drain Current (V_{DS} = 0.85 BV_{DSS}, V_{GS} = 0) T_J = 100°C	I_{DSS}	— —	0.25 2.5	mAdc
Gate-Body Leakage Current (V_{GS} = 20 Vdc, V_{DS} = 0)	I_{GSS}	—	500	nAdc
ON CHARACTERISTICS*				
Gate Threshold Voltage (I_D = 1.0 mA, V_{DS} = V_{GS}) T_J = 100°C	$V_{GS(th)}$	2.0 1.5	4.5 4.0	Vdc
Drain-Source On-Voltage (V_{GS} = 10 V) (I_D = 3.0 Adc) (I_D = 6.0 Adc) (I_D = 3.0 Adc, T_J = 100°C)	$V_{DS(on)}$	— — —	4.5 10 9.0	Vdc
Static Drain-Source On-Resistance (V_{GS} = 10 Vdc, I_D = 3.0 Adc)	$r_{DS(on)}$	—	1.5	Ohms
Forward Transconductance (V_{DS} = 15 V, I_D = 3.0 A)	g_{fs}	2.0	—	mhos

The FET gain g_m (called g_{FS} on the spec sheet) is 2 mho minimum. Therefore, to get the output current (I_{DS}), rearrange Equation 4.5; that is, multiply the active gate voltage by the gain:

$$I_{DS} = V_{GS} \times g_{FS} = 3\,\text{V} \times 2\,\text{mho} = 6\,\text{A (minimum)}$$

Now consider the case if the gate threshold voltage was at its maximum of 4.5 V. Then, the active gate voltage is

$$V_{GS} = 5\,\text{V} - 4.5\,\text{V} = 0.5\,\text{V}$$

Because the minimum FET gain (g_{FS}) is still 2 mho, we recalculate the output current:

$$I_{DS} = V_{GS} \times g_{FS} = 0.5\,\text{V} \times 2\,\text{mho} = 1\,\text{A (minimum)}$$

Thus, we have found that the minimum value of I_{DS} for the MTM6N55 FET with a gate voltage of 5 V is 1 A. Therefore, the measured value of 4 A is within specifications. ◆

4.4 SILICON-CONTROLLED RECTIFIERS

The **silicon-controlled rectifier** (SCR) is a three-terminal power-control device that is a member of a group known as **thyristors**. Thyristors are four-layer semiconductor devices that behave like a switch; that is, they are either conducting or nonconducting (on or off). Constructed from four layers of semiconductor material (PNPN), as shown in Figure 4.25(a), the SCR has three terminals: the **anode** (A), the **cathode** (K), and the **gate** (G), as shown in the schematic symbol [Figure 4.25(b)].

The SCR is usually connected in series with the load as shown in Figure 4.26. Notice that this circuit is similar to a transistor or FET power-driving circuit. In this case, the load current enters the anode and leaves through the cathode. The gate is used to switch

Figure 4.25

A silicon-controlled rectifier.

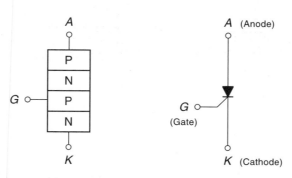

(a) SCR construction   **(b)** SCR schematic symbol

the SCR from its nonconduction state (off) to its conduction state (on). Unfortunately *the gate cannot be used to turn the SCR off*, as will be addressed later in this section. Also, current can travel in only one direction through the SCR (as it does in a diode).

Figure 4.26
The basic SCR circuit.

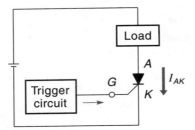

The best way to explain the operation of the SCR is with the characteristic curve, which is given in Figure 4.27. Take time to consider this graph. First recognize that voltage across the SCR is on the horizontal axis, and current through the SCR is on the vertical axis. The left side of the graph corresponds to the SCR being reversed-biased, which explains why there is no current flow (until reverse breakdown occurs at some high negative voltage). The interesting action takes place on the right half of the graph, where the *solid line shows the action of the SCR when there is no gate voltage*. As the voltage across the SCR gets more positive, the current is held off until the voltage reaches a value known as the **forward breakover voltage**, at which point the SCR turns on and current starts to flow (and the blocking voltage shrinks back). The SCR is now in the **forward conduction region** and is acting like a low-resistance closed switch. If the current is now reduced below the value known as the **holding current** (I_H), the SCR will immediately switch back into its high-resistance nonconducting state.

Figure 4.27
An SCR characteristic curve (sample).

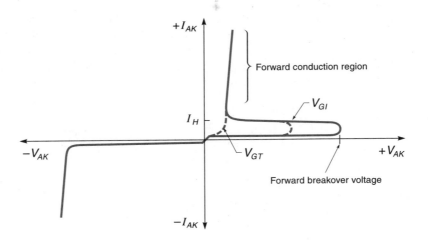

We now turn our attention to how the gate voltage affects the operation of the SCR. When the gate voltage is V_{G1}, the forward breakover voltage is significantly reduced—that is, the SCR will turn on at a lower anode–cathode voltage (indicated by the dashed line in Figure 4.27). If the gate voltage is high enough (called V_{GT}), the SCR starts conducting right away and requires almost no forward breakover voltage whatsoever. This is the turn-on mode most practical SCR circuits use. Once the SCR starts to conduct, the gate voltage can be removed, and the SCR will remain in the conduction state. Therefore, in many SCR circuits, the gate signal is actually a short "trigger" pulse, just long enough to ensure that switching has occurred.

Turning on the SCR with the gate is relatively easy, but turning it off is another matter. *There is no way to command the SCR to turn off*; the only way to turn it off is to reduce the load current below the holding current (I_H) value. For DC circuits, this can be done in a number of ways. The first method, illustrated in Figure 4.28(a), is to open the load circuit with some other device—say, a switch. The second method, called **forced commutation**, involves momentarily forcing the voltage across the SCR (V_{AK}) to zero (or below), thereby stopping the current. One way to do this is illustrated in Figure 4.28(b), where a "shorting transistor" momentarily reroutes the current around the SCR. Another way is shown in Figure 4.28(c) and uses a second SCR$_2$ to turn off the main SCR$_1$. The circuit works as follows: When the main SCR$_1$ is on, its anode voltage is practically 0 V; therefore, capacitor C can charge up through resistor R. To turn off the main SCR$_1$, SCR$_2$ is triggered into conduction which "pulls down" the voltage at the positive ($+$) end of C to 0 V, temporarily causing the negative ($-$) end of C to drop as well. The negative voltage on the anode of SCR$_1$ turns it off. Finally, when the capacitor discharging current wanes, SCR$_2$ turns off by itself.

Figure 4.28

Turning off an SCR for a DC load.

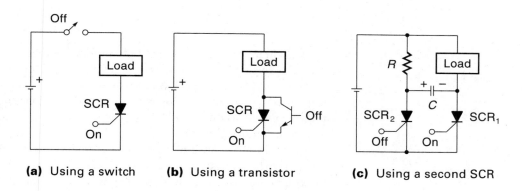

(a) Using a switch   **(b)** Using a transistor   **(c)** Using a second SCR

Turning off the SCR for AC circuits is not a problem because it will naturally turn itself off at each negative half-cycle (remember, the SCR can only conduct in one direction). So for AC applications, the SCR must be triggered on once for each positive half-cycle, and then it will turn itself off at the end of that positive half-cycle.

Because SCRs work so well with AC, they are often used as the rectifier in an AC-to-DC converter—for example, to control the speed of a DC motor. A simple SCR speed-control circuit is shown in Figure 4.29(a). As you can see, the gate voltage is supplied by the AC itself through potentiometer R and diode D. When the level of the gate voltage gets high enough, the SCR turns on, and it stays on for the remainder of that positive half-cycle. By increasing the resistance of R, the time at which the SCR fires can be delayed, thus causing a shorter conduction period. These power bursts are averaged out by the load (such as a motor) into an effective DC voltage.

Figure 4.29

Turning on an SCR for an AC load: (a) Trigger circuit uses a resistor only and (b) trigger circuit uses a resistor and capacitor.

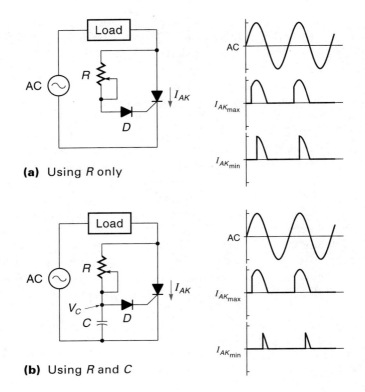

(a) Using R only

(b) Using R and C

The circuit just described can only control the trigger time from 0 to 90° of the sine wave. To delay the trigger point beyond 90°, a capacitor can be added to the gate circuit as shown in Figure 4.29(b). As the capacitor C charges up through R, V_C lags behind (remember, this happens for each cycle). Using this circuit, known as a *half-wave* **phase-control circuit**, the SCR firing can be delayed up to about 170°, resulting in such a narrow "on time" that the DC equivalent is practically 0 V.

The AC-to-DC converter circuits described in the last two paragraphs are half-wave rectifiers, which means only half of the potential AC power is used. The SCR can be used in conjunction with a full-wave rectifier to make use of more of the AC power by giving two power pulses per cycle. Such a circuit is shown in Figure 4.30 and is known as a *full-wave phase-control circuit*. The waveform (Figure 4.30) for the SCR current (I_{AK}) depicts the situation when conduction starts at about 90°. On real systems, voltage is easier to measure than current, so it is important to know what the voltage waveforms look like. Notice that during conduction the SCR voltage (V_{AK}) is practically zero, and when the SCR is off, it assumes the entire line voltage.

Some other conditions (besides those already mentioned) could cause the SCR to turn on:

1. If the voltage across the SCR (V_{AK}) rises too quickly, the SCR may turn on, even if the voltage is below the forward breakover voltage. This is called the **dv/dt**

effect. To eliminate this possibility, R_C **snubber** circuits are connected across the SCR to prevent the voltage from rising too fast (see Figure 4.30).

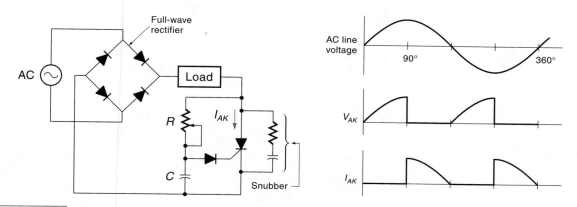

Figure 4.30 An SCR used with a full-wave rectifier.

2. After turning off, V_{AK} and I_H must be held low for a given period of time or the SCR will turn on again.

3. As the SCR temperature increases, the required breakover voltage decreases, thus possibly causing the SCR to turn on.

4. A type of SCR, known as a LASCR (light-activated SCR), uses light energy to trigger the conduction state. These devices have a little window that allows the light to hit the necessary PN junction.

SCRs are available in a wide range of sizes, with current capacities from 0.5 to over 1000 A. A specification sheet of some Motorola SCRs is given in Figure 4.31. An explanation of the parameters used is as follows:

V_{DRM}:   peak forward blocking voltage

V_{RRM}:   peak reverse blocking voltage

I_{TSM}:   peak surge current

I_{GT}:   gate trigger current, continuous DC

V_{GT}:   gate trigger voltage, continuous DC

I_H: holding current

For example, a 2N6401 (Figure 4.31, third column, third row) can carry 16 A of load current, should be used with supply voltages under 100 V_{peak} (+ or −), can handle brief surge load currents of up to 160 A, and requires 30 mA of gate current to turn on, which can be achieved with 1.5 V on the gate, and its holding current is 40 mA.

♦ **EXAMPLE 4.7**

A 10-A 120-Vdc motor is to be powered from a full-wave SCR circuit similar to that shown in Figure 4.30. The source of power is 120 Vac. Select an SCR to do the job from the list given in Figure 4.31.

		ON-STATE (RMS) CURRENT						
	12.5 AMPS	**16 AMPS**		**20 AMPS**				
	 Case 54 Style 2	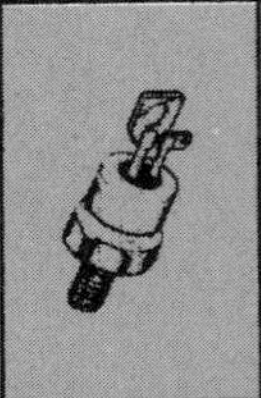 Case 263-02 Style 1	 Case 221A-02 TO-220AB Style 3	 Case 310 Style 1	 Case 263 Style 1	 Case 311-01 Style 1	 Case 54 Style 2	 Case 174-03 T0203AA Style 1
25 V		2N1842 2N1482A					MCR649AP1	MCR3818-1
50 V		2N1843 2N1843A	2N6400	2N5164	2N5168		MCR649AP2	MCR3818-2
100 V	2N3668	2N1844 2N1844A	2N6401	S6200A	S6210A	2N6167 S6220A	MCR649AP3	MCR3818-3
200 V	2N3669	2N1846 2N1848A	2N6402	2N5165 S6200B	2N5169 S6210B	2N6168 S6220B	MCR649AP4	MCR3818-4
300 V		2N1848 2N1848A	MCR221-5				MCR649AP5	MCR3818-5
400 V	2N3670	2N1849 2N1849A	2N6403	2N5166 S6200D	2N5170 S6210D	2N6169 S6220D	MCR649AP6	MCR3818-6
500 V	2N4103	2N1850 2N1850A	MCR221-7				MCR649AP7	MCR3818-7
600 V			2N6404	2N5167 S6200M	2N5171 S6210M	2N6170 S6220M	MCR649AP8	MCR3818-8
700 V			MCR221-9				MCR649AP9	MCR3818-9
800 V			2N6405				MCR649AP10	MCR3818-10
I_{TSM} (Amps)	200	125	160	240			260	240
I_{GT} (mA)	40	80	30	40			40	40
V_{GT} (V)	2.0	2.0	1.5	1.5			3.5	1.5
I_H (mA)	20 Typ	20 Typ	40	50			20 Typ	50

The leftmost label column reads V_{DRM} / V_{RRM} for the voltage rows and "MAXIMUM ELECTRICAL CHARACTERISTICS" for the characteristics rows.

Figure 4.31 Specification sheet for Motorola SCRs.

Solution   The concern here is the current and the voltage capability of the SCR. Any of the SCRs could handle the current, but, for economy of design, we might select one from the first column rated at 12.5 A. The peak voltage of 120 Vac can be calculated as follows:

$$V_{PEAK} = \frac{120\ V}{0.707} = 170\ Vdc$$

With a peak voltage of 170 Vdc, we must select at least a 200-V unit such as the 2N3669 (12.5 A, 200 V). Of course, many other SCRs on the list would work as well.

◆

4.5   TRIACS

The **triac** is a three-terminal device that is similar to the SCR except that the triac can conduct current in both directions. Its primary use is to control power to AC devices such as motors. As indicated on schematic symbol [Figure 4.32(a)], the three terminals are MT_1, MT_2* (main terminals), and the **gate**. As the name suggests, the load current passes through the main terminals, and the gate controls the flow. Figure 4.32(b) shows the equivalent circuit of the triac, which consists of two back-to-back SCRs with a common gate. The (conventional) current flowing down (from MT_2) would go through SCR_1, while current flowing up (from MT_1) would go through SCR_2.

Figure 4.32
Triac.

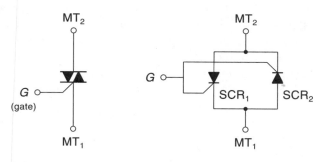

(a) Triac symbol   **(b)** Triac equivalent circuit

The operating characteristics of the triac are best explained using the characteristic curve shown in Figure 4.33. Notice that the right half of the curve looks just like the SCR curve; no current flows until either the breakover voltage is reached or the gate is

*Sometimes called *anode 1* and *anode 2*.

triggered (indicated by dashed line). This same pattern is repeated on the left side of the curve (for voltage and current of the opposite direction). Also, like the SCR, once the triac is triggered on, it will remain on by itself until the load current drops below the holding current value (I_H).

Figure 4.33

A triac characteristic curve (sample).

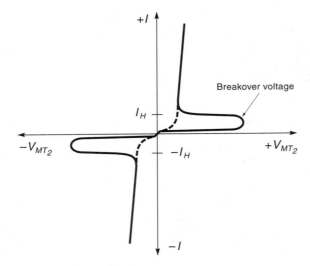

A single cycle of AC has a positive and a negative half-cycle. *The triac requires a trigger pulse at the gate for each half-cycle*, and the trigger should be positive for the positive half-cycle and negative for the negative half-cycle (although other possibilities exist).

A triac can be used as an on–off solid-state switch for AC loads or to regulate power to an AC load, such as a dimmer switch. A typical dimmer switch circuit is shown in Figure 4.34. The action of this circuit is similar to that of the SCR circuit in Figure 4.29(b). For each half-cycle, the capacitor C starts to charge up through resistor R. When the voltage V_C gets large enough, the triac triggers into conduction and stays on for the remainder of the half-cycle. By changing the resistance in R, the trigger point can be delayed, thus reducing the power to the load.

Figure 4.34

A phase-control circuit using a triac (as might be used for a dimmer switch).

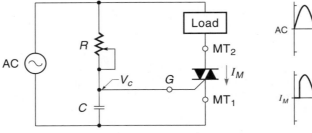
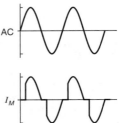

Triacs are available in various packages, some of which can handle currents up to 50 A (which is considerably less than the SCR). Figure 4.35 shows a sample selection of triacs. An explanation of the specifications is as follows on page 118:

Figure 4.35   A selection of triacs. (Courtesy of Motorola Inc.)

		ON-STATE (RMS) CURRENT								
		4.0 AMPS 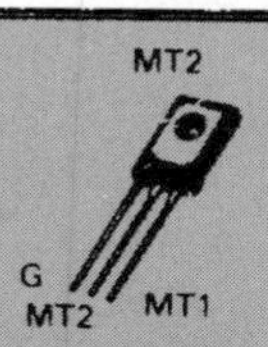			**6.0 AMPS** 			**8.0 AMPS** 		
		Sensitive Gate Case 77-04 Style 5			Case 221A-02 TO-220AB Style 4					
V_{DRM}	25 V	2N6068	2N6068A	2N6068B					MAC222-1 MAC222A-1	
	50 V	2N6069	2N6069A	2N6069B					MAC222-2 MAC222A-2	MAC220-2 MAC221-2
	100 V	2N6070	2N6070A	2N6070B	T2500A	T2801A	SC141A		MAC222-3 MAC222A-3	MAC220-3 MAC221-3
	200 V	2N6071	2N6071A	2N6071B	T2500B	T2801B	SC141B	MAC218-4 MAC218A-4	MAC222-4 MAC222A-4	2N6342 2N6346
	300 V	2N6072	2N6072A	2N6072B	T2500C	T2801C	SC141C	MAC218-5 MAC218A-5	MAC222-5 MAC222A-5	MAC220-5 MAC221-5
	400 V	2N6073	2N6073A	2N6073B	T2500D	T2801D	SC141D	MAC218-6 MAC218A-6	MAC222-6 MAC222A-6	2N6343 2N6347
	500 V	2N6074	2N6074A	2N6074B	T2500E	T2801E	SC141E	MAC218-7 MAC218A-7	MAC222-7 MAC222A-7	MAC220-7 MAC221-7
	600 V	2N6075	2N6075A	2N6075B	T2500M	T2801M	SC141M	MAC218-8 MAC218A-8	MAC222-8 MAC222A-8	2N6344 2N6348
	700 V				T2500S	T2801S	SC141S	MAC218-9 MAC218A-9	MAC222-9 MAC222A-9	MAC220-9 MAC221-9
	800 V				T2500N	T2801N	SC141N	MAC218-10 MAC218A-10	MAC222-10 MAC222A-10	2N6345 2N6349
MAXIMUM ELECTRICAL CHARACTERISTICS	I_{TSM} (Amps)	30	30	30	60	80	80	100	80	100
	I_{GT} @ 25°C (mA)									
	MT2(+)G(+)	30	5.0	3.0	25	80	50	50	50	50
	MT2(+)G(-)	—	5.0	3.0	60	80	50	50	50	75#
	MT2(-)G(-)	30	5.0	3.0	25	80	50	50	50	50
	MT2(-)G(+)	—	10	5.0	60	80	—	—	75#	75#
	V_{GT} @ 25°C (V)	@ —40°C	@ —40°C	@ —40°C						
	MT2(+)G(+)	2.5	2.5	2.5	2.5	4.0	2.5	2.5	2.0	2.0
	MT2(+)G(-)	—	2.5	2.5	2.5	4.0	2.5	2.5	2.0	2.5#
	MT2(-)G(-)	2.5	2.5	2.5	2.5	4.0	2.5	2.5	2.0	2.5
	MT2(-)G(+)	—	2.5	2.5	2.5	4.0	—	—	2.5#	2.5#

\# Denotes 2N6346-49, MAC221, and MAC222A series only

V_{DRM}:   maximum off-state voltage allowed

I_{TSM}:   maximum load surge current

I_{GT} and V_{GT}:   gate current and voltage required to trigger triac into conduction

The two most common trigger conditions are MT2(+)G(+) and MT2(−)G(−), as used in the circuit of Figure 4.34.

Calculation of Delay and Conduction Periods

As we have seen in the preceding sections, phase-control circuits (using SCRs and triacs) regulate power to the load by turning on for a portion of the cycle period. As shown in Figure 4.36, for each half-cycle, the SCR (or triac) delays for a period of time and then switches into conduction for the rest of the half-cycle. Delay and conduction measurements are commonly expressed in units of time or as an angle. It is useful to be able to make conversions between these two measurements.

Figure 4.36

Delay and conduction periods in SCR and triac waveforms.

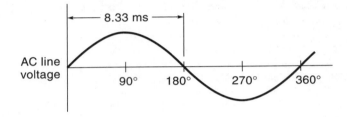

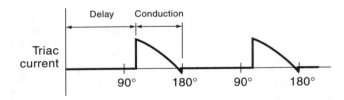

In North America, AC power has a frequency of 60 Hz. Therefore, the time for each half-cycle is

$$0.5 \times \frac{1}{60 \text{ Hz}} = 8.33 \text{ ms}$$

Therefore,

$$\text{Delay time} + \text{conduction time} = 8.33 \text{ ms}$$

So, for example, if an SCR had a conduction time of 3 ms, the delay time would be calculated as follows: 8.33 ms − 3 ms = 5.33 ms.

When dealing with angles, recall that each half-cycle is 180°. Therefore,

$$\text{Delay angle} + \text{conduction angle} = 180°$$

So, for example, if a triac had a delay angle of 30°, then

$$\text{Conduction angle} = 180° - 30° = 150°$$

Converting between time and angle measurements requires knowing the relationship between time and degrees. For a 60-Hz system, there are 8.33 ms per half-cycle of 180°. Therefore,

$$\frac{8.33 \text{ ms}}{180°} = 46.3 \text{ μs/deg}$$

So, for example, if a triac had a conduction angle of 120°, we could calculate the conduction time as follows: 46.3 μs/deg × 120° = 5.56 ms.

4.6   TRIGGER DEVICES

The SCR and triac devices discussed in the previous sections both require **trigger circuits** to provide the gate turn-on signal. Ideally, this signal is a strong pulse delivered at the right time. Various semiconductor devices are particularly suitable for this job because of their bistable nature. This section gives a brief introduction to some of the most common triggering devices. For more information, consult a semiconductor devices text.

Unijunction Transistors

The **unijunction transistor** (UJT) symbol and characteristic curve are shown in Figure 4.37. Notice that it has three terminals: *base 1* (B_1), *base 2* (B_2), and the *emitter* (E). The UJT operates as follows: For low emitter voltages (V_{EB_1}), there is no emitter current (I_E). Once V_{EB_1} reaches V_P (which is about two-thirds of the voltage between B_1 and B_2), the transistor fires, and current is allowed to surge through the emitter and B_1. UJTs are often used to trigger SCRs, as shown in the simplified circuit of Figure 4.38. For each positive half-cycle of the AC, the capacitor C charges up through R_1. When V_C gets high enough, the UJT fires, and the resulting current through B_1 develops a large voltage across R_3 and triggers the SCR.

Figure 4.37

The unijunction transistor (UJT).

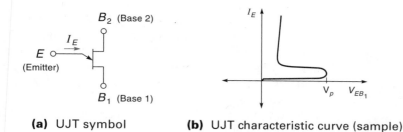

(a) UJT symbol   **(b)** UJT characteristic curve (sample)

Figure 4.38
Using a UJT to trigger on SCR.

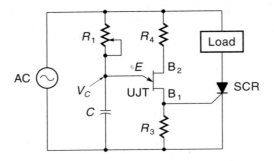

Programmable Unijunction Transistors

The **programmable unijunction transistor** (PUT) behaves like the UJT except that the firing voltage is adjustable via a resistor-divider network. The symbol and simplified trigger circuit are shown in Figure 4.39. Notice that it has three terminals: the **anode** (A), **cathode** (K), and **gate** (G). Like the SCR, the main current from A to K is held off until the anode voltage reaches V_P (peak voltage), at which point the PUT is triggered into conduction. The DC voltage on the gate determines what V_P will be. In the circuit shown in Figure 4.39(b), R_3 and R_4 form a resistor divider that establishes the gate voltage and hence the trigger point. The rest of the circuit is similar to other trigger circuits already discussed.

Figure 4.39
The programmable unijunction transistor (PUT).

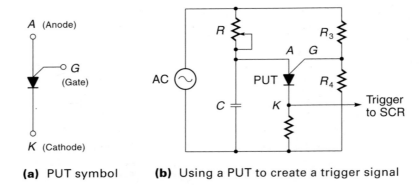

(a) PUT symbol   **(b)** Using a PUT to create a trigger signal

Diac

The **diac** symbol and characteristic curve are shown in Figure 4.40. It is a two-terminal bidirectional device, and it behaves essentially like a triac without a gate. As the curve indicates, if the voltage across the diac exceeds the breakover voltage (in either direction), it triggers into conduction mode and stays there until the the load current drops below the holding current. Diacs are frequently used to trigger triacs in full-wave AC applications, as shown in the simplified circuit of Figure 4.41. The *silicon bidirectional switch* (SBS) is similar to the diac except that it has an extra gate terminal for external synchronization.

Figure 4.40
The diac.

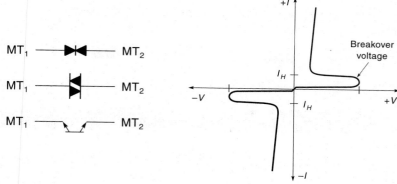

(a) Diac symbols   **(b)** Diac characteristic curve

Figure 4.41
Using a diac to trigger a triac.

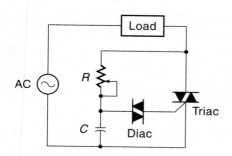

SUMMARY

Switches are manually operated contacts that open and close. Many sizes and styles of switches are available, the most common being toggle, pushbutton, slide, DIP, rotary, thumbwheel, and membrane. Many of these styles are available with multiple contact configurations.

The traditional electromechanical relay (EMR) uses an electromagnet to open and close electrical contacts. Relay contacts are either normally open or normally closed, which refers to their unenergized state. As with switches, relays come in many sizes and contact configurations. Solid-state relays (SSRs) are solid-state switching circuits packaged so they can be used in place of EMRs in some applications.

The bipolar junction transistor (BJT), or simply *transistor*, is a three-terminal solid-state device that works like a valve to control the flow of current through a load (such as a motor). The transistor is a current amplifier, which means that the load current will be a multiple of the input base current. Another kind of transistor, called the field effect transistor (FET), performs the same job as the BJT, the main difference being that (for the FET) the load current is a multiple of the input gate voltage.

The silicon-controlled rectifier (SCR) is a three-terminal, bistable, solid-state device capable of switching very large DC currents. The SCR is easily switched on with a trigger pulse (to its gate terminal); however, the only way to turn it off is by reducing the load current to a very low value. Turning off the SCR requires additional circuitry for DC loads, but is not a problem in AC applications.

A triac is a three-terminal, bistable, solid-state device capable of switching AC current. The triac must be triggered into conduction each half-cycle with a trigger circuit. By delaying the trigger pulses, the amount of load power can be reduced, as in a lamp dimmer switch.

There are a number of smaller, bistable, solid-state devices that find uses in triggering SCRs and triacs. Examples of these devices are the unijunction transistor (UJT), the programmable unijunction transistor (PUT), and the diac.

GLOSSARY

anode   One of three terminals of an SCR and PUT.

base   One of three terminals of a transistor.

biasing circuit   The part of a transistor circuit that generates the forward bias voltage.

bipolar junction transistor (BJT)   Known as a *transistor*, a solid-state device that can be used as a switch or an amplifier.

BJT   *See* **bipolar junction transistor**.

cathode   One of three terminals of an SCR and PUT.

class A operation   The event when a transistor is biased so the collector current is approximately half of its maximum value.

class B operation   The event when a transistor is biased so that the collector current is just turning on.

class C operation   The event when a transistor is biased below cutoff and used as an on–off switch.

collector   One of the three terminals of a transistor.

contactor   A heavy-duty relay that switches power directly to motors and machinery.

diac   A bistable, two-terminal, solid-state device that is used to trigger the triac.

DIP switch   A set of SPST switches built into the shape of an IC (DIP stands for dual in-line package).

double-pole/double-throw switch (DPDT)   A switch contact configuration.

DPDT   *See* **double-pole/double-throw switch**.

drain   One of three terminals in an FET.

dv/dt effect   The event when the SCR turns on if the anode–cathode voltage rises too quickly.

electromechanical relay (EMR)   A device that uses an electromagnet to open or close electrical contacts.

emitter   One of three terminals of a transistor.

EMR   *See* **electromechanical relay**.

enhancement mode   Property of some MOSFETs where the gate voltage can always be positive (for N-channel).

FET   *See* **field effect transistor**.

field effect transistor (FET)   A three-terminal solid-state amplifying device that uses voltage as its input signal.

forced commutation   The process of turning off an SCR by momentarily forcing the anode–cathode voltage to 0 V (or below).

forward bias voltage   The DC offset base voltage required to start the transistor conducting.

forward breakover voltage   The voltage across a thyristor that causes it to switch into its conduction state.

forward conduction region   The operating range of a thyristor when it is conducting.

forward current gain (h_{FE})   The gain of a transistor; the collector current divided by the base current.

gain-bandwidth product (f_T)   A constant for an amplifier that is the product of the open-loop gain and the frequency; can be read as the frequency when a transistor's gain has been reduced to 1.

gate   One of three terminals in an SCR, PUT, FET, and triac.

heat sink   A piece of metal, possibly with cooling fins, used to dissipate heat from a power device to the air.

holding current (I_H)   The minimum anode current required to keep the thyristor in the conduction state.

hybrid solid-state relay   A device that uses a reed relay to activate a triac.

I_{DSS} The highest possible drain current for a particular JFET; occurs when the gate voltage is 0 V.

JFET   *See* **junction FET**.

junction FET   One type of an FET.

limit switch   A switch used as a proximity sensor.

membrane switch   Usually a keypad with a flexible membrane over the top.

metal-oxide semiconductor FET (MOSFET)   One type of an FET.

microswitch A small push-button switch with a very short throw distance.

minimum holding voltage Minimum voltage needed to keep a relay activated.

momentary-contact switch A switch position that is spring-loaded.

MOSFET *See* **metal-oxide semiconductor FET.**

MT$_1$ One of three terminals of a triac.

MT$_2$ One of three terminals of a traic.

normally open (NO), normally closed (NC) The state of momentary switch or relay contacts when the device is not activated.

NPN, PNP The two basic types of transistors, the difference being the direction of current and voltage within the device.

phase-control circuit A circuit that can delay conduction for part of the AC cycle for the purpose of reducing average output voltage.

power transistor A transistor designed to carry a large current and dissipate a large amount of heat.

programmable unijunction transistor (PUT) A solid-state device that performs the same function as a UJT except that the trigger voltage is adjustable.

pull-in current Minimum current needed to activate a relay.

pull-in voltage Minimum voltage needed to activate a relay.

push-button switch A momentary switch activated by pushing.

PUT *See* **programmable unijunction transistor.**

reed relay A small relay with contacts sealed in a tube and activated by a magnetic field.

rotary switch A rotating knob that activates different switch contacts.

SCR *See* **silicon-controlled rectifier.**

sealed current Current required to keep a relay energized.

shaded pole A small metal ring around the end of the electromagnetic pole of an AC relay, for the purpose of keeping the relay from "buzzing" at 60 Hz.

silicon-controlled rectifier (SCR) A bistable, three-terminal, semiconductor device used to switch power to a load.

single-pole/double-throw switch (SPDT) A switch contact configuration.

single-pole/single-throw switch (SPST) A switch contact configuration.

slide switch Similar to a toggle switch except the handle slides back-and-forth.

snubber A circuit that prevents a fast voltage rise across an SCR, for the purpose of keeping it from false firing.

solid-state relay (SSR) A solid-state switching device used as a relay.

source One of three terminals in an FET.

SPDT *See* **single-pole/double-throw switch.**

SPST *See* **single-pole/single-throw switch.**

SSR *See* **solid-state relay.**

switch wafer Part of a rotary switch.

three-position switch Switch with a center position.

thumbwheel switch Switch that rotates a drum to select numeric data.

thyristor A class of four-layer semiconductor devices (such as PNPN) that are inherently bistable; the SCR and triac are thyristors.

toggle switch A manually operated device that connects or disconnects power.

transconductance (g_m) The gain of an FET, which is the change in drain current divided by the change in gate voltage.

triac A bistable, three-terminal, solid-state device that switches power.

trigger circuit A circuit used to generate the turn-on pulse for an SCR or a triac.

two-position switch A toggle or slide switch with two positions.

UJT *See* **unijunction transistor.**

unijunction transistor (UJT) A bistable, three-terminal solid-state device that primarily triggers an SCR.

V$_{GS(off)}$ The gate voltage necessary to turn the drain current off (for a JFET).

wiper The center and/or moveable contact in a switch.

zero-voltage switching As applied to SSRs, the delay in switching until the AC voltage crosses the zero point.

EXERCISES

Section 4.1

1. Draw the symbol of a DPDT toggle switch and explain how it operates.
2. Draw the symbol of the 1SCY191 switch (Table 4.1) and explain how it operates.
3. Draw the symbol of the 2SHX191 switch (Table 4.1) and explain how it operates.
4. For a rotary switch, draw a picture of the contact configuration for a one-pole/four-throw switch (terminal C makes contact with one of four terminals).

Section 4.2

5. Draw the schematic symbol of a DPDT relay and explain how it works.
6. Draw the schematic symbol for the Y4-V52 relay (Table 4.2).
7. For relay Y4-V52 (Table 4.2), what are the voltage and current specifications for the coil and contacts?
8. For the relay Y6-V430 (Table 4.2), what are the voltage and current specifications for coil and contacts?
9. From the list of relays (Table 4.2), select a model to be used on an automobile to turn on a small spotlight (1.5 A).
10. List the advantages and disadvantages of an SSR versus an EMR.

Section 4.3

11. A transistor has a current gain of 60 and a collector current of 5 A. Find the base current.
12. A transistor has a current gain of 40 and a base current of 25 mA. Find the collector current.
13. Calculate the exact and approximate emitter currents for the transistor described in Exercise 12.
14. A transistor with a gain of 35 has an input curve similar to that shown in Figure 4.16. The base voltage (V_{BE}) is measured to be 0.8 V.
    a. Find the base current.
    b. Find the collector current.
15. For the transistor circuit shown in Figure 4.19, calculate I_C if the resistor R_1 is changed to 13 kΩ. (Assume input curve Figure 4.16 applies.)

16. For the transistor circuit shown in Figure 4.19, calculate the I_C if the resistor R_2 is changed to 1.2 kΩ. (Assume input curve Figure 4.16 applies.)
17. In regards to transistor amplifiers, explain the difference between class A, class B, and class C operations.
18. A 100-W power transistor is operating at a temperature of 60°C. What is the actual power capacity? (Assume a derate 0.57 W/°C above 25°C.)
19. Select a transistor from the list in Figure 4.21 to meet at least the following specifications: Load current = 5 A, power dissipation = 80 W, and minimum gain = 10.
20. Draw a simple series circuit where an FET is used to control the power to a load and explain how it works. Include in your explanation the terms *gate voltage*, *drain–source current*, and g_m.
21. List the differences between a *JFET* and a *MOSFET*.
22. For the FET circuit shown in Figure 4.23, find the load current if the gate voltage is 6.5 V.
23. A circuit uses the MTM6N55 MOSFET (Figure 4.24) and has a gate voltage (V_{GS}) of 6 V. Find the minimum and maximum drain current that could be expected in this circuit. (Assume g_m = 2.5 mho.)

Section 4.4

24. For an SCR, sketch the characteristic curve and use it to explain how the device operates.
25. List the three ways to turn an SCR off (for a DC circuit).
26. Explain why turning off the SCR is not a problem in AC circuits.
27. For a half-wave phase-control circuit, draw the waveforms for I_{AK} and V_{AK} if the SCR starts conducting at 45°.
28. For a full-wave phase-control circuit, draw the waveforms for I_{AK} and V_{AK} if the SCR starts conducting at 135°.
29. Select an SCR from the list in Figure 4.31 to meet the following specifications: Load current = 15 A and load voltage = 240 Vac. What is the trigger voltage required for your selected SCR?

Section 4.5

30. Sketch the characteristic curve for the triac and use it to explain the operation of the device.

31. Draw the output waveform (I_m) for a triac circuit that triggers at 60°.

32. Select a triac from the selection given in Figure 4.35 to meet the following specifications: Load current = 7 A and load voltage = 120 Vac. What is the trigger voltage required for your selected triac?

33. For a 60-Hz system, a triac has a conduction time of 2 ms. Find the delay time.

34. For a 60-Hz system, a triac has a conduction angle of 45°. Find the delay angle.

35. For a 60-Hz system, a triac has a delay time of 3.52 ms. Find the conduction time.

36. For a 60-Hz system, a triac has a conduction angle of 85°. Find the conduction time.

Section 4.6

37. What is the function of R_1 in the UJT circuit of Figure 4.38?

38. What is the purpose of R_3 and R_4 in the PUT circuit of Figure 4.39?

39. Explain how the diac triggers the triac in the circuit of Figure 4.41.

5 Mechanical Systems

OBJECTIVES

After studying this chapter, you should be able to:

- ❑ Understand the properties of static, sliding, and viscous friction.
- ❑ Differentiate among the various types of springs and calculate the force that a spring exerts.
- ❑ Use the basic equations of linear and rotational motion to calculate the distance, velocity, and acceleration of an object acted on by a force.
- ❑ Convert the equivalent amounts of energy used in chemical, thermal, mechanical, and electrical systems and calculate energy-conversion efficiency.

- ❑ Understand the concept of heat conduction and perform simple heat-conduction calculations.
- ❑ Understand the properties of underdamped, overdamped, and critically damped mechanical systems.
- ❑ Calculate mechanical resonant frequencies.
- ❑ Understand the use of the various gear types and their terminology and perform gear train calculations.
- ❑ Know the characteristics of belts and roller chains used for power transfer.

INTRODUCTION

To understand the total electromechanical system, we need to understand some basic mechanical principles. In some ways, understanding the mechanics is easier than the electronics because at least you can see what the mechanical parts are doing. People have been watching things move all their lives—motors, gears, levers, springs, and so on, and have some feel for what is going on. In this chapter, we will first review the basic principles that dictate how mechanical systems respond to various forces and movements. Then the concepts of energy transfer, efficiency, and heat conduction will be introduced. We will not examine how strong a part is or how to design it—that is left to mechanical engineering textbooks.

An important part of any mechanical system is how power is transmitted from the power source to what is being driven. We will examine three ways this is done: gears, belts, and roller chains.

You may notice that many of the mechanical concepts have electrical analogs. For example, in a mechanical system, force causes movement; in an electrical system, voltage causes current. Friction is analogous to resistance, and mechanical resonance is similar to electrical resonance. The same set of principles can be applied in many areas.

A note about units: The United States is in the middle of the long slow process of switching from the English (or customary) units (inches, feet, pounds, etc.) to the SI (or metric) units (centimeters, meters, kilograms, etc.) Today, many machines in service, as well as many new machines, are based on the English system, but the trend is toward machines based on the metric system. This text uses primarily English units, but some examples are repeated using the SI units so that students can be exposed to both systems.

5.1 BEHAVIOR OF MECHANICAL COMPONENTS

Overview

An electronic controller outputs signals that are either analog or digital. When these signals are converted into mechanical motion, it becomes an electromechanical system, and we need to deal with a whole new set of conditions. Mechanical systems are subject to friction, flexing of parts, backlash, and effects of weight and inertia. The general effect of these factors is to slow the reaction time and/or make it difficult to move to a certain position with precision. Furthermore, mechanical parts tend to vibrate under certain conditions, which can cause damage to the parts as well as render the system useless.

Friction

Sliding friction, the drag force that is always present when parts slide against each other, is caused by high spots on the sliding surfaces interfering with each other. Figure 5.1 illustrates this; the high spots tend to "catch" on each other and resist motion.

Figure 5.1

Friction is caused by uneven surfaces "catching."

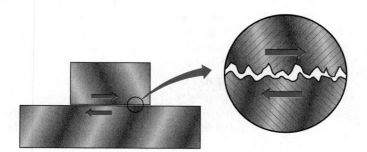

Friction force is proportional to the **normal force** (F_n), which is the force pushing the parts together. The normal force could simply be the weight of the sliding part [Figure 5.2(a)], or the force could come from a bolt or spring, squeezing the sliding surfaces together, as in the partially tightened bolt in Figure 5.2(b). Often, parts need to move freely against each other, and this can present a dilemma. A snug fit is desirable for reducing rattles, but will have more friction than if the parts fit loosely.

Friction is greatest at startup. It takes more force to start a part moving than it does to keep it moving, sometimes twice as much. Startup friction is known as **static friction**; it is particularly troublesome for control systems because, once enough force is applied to overcome the static friction, the resistance immediately drops and the part tends to overshoot its destination. We will discuss how to deal with this problem in Chapter 11.

Figure 5.2
Friction is proportional to normal force (F_n) caused by (a) a weight and (b) bolt tightening.

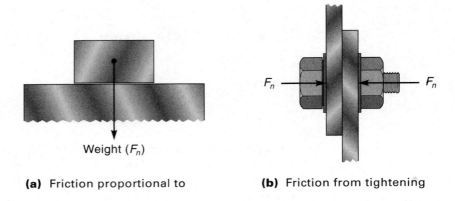

Weight (F_n)

(a) Friction proportional to   **(b)** Friction from tightening

Friction is different for different materials. Generally, the friction force is lower for harder materials such as metals and higher for softer materials such as wood or rubber. The surface condition also makes a difference: The smoother the surface, the less friction.

Lubrication between the parts can greatly reduce the friction and change its character. Ideally, the lubricant keeps the parts from actually touching each other, and the movement is supported by layers of lubricant slipping over each other. The drag on the part from the lubricant is known as **viscous friction**, and its force is directly proportional to the relative velocity between the moving parts. (This is different from sliding friction where the drag force is relatively constant.) A good lubricant must be sticky enough to adhere to the material, but not so sticky as to impede motion. A potential problem is that contaminants in the area such as dirt and sand will tend to stick in the lubricant, which will increase the friction and possibly damage the parts. Some applications use a "dry" lubricant, which is available in a spray can and is applied as a thin, nonstick coating on a part. Some mechanical joints are designed to run "dry" without any lubricant; therefore, check the manual before you lubricate with oil every moving joint you see.

Rolling friction produces considerably less drag than sliding friction. For example, using a wagon to carry something requires less "pulling force" than dragging it. Ball bearings and roller bearings (Figure 5.3) make use of this property to achieve very low friction values.

Figure 5.3

A ball bearing and a roller bearing. (Ball bearing courtesy of INA Bearing Co., Inc., roller bearing courtesy of The Timken Co.)

(a) Ball bearing

(b) Roller bearing

Energy expended overcoming friction is converted to heat. Any system that transports mechanical energy, such as a ball bearing or a gearbox, has a certain efficiency. The efficiency tells what percentage of the mechanical power going in actually comes out, with the rest being lost to friction. A ball bearing can be 99% efficient, a bicycle chain is about 96% efficient, but a worm gear may be only 60% efficient. (Energy and efficiency are discussed later in this chapter.)

Springs

Common components in mechanical systems, springs absorb shock (for example, automobile chassis springs), store energy in spring-wound motors, and provide a constant pressure (for example, in a clothespin).

The force that a spring produces can be predicted using **Hooke's law**, which the following states mathematically:

$$F = kx \qquad (5.1)$$

where

F = push or pull force of the spring
k = spring constant, different for each size of spring
x = distance spring is extended from its "rest length"

Equation 5.1 tells us that the more a spring is extended or compressed, the more force it produces. Because force increases linearly, to get twice the force, you must extend it twice as far. This relationship has a limit, of course. At some point, the spring will deform and never return to its original shape. The use of Hooke's law is illustrated in Example 5.1.

◆ **EXAMPLE 5.1**

A coil spring with a spring constant of 10 lb/in. has a rest length of 2.0 in. In a machine, it provides a constant pressure on a belt-tensioning pulley (Figure 5.4). The spring length in the machine is 2.5 in. How much force is the spring providing?

Solution

First, we calculate the distance the spring is extended by finding the difference between the extended length and the rest length:

$$x = 2.5 \text{ in.} - 2.0 \text{ in.} = 0.5 \text{ in.}$$

Now use the spring constant (k) of 10 lb/in. in Equation 5.1 to calculate the force the spring exerts on the pulley:

$$F = kx$$

$$= \frac{10 \text{ lb}}{\text{in.}} \times 0.5 \text{ in.} = 5 \text{ lb}$$

The spring is exerting a pressure of 5 lb on the tensioning pulley.

Figure 5.4

A spring used to provide tension on a belt-tensioning pulley (Example 5.1).

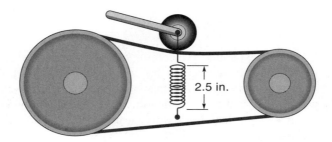

2.5 in.

◆

Many different types of springs are in use. Perhaps the most common is the *coil spring*; coil springs can be either extension or compression types [Figure 5.5(a)]. *Extension springs* may have completely closed coils at rest. They are used to provide a pull as described in Example 5.1. *Compression springs* usually have flat ends and are intended to be compressed. Examples of these are valve springs in an engine and the small springs under each key in a keyboard.

Figure 5.5(b) illustrates a *motor spring*, a flat strip rolled in a coil with one end fixed. The energy-storage capacity of a spring is not as high as a battery for equal weights; however, a spring will not discharge "on the shelf," and its output is direct mechanical motion. *Torsion springs* are used to provide torque or twist. Figure 5.5(c) illustrates one example. A typical clothespin uses a torsion spring. A *flat spring*, a thin flat piece of metal [Figure 5.5(d)], is a simple device used extensively in applications where a constant pressure is needed and the deflection distance is low. A *leaf spring* [Figure 5.5(e)] is made from several large flat springs. The leaf spring has the combined load strength of the individual flat springs.

Finally, it is important to realize that *every* mechanical part is a spring, whether it was intended to be or not. Any piece of material (particularly metal) will deflect under

Figure 5.5

Types of springs.

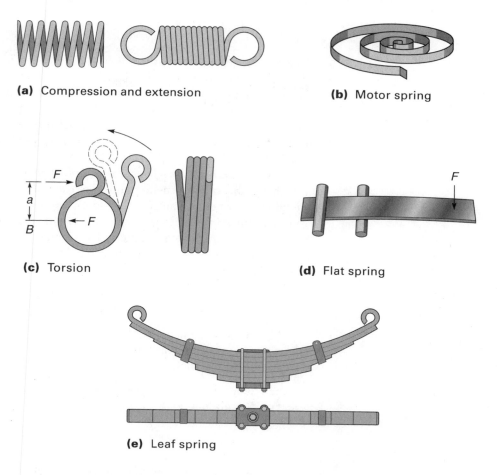

(a) Compression and extension

(b) Motor spring

(c) Torsion

(d) Flat spring

(e) Leaf spring

pressure and then spring back when the pressure is removed. The actual spring constant of a part depends on its composition, its shape, and how the force is applied. In general, a long thin part is much more springy than a shorter thicker part (Figure 5.6). As we will see later in this chapter, the springiness of system components can adversely affect the performance of a machine.

Figure 5.6

A long thin part is more springy than short thick parts.

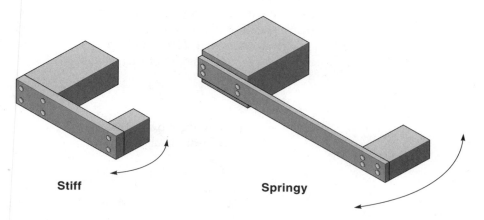

Stiff

Springy

Mass and Inertia

Every mechanical part has a **mass**, which is a measure of the amount of matter present. The primary unit of mass is the kilogram (kg), (there is no common unit of mass in customary units). Related to mass is the familiar concept of weight. **Weight** is the downward force that an object exerts because of gravitational attraction. Weight is measured in units of force, such as the pound (lb). In the SI system, force is measured in **Newtons** (N), where 1 N is equivalent to $1 \text{ kg} \cdot \text{m/s}^2$. The Newton is equal to a force of .224 lb.

On Earth, mass and weight are directly proportional to each other and are often used interchangeably by the general public. For example, in Europe, people buy oranges by the kilogram, whereas oranges are sold by the pound in the United States. However, from an engineering standpoint, mass and weight (force) are different concepts and must be treated as such. Newton's second law of motion (Equation 5.2) specifies the relationship between mass and force, telling us that when a force is applied to a mass, the mass will accelerate:

$$F = ma \tag{5.2}$$

where

F = force applied
m = mass of the object being moved
a = acceleration of the object

The relationship between weight and mass can be expressed as a special case of Equation 5.2, given as

$$w = mg \tag{5.2a}$$

where

w = weight (downward force due to gravity)
m = mass
g = acceleration of gravity (32 ft/s^2 or 9.8 m/s^2)

In Equation 5.2a, weight has been substituted for force, and the gravitational constant (g) has been substituted for acceleration. As you will see, Equation 5.2a will be used when solving numerical problems.

Momentum is a property of a moving object. It takes more effort to stop a rolling car than a child's wagon—we say that the moving car has more momentum. **Momentum** is defined as mass times velocity and is a measure of how much energy it takes to get a part moving. This same amount of energy must be dissipated when the part is stopped.

$$\text{Momentum} = mv \tag{5.3}$$

where

m = mass of the moving object
v = velocity of the object

Inertia, a basic characteristic of mass, is the tendency of an object to remain at rest or in motion unless acted on by an external force. An example would be throwing a ball. Once the ball leaves your hand, it continues to move through the air because of inertia. Figure 5.7 uses a car to illustrate the basic concepts of linear motion (moving in a straight line).

Figure 5.7   A demonstration of force and inertia.

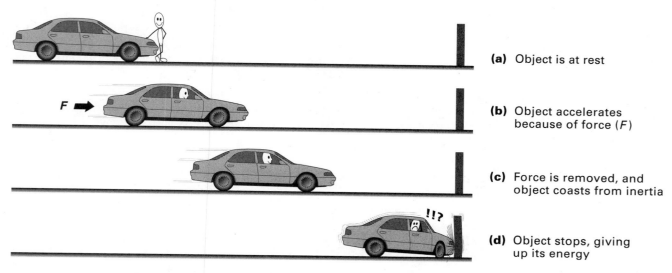

(a) Object is at rest

(b) Object accelerates because of force (F)

(c) Force is removed, and object coasts from inertia

(d) Object stops, giving up its energy

Inertia is also demonstrated when an object rotates about an axis. Consider a bicycle wheel and a car wheel. Both are about the same diameter, but the heavier car wheel has much more inertia and therefore takes more energy to get spinning. However, the inertia of a rotating part is a function of *shape and mass*, not just mass alone, and is given a special name: **moment of inertia** (I). Figure 5.8 shows a few shapes. In general, parts that have the bulk of their mass at a distance from the axis have more rotational inertia than parts where the bulk of the mass is close to the axis. Equations are available to calculate

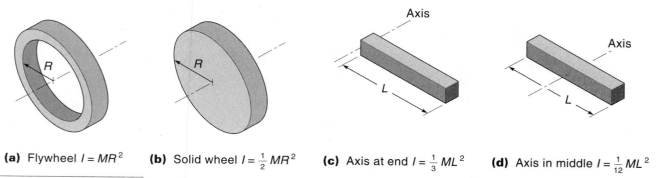

(a) Flywheel $I = MR^2$ (b) Solid wheel $I = \frac{1}{2}MR^2$ (c) Axis at end $I = \frac{1}{3}ML^2$ (d) Axis in middle $I = \frac{1}{12}ML^2$

Figure 5.8   Moment of inertia for four shapes.

the moment of inertia for different shapes. Notice that the flywheel in Figure 5.8(a) has far more rotational inertia than the solid wheel of Figure 5.8(b), even though both wheels are the same mass. This is because most of the mass in the flywheel is concentrated as far from the axis as possible. Figure 5.8(c) shows the rotational inertia in a long beam. A part does not have to look like a wheel to have rotational inertia; the only requirement is that it rotates about an axis. In Figure 5.8(c), most of the mass is farther from the axis, so the moment of inertia is higher than the situation in Figure 5.8(d) where the axis is through the middle.

In some situations, high rotational inertia is desirable, in which case a flywheel is used. For example, in a car engine, the flywheel smooths out the power impulses from the individual pistons. However, in most cases, inertia creates problems for electromechanical control systems because a high-inertia part takes more force to get moving and more force to stop. This results in slower response times and positional inaccuracies. The question is, Why not design parts to have less inertia? The problem is that inertia is a function of weight, and weight is a function of strength. Obviously, a part must be strong enough to carry the load, so a certain amount of inertia is unavoidable.

Basic Equations of Motion for Linear Systems

A set of basic equations allows you to calculate the position, velocity, and the time it takes to get someplace, for an object being acted on by a force. Perhaps you have seen and used these before in a physics class, but it is worthwhile to repeat here because we will be using them from time to time in the text to explain mechanical behavior.

We will start by repeating Equation 5.2:

$$F = ma$$

This equation is used to calculate the force required to accelerate an object. The velocity of an object that is being uniformly accelerated can be computed from Equation 5.4. Remember that a uniform **acceleration** means that the velocity is increasing at a constant rate:

$$v = at \tag{5.4}$$

where

v = velocity
a = acceleration
t = time it takes to go from rest to velocity v

The distance that an object with a constant velocity has traveled in a certain time can be computed as follows:

$$d = vt \tag{5.5}$$

where

d = distance the object has moved
v = velocity of the object
t = time the object has been moving

The distance that an object under uniform acceleration has traveled in a certain time is computed as follows:

$$d = \frac{1}{2}at^2 \tag{5.6}$$

where

d = distance the object has moved
a = acceleration
t = time since acceleration started

Finally, we can use Equation 5.7 to calculate the velocity if the distance over which an object has been accelerating is known:

$$v = \sqrt{2ad} \tag{5.7}$$

where

v = velocity
a = acceleration
d = distance the object has moved under acceleration

◆ **EXAMPLE 5.2**

A pneumatic ram pushes defective parts off a conveyer belt (Figure 5.9). A part could weigh up to 10 lb, and it must be pushed a distance of 1 ft in 1 s or less. What is the force of the ram?

Solution

We are given the distance a part has to be moved and the time to do it. None of the preceding equations relate distance, time, and force directly, but Equation 5.6 relates time, distance, and acceleration. Once we know the acceleration required, we can calculate the force required, using Equation 5.2. So, starting with Equation 5.6,

$$d = \frac{1}{2}at^2$$

We first solve for a:

$$a = \frac{2d}{t^2} = \frac{2 \times 1 \text{ ft}}{1 \text{ s}^2} = 2 \text{ ft/s}^2$$

Knowing the acceleration, we now calculate the force required to give that acceleration, using Equation 5.2:

$$F = ma$$

But first we must calculate mass from weight using Equation 5.2a.

$$m = \frac{w}{g} = \frac{10 \text{ lb}}{32 \text{ ft/s}^2} = 0.31 \frac{\text{lb} \cdot \text{s}^2}{\text{ft}}$$

Finally, we calculate the force of the ram from Equation 5.2:

$$F = ma = \frac{0.31 \text{ lb} \cdot \text{s}^2}{\text{ft}} \times \frac{2 \text{ ft}}{\text{s}^2} = 0.62 \text{ lb}$$

The ram must push with a force of at least 0.62 lb.

Figure 5.9
The pneumatic ram must push a defective part off conveyer belt in 1 s or less (Example 5.2).

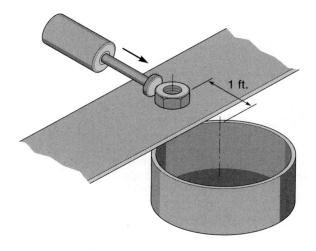

1 ft.

EXAMPLE 5.2 (Repeated with SI Units)

A pneumatic ram pushes defective parts off a conveyer belt (Figure 5.9). A part could have a mass of up to 5 kg, and it must be pushed a distance of 25 cm in 1 s or less. What is the force of the ram?

Solution   Calculate the required acceleration of the part, using Equation 5.6:

$$d = \frac{1}{2}at^2$$

Next, solve for a:

$$a = \frac{2d}{t^2} = \frac{2 \times 25 \text{ cm}}{1 \text{ s}^2} = \frac{50 \text{ cm}}{\text{s}^2} = \frac{0.5 \text{ m}}{\text{s}^2}$$

Knowing the acceleration, we now calculate the force required to give that acceleration, using Equation 5.2:

$$F = ma$$
$$= 5 \text{ kg} \times \frac{0.5 \text{ m}}{\text{s}^2} = 2.5 \text{ N}$$

The ram must push with a force of at least 2.5 N.   ◆

◆ **EXAMPLE 5.3**

An electric solenoid is used to drive the print hammer in a high-speed printer. The hammer presses the print type into the inked ribbon, which prints on the paper (Figure 5.10). The hammer weighs 0.1 lb. It must strike the type with a velocity of 60 in./s and moves through a distance of 0.5 in. How much force must the solenoid provide to the hammer?

Solution   In this problem, we are given distance and velocity and need to find the force pushing the hammer. As in Example 5.2, this is a two-step problem: We first need to find the acceleration required and then the force to create that acceleration. We will modify Equation 5.7 to compute the acceleration, but we first need to convert weight to mass:

$$m = \frac{w}{g} = \frac{0.1 \text{ lb}}{32 \text{ ft/s}^2} = 0.00313 \frac{\text{lb} \cdot \text{s}^2}{\text{ft}}$$

Rearranging Equation 5.7 to solve for acceleration, we get

$$v = \sqrt{2ad}$$

$$a = \frac{v^2}{2d} = \frac{(60 \text{ in./s})^2}{2 \times 0.5 \text{ in.}} = \frac{3600 \text{ in.}^2}{1 \text{ in.} \cdot \text{s}^2} = 3600 \frac{\text{in.}}{\text{s}^2}$$

Now, from Equation 5.2, we can calculate the force to push the hammer:

$$F = ma = \frac{0.00313 \text{ lb} \cdot \text{s}^2}{\text{ft}} \times \frac{3600 \text{ in.}}{\text{s}^2} \times \frac{1 \text{ ft}}{12 \text{ in.}} = 0.939 \text{ lb}$$

The solenoid would have to provide a force on the hammer of at least 0.939 lb.

Figure 5.10

The print hammer must hit type at a specified velocity (Example 5.3).

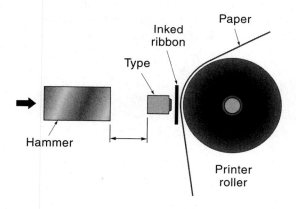

EXAMPLE 5.3 (Repeated with SI Units)

An electric solenoid is used to drive the print hammer in a high-speed printer. The hammer presses the print type into the inked ribbon, which prints on the paper (Figure 5.10). The mass of the hammer is 0.05 kg (50 g). It must strike the type with a velocity of 1.5 m/s and moves through a distance of 1 cm. How much force must the solenoid provide to the hammer?

Solution   In this problem, we are given distance and velocity and need to find the force pushing the hammer. As in Example 5.2, this is a two-step problem: We first need to find the acceleration required and then the force to create that acceleration. To solve for acceleration, rearrange Equation 5.7:

$$v = \sqrt{2ad}$$

$$a = \frac{v^2}{2d} = \frac{(1.5 \text{ m/s})^2}{2 \times 1 \text{ cm}} = \frac{2.25 \text{ m}^2}{2 \text{ cm} \cdot \text{s}^2} \times \frac{100 \text{ cm}}{\text{m}} = \frac{112.5 \text{ m}}{\text{s}^2}$$

Now, from Equation 5.2, we can calculate the required hammer force:

$$F = ma = 0.05 \text{ kg} \times 112.5 \frac{\text{m}}{\text{s}^2} = 5.6 \text{ N}$$

The solenoid would have to provide a force on the hammer of at least 5.6 Newtons.   ◆

Basic Equations of Motion for Rotational Systems

Rotational motion refers, of course, to things that go around. Wheels, gears, axles, and motors all follow the laws of rotational motion. The basic equations for rotational systems are almost the same as those for linear motion. Every quantity used in linear motion has an analog in rotational motion. Instead of force, we have torque. Instead of acceleration and velocity, we have angular acceleration and angular velocity; instead of mass, we have moment of inertia (I). Moment of inertia, as discussed earlier, is dependent on mass, shape, and location of the axis. Equations for calculating I for a few standard shapes are given in Figure 5.8. Equations for I for other shapes are readily available in handbooks.

Torque is the kind of force motors produce. **Torque** is a twisting force acting at a certain radial distance from the center of rotation (Figure 5.11). The equation for torque follows:

$$T = Fd \tag{5.8}$$

where

T = torque
F = force
d = distance of force from the axis

The basic law of rotational motion is given in Equation 5.9 and has the same form as $F = ma$. It tells us that if a constant torque is applied to a wheel, that wheel will accelerate its rotation:

$$T = I\alpha \tag{5.9}$$

where

T = torque applied to a rotating object
I = moment of inertia for the object
α = angular acceleration of the object

Figure 5.11

Torque is caused by a force (*F*) acting at a distance from the axis.

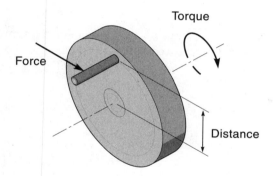

The angular position of an object rotating at a constant angular velocity can be calculated from Equation 5.10:

$$\theta = \omega t \tag{5.10}$$

where

θ = angular position in radians
ω = angular velocity
t = time the object has been rotating

The angular velocity of an object under constant angular acceleration can be calculated from Equation 5.11:

$$\omega = \alpha t \tag{5.11}$$

where

ω = angular velocity
α = angular acceleration
t = time the object has been accelerating

The angular position of an object under constant acceleration can be calculated from Equation 5.12:

$$\theta = \frac{1}{2}\alpha t^2 \tag{5.12}$$

The angular velocity of an object that has moved to a certain position with a certain angular acceleration can be calculated from Equation 5.13:

$$\omega = \sqrt{2\alpha\theta} \tag{5.13}$$

◆ **EXAMPLE 5.4**

A motor is used to open and close a damper in an air duct (Figure 5.12.) The motor can produce a torque of 30 in. · lb, and the moment of inertia (*I*) of the damper is 0.8 lb · s² · in. How long will it take the damper to open, and what is the maximum angular velocity of the damper? In this simple system, the motor comes on full torque and stays on until the damper hits a limit switch.

Solution   Because we know how much torque the motor can supply and the *I* of the load, we first calculate the angular acceleration of the damper from Equation 5.9:

$$T = I\alpha$$

Solving for α, we get

$$\alpha = \frac{T}{I} = \frac{30 \text{ in.} \cdot \text{lb}}{0.8 \text{ lb} \cdot \text{s}^2 \cdot \text{in.}} = 37.5 \frac{\text{rad}}{\text{s}^2}$$

We now have the acceleration, but we are looking for the time the damper takes to open (that is, time for the damper to rotate 90°). Equation 5.12 relates time, position, and acceleration, but position must be in radians, so we first convert 90° to radians. Recalling that there are π radians in 180°, we find that

$$90° = 90° \times \frac{\pi \text{ rad}}{180°} = 1.57 \text{ rad}$$

Now we use Equation 5.12, $\theta = (1/2)\alpha t^2$, and solve for *t*:

$$t = \frac{2\theta}{\alpha} \qquad \sqrt{} = \sqrt{\frac{2 \times 1.57 \text{ rad} \cdot \text{s}^2}{37.5 \text{ rad}}} = 0.29 \text{ s}$$

The calculations show that the damper will open in about one-third of a second. We were also asked to find the maximum angular velocity of the damper. Because the motor provides a constant torque, the damper is accelerating all the time it is opening, and the maximum velocity will occur at 0.29 s. We already calculated the acceleration and time, so we will use Equation 5.11 because it relates velocity, acceleration, and time:

$$\omega = \alpha t = 37.5 \frac{\text{rad}}{\text{s}^2} \times 0.29 \text{ s} = 10.87 \frac{\text{rad}}{\text{s}}$$

Converting to degrees/second gives us

$$\frac{10.87 \text{ rad}}{\text{s}} \times \frac{180°}{\pi \text{ rad}} = \frac{623°}{\text{s}}$$

Therefore, the fastest angular velocity of the damper is 623°/s.

Figure 5.12

A damper in an air duct (Example 5.4).

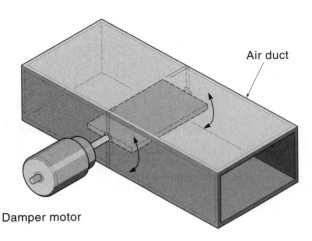

Air duct

Damper motor

EXAMPLE 5.4 (Repeated with SI Units)

A motor is used to open and close a damper in an air duct (Figure 5.12). The motor can produce a torque of 3 N · m, and the moment of inertia (I) of the damper is 0.08 kg · m². How long will it take the damper to open, and what is the maximum angular velocity of the damper? In this simple system, the motor comes on full torque and stays on until the damper hits a limit switch.

Solution

First, calculate the angular acceleration of the damper from Equation 5.9:

$$T = I\alpha$$

Solving for α, we get

$$\alpha = \frac{T}{I} = \frac{3 \text{ N} \cdot \text{m}}{0.08 \text{ kg} \cdot \text{m}^2} = 37.5 \frac{\text{rad}}{\text{s}^2}$$

Now, use Equation 5.12 to calculate the time it takes the damper to open (rotate 90°). However, we must first convert 90° to radians. Recalling that there are π radians in 180°, we find that

$$90° = 90° \times \frac{\pi \text{ rad}}{180°} = 1.57 \text{ rad}$$

Now use Equation 5.12, $\theta = (1/2)\alpha t^2$, and solve for t:

$$t = \sqrt{\frac{2\theta}{\alpha}} \quad = \sqrt{\frac{2 \times 1.57 \text{ rad} \cdot \text{s}^2}{37.5 \text{ rad}}} = 0.29 \text{ s}$$

The calculations show that the damper will open in about one-third of a second. Use Equation 5.11 to calculate the maximum rotational velocity, which will occur at 0.29 s:

$$\omega = \alpha t = 37.5 \frac{\text{rad}}{\text{s}^2} \times 0.29 \text{ s} = 10.87 \frac{\text{rad}}{\text{s}}$$

Converting to degrees/second gives us

$$\frac{10.87 \text{ rad}}{\text{s}} \times \frac{180°}{\pi \text{ rad}} = \frac{623°}{\text{s}}$$

Therefore, the fastest angular velocity of the damper is 623°/s.   ◆

5.2   ENERGY

Energy Conversion

Energy commonly exists in four different forms; thermal (heat), mechanical (motion), electrical, and chemical (fuels). These forms of energy can be converted from one type to another using various processes or machines. The process of combustion releases chemical energy into thermal energy. The resulting heat may be used directly—say, to heat your home—or may be converted into mechanical energy, as with a steam turbine (or gasoline engine). Mechanical energy can be converted into electrical energy with a generator. Electrical energy can be converted back into mechanical energy with an electric motor and into thermal energy with resistance coils.

Thermal energy is measured in British thermal units (Btu), where 1 Btu is the amount of heat it takes to heat 1 lb of water 1°F. In SI units, thermal energy is measured in joules (J), where 1 J is the amount of heat it takes to heat 0.239 gram of water 1°C.

Mechanical energy is measured in ft · lb, where 1 ft · lb is the energy it takes to lift 1 lb vertically 1 ft. In SI units, mechanical energy is measured in N · m, where 1 N · m is the energy it takes to lift an object weighing 1 N vertically 1 m.

To find the equivalent amount of different types of energy, we can use standard conversion factors:

Chemical → Thermal
1 gal fuel oil → 142,500 Btu (or 1.5×10^8 J)

Thermal → Mechanical
1 Btu → 778 ft · lb (or 1055 N · M)
1 J (or 0.239 calories) → 1 N · m

Mechanical → Electrical
1 ft · lb → 3.76×10^{-4} W · h
1 N · m → 2.78×10^{-4} W · h

Converting between energy types is simply a matter of using these conversion factors, but remember that energy and power are related but different. **Energy** is the *amount of work it takes to do a job*, and **power** is the *rate at which the energy is used*. A more powerful motor allows you to do a particular job in less time than a weaker motor, but the amount of energy expended is the same in both cases. Examples of energy–power conversion are given next.

◆ EXAMPLE 5.5

A small 2000-Btu/h oil-fired furnace is to be replaced by an electric furnace operating on 120 Vac. How much current will the electric furnace require?

Solution

First, convert the thermal energy per hour into its electrical equivalent:

$$\frac{2000 \text{ Btu}}{h} \times \frac{778 \text{ ft·lb}}{\text{Btu}} \times \frac{3.76 \times 10^{-4} \text{ W·h}}{\text{ft·lb}} = 584 \text{ W}$$

The result is measured in watts, which is a unit of power because the 2000 Btu/h is a "power," being energy per unit time. Knowing the voltage to be 120 V, we can find the current:

$$P = VI$$

$$I = \frac{P}{V} = \frac{584 \text{ W}}{120 \text{ V}} = 4.86 \text{ A}$$

Therefore, the electric furnace would require a current of 4.86 A.

♦ **EXAMPLE 5.6**

A high-efficiency electric motor used on a winch draws 2 A at 24 Vdc. How fast can it lift a 100-lb weight?

Solution

First, calculate the electric power in the motor:

$$P = VI = 24 \text{ V} \times 2 \text{ A} = 48 \text{ W}$$

Now convert electrical power into mechanical power:

$$48 \text{ W} \times \frac{1 \text{ ft} \cdot \text{lb}}{3.76 \times 10^{-4} \text{ W} \cdot \text{h}} = 127{,}700 \text{ ft} \cdot \text{lb/h}$$

To see how fast the motor can lift 100 lb, we divide by 100 lb (so the units of pounds cancel):

$$\frac{127{,}700 \text{ ft} \cdot \cancel{\text{lb}}/\text{h}}{100 \ \cancel{\text{lb}}} = 1277 \text{ ft/h, or } 21.3 \text{ ft/min}$$

Therefore, the winch could lift 100 lb at a rate of 21.3 ft/min.

EXAMPLE 5.6 (Repeated with SI Units)

A high-efficiency electric motor used on a winch draws 2 A at 24 Vdc. How fast can it lift a 50-kg mass?

Solution

First, calculate the electric power in the motor:

$$P = VI = 24 \text{ V} \times 2 \text{ A} = 48 \text{ W}$$

Now convert electrical power into mechanical power:

$$48 \text{ W} \times \frac{1 \text{ N} \cdot \text{m}}{2.78 \times 10^{-4} \text{ W} \cdot \text{h}} = 172{,}660 \text{ N} \cdot \text{m/h}$$

To see how fast the motor can lift 50 kg, we first find the force required to lift 50 kg in the presence of gravity:

$$F = ma = mg = 50 \text{ kg} \times 9.8 \text{ m/s} = 490 \text{ N}$$

Now divide the power by 490 N (so the units of Newtons cancel):

$$\frac{172{,}660 \ \cancel{\text{N}} \cdot \text{m/h}}{490 \ \cancel{\text{N}}} = 352 \text{ m/h, or } 5.9 \text{ m/min}$$

Therefore, the winch could lift 50 kg at a rate of 5.9 m/min. ♦

Two general principles govern energy conversion: First, you can never get *more* energy out of a conversion than you put in; second, no energy conversion is perfect—there is always some waste, which is typically in the form of heat. Both concepts are expressed in the following equation:

$$E_{\text{in}} = E_{\text{out}} + E_{\text{waste}} \tag{5.14}$$

We account for this waste energy with the concept of efficiency. The **efficiency** of an energy conversion is the percentage of useful energy out. For example, the typical automobile engine is only about 20% efficient, meaning that only about 20% of the chemical energy available in the gasoline is converted into useful mechanical energy. The remaining 80% is lost as heat (both through the radiator and exhaust) and internal friction.

♦ **EXAMPLE 5.7**

An electric motor operates on 120 V and draws 5 A. If it is 90% efficient, how much power is lost to heat?

Solution

First, we compute the amount of power the motor is using:

$$\text{Power} = VI$$

$$= 120 \text{ V} \times 5 \text{ A} = 600 \text{ W}$$

Using the rated efficiency, we can compute how this power is converted:

$$0.9 \times 600 \text{ W} = 540 \text{ W of mechanical power}$$

$$0.1 \times 600 \text{ W} = 60 \text{ W of waste heat}$$

With 60 W of waste heat, this motor would give off as much heat as a 60-W light bulb.   ♦

Heat Transfer

Because waste heat is generated in many electrical and mechanical systems, there must be some provision to dispose of it. In many cases, the waste heat is simply transferred to the surrounding air. For example, consider the power transistor shown in Figure 5.13. The transistor is mounted on a *heat sink*, which is a piece of aluminum with "fins" (such as found on an air-cooled engine). The fins provide a larger surface area for the heat to be transferred to the air; if the air is moving (say, from a fan), you get even better heat transfer. In Figure 5.13, the heat must travel from its source (the NPN junction) to the transistor case, to the heat-sink base, and out to the fins.

Figure 5.13

Heat conducting from a transistor, through a heat sink to the air.

Transistor

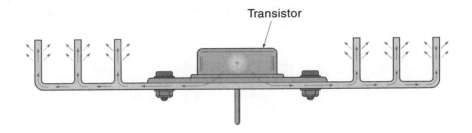

The amount of heat that conducts through a material is dependent on three factors; type of material, its shape, and the temperature difference between the ends. Figure 5.14 shows heat conducting through a rectangular solid of length L and cross-sectional area A. Heat "flows" because the temperature of the hot end (T_2) is higher than the temperature of the cool end (T_1), and the actual amount of that conduction can be calculated with Equation 5.15:

$$H = \frac{KA(T_2 - T_1)}{L} \tag{5.15}$$

where

H = amount of heat being conducted
K = constant for different materials
A = cross-sectional area of the material
T_2, T_1 = high and low temperatures at each end of the material
L = length of the material

Figure 5.14

Heat conduction through a solid.

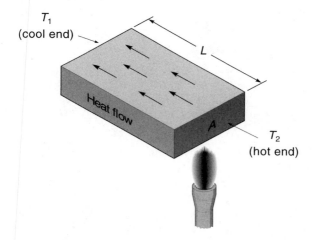

Table 5.1 gives a sample of the heat conduction constant K for several materials. Table 5.1 clearly shows that metals—in particular, copper and aluminum—are the best

TABLE 5.1 Heat Conduction of Various Substances

Material	$K[(Btu \cdot ft)/(h \cdot ft^2 \cdot °F)]$	$K[(cal \cdot cm)/(s \cdot cm^2 \cdot °C)]$
Copper	223	0.92
Aluminum	119	0.49
Steel	29	0.12
Glass	0.48	0.002
Wood	0.048	0.0002
Cork, felt	0.024	0.0001
Air	0.014	0.000057

conductors of heat. The poorest conductor is air, which explains why it is a challenge to conduct away large amounts of waste heat to the air. A simple heat-conduction problem is worked in Example 5.8.

♦ **EXAMPLE 5.8**

A power transistor is mounted on one end of an aluminum bracket that is 2 in. wide and ⅛ in. thick. The other end of the bracket is mounted on a large chassis (Figure 5.15). The transistor is about 1 in. from the chassis. The transistor case can be about 100°C (180°F) above the chassis temperature. How much power can the bracket transfer?

Solution

We will use Equation 5.15 to calculate the heat flow, but first we need to calculate the length and cross-sectional area of the heat conductor (in feet):

$$\text{Length} = 1 \text{ in.} = 0.0833 \text{ ft}$$

$$\text{Area} = \frac{1}{8} \text{ in.} \times 2 \text{ in.} = 0.25 \text{ in.}^2 = 0.00174 \text{ ft}^2$$

Now apply Equation 5.15:

$$H = \frac{KA(T_2 - T_1)}{L}$$

$$= \frac{119 \text{ Btu} \cdot \text{ft}}{\text{h} \cdot \text{ft}^2 \cdot °\text{F}} \times \frac{0.00174 \text{ ft}^2 \times (180°\text{F})}{0.0833 \text{ ft}} = 446 \text{ Btu/h}$$

Next convert 446 Btu/h to watts using the conversion factors from the previous section:

$$\frac{446 \text{ Btu}}{\text{h}} \times \frac{778 \text{ ft} \cdot \text{lb}}{1 \text{ Btu}} \times \frac{3.76 \times 10^{-4} \text{ W} \cdot \text{h}}{1 \text{ ft} \cdot \text{lb}} = 130 \text{ W}$$

The aluminum bracket can transfer 130 W of thermal energy. In practice, this number would be derated to account for the thermal resistance across the transistor and chassis-mounting areas.

Figure 5.15

A power transistor on a bracket (Example 5.8).

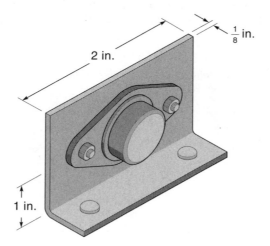

♦

5.3   RESPONSE OF THE WHOLE MECHANICAL SYSTEM

Consider the case of a rotating antenna system (Figure 5.16). A motor drives a gear train, which rotates the antenna. It all seems simple enough; if the antenna needs to go to a certain position, the motor comes on, rotates until the antenna is in position, and then stops. This is *not* a simple situation, however. First, the antenna has inertia, so it takes more force to get it moving than it does to keep it moving. Second, the power train—consisting of gears, shafts, and other metal parts—has a certain flexibility. For example, if the antenna were clamped immobile and the motor tried to turn, all it could do is twist the input gear a little bit, like winding up a stiff spring. This springiness comes about because all parts in the power train will bend a little under stress. The other real-life consideration is friction. All moving parts in the system are turning on bearings, which have friction. The friction force resists motion in either direction.

Figure 5.16

A rotating antenna system.

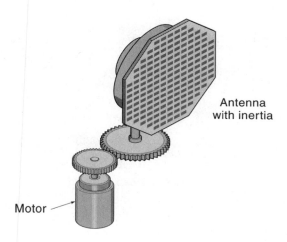

Antenna
with inertia

Motor

Underdamped, Critically Damped, and Overdamped Mechanical Systems

Any mechanical system, such as an antenna (see Figure 5.16), can be diagrammed (Figure 5.17). In the diagram, the load that must be moved is represented by a box. Because this load has mass (inertia for rotational systems), it resists being accelerated or decelerated. Further, there is always some external drag on the load from friction and other sources (such as wind resistance or intentional hydraulic dampers). This drag is called **damping**. In Figure 5.17, the damping force is represented by a shock absorber. The connecting link between the load and the actuator (the hydraulic cylinder) is represented by a spring. The spring is shown because we want to represent the flexibility of the power train (flexing of levers, gears, etc.).

Refer to Figure 5.17(a) and notice that the load is at rest in position A and needs to move to position B. First, the hydraulic cylinder extends to the right (pushing on the spring). In Figure 5.17(b), both friction and inertia prevent the load from responding immediately, so at first the spring just compresses. At some point, the pressure on the

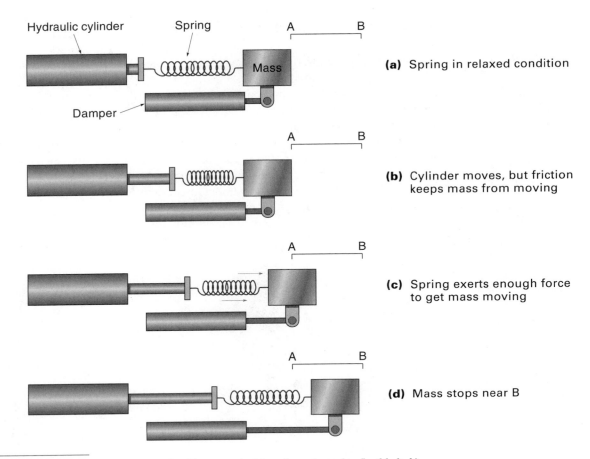

Figure 5.17 Model of a mechanical system (load being pushed by a force through a flexible link).

load from the spring is large enough to start the load moving [Figure 5.17(c)]. Eventually, the load will come to rest somewhere near point B [Figure 5.17(d)]. The motion of the load in response to the force is said to be either **underdamped**, **critically damped**, or **overdamped** as described next.

Figure 5.18(a) shows the response of an underdamped system. Recall that the load was at rest at position A, and then a hydraulic cylinder gave it a push. The dashed line shows the movement of the hydraulic cylinder. At first, the load lags behind a little, but if the connecting link is stiff enough (and the load is not too heavy), the load will speed up quickly. In fact, the inertia of the fast-moving load carries it past point B; this is called **overshoot**. The load may oscillate back-and-forth several times before coming to rest. You can see now why this system is called underdamped because the friction (damping force) is not enough to keep the load from overshooting its destination.

Figure 5.18(b) shows the response of a critically damped system. In this case, the damping is just enough to prevent overshoot. Notice that the time to get from position A to B is a little longer, but the absence of oscillations is desirable in many cases. In fact, the time required to actually settle out is the shortest for a critically damped system.

Figure 5.18
The effect of damping on a
mechanical system response.

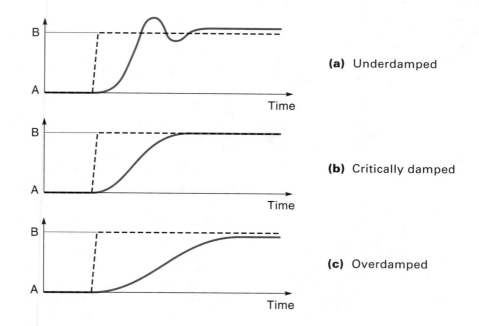

(a) Underdamped

(b) Critically damped

(c) Overdamped

Figure 5.18(c) shows the response of an overdamped system. Overdamped is any damping more than critically damped, so the response time is even slower, one might say sluggish. Also notice that the final resting place of the load may fall considerably short of position B.

In real control systems, the controller can compensate somewhat for the natural mechanical tendencies that have been discussed here. For example, overshoot [Figure 5.18(a)] could be greatly reduced with the proper control algorithm. (These concepts are discussed in Chapter 11.)

Mechanical Resonance

Figure 5.19 shows a mass that is supported by a spring. If you pull the mass down a little and let it go, it will oscillate up-and-down. The frequency of the oscillations is called the **natural resonant frequency** and can be calculated by Equation 5.16:*

$$f = \frac{1}{2\pi}\sqrt{\frac{k}{M}} \tag{5.16}$$

where

f = resonant frequency (in Hz)
k = spring force constant
m = mass

---

*Notice how similar this equation is to the equation for electrical resonance: $f_c = 1/(2\pi\sqrt{LC})$.

Figure 5.19

Mass (*m*) will oscillate
up-and-down on a spring.

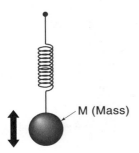

M (Mass)

With the mass and spring setup of Figure 5.19, it's easy to visualize the mass oscillating up-and-down. The point is that all mechanical parts have mass and all are connected to other parts through a power train or a superstructure that has some spring constant *k*. If all mechanical parts in a system have mass and are connected by "springs," then all parts will have resonant frequencies. Figure 5.20 shows two examples. In Figure 5.20(a), a mass is in the middle of a steel support bracket. The bracket is the "spring." If you push down on the mass, the bracket will bend (shown with dashed line). When you let up, the bracket will return to its normal position. This is a stiff spring, meaning the spring constant (*k* in Equation 5.16) will be relatively high, which in turn causes the mechanical resonant frequency to be high, perhaps 500 Hz. So although you couldn't actually see it vibrate, you could probably hear it because 500 Hz is in the audio range. The setup in Figure 5.20(b) has the same mass, but this time the support bracket is anchored at only one end. It will take much less force to bend the bracket in this configuration; thus, the spring constant is less, which (again, according to Equation 5.16) causes the resonant frequency to be lower, perhaps 25 Hz.

Figure 5.20

A support bracket acts as a spring: (a) stiff support and (b) much less stiff because of support on only one end.

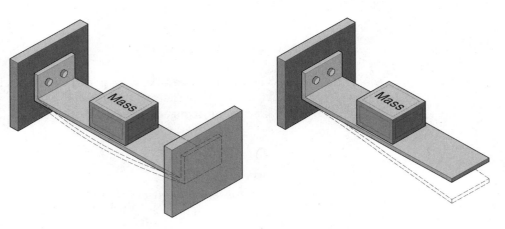

(a) Supported on both ends
makes a stiffer support

(b) Supported on one end
makes a springier support

The concept of resonant frequency is important because it explains why mechanical parts will vibrate, buzz, or develop cracks and break under certain conditions. These events happen because the part is oscillating back-and-forth at its resonant frequency. It takes surprisingly little energy to keep a part vibrating *if the energy itself is applied in the form of a vibration at or near the resonant frequency.* You can demonstrate this easily with a length of string and a mass (say, a small rock), as illustrated in Figure 5.21. First, get the rock swinging [Figure 5.21(a)] and notice how little energy it takes to keep the rock oscillating; you barely have to move your hand at all. Next slow the frequency way down so that you are slowly moving your hand back-and-forth [Figure 5.21(b)]. What happens to the rock? It slowly follows under your hand, with no wild oscillations. Finally, increase your hand frequency way above the resonant frequency [Figure 5.21(c)]. The rock pretty much stays still, and the string absorbs the motion. The important conclusion: A mechanical part may make wild movements if it is driven by a vibration at or near its resonant frequency.

Figure 5.21

A small rock and string demonstrate (a) natural resonant frequency, (b) below-resonant frequency, and (c) above-resonant frequency.

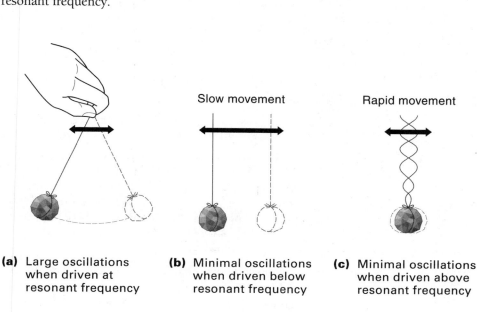

Slow movement

Rapid movement

(a) Large oscillations when driven at resonant frequency

(b) Minimal oscillations when driven below resonant frequency

(c) Minimal oscillations when driven above resonant frequency

Sources of mechanical vibrations abound in machinery—from motors, gears, and bearings or from *anything* that moves in a repeated motion. Mechanical engineers sometimes use special computer programs that analyze and predict the resonant frequency of each part of the mechanical design. If the design is something that must be right the first time, such as a space satellite, a prototype of the device is mounted on a test machine called a shake table, which vibrates the new design at various frequencies while high-speed cameras record the response. Any wild mechanical oscillations can then be identified and corrected.

♦ **EXAMPLE 5.9**

The electrical control box for an electric motor is attached to the motor case with a bracket (Figure 5.22). The bracket is developing cracks even though it is clearly strong enough to support the weight of the box. What is the problem here?

Solution

The likely problem is that the box is vibrating back-and-forth on the bracket when the motor is running, thus putting a strain on the bracket. That is, we suspect that the resonant frequency of the box swinging on the bracket is at or near the motor rpm. To test this theory, first determine the spring constant of the bracket by noticing that it defects 0.1 in. when pushed with 10 lb:

$$k = \frac{\text{force}}{\text{deflection}} = \frac{10 \text{ lb}}{0.1 \text{ in.}} = 100 \frac{\text{lb}}{\text{in.}}$$

Next determine the mass of the box. It is found to weigh 1 lb. We then find the mass by dividing by the gravity term, which is 32 ft/s², or 384 in./s²:

$$m = \frac{w}{g} = \frac{1 \text{ lb}}{384 \text{ in./s}^2} = 0.0026 \frac{\text{lb} \cdot \text{s}^2}{\text{in.}}$$

Now use Equation 5.16 to calculate the resonant frequency of the box swinging on the bracket:

$$f = \frac{1}{2\pi} \sqrt{\frac{k}{m}} = \frac{1}{6.28} \sqrt{\frac{100 \text{ lb/in.}}{0.0026 \text{ lb} \cdot \text{s}^2/\text{in.}}} = 31 \text{ Hz}$$

From the motor plate, we read that it operates at 1750 rpm, which works out to 29 rps (dividing by 60). The motor is vibrating at 29 Hz, which is very close to the resonant frequency of the box: 31 Hz. No doubt, the box has been swinging back-and-forth on the bracket. To solve the problem, we need to change the resonant frequency of the box; up or down would be sufficient. This could be done by making the bracket shorter or longer or changing its shape. ♦

One last point on resonant frequency: If free movement of a part is restricted because of friction or some other form of damping, the magnitude of the oscillations will be reduced. In fact, if you look at Figure 5.18(a) (underdamped case), the overshoot and undershoot are the beginnings of resonant frequency oscillations, which are quickly damped out. If the part is critically damped [Figure 5.18(b)], no oscillations will occur at all—even when driven at its resonant frequency.

Figure 5.22

Control-box vibration on a motor (Example 5.9).

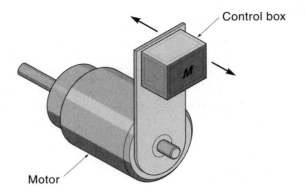
Control box

M

Motor

5.4   GEARS

Gears allow us to change the rotational velocity and the torque to suit the motor and load conditions. Gears also can be used to simply transport power from one shaft to a parallel shaft. The concept of gearing is certainly not new. The ancient Greeks were using wooden gears for such things as transferring power from the waterwheel to the millstone (a practice still in use in existing old water-powered mills). Modern precision-cut metal and plastic gears provide an efficient, smooth, and durable means of power transmission. Figure 5.23 shows a selection of gears.

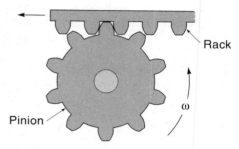

(a)   Rack and pinion

(b)   Straight bevel gears

(c)   Spur gears

(d)   Worm and worm gears

(e)   Spiral bevel gears

Figure 5.23   A selection of gear types. (Photos courtesy of Boston Gear)

Spur Gears

Figure 5.24 shows two meshed **spur gears**. Power is transmitted by a tooth of one gear pushing against the tooth of the mating gear, one at a time. You might think that this would cause a jerky motion; if the teeth have the proper shape, however, they actually roll on each other, handing the power smoothly over to the next tooth (Figure 5.25). Because the teeth are actually rolling and not sliding on each other, there is very little friction in a **gear pass**, which is one set of mating gears.

When two gears of different diameters are meshed, they rotate at different velocities. The motion is similar to two wheels of different diameters rolling against each other, where the size of each wheel is a theoretical circle called the **pitch circle** (see Figure 5.24). The diameter of this pitch circle is called the **pitch diameter**. Notice that the pitch diameter is smaller than the overall diameter of the gear because the teeth overlap in the mesh.

The distance along the pitch circle of one tooth (and "valley") is called the **circular pitch**. Of course, an integer number of teeth must be on a gear (you cannot have a 5½-tooth gear!). The number of teeth on a gear can be calculated from Equation 5.17:

$$N = \frac{\text{circumference}}{\text{distance between teeth}} = \frac{\pi D}{P_c} \tag{5.17}$$

where

N = total number of teeth
D = pitch diameter
P_c = circular pitch

Finally, there is the **diametral pitch** or simply *pitch*, which is the ratio of the number of teeth per inch of pitch diameter, as shown in Equation 5.18:

Figure 5.24
Meshed spur gears.

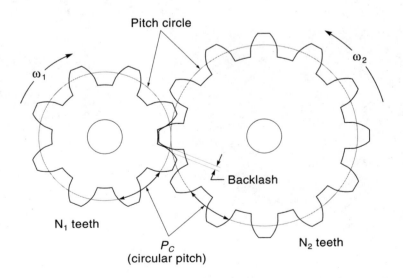

Pitch circle

ω_1

ω_2

Backlash

N_1 teeth

N_2 teeth

P_c
(circular pitch)

Figure 5.25
Showing different positions of a single involute tooth rolling on its mating tooth (follow contact point A–B–C–D).

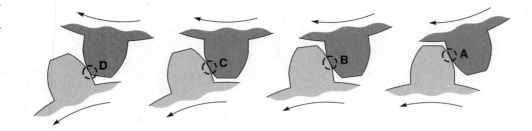

$$\text{Diametral pitch} = \text{pitch} = \frac{N}{D} \qquad (5.18)$$

where

N = total number of teeth
D = pitch diameter (in inches)

Diametral pitch is the parameter most often used to specify the tooth size in a gear. Figure 5.26 illustrates the size of teeth with pitches ranging from 4 to 48. Notice that a larger tooth has a smaller pitch. In a gear mesh, only one tooth is in contact at any time, so the larger the tooth, the more power can be transmitted. Lightly loaded machines such as printers and plotters typically use gears with a pitch between 48 and 32; these tend to be known as **instrument gears**. Automotive transmissions use a pitch range of 10–5. Of course, a bottom-line requirement is that mating gears have the same pitch (tooth size) so that they will fit together.

Figure 5.26
Gear-teeth sizes for different diameter pitches. (Courtesy of Boston Gear)

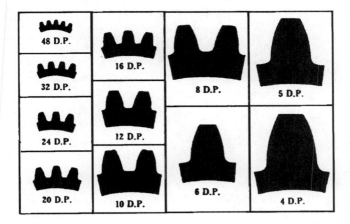

Using Gears to Change Speed

The **gear ratio** is a ratio of the number of teeth of two gears. The two gears in Figure 5.27 have 40 and 20 teeth, respectively, so the gear ratio is 40/20 = 2. However, there is more to it. For the gears to mesh, their tooth sizes must be the same, which means they both have the same number of teeth per inch of circumference. So the gear ratio is also the ratio of circumferences. Finally, the diameter of any circle is directly proportional to its circumference, so the gear ratio is also a ratio of diameters, which makes it convenient to find the ratio of an existing gear mesh; all you do is measure the diameters and divide.* This is presented in Equation 5.19:

Figure 5.27

Two gears in mesh; gear ratio is 40/20 = 2.

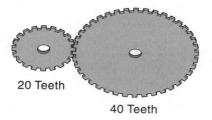

20 Teeth

40 Teeth

$$N_g = \frac{N_2}{N_1} = \frac{\mathrm{Cir}_2}{\mathrm{Cir}_1} = \frac{\pi \, \mathrm{Dia}_2}{\pi \, \mathrm{Dia}_1} = \frac{\mathrm{Dia}_2}{\mathrm{Dia}_1} \qquad (5.19)$$

where

N_g = gear ratio between gears 1 and 2
N = number of teeth
Cir = circumference of the gear
Dia = diameter of the gear

The assumption in Equation 5.19 is that gear 1 is supplying the power (called the driver gear), and gear 2 is receiving the power (called the driven gear).

*Technically, you should divide the pitch diameters, but with most small-toothed instrument gears, the overall diameter and the pitch diameter are practically the same.

♦ **EXAMPLE 5.10**

What is the gear ratio of the gear mesh shown in Figure 5.28?

Solution   The driver gear has a diameter of 0.75 in., and the driven gear has a diameter of 2.25 in. Find the ratio by dividing the diameters:

$$N_g = \frac{\text{Dia}_2}{\text{Dia}_1} = \frac{2.25 \text{ in.}}{0.75 \text{ in.}} = 3$$

♦

Figure 5.28

A 3:1 gear mesh (Example 5.10).

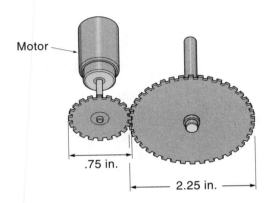

Motor

.75 in.

2.25 in.

Looking at the 3 : 1 gear mesh in Figure 5.28, you can see that the small gear (called the **pinion**) makes three revolutions in the same time it takes the big gear to make one revolution. In other words, the small gear must rotate three times faster than the big gear. This kind of logic leads us to conclude that the ratio of position, velocity, and acceleration are inversely proportional to the gear ratio N_g:

$$N_g = \frac{N_2}{N_1} = \frac{\theta_1}{\theta_2} = \frac{\omega_1}{\omega_2} = \frac{\alpha_1}{\alpha_2} \qquad (5.20)$$

where

N_g = gear ratio
N = number of teeth
θ = gear position (angle in degrees)
ω = angular velocity of the gear
α = angular acceleration

In applying this equation, sometimes confusion arises about how to plug in the gear ratio; if the ratio is 3, does this go into the equation: as 3/1 or 1/3? The rule that never fails is this: Look at the gears (or a sketch); the bigger gear will always rotate *less* of an angle and at *less* velocity than the small gear. Plug in the gear ratio such that this relationship holds. The following examples illustrate the principles of Equation 5.20.

◆ EXAMPLE 5.11

A radio tuner is connected to the tuning knob through a 3 : 1 gear mesh (Figure 5.29). If the knob is turned 70°, how many degrees does the tuner rotate?

Solution   The gear attached to the tuner is the bigger of the two; therefore, it must turn *less* than 70°. We will use the part of Equation 5.20 that deals with position, solving for the angle of the tuner gear, θ_2:

$$N_g = \frac{3}{1} = \frac{70°}{\theta_2}$$

so

$$\theta_2 = \frac{70°}{3} = 23.33°$$

The tuner will rotate 23.33° when the knob rotates 70°.

Figure 5.29
A gear setup (Example 5.11).

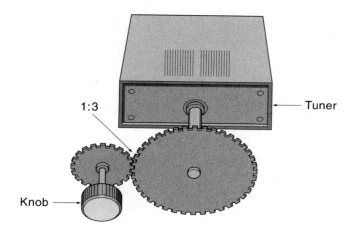

◆ EXAMPLE 5.12

A small motor running at 200 rpm drives a paper roller in a business machine (Figure 5.30). The gear on the motor has 20 teeth, and the gear on the roller has 50 teeth. How fast is the roller turning?

Solution   First, note that the roller gear is larger than the motor gear, so it will be turning slower than the motor. To calculate the speed, apply Equation 5.20, solving for velocity:

$$\frac{50 \text{ teeth}}{20 \text{ teeth}} = \frac{200 \text{ rpm}}{\omega_2}$$

so

$$\omega_2 = \frac{200 \text{ rpm} \times 20 \text{ teeth}}{50 \text{ teeth}} = 80 \text{ rpm}$$

◆

Figure 5.30

Gear drive for a paper roller (Example 5.12).

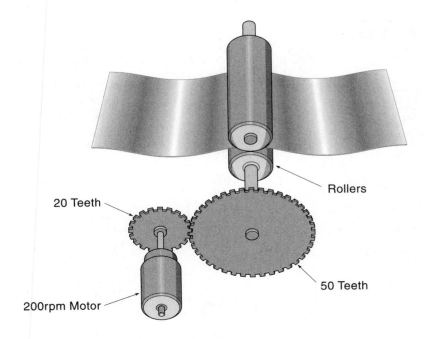

Rollers

20 Teeth

50 Teeth

200rpm Motor

A **gear train** consists of more than a single gear pass. Figure 5.31 shows a system with three gears. Notice that the middle gear is actually two gears fastened together. The gear ratio of the total system is the product of the individual gear ratios:

$$N_{g(\text{tot})} = N_{g1}N_{g2}N_{g3}\cdots \tag{5.21}$$

where

$N_{g(\text{tot})}$ = overall gear ratio
N_{g1} = gear ratio of first pass
N_{g2} = gear ratio of second pass, and so on

Figure 5.31

A two-pass gear train (Examples 5.12 and 5.13).

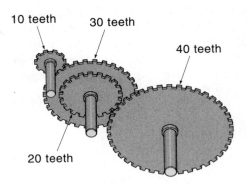

10 teeth   30 teeth

40 teeth

20 teeth

♦ **EXAMPLE 5.13**

Find the overall gear ratio of the gear train pictured in Figure 5.31.

Solution   There are actually four gears here (10, 20, 30, and 40 teeth) and two gear passes. The 20- and 30-teeth gears are fastened together and so rotate at the same velocity. The overall gear ratio is found by multiplying the two individual gear ratios:

$$N_{g1} = \frac{30 \text{ teeth}}{10 \text{ teeth}} = 3 \qquad \text{Ratio of first pass}$$

$$N_{g2} = \frac{40 \text{ teeth}}{20 \text{ teeth}} = 2 \qquad \text{Ratio of second pass}$$

$$N_{g1}N_{g2} = 3 \times 2 = 6 \qquad \text{Overall ratio}$$

♦

Using Gears to Transfer Power

For precision-made spur gears, the assumption is that power is conserved across the gear pass. Except for a small loss through friction, the gear pass can neither create nor destroy power:

$$\text{Power in, gear } 1 = \text{Power out, gear } 2 \tag{5.22}$$

Therefore, because power = torque × velocity, we rewrite Equation 5.22 as follows:

$$T_1\omega_1 = T_2\omega_2 \tag{5.23}$$

where

T = torque (such as ft · lb or N · m)
ω = angular velocity (such as deg/s)

Rearranging Equation 5.23, we get

$$\frac{T_2}{T_1} = \frac{\omega_1}{\omega_2} = N_g \tag{5.24}$$

Equation 5.23 tells us that the product of torque and velocity is the same on each side of the gear pass. This means that the faster turning gear has less torque, and the gear that is turning slower has more torque.* This is demonstrated every time you drive a car; the lower gears give more torque for start up, and the higher gears give more speed but less torque. Equation 5.24 shows that the ratio of torques in a gear pass is inversely proportional to the gear ratio.

---

*Note the similarity between a gear pass and an ideal electrical transformer, where the product of $V \times I$ is maintained from primary to secondary windings.

◆ **EXAMPLE 5.14**

An electric motor supplies 60 in. · oz of torque while running at 100 rpm and it is driving a load through a 1 : 5 gear ratio (Figure 5.32). Find the output torque and velocity.

Solution

Equation 5.24 relates torque and velocity to gear ratio. Solving first for torque on the driven (or load) gear, we have $T_2/T_1 = N_g$, so

$$T_2 = T_1 N_g = 60 \text{ in.} \cdot \text{oz} \times 5 = 300 \text{ in.} \cdot \text{oz}$$

Solving for the velocity of the load gear we have $\omega_1/\omega_2 = N_g$, so

$$\omega_2 = \frac{\omega_1}{N_g} = \frac{100 \text{ rpm}}{5} = 20 \text{ rpm}$$

This is a typical application for a gear pass. Electric motors tend to be high-speed and low-torque, whereas many applications require higher torques and lower speeds. In this case, the torque was increased by a factor of 5, to 300 in. · oz, but the speed was reduced by a factor of 5, to 20 rpm.

Figure 5.32

A motor driving a load (Example 5.14).

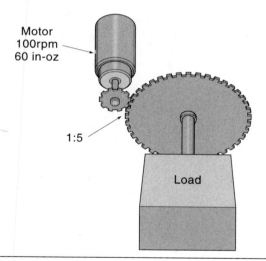

Motor
100rpm
60 in-oz

1:5

Load

◆

Recall from Section 5.1 that every load rotating on an axis has a moment of inertia (I). The more I it has, the more energy it takes to get it spinning. Ultimately, the motor has to supply this energy; thus, the question is, What moment of inertia does the motor "see" if the load is driven through a set of gears (Figure 5.33)? It turns out that the reflected moment of inertia is inversely proportional to the gear ratio squared, as given in Equation 5.25:

$$I \text{ reflected} = \frac{I_{\text{load}}}{N_g^2} \tag{5.25}$$

Figure 5.33

The moment of inertia of a load is "reflected back" to the motor.

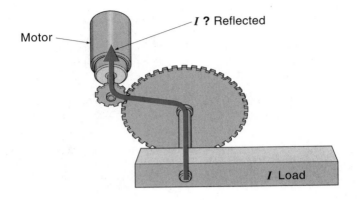

This means that in most cases the motor "sees" a much lower inertia than the load actually has. The comprehensive motor example at the end of Chapter 7 gives a practical application of the reflected inertia equation.

Long Gear Trains

Large gear reductions (ratios) can be made with multiple gear passes, but this approach presents special problems. For example, a reduction of 1 : 256 might take as many as four passes using conventional spur gears. Such a gear train is shown in Figure 5.34. Undesirable qualities such as inefficiency and backlash tend to magnify with each gear pass. **Backlash** is the small free clearance between mating teeth. A small amount of backlash is necessary to prevent binding of the gears if they are slightly out of round. However, in a long gear train, the effect of the backlash is amplified along with the gear ratio. In the case of Figure 5.34, backlash of 1° in the first pass would be amplified by the three other gear passes (4 × 4 × 4 = 64) to become 64° of backlash at the end—clearly unacceptable. Efficiency is another problem. If the efficiency of each pass in Figure 5.34 was a respectable 96%, the overall efficiency for four passes would be 96% × 96% × 96% × 96% = 85%.

Figure 5.34

A gear train with four passes; total gear ratio is 256.

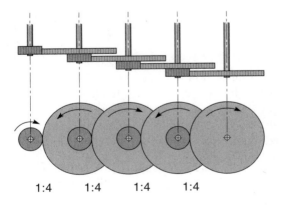

1:4   1:4   1:4   1:4

Worm Gears

Worm gears yield very high gear ratios on a single pass with little or no backlash, but they can be highly inefficient. Figure 5.35 shows a worm gear mesh and consists of the **worm** and the **worm gear**, or pinion. For each revolution of the worm, the pinion advances one tooth. This is what makes the high gear ratios possible; a pinion with 50 teeth would yield a gear ratio of 50. Notice, however, that as the worm rotates it is really sliding across the pinion's teeth. The friction from this sliding is the cause of the inefficiency, which is typically in the 50–80% range.

Figure 5.35
Worm gearing.

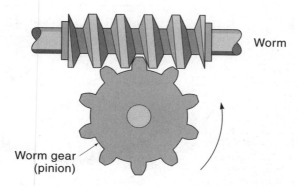

Another advantage of worm gearing is the **lockup property**. Lockup occurs in higher-ratio worm gears, and it means that power flows only one way—from the worm to the pinion. If you try to turn the pinion directly, the friction is so great that it locks up. A common use of this feature is the little worm gears used to tighten guitar strings. You can twist the worm and tighten the string, but the tension of the string cannot spin the worm gears backward.

Harmonic Drive

The **harmonic drive** is a unique gear system that gives high efficiency, no backlash, and high gear ratio. Illustrated in Figure 5.36, the drive looks like a cup with the input shaft going through the center. The rigid cup has an internal set of teeth all the way around known as the *rigid circular spline*. Inside the rigid cup is a slightly smaller flexible cup called the *flex spline*. The flex spline typically has two teeth less than the outer cup. A rotating device called the *wave generator* pushes out on the flex spline so that the teeth mesh only on opposite sides. The input shaft is connected to the wave generator, and as it rotates, the two splines (inner and outer) mesh with each other. Because they have a different number of teeth, the outer cup (which is connected to the output shaft) will advance a little for each revolution of the flex spline. The actual gear ratio can be calculated from the following equation:

$$N_h = \frac{\text{number of teeth on outer rigid cup}}{\text{difference in number of teeth on splines}} \qquad (5.26)$$

Figure 5.36
The harmonic drive.

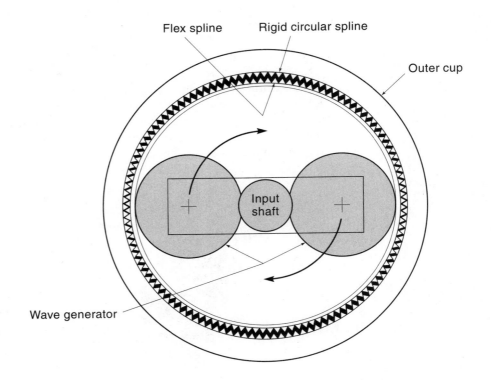

where N_h is the gear ratio (more correctly *velocity ratio*) of the harmonic drive. For example, in the drive illustrated in Figure 5.36, the number of teeth in the outer cup is 100, and the number of teeth on the flex spline is 98; thus, the gear ratio is 100/2 = 50.

5.5   OTHER POWER-TRANSMITTING TECHNIQUES

The most direct way to transmit rotary power is through a shaft, such as a drive shaft or an axle shaft. Gears can transmit power between parallel or perpendicular shafts and change the velocity and torque in the process. In this section, we examine some other techniques for transmitting power from shaft to shaft.

Belts

Under the correct circumstances, belts are a good means of transmitting power. Usually made of rubber and hence very flexible, they come in various types (Figure 5.37). The advantages of belts include low cost, quiet running, low maintenance (no lubrication), shock absorbtion, and tolerance of nonparallel shafts, and they allow for "gearing" by having one pulley larger than the other.

Figure 5.37

Different types of power-transmission belts.

V belt   Flat belt   Toothed belt

Figure 5.38 shows a standard belt setup, which includes a driver pulley and a driven pulley. Notice that the power is carried entirely by the lower belt segment. The ratio of diameters of the pulleys provides exactly the same change in velocity and torque as for gears. In fact, all equations given for spur gears (5.19–5.25) apply to belts. One difference is that meshed gears are rotating in opposite directions, whereas both pulleys in a belt system rotate in the same direction (compare Figure 5.24 with Figure 5.38).

Figure 5.38

Power transmission using a belt (Example 5.15).

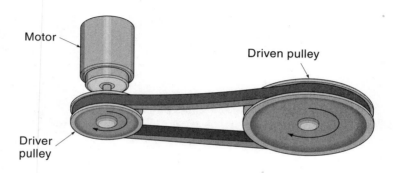

Motor

Driven pulley

Driver pulley

◆ **EXAMPLE 5.15**

For the belt system shown in Figure 5.38, the diameter of the small pulley is 3 in., and the diameter of the large pulley is 8 in. The motor is rotating at 1000 rpm, providing a torque of 10 in. · lb. Find the velocity and available torque of the large pulley shaft.

Solution

The velocity ratio N_v (similar to gear ratio) is the ratio of their diameters:

$$N_v = \frac{\omega_1}{\omega_2} = \frac{\text{Dia}_2}{\text{Dia}_1} = \frac{8 \text{ in.}}{3 \text{ in.}} = 2.67 \qquad (5.27)$$

This means the big pulley will rotate 2.67 times slower than the small pulley. We can solve this problem by simply dividing 1000 rpm by 2.67 to get 375 rpm, or we can do it more formally by equations. From equation 5.27, $N_v = \omega_1/\omega_2$, where ω_1 and ω_2 are pulley rpms. Solving for ω_2, we get

$$\omega_2 = \frac{\omega_1}{N_v} = \frac{1000 \text{ rpm}}{2.67} = 375 \text{ rpm}$$

The torque of the large pulley will be 2.67 larger than the small pulley. Again, we could find the output torque by simply multiplying 10 in. · lb by 2.67 to get 26.7 in. · lb, or do it by Equation 5.24 (which also applies to belts):

$$N_v = \frac{T_2}{T_1}$$

where T_1 and T_2 are torques. Solving for T_2, we get

$$T_2 = T_1 N_v = 10 \text{ in.} \cdot \text{lb} \times 2.67 = 26.7 \text{ in.} \cdot \text{lb} \qquad ◆$$

With the exception of the toothed variety, belts rely on friction to transmit power. As discussed in Section 5.1, friction is proportional to how tightly the parts are pressed together. For belts, this is usually accomplished by keeping the belt in tension. A V belt increases the friction by using a wedging action (Figure 5.39). Just as a wedge produces large forces when splitting a log, the V belt produces large sideways forces that increases its grip on the pulley. Even so, a general rule is that a belt should have at least 120° of contact to give it enough gripping area (Figure 5.40). This requirement prevents really large velocity ratios.

Figure 5.39
Belt-wedging action produces large sideways forces, increasing friction.

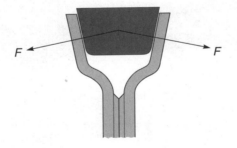

Figure 5.40
Minimum pulley contact is 120°.

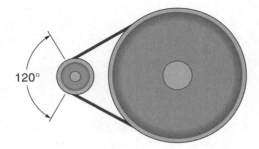

Untoothed belts tend to **creep** over time and cannot be used to maintain position relationships. Creep is *not* the same as slipping; slipping is when the pulley rotates under the belt causing a squealing noise. Creep and slip occur because there is no positive interlocking between belt and pulley, a fact that allows for some special capabilities such as variable-speed transmission and belt clutches.

Figure 5.41 shows the variable-speed transmission. It consists of two specially made pulleys, designed so the sides can go in and out, that can effectively change their diameter while the belt is running. When the sides are close together, the belt is forced to the outside, thus making a larger diameter; when the pulley sides move apart, the belt drops down to a smaller diameter. If both pulleys (A and B) change their effective diameters at the same time but in opposite directions, then the belt stays tight even while the velocity ratio between the pulleys is changing. This is a mechanical engineer's dream: an infinite ratio transmission that can be changed "on the fly." This design is used in industry, for example, where feed speeds need to be exactly matched to other machines.

Figure 5.41

A variable-speed belt drive.

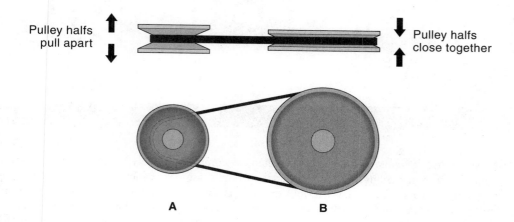

A belt clutch is shown in Figure 5.42. The belt around the two pulleys is intentionally loose so that the driver pulley will slip when the idler wheel is up [Figure 5.42(a)]. When the idler wheel is lowered [Figure 5.42(b)], the belt tightens around both pulleys, and power is transmitted.

Figure 5.42

A belt clutch; (a) disengaged— belt loose; (b) engaged— idler wheel making belt tight.

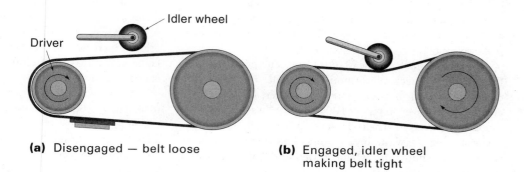

(a) Disengaged — belt loose   **(b)** Engaged, idler wheel making belt tight

Roller Chain

Another way that rotary power is transmitted is with roller chain, the kind used on bicycles. Shown in Figure 5.43, it consists of a series of links connected by pins, with a roller around each pin. The rollers engage teeth on the sprocket, so unlike belts, there is no slippage, and the "gear ratio" is a function of the number of teeth on each sprocket. Chain drives do not rely on friction as belts do, so there is no need to have the chain tight; in fact, there should always be some small slack to ensure there will be no binding. Like belts, all the load is carried by one side of the chain, and both sprockets rotate in the same direction. Properly lubricated roller chain drives can be very efficient.

Figure 5.43
A roller chain.

SUMMARY

Most control systems include some mechanical components, so it is important to have some idea of how these components behave. Friction is the drag force that is always present when parts slide against each other. There are three classifications of friction: static (moving an object from rest), sliding (object is moving), and viscous (sliding over a thick lubricant).

Springs come in many sizes and shapes and find many applications in mechanical systems. All springs obey Hooke's law, which states that the deformation of a spring is proportional to the applied force.

Physical objects have mass, which means they tend to resist being moved. Two basic sets of equations (from physics) relate how an object responds to a force: one for linear systems and one for rotational systems. With the appropriate givens, these equations can be used to calculate the position, velocity, or acceleration of an object.

Energy is usually in one of four forms: chemical (fuels), thermal (heat), mechanical (motion), and electrical. Energy can be converted from one form to another, but the efficiency of the conversion is not 100% and the "waste energy" is usually in the form of heat, as when an electric motor heats up. Managing the waste heat requires knowing something about heat conduction, which is the study of how heat moves through materials.

A real mechanical system is made up of parts, each of which has some flexibility, some inertia, and probably some frictional resistance to movement. The response of a system to a force is described as being either underdamped (too flexible), critically damped (just right), or overdamped (sluggish). Any mechanical structure has a natural resonant frequency at which it will tend to vibrate. Care should be taken to ensure that an external exciting force does not have the same frequency as the mechanical resonant frequency, or wild oscillations and breakage could result.

Gears, available in different sizes and types, are a very efficient way to transport and reconfigure rotational power. The gear ratio is the ratio of the effective gear diameters in a gear pass. The rotational velocity across a gear pass will be increased or decreased by the gear ratio. Also important is that power is conserved across the gear pass, so if the velocity increases (across a gear pass), the torque decreases. Large gear ratios can be built from multiple gear passes, called gear trains.

Two other power-transmitting devices are belts and roller chains. Belts have the advantage of running smooth and quiet and are tolerant to slight misalignments. They also allow for interesting designs such as clutches and variable-speed transmissions. Roller chains are more rigid than belts and can be a very efficient way to transport power.

GLOSSARY

acceleration   The process of increasing velocity; uniform acceleration is when the velocity is increasing at a constant rate.

backlash   In a gear pass, the small amount of free play between gears so that the teeth don't bind on each other.

circular pitch   The distance along the pitch circle of a gear of one tooth and valley.

creep   The property that explains why two pulleys connected with a belt will not stay exactly synchronized.

critically damped   A system that is damped just enough to prevent overshoot.

damping   A drag force on a system from sliding or viscous friction, which acts to slow movement (make it sluggish).

diametral pitch   The ratio of the number of gear teeth per inch of pitch diameter; in practice, the number that describes the tooth size when specifying gears.

efficiency   When describing energy conversions, the percentage of input energy that is converted to useful output energy.

energy   The amount of work it takes to do a job; energy has different units for chemical, thermal, mechanical, and electrical systems.

gear pass   Two gears in mesh.

gear ratio   The ratio of the number of teeth of two gears in mesh. (Also the ratio of pitch diameters.)

gear train   A gear system consisting of more than one gear pass.

harmonic drive   A unique device using a flexible gear that can provide a large gear ratio with virtually no backlash.

Hooke's law   The "spring law" that states the amount a spring deflects is proportional to applied force.

inertia   The property that explains why an object in motion will tend to stay in motion. Inertia is directly related to mass; the more mass an object has, the more energy it takes to get it moving or to stop it.

instrument gears   Smaller gears with pitch in the 48–28 range, found in office machines and smaller industrial machines.

lockup property   The property of high-ratio worm gears that cannot be driven backward.

mass   The amount of material in an object; mass is related to weight in that the more mass it has, the more it will weigh.

moment of inertia   A mechanical property of an object that is based on its shape and mass and the axis of rotation; the larger the moment of inertia, the more torque it takes to spin the object about the designated axis.

momentum   The property of a moving object that tends to keep it moving in the same direction.

natural resonant frequency   In mechanical systems, the frequency at which a part or parts will vibrate. The resonant frequency is a function of the mass and spring constant.

Newton   Unit of force in the SI system (1 N = 0.224 lb)

normal force   In friction calculations, the force pushing the sliding surfaces together.

overdamped   A system that has so much drag (from static or viscous friction) that its response is sluggish.

overshoot   The event when a mechanical (or electrical) output approaches its destination too fast and goes beyond; underdamped systems tend to overshoot.

pinion   The small driver gear in a gear pass.

pitch circle   If gears were solid disks, the pitch circle would be the theoretical circle that meshed gears roll on.

pitch diameter   The diameter of the pitch circle.

power   A property that describes how fast energy is being used; in other words, power is energy per unit time.

sliding friction   The frictional drag force on two dry sliding objects.

spur gear   A type of circular gear with radial teeth (the most common type of gear).

static friction   The friction force that must be overcome to get an object at rest to move; for a particular object, static friction is greater than sliding friction.

torque   A twisting force such as would come from a motor; torque is used in rotational systems just as force is used in linear systems.

underdamped   A system that has relatively little damping so that it responds quickly and tends to overshoot.

viscous friction   A drag force experienced when a lubricant is used between the objects so that the objects do not actually touch; the drag force, which is proportional to velocity, comes from the layers of lubricant slipping over each other.

weight   Technically, a downward force exerted by a mass, caused by gravity.

worm   The spiral-looking gear in a worm gearbox.

worm gear   The circular gear that the worm meshes with in a worm gearbox.

EXERCISES

Section 5.1

1. What is the difference between *static*, *sliding*, and *viscous friction*?

2. A coil spring has a spring constant of 5 lb/in. and a rest length of 3 in. In a machine, the spring is stretched to 4.5 in. and is used to exert a constant force on a lever. How much force does the spring exert?

3. The keys of a certain keyboard depress ¼ in. under 1 oz of force. What is the spring constant?

4. A robot arm must be able to extend 24 in. horizontally in 2 s while carrying a load of 10 lb. How much force is required, assuming the arm is accelerating all the way?

5. Rework Exercise 4 with the more realistic assumption that the arm accelerates for the first 12 in., and then decelerates the last 12 in. The total time should still be 2 s.

6. A robot arm must be able to extend 50 cm horizontally in 2 s while carrying a load of 5 kg. How much force is required, assuming the arm is accelerating all the way?

7. Rework Exercise 6 with the more realistic assumption that the arm accelerates for the first 25 cm and then decelerates the last 25 cm. The total time should still be 2 s.

8. A linear actuator can provide a force of 2 lb. How long will it take to accelerate a 6-lb load to a speed of 5 ft/s? (Assume no friction.)

9. Two rotating parts, one heavy and one light, need to be spun up to the same speed in the same period of time. If they end up at the same speed, why does the heavy part require a bigger motor?

10. A torque of 10 in. · lb is applied to a rotating part. The part has a moment of inertia of 0.5 lb · s² · in. How much time will it take to rotate the part 180°?

11. A wheel is attached to the end of an electric motor shaft. The wheel has a moment of inertia of 1.5 lb · s² · in. If the motor produces a constant torque of 2 in. · lb, how long will it take the motor to spin the wheel up to 100 rpm?

12. A torque of 18 N · m is applied to a rotating part. The part has a moment of inertia of 0.078 kg · m². How much time will it take to rotate the part 180°?

13. A wheel is attached to the end of an electric motor shaft. The wheel has a moment of inertia of 0.23 kg·m². If the motor produces a constant torque of 0.035 N · m, how long will it take the motor to spin the wheel up to 100 rpm?

14. A new disk drive must rotate at 300 rpm. How much torque must the motor supply to go from 0 to 300 rpm in 3 s? The moment of inertia of the drive is 0.04 lb · s² · in.

Section 5.2

15. An electric heater is rated at 220 Vac at 15 A. How many Btu/h of heat does it put out?

16. A small gasoline engine can put out 12,000 ft · lb/min of mechanical power. The engine is used to drive an electric generator that is 80% efficient. Find the output of the generator in watts.

17. A 120-V electric motor is needed that can do the equivalent of lifting a 100-lb weight 10 ft in 1 min. Assuming the motor is 85% efficient, find the motor current.

18. A small gasoline engine can put out 18,000 N · m/min of mechanical power. The engine is used to drive an electric generator that is 80% efficient. Find the output of the generator in watts.

19. A 120-V electric motor is needed that can do the equivalent of lifting a 50-kg mass 3 m in 1 min. Assuming the motor is 85% efficient, find the motor current.

20. An electronic package is to be mounted 6 in. above a diesel engine on a steel bracket. The bracket is ¼ in. thick and 2 in. wide. Assume the engine block is 175°F and the package is at 100°F. How much heat (Btu/h) will come through the bracket (neglect heat dissipated by the bracket itself)?

Section 5.3

21. The shock absorbers on a car dampen out bouncing when the car goes over a bump. Use this example to explain the concept of underdamped, critically damped, and overdamped systems. Specifically, how would the car react to a bump in each case?

22. What may happen if the vibrations in a system—say, from a rotating motor—are at the same frequency as the resonant frequency of one of the system parts?

23. A 2-lb control box is connected by a bracket to a machine. If you push on the control box with 20 lb of

force, the bracket bends a little, and the box moves about 0.25 in. Find the resonant frequency of the control box.

Section 5.4

24. What is *diametral pitch*, and why is it an important parameter in a gear mesh?

25. Find the gear ratios of the gears shown in Figure 5.44.

26. For each gear pass shown in Figure 5.44, how many degrees must the small gear turn for the big gear to turn 90°?

27. For each gear pass shown in Figure 5.44, how many degrees must the big gear turn for the small gear to make two complete revolutions?

28. For each gear pass shown in Figure 5.44, the pinion is rotating at 50 rpm. Find the rpm of the big gear.

29. Find the overall gear ratio of the gear train shown in Figure 5.45. If the motor is rotating at 500 rpm, what is the rotational velocity of the output gear?

30. The motor in Figure 5.46 is turning at 600 rpm with a torque of 2 in. · lb. What is the velocity and torque available at the output shaft?

31. A roller needs to turn at 60 rpm and requires 10 in. · lb of torque. A motor is available with a maximum velocity of 500 rpm at a torque of 1.3 in. · lb. Can the motor be used?

32. How is the moment of inertia of the load affected by a gear pass?

33. Explain the term *backlash*. What is the effect on backlash from long gear trains?

34. Give an advantage and a disadvantage of worm gear drives.

35. What does the term *lockup* refer to in worm gear drives?

36. What are the advantages of the harmonic drive?

37. What is the gear ratio of a harmonic drive with 200 teeth in the outer cup and 196 teeth on the flex spline?

Section 5.5

38. What are some advantages of transmitting power by belts?

39. A "squirrel-cage" fan is driven by belt from a motor that turns at 1750 rpm (Figure 5.47). The motor pulley is 3 in. in diameter, and the fan pulley is 12 in. At what speed does the fan rotate?

40. The pulleys in a variable-speed transmission expand from 4 to 8 in. in diameter. Find the range of the velocity ratio.

Figure 5.45

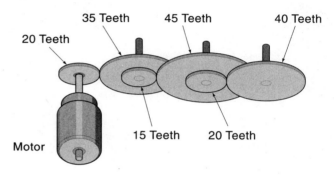

Figure 5.46

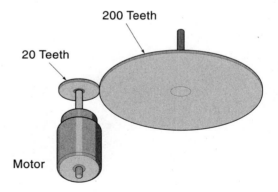

Figure 5.44

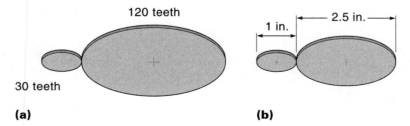

(a)   (b)

41. A "squirrel-cage" fan is driven by belt from a motor that turns at 1750 rpm (Figure 5.47). The motor pulley is 8 cm in diameter, and the fan pulley is 30 cm. At what speed does the fan rotate?

42. The pulleys in a variable-speed transmission expand from 6 to 20 cm in diameter. Find the range of the velocity ratio.

43. What are some similarities and differences between belts and roller chains as used in power transmission?

Figure 5.47

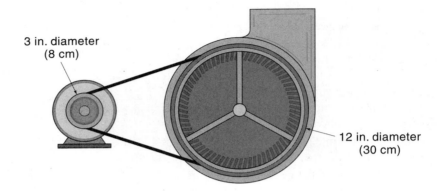

3 in. diameter
(8 cm)

12 in. diameter
(30 cm)

6 Sensors

OBJECTIVES

After studying this chapter, you should be familiar with the characteristics and operation of such sensors as:

❑ Position sensors including potentiometers, optical rotary encoders, and linear variable differential transformers.
❑ Velocity sensors including optical and direct current tachometers.
❑ Proximity sensors including limit switches, optical proximity switches, and Hall-effect switches.
❑ Load sensors including bonded-wire strain gauges, semiconductor force strain gauges, and low-force sensors.

❑ Pressure sensors including Bourdon tubes, bellows, and semiconductor pressure sensors.
❑ Temperature sensors including bimetallic temperature sensors, thermocouples, resistance temperature detectors, thermistors, and IC temperature sensors.
❑ Flow sensors including orifice plates, venturis, pitot tubes, turbines, and magnetic flowmeters.
❑ Liquid-level sensors including discrete and continuous types.

INTRODUCTION

The devices that inform the control system about what is actually occurring are called **sensors** (also known as transducers). As an example, the human body has an amazing sensor system that continually presents our brain with a reasonably complete picture of the environment—whether we need it all or not. For a control system, the designer must ascertain exactly what parameters need to be monitored—for example, position, temperature, and pressure—and then specify the sensors and data interface circuitry to do the job. Many times a choice is possible. For example, we might measure fluid flow in a pipe with a flowmeter, or we could measure the flow indirectly by seeing how long it takes for the fluid to fill a known-sized container. The choice would be dictated by system requirements, cost, and reliability.

Most sensors work by converting some physical parameter such as temperature or position into an electrical signal. This is why sensors are also called **transducers**, which are devices that convert energy from one form to another.

6.1   POSITION SENSORS

Position sensors report the physical position of an object with respect to a reference point. The information can be an angle, as in how many degrees a radar dish has turned, or linear, as in how many inches a robot arm has extended.

Potentiometers

A **potentiometer** (pot) can be used to convert rotary or linear displacement to a voltage. Actually, the pot itself gives resistance, but as we will see, this resistance value can easily be converted to a voltage. Pots used for position sensors are the same in principle as a standard "volume-control" type but are made to more exact standards.

Figure 6.1(a) illustrates how the pot works. A resistive material, such as conductive plastic, is formed in the shape of a circle (terminating at contacts A and C). This material has a very uniform resistivity so that the ohms-per-inch value along its length is a constant. Connected to the shaft is the **slider**, or **wiper**, which slides along the resistor and taps off a value [contact B in Figure 6.1(a)]. Figure 6.1(b) shows the circuit symbol. The pot just described is the single-turn type, which actually has only about 350° of useful range. A single-turn pot may have "stops" at each end of its travel. Obviously, such a pot could only be used where the rotation never exceeds 350°. A single-turn pot without stops has a small "dead zone" when the wiper crosses the end of the resistor. Multiturn pots are available with a wiper that moves in a helix motion, allowing for up to 25 or more revolutions of the shaft from stop to stop. Figure 6.1(c) illustrates a linear-motion potentiometer. In this case, the wiper can move back-and-forth in a straight line. Linear-motion pots are useful for sensing the position of objects that move in a linear fashion.

Figure 6.1
Potentiometer.

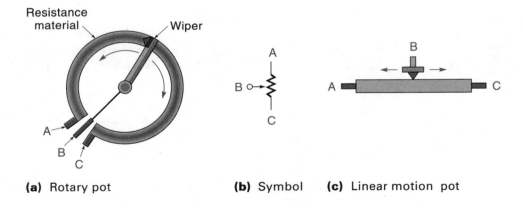

(a) Rotary pot   **(b)** Symbol   **(c)** Linear motion pot

Figure 6.2(a) shows a pot that detects the angular position of a robot arm. In this case, the pot body is held stationary, and the pot shaft is connected directly to the motor shaft. Ten volts is maintained across the (outside) terminals of the pot. Look at Figure 6.2(b) and imagine how the voltage is changing evenly from 0 to 10 Vdc along the resistive element. The wiper merely taps off the voltage drop between its contact point and ground. For example, if the wiper is at the bottom, the output is 0 V corresponding to 0°. When the wiper is at the top, the output is 10 V corresponding to 350°; in the exact middle, a 5-V output indicates 175° (350°/2 = 175°). Example 6.1 demonstrates how to calculate the pot voltage for any particular angle.

Figure 6.2

Potentiometer as a position sensor.

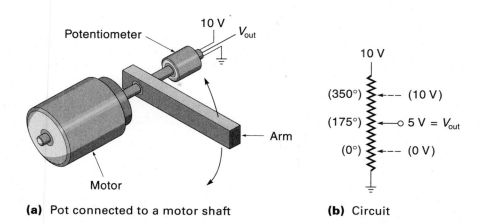

(a) Pot connected to a motor shaft   **(b)** Circuit

◆ **EXAMPLE 6.1**

A pot is supplied with 10 V and is set at 82° [similar to Figure 6.2(b)]. The range of this single-turn pot is 350°. Calculate the output voltage.

Solution   If the pot is supplied 10 V, then the maximum angle of 350° will produce a 10-V output. Using these values, we can set up a ratio of output to input and use that ratio to calculate the output for any input [this ratio is an example of a simple transfer function (TF) discussed in Chapter 1]:

$$\text{TF}_{\text{pot}} = \frac{\text{output}}{\text{input}} = \frac{10 \text{ Vdc}}{350°}$$

To find the output voltage for a particular angle, multiply the angle with the transfer function (and as always, be sure the units work out correctly—in this case, *degrees* cancel, leaving *volts* as the unit):

$$\text{Pot voltage (at 82°)} = \frac{10 \text{ Vdc}}{350°} \times 82° = 2.34 \text{ Vdc}$$

◆

The potentiometer circuit being discussed here is actually a voltage divider, and to work properly the same current must flow through the entire pot resistance. A **loading error** occurs when the pot wiper is connected to a circuit with an input resistance that is *not* considerably higher than the pot's resistance. When this happens, current flows out through the wiper arm, robbing current from the lower portion of the resistor and causing the reading to be low (see Example 6.2). To solve this problem, a high-impedance buffer circuit such as the voltage follower (discussed in Chapter 3) can be inserted between the pot and the circuit it must drive. Loading error is the difference between the unloaded and loaded output as given in Equation 6.1a:

$$\text{Loading error} = V_{NL} - V_{L} \qquad (6.1a)$$

where

V_{NL} = output voltage with no load
V_{L} = output voltage with load applied

◆ **EXAMPLE 6.2**

A 10-kΩ pot is used as a position sensor (Figure 6.3). Assume that the wiper is in the middle of its range. Find the loading error when

    a. The interface circuit presents an infinite resistance.

    b. The interface circuit presents a resistance of 100 kΩ.

Solution

    a. Figure 6.3(a) shows the ideal situation where the interface circuit resistance is so high that there is virtually no current in the pot wiper wire. The pot will behave like two 5-kΩ resistors in series, and we can use the voltage-divider rule to calculate the pot voltage:

$$V_{pot} = 10\text{ V} \times \frac{5\text{ k}\Omega}{5\text{ k}\Omega + 5\text{ k}\Omega} = 5\text{ V}$$

As we would expect, the pot voltage is exactly half of the 10-V supply voltage. There is no loading error in this case.

    b. Now consider the case where the input resistance of the interface circuit is 100 kΩ, as shown in Figure 6.3(b). We will use the voltage-divider rule again to compute the pot voltage, but this time the lower resistance is the parallel combination of 5 kΩ and 100 kΩ (as shown in [Figure 6.3(c)]:

$$5\text{ k}\Omega//100\text{ k}\Omega = \frac{1}{\dfrac{1}{5\text{ k}\Omega} + \dfrac{1}{100\text{ k}\Omega}} = 4.76\text{ k}\Omega$$

which is the equivalent lower resistance. Using this value in the voltage divider, we now recalculate the pot voltage:

$$V_{pot} = 10\text{ V} \times \frac{4.76\text{ k}\Omega}{5\text{ k}\Omega + 4.76\text{ k}\Omega} = 4.88\text{ V}$$

Thus, the actual pot voltage is only 4.88 V when it should be 5 V. The loading error is

$$5\text{V} - 4.88\text{V} = .12\text{V}$$

The maximum loading error occurs when the pot is ⅔ of full range. If you were to rework this problem for a pot voltage of 2.5 V, you would find the error is only 0.045 V. Therefore, the effect of loading errors is not linear.

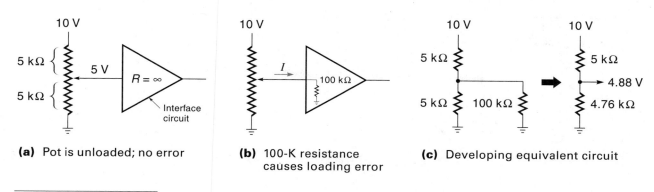

(a) Pot is unloaded; no error

(b) 100-K resistance causes loading error

(c) Developing equivalent circuit

Figure 6.3   Loading errors.

In many applications, the total rotary movement to be measured is less than a full revolution. Consider the arm in Figure 6.4 that moves through an angle of only 90°. Using as much of the pot's range as possible in order to get a lower average error rate is advantageous, so we might use a 3 : 1 gear ratio that causes the pot to turn through 270°. (In Figure 6.4, the small pot gear must make three revolutions for each revolution of the motor gear.) The controller will be programmed to understand that 3° of the pot corresponds to only 1° of the actual arm.

Figure 6.4

When motor shaft is restricted to 90°, the 3 : 1 gear pass turns the pot through 270°.

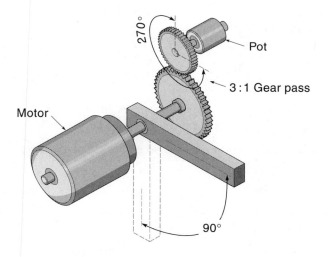

As in all physical systems, we must be aware of certain errors that creep in. In this case, carbon pots cannot be made perfectly linear, so we define **linearity error** as the difference between what the angle really is and what the pot reports it to be. The graph of Figure 6.5 shows the ideal versus actual resistance (R) for a pot position sensor. The error is the difference in resistance between these two lines. Notice that the error is not the same everywhere, but the maximum error is designated as ΔR. Linearity error is defined in percentage, as shown below, and ranges between 1.0 and 0.1% (but higher precision costs more, of course):

Figure 6.5
Linearity error of a pot: ideal vs. actual.

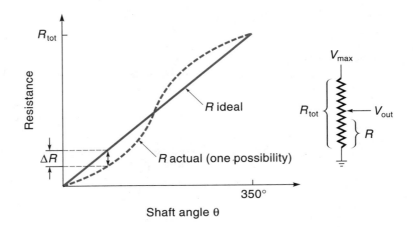

$$\text{Linearity error} = \frac{\Delta R \times 100}{R_{tot}} \qquad (6.1b)$$

where

ΔR = maximum resistance error
R_{tot} = total pot resistance

When the potentiometer is used as a position sensor, the output voltage is directly proportional to the shaft angular position, so linearity error can also be expressed in terms of angle:

$$\text{Linearity error} = \frac{\Delta\theta \times 100}{\theta_{tot}} \qquad (6.1c)$$

where

$\Delta\theta$ = maximum angle error (in degrees)
θ_{tot} = total range of the pot (in degrees)

(*Note*: Loading effects will also contribute to the error.)

♦ **EXAMPLE 6.3**

A single-turn pot (350°) has a linearity error of 0.1% and is connected to a 5 Vdc source. Calculate the maximum angle error that could be expected from this system.

Solution

To calculate the maximum possible angle error, rearrange Equation 6.1c and solve:

$$\Delta\theta = \frac{\text{linearity} \times \theta_{tot}}{100} = \frac{0.1 \times 350°}{100} = 0.35°$$

If this pot were in a control system, the controller would only know the position to within 0.35° or about one-third of a degree. ♦

Linearity error determines the accuracy of a sensor. A related but different measurement concept is resolution. **Resolution** refers to the smallest increment of data that can be detected and/or reported. In digital systems, the resolution usually refers to the value of the **least significant bit** (LSB) because that is the smallest change that can be reported. For example, a 2-bit number has four possible states (00, 01, 10, 11). If we used this 2-bit number to quantify the gas level in your tank, we could specify the amount of gas only to the nearest fourth, that is, one-quarter, one-half, three-quarters, and full. Thus, the resolution would be one-fourth of a tank of gas. The "accuracy" of a digital system should be ± ½ LSB (although it could be worse, in which case the LSB wouldn't mean very much). In the gas-gauge example, ± ½ LSB accuracy corresponds to ± ⅛ tank, so if the gauge reads half full, you would know that the actual level was between three-eights and five-eighths full.

For an analog device such as a potentiometer, resolution refers to the smallest change that can be measured. It is usually expressed in percentage:

$$\% \text{ resolution} = \frac{\text{smallest change in resistance} \times 100}{\text{total resistance}}$$

$$= \frac{\Delta R}{R_{\text{tot}}} \times 100 \tag{6.2}$$

Let's examine resolution in conjunction with the **wire-wound potentiometer**. A wire-wound pot uses a coil of resistance wire for the resistive element (see Figure 6.6). The wiper bumps along on the top of the coil. Clearly, the resolution in this case is the resistance of one loop of the coil. This concept is illustrated in Example 6.4.

Figure 6.6
Resolution in a wire-wound pot.

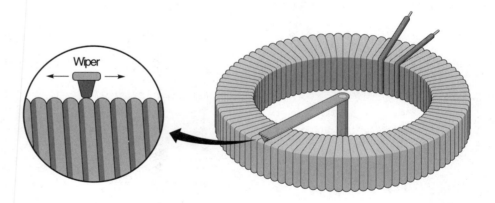

♦ EXAMPLE 6.4

The resistive element of a wire-wound pot is made from 10 in. of 100 Ω/in. resistance wire and is wound as a coil of 200 loops. The range of the pot is 350°. What is the resolution of this pot?

Solution

$$R_{\text{tot}} = 10 \text{ in.} \times \frac{100 \ \Omega}{\text{in.}} = 1 \text{ k}\Omega$$

The pot coil has 200 turns of wire. Therefore, the smallest increment of resistance corresponds to one loop of the coil. The resistance of one loop of wire is

$$\text{Resistance/Loop} = \frac{1 \text{ k}\Omega}{200 \text{ loops}} = 5 \text{ }\Omega/\text{loop}$$

Thus, resolution is

$$\frac{\Delta R \times 100}{R_{\text{tot}}} = \frac{5 \text{ }\Omega \times 100}{1 \text{ k}\Omega} = 0.5\%$$

If this pot were to be used as a position sensor, it would be useful to know what the resolution is in degrees. The smallest measurable change corresponds to one loop of the resistance coil, and this pot divides 350° into 200 parts; therefore, the resolution in degrees would be 350°/200 loops = 1.75°.   ♦

The output of a position sensor should be a continuous DC voltage, but the slider action of pots can sometimes cause voltage transients. This is particularly true for wire-wound pots because the slider may momentarily break contact as it bumps from wire to wire. If this is a problem, it can usually be resolved with a low-pass filter, which is simply a capacitor to ground (Figure 6.7). The capacitor stays charged up to the average pot voltage and resists momently voltage changes.

Figure 6.7   Pot sensor position system for robot arm (Example 6.5).

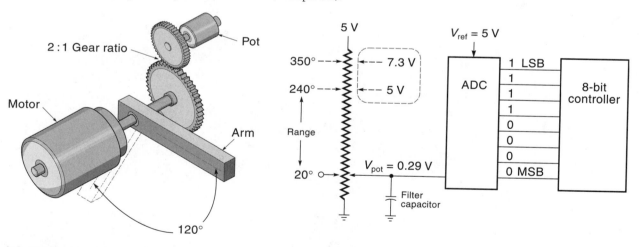

(a) Hardware setup

(b) Sensor circuit

Example 6.5 uses a potentiometer as the position sensor for a digital feedback control system. The main consideration here is resolution from the analog to the digital.

♦ **EXAMPLE 6.5**

The robot arm illustrated in Figure 6.7 rotates 120° stop-to-stop and uses a pot as the position sensor. The controller is an 8-bit digital system and needs to know the actual position of the arm to within 0.5°. Determine if the setup shown in Figure 6.7 will do the job.

Solution

To have 0.5° resolution means that the entire 120° will be divided into 240 increments, each increment being 0.5°. An 8-bit number has 255 levels (from 00000000 to 11111111), so it has more than enough to do the job. (That's good!)

The pot is supplied with 5 V. Therefore, the output of the pot would be 5 V for the maximum pot angle of 350° (if it could rotate that far). Notice that the reference voltage of the ADC (analog-to-digital converter) is also set at 5 V; thus, if the pot voltage (V_{pot}) is 5 V, the digital output would be 255 (11111111_{bin}). Table 6.1 summarizes this (see last three columns).

A single-turn pot has a range of 350°, but the robot arm only rotates 120°, hence the 2 : 1 gear ratio between the pot and the arm. With this arrangement, the pot rotates 240° when the arm rotates 120°. By doubling the operating range of the pot, the linearity and resolution errors (from the pot) are reduced by half.

Consider the case when the robot arm is at 10° (second line in Table 6.1). Because of the 2 : 1 gear ratio, the pot would be at 20°. To calculate the pot voltage at 20°, we use the transfer function of the pot (5 V/350°):

$$V_{pot} = \frac{5\,\text{V}}{350°} \times 20° = 0.29\,\text{V}$$

This 0.29 V is then converted into binary with the ADC (see Figure 6.7). To calculate the binary output, first form the ADC transfer function:

$$\frac{\text{output}}{\text{input}} = \frac{255\,\text{states}}{5\,\text{V}}$$

Now calculate the ADC binary output using the 0.29 V (pot voltage) as the input:

$$\frac{255\,\text{states}}{5\,\text{V}} \times 0.29\,\text{V} = 14.8 \approx 15_{\text{states}} = 00001111_{bin}$$

We now turn our attention to the system resolution, which is the smallest measurable change. In a digital system, this usually corresponds to the value assigned

TABLE 6.1   **System Values for Various Angles of Robot Arm**

Arm angle (degrees)	Pot angle (degrees)	Pot voltage (V)	ADC output (binary states)
0	0	0	00000000
10	20	0.29	00001111
120	240	3.43	10110000
175	350	5	11111111

(Actual range of pot

to the LSB (you can't change half a bit!). We can find the resolution by calculating the pot angle corresponding to a single binary state. This is done by multiplying the transfer functions of each of the system elements together to get an overall system transfer function (you may notice that we actually used the inverse of the transfer functions in order to get the desired units):

$$\underbrace{\frac{1°_{\text{arm}}}{2°_{\text{pot}}}}_{\text{Gears}} \times \underbrace{\frac{350°_{\text{pot}}}{5\text{ V}}}_{\text{Pot}} \times \underbrace{\frac{5\text{ V}}{255\text{ states}}}_{\text{ADC}} = \frac{0.686°_{\text{arm}}}{\text{state}}$$

This result tells us that the LSB of the ADC is 0.686°, which is too big! We need the LSB to be 0.5°. As it stands, this design does not meet the specification. Can it be fixed? Yes, looking back, you can see that at 350° the pot sends 5 V to the ADC, but this will *never* happen because the pot is constrained to 240°. To get maximum resolution from the ADC, the pot should send 5 V to the ADC when the pot is 240°. This will require raising the pot supply voltage to 7.3 V [by ratio, 5 V × (350°/240°) = 7.3 V].

The revised voltages are shown in the dashed circle in Figure 6.7. Now the resolution is recalculated to be

$$\frac{1°_{\text{arm}}}{2°_{\text{pot}}} \times \frac{350°}{7.3\text{ V}} \times \frac{5\text{ V}}{255\text{ states}} = 0.470°/\text{state}$$

This result is within the 0.5° specification for resolution.   ♦

Optical Rotary Encoders

An **optical rotary encoder** produces angular position data directly in digital form, eliminating any need for the ADC converter. The concept is illustrated in Figure 6.8, which shows a slotted disk attached to a shaft. A light source and photocell arrangement are mounted so that the slots cut the light beam as the disk rotates. The angle of the shaft is deduced from the output of the photocell. There are two types of optical rotary encoders: the absolute encoder and the incremental encoder.

Figure 6.8

An optical rotary encoder.

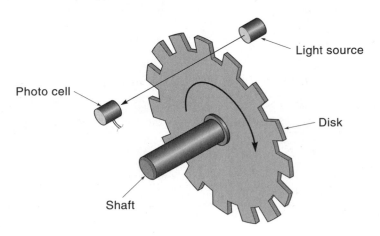

Absolute Optical Encoders

Absolute optical encoders use a glass disk marked off with a pattern of concentric tracks (Figure 6.9). A separate light beam is sent through each track to individual photo sensors. Each photo sensor contributes 1 bit to the output digital word. The encoder in Figure 6.9 outputs a 4-bit word with the LSB coming from the outer track. The disk is divided into 16 sectors, so the resolution in this case is $360°/16 = 22.5°$. For better resolution, more tracks would be required. For example, eight tracks (providing 256 states) yield $360°/256 = 1.4°$/state, and ten tracks (providing 1024 states) yield $360°/1024 = 0.35°$/state.

An advantage of this type of encoder is that the output is in straightforward digital form and, like a pot, always gives the absolute position. This is in contrast to the incremental encoder that, as will be shown, provides only a relative position. A disadvantage of the absolute encoder is that it is relatively expensive because it requires that many photocells be mounted and aligned very precisely.

If the absolute optical encoder is not properly aligned, it may occasionally report completely erroneous data. Figure 6.10 illustrates this situation, and it occurs when more than 1 bit changes at a time, say, from sector 7 (0111) to 8 (1000). In the figure,

Figure 6.9

An absolute optical encoder using straight binary code.

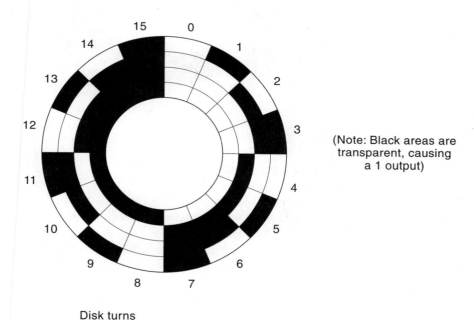

(Note: Black areas are transparent, causing a 1 output)

Figure 6.10

An absolute optical encoder showing how an out-of-alignment photocell can cause an erroneous state. (*Note:* Dark areas produce a 1, and light areas produce a 0.)

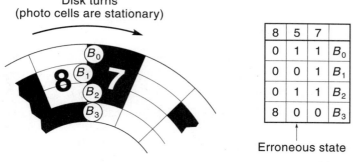

Disk turns
(photo cells are stationary)

8	5	7	
0	1	1	B_0
0	0	1	B_1
0	1	1	B_2
8	0	0	B_3

↑
Erroneous state

the photo sensors are not exactly in a straight line. In this case, sensor B_1 is out of alignment and switches from a 1 to a 0 before the others. This causes a momentary erroneous output of 5 (0101). If the computer requests data during this "transition" time, it would get the wrong answer. One solution is to use the **Grey code** on the disk instead of the straight binary code (Figure 6.11). With the Grey code, only 1 bit changes between any two sectors. If the photocells are out of line, the worst that could happen is that the output would switch early or late. Put another way, the error can never be more than the value of 1 LSB when using the Grey code.

Figure 6.11

An absolute optical encoder using a Grey code.

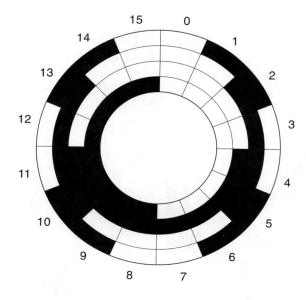

Incremental Optical Encoders

The **incremental optical encoder** (Figure 6.12) has only one track of equally spaced slots. Position is determined by counting the number of slots that pass by a photo sensor, where each slot represents a known angle. This system requires an initial reference point, which may come from a second sensor on an inner track or simply from a mechanical stop or limit switch. In many applications, the shaft being monitored will be cycling back-and-forth, stopping at various angles. To keep track of the position, the controller must know which direction the disk is turning as well as the number of slots passed. Example 6.6 illustrates this.

♦ **EXAMPLE 6.6**

An incremental encoder has 360 slots. Starting from the reference point, the photo sensor counts 100 slots clockwise (CW), 30 slots counterclockwise (CCW), then 45 slots CW. What is the current position?

Solution   If the disk has 360 slots, then each slot represents 1° of rotation. Starting at the reference point, we first rotated 100° CW, then reversed 30° to 70°, and finally reversed again for 45°, bringing us finally to 115° (CW) from the reference point. ♦

Figure 6.12

An incremental optical encoder.

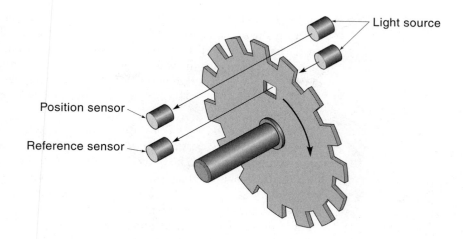

A single photo sensor cannot convey which direction the disk is rotating; however, a clever system using two sensors can. As Figure 6.13(a) illustrates, the two sensors, V_1 and V_2, are located slightly apart from each other on the same track. For this example, V_1 is initially off (well, almost—you can see it is half-covered up), and V_2 is on. Now imagine that the disk starts to rotate CCW. The first thing that happens is that V_1 comes completely on (while V_2 remains on). After more rotation, V_2 goes off, and slightly later V_1 goes off again. Figure 6.13(b) shows the waveform for V_1 and V_2. Now consider what

Figure 6.13

An incremental optical encoder.

(a) Two photo-sensor arrangement to determine direction

(b) CCW—Photo cell waveforms for counter-clockwise

(c) CW—Photo cell waveforms for clockwise

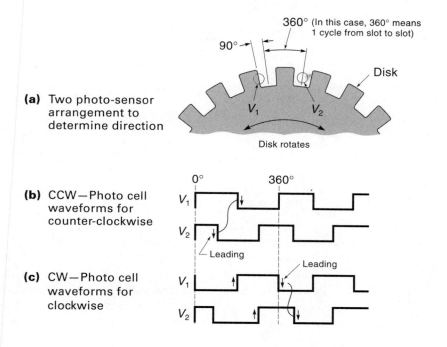

happens when the disk is rotated in the CW direction [starting again from the position shown in Figure 6.13(a)]. This time V_1 goes off immediately, and V_2 stays on for half a slot and then goes off. Later V_1 comes on, followed by V_2 coming on. Figure 6.13(c) shows the waveforms generated by V_1 and V_2. Compare the two sets of waveforms—notice that *in the CCW case* V_2 *leads* V_1 *by 90°, whereas for the CW case* V_1 *is leading* V_2 *by 90°.* This difference in phase determines which direction the disk is turning.

Decoding V_1 and V_2   The hardware of the incremental encoder is simpler than for the absolute type. The price paid for that simplicity is that we do not get direct binary position information from V_1 and V_2. Instead, a decoder circuit must be employed to convert the signals from the photo sensors into a binary word. Actually, the circuit has two parts: The first part extracts direction information, and the second part is an up–down counter, which maintains the slot count. The block diagram of Figure 6.14 shows this. Referring to the diagram, we see that V_1 and V_2 are converted into two new signals denoted by "count-down" and "count-up". The count-down signal gives one pulse for every slot passed when the disk is going counterclockwise. The count-up signal gives one pulse for each slot when the disk is rotating clockwise. These signals are then fed to an up–down counter such as the TTL 74193. This counter starts out at 0 (it is usually reset by the reference sensor) and then proceeds to maintain the position by keeping track of the CCW and CW counts. Referring again to Example 6.6, the counter would start at 0, count up to 100, count down 30 pulses to 70, and then count up 45 pulses to 115. Thus, the accumulated total on the counter always represents the current absolute position.

Figure 6.14

Block diagram of an incremental encoder system.

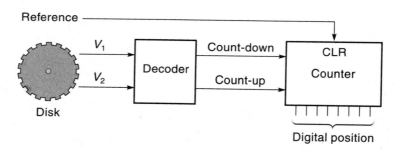

The simplest way to perform the decoding is with a single D-type flip-flop and two AND gates (Figure 6.15). To understand how this circuit functions, we need to examine the waveforms of V_1 and V_2 (Figure 6.16). In the CCW case, every time V_2 goes low, V_1 is high; in the CW case, when V_2 goes low, V_1 is low. This fact is used to separate CCW and CW rotation. V_2 is connected to the negative-going clock of the flip-flop, and V_1 is connected to the D input. Every time V_2 goes low, V_1 is latched and appears at the output. Thus, as long as the disk is rotating CCW, the output will be high; and as long as it rotates CW, the output will be low. These direction signals can be gated with V_2 to produce the required counter inputs count-up and count-down. The count-up signal pulses once per slot when the disk is turning clockwise, and the count-down signal pulses when the disk is turning counterclockwise.

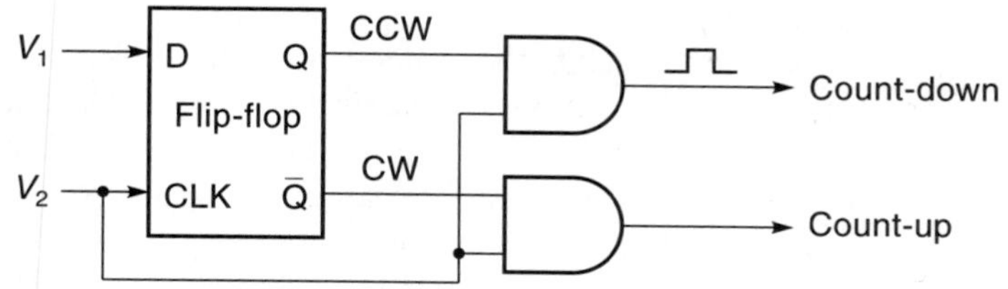

Figure 6.15

A decoder for an incremental optical encoder.

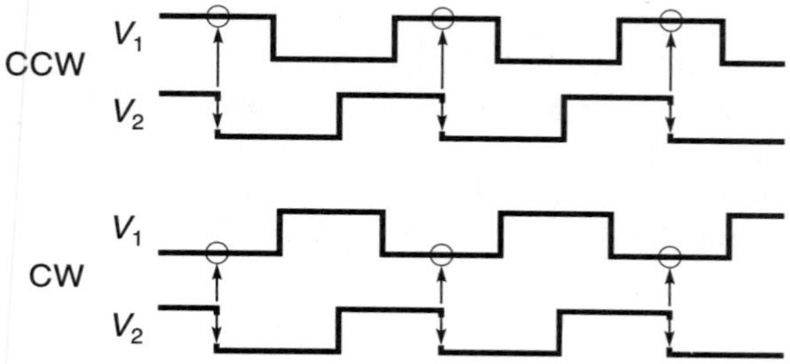

Figure 6.16

Decoding direction from V_1 and V_2.

♦ **EXAMPLE 6.7**

A position-sensor system (Figure 6.14) uses a 250-slot disk. The current value of the counter is 00100110. What is the angle of the shaft being measured?

Solution

For a 250-slot disk, each slot represents $360°/250 = 1.44°$, and a count of $00100110 = 38$ decimal, so the position is $38 \times 1.44° = 54.72°$. ♦

The decoding described so far is the straightforward low-resolution approach. Getting a resolution four times better with more sophisticated decoding is possible because the signals V_1 and V_2 cycle through four distinct states each time a slot passes the sensors. These states can be seen in Figure 6.17. If we were to decode each of these states in Example 6.7, then we would know the angle to the nearest $0.36°$ ($1.44°/4 = 0.36°$) instead of $1.44°$.

Figure 6.17

An incremental encoder showing four unique states for each slot cycle.

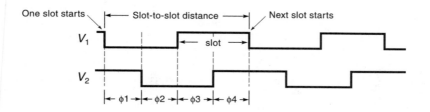

Interfacing the Incremental Encoder to a Computer   There is a special problem when attempting to pass data to a computer from a standard ripple-type digital counter.* The counter is counting real-world events and so is not synchronized with the computer. If the computer requests position data while the counter is changing, it may very well get meaningless data. Because this is a remote possibility, the resulting errors are infrequent and in many applications can be ignored. But other situations require that data *always* be accurate.

One approach to the problem might be to disable or "freeze" the counter during the time when the computer is receiving data. But if a count-pulse occurs while the counter is frozen, it will be lost. The solution is to put a *latch* (a temporary holding register) between the counter and the computer (Figure 6.18). With this setup, the counter is never disabled and always holds the correct count. The latch is connected so that it ordinarily contains the same value as the counter. During those brief times when the counter is counting, the latch is inhibited from changing. With this system, a count is never permanently lost. The worst situation would be if a count came in while a computer exchange was in progress; in this case, the new count would not be reported with the current exchange because the latch is frozen. As soon as the counter finished updating, however, the latch would be updated, and the count would be reported with the next computer exchange.

Figure 6.18

An incremental encoder interface circuit showing how the latch is inhibited from changing when the counter is updated.

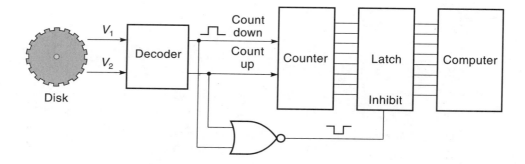

♦ **EXAMPLE 6.8**

The angular position of a shaft must be known to a resolution of 0.5°. A system that uses a 720-slot encoder (Figure 6.19) is proposed. The controller uses a 8051 microcontroller which has 8 bit ports. Will this design meet the specifications?

Solution   For 0.5° resolution, the encoder must have a slot every 0.5° as a minimum. First, calculate the number of slots required:

---

*The common ripple counter takes a finite time to settle out because a new count may cause a "carry" to ripple up through all bits.

$$\frac{360°}{0.5°/\text{slot}} = 720 \text{ slots}$$

The 720-slot encoder will work just fine. Being a digital system, the resolution is determined by the LSB, which in this case should correspond to 1 slot on the disk (0.5°). The binary output should have a range of 0–719 (for 720 states), so the circuit must have the capacity to handle 10-bit data:

$$719 \text{ (decimal)} = 1011001111 \text{ (binary)}$$

which is 10 bits. Because the controller is an 8-bit microcontroller, it will require two ports to input the entire 10 bits. As shown in Figure 6.19, the counter consists of three 74193 4-bit up–down counters. The outputs of the counter are constantly updating the 10-bit latch made from two 74373s. The outputs of the latch are connected to ports 1 and 3 of the 8051.

Figure 6.19

A circuit diagram (Example 6.8).

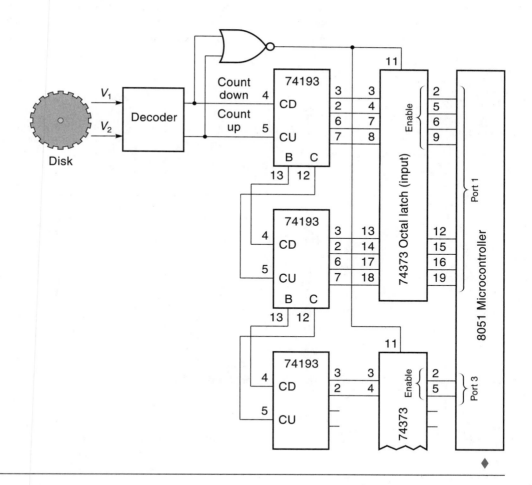

Linear Variable Differential Transformers

The **linear variable differential transformer** (LVDT) is a high-resolution position sensor that outputs an AC voltage with a magnitude proportional to linear position. It has a relatively short range of about 2 in., but it has the advantage of no sliding contacts. Figure 6.20(a) illustrates that the unit consists of three windings and a movable iron core. The center winding, or *primary*, is connected to an AC reference voltage. The outer two windings, called *secondaries*, are wired to be out of phase with each other and are connected in series. If the iron core is exactly in the center, the voltages induced on the secondaries by the primary will be equal and opposite, giving a net output (V_{net}) of 0 V [as shown in Figure 6.20(b)]. Consider what happens when the core is moved a little to the right. Now there is more coupling to secondary 2 so its voltage is higher, while secondary 1 is lower. Figure 6.20(c) illustrates the waveforms of this situation. The algebraic sum of the two secondaries is in phase with secondary 2, and the magnitude is proportional to the distance the core is off center. If the core is moved a little left of center, then secondary 1 has the greater voltage, producing a net output that is in phase with secondary 1 [Figure 6.20(d)]. In fact, the only way we can tell from the output which direction the core moved is by the phase. Summarizing, the output of the LVDT is an AC voltage with a magnitude and phase angle. The magnitude represents the distance that the core is off center, and the phase angle represents the direction of the core (left or right.)

Figure 6.20

A linear variable differential transformer (LVDT).

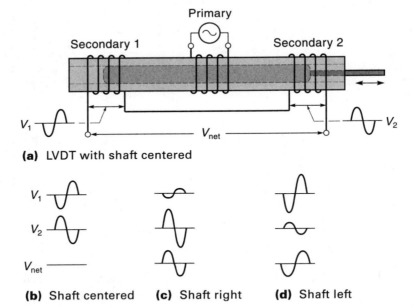

(a) LVDT with shaft centered

(b) Shaft centered   **(c)** Shaft right   **(d)** Shaft left

Figure 6.21 illustrates an LVDT with its support electronics. An oscillator provides the AC reference voltage to the primary—typically, 50–10 KHz at 10 V or less. The output of the LVDT goes first to a phase-sensitive rectifier. This circuit compares the phase of LVDT output with the reference voltage. If they are in phase, the rectifier outputs only the positive part of the signal. If they are out of phase, the rectifier outputs only the negative parts. Next, a low-pass filter smoothes out the rectified signal to produce DC. Finally, an amplifier adjusts the gain to the desired level. The output of the LVDT interface circuit is a DC voltage that is proportional to the linear distance that the core is offset from the center. Some integrated circuits, such as the NE 5520 (Signetics), combine all the functions shown within the box (of Figure 6.21) on a single chip.

Figure 6.21

An interface circuit for an LVDT.

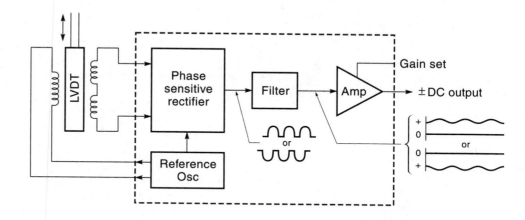

6.2 VELOCITY SENSORS

Velocity sensors, or *tachometers*, are devices that give an output proportional to angular velocity. These sensors find wide application in motor-speed control systems. They are also used in position systems to improve their performance.

Velocity from Position Sensors

Velocity is the rate of change of position. Expressed mathematically,

$$\text{Velocity} = \frac{\Delta d}{\Delta t} = \frac{d_2 - d_1}{t_2 - t_1} \qquad (6.3)$$

where

$$\Delta d = \text{change in distance}$$
$$\Delta t = \text{change in time}$$
$$d_2, d_1 = \text{position samples}$$
$$t_2, t_1 = \text{times when samples were taken}$$

Because the only components of velocity are position and time, extracting velocity information from two sequential position data samples should be possible (if you know the time between them). This concept is demonstrated in Example 6.9. The math could be done with hard-wired circuits or software. If the system already has a position sensor, such as a potentiometer, using this approach eliminates the need for an additional (velocity) sensor.

◆ **EXAMPLE 6.9**

A rotating machine part has a pot position sensor connected through an ADC such that LSB = 1°. Determine how to use this setup to get velocity data.

Solution

Velocity can be computed from two sequential position samples—d_1 taken at time t_1 and d_2 taken at t_2, as specified in Equation 6.3:

$$\text{Velocity} = \frac{\Delta d}{\Delta t} = \frac{d_2 - d_1}{t_2 - t_1}$$

If we took a data sample exactly every second, then the denominator of Equation 6.3 would be 1. In that case, velocity would just equal ($d_2 - d_1$), but 1 s is probably too long a time for the controller to wait between samples. Instead, select $\frac{1}{10}$ s (100 ms) as the time between samples. Now,

$$\frac{\Delta d}{\Delta t} = \frac{d_2 - d_1}{1/10} = 10(d_2 - d_1)$$

Thus, all the software has to do to calculate velocity is

1. Take two position samples exactly 1/10 apart.
2. Subtract the values of the two samples.
3. Multiply the result by 10.

◆

Velocity data can be derived from an optical rotary encoder in two ways. The first would be the method just described for the potentiometer; the second method involves determining the time it takes for each slot in the disk to pass. The slower the velocity, the longer it takes for each slot to go by. The digital counter circuit shown in Figure 6.22 can be used as a timer to time how long it takes for one slot to pass. The idea is to count the cycles of a known high-speed clock for the duration of one *slot period*. The final count would be proportional to the time it took for the slot to pass.

The operation of the circuit (Figure 6.22) is as follows. One of the outputs of the optical encoder (say, V_1) is used as the input to the timer. V_1 triggers a one-shot to produce V_1', which is a brief negative-going pulse to clear the counter. When V_1' returns high (removing the clear), a high-speed clock is counted by the counter. When the next slot triggers the one-shot, the counter data are transferred into a separate latch, and the counter is cleared so it can start over again. The controller reads the count from the latch. The value of the count is proportional to the *reciprocal* of the angular velocity. The slower the velocity, the larger the count. This means that for very slow velocities the counter might overflow and start counting up from 0 again (such as your car odometer turning over from 99,999 to 00,000). In fact, when the disk comes to a dead stop, *any* counter would overflow eventually. To solve this problem, a special circuit using another one-shot has been added. Every time the counter fills up, the one-shot fires and reloads 1s into all bits. This action prevents the full counter from ever rolling over to 0. The result is that a full counter is interpreted by the controller as meaning "velocity too low to measure."

Figure 6.22

Circuit for counting slot-cycle time (for determining velocity from an incremental encoder).

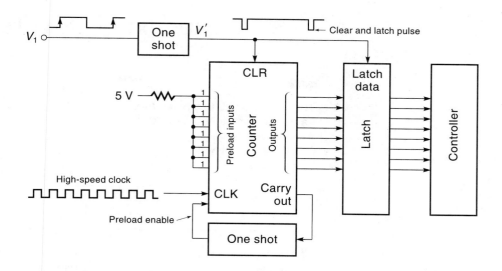

Tachometers

Optical Tachometers

The **optical tachometer**, a simple device, can determine a shaft speed in terms of revolutions per minute (rpm). As shown in Figure 6.23, a contrasting stripe is placed on the shaft. A photo sensor is mounted in such a way as to output a pulse each time the stripe goes by. The period of this waveform is inversely proportional to the rpm of the shaft and can be measured using a counter circuit like that described for the optical shaft encoder (Figure 6.22). Notice that this system cannot sense position or direction. However, if two photo sensors are used, the direction could be determined by phasing, similar to the incremental optical shaft encoder.

Figure 6.23

An optical tachometer.

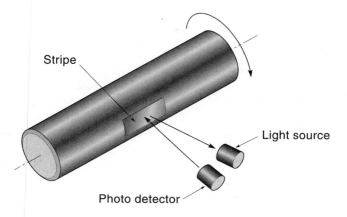

Direct Current Tachometers

A **direct current tachometer** is essentially a DC generator that produces a DC output voltage proportional to shaft velocity. The output polarity is determined by the direction of rotation. Typically, these units have stationary permanent magnets (discussed in

Figure 6.24

The CK20 DC tachometer.

CK20 DC TACHOMETER

The model CK20 is a moving coil tachometer designed for use in applications requiring velocity feedback with minimum system inertia load.

Parameter	Value	Units
Linerity	.2	% max. deviation
Ripple	1.5	max, % peak to peak AC
Ripple Frequency	19	Cycles per revolution
Speed Range	1-6000	RPM
Armature Inertia	9×10^{-5}	in-oz-sec^2
Friction Torque	.25	in-oz, max.
Rated Life	10,000	Hours at 3000 RPM

WINDING VARIATIONS

	CK20-A	CK20-B	CK20-C
Output Voltage Gradient (V/KRPM)	3.0	2.5	1

Chapter 7), and the rotating part consists of coils. Such a design keeps the inertia down but requires the use of brushes, which eventually wear out. Still, these units are useful because they provide a direct conversion between velocity and voltage.

Figure 6.24 gives the specifications of the CK20 tachometer. The housing of this unit is constructed so that it can mount "piggyback" on a motor, providing direct feedback of the motor velocity. The transfer function for the tachometer has units of volts/1000 rpm. We can use the transfer function to calculate the output voltage for a particular speed. Looking at the bottom of Figure 6.24, you can see that the CK20 comes in three models. For example, the CK20-A outputs 3 V for 1000 rpm (3 V/Krpm). It has a speed range of 0–6000 rpm, so the maximum voltage would be 18 V at 6000 rpm. This information can be displayed as a linear graph (Figure 6.25). From the graph, we can easily find the output voltage for any speed. The "linearity" of the motor is given as 0.2%, which means that the actual velocity may be as much as 0.2%, different from what it should be. For example, if the output is 9 V, the velocity should be 3000 rpm; however, because 0.2% × 3000 = 6 rpm, the actual velocity could be anywhere from 2994 to 3006 rpm.

Figure 6.25

Graph of speed versus output DC volts for the CK20-A tachometer.

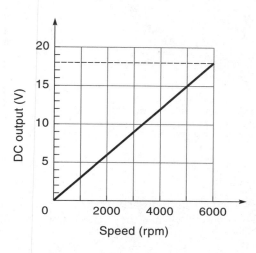

Velocities of thousands of rpm are much higher than you would normally find for actual heavy mechanical parts. Therefore the tachometer is frequently attached to the motor, and the motor is geared down to drive the load. Example 6.10 demonstrates this.

♦ EXAMPLE 6.10

As shown in Figure 6.26, a motor with a piggyback tachometer has a built-in gear box with a ratio of 100 : 1 (that is, the output shaft rotates 100 times slower than the motor). The tachometer is a CK20-A with an output of 3 V/Krpm. This unit is driving a machine tool with a maximum rotational velocity of 60°/s.

a. What is the expected output of the tachometer?

b. Find the resolution of this system if the tachometer data were converted to digital with an 8-bit ADC as illustrated in Figure 6.26.

Solution

a. A maximum tool velocity of 60°/s can be converted to rpm as follows:

$$\frac{60°}{s} \times \frac{1 \text{ rev}}{360°} \times \frac{60 \text{ s}}{\text{min}} = 10 \text{ rpm}$$

Because of the gear ratio, the tachometer is turning 100 times faster than the tool. Calculating the overall transfer function of the velocity sensor, we find

$$\underbrace{\frac{3 \text{ V}}{1000 \text{ rpm}_{motor}}}_{\text{Tachometer}} \times \underbrace{\frac{100 \text{ rpm}_{motor}}{1 \text{ rpm}_{tool}}}_{\text{Gearbox}} = 0.3 \text{ V/rpm}_{tool}$$

Figure 6.26

A tachometer interface circuit (Example 6.10).

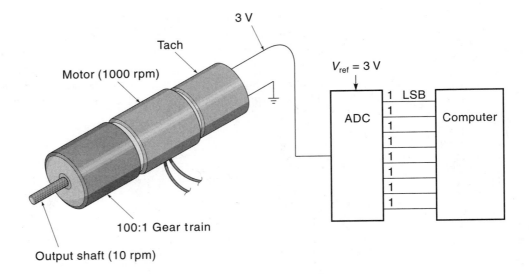

Now, using this transfer function, we can calculate what the tachometer voltage would be when the tool is rotating at 10 rpm:

$$V_{tach} = \frac{0.3\,V}{rpm_{tool}} \times 10\,rpm_{tool} = 3\,V$$

b. To get the best resolution, we would reference the ADC to 3 V so that 3 V = 11111111_{bin} (255 decimal). Because we know that the tachometer is producing 3 V when the shaft is rotating at 10 rpm and that 8 bits represent 255 levels, we can calculate the rpm represented by each binary state:

$$\text{Resolution (LSB)} = \frac{10\,rpm}{255\,states} = 0.04\,rpm/state$$

This means that the digital controller will know the shaft velocity to within 0.04 rpm. Therefore, the resolution is 0.04 rpm. ♦

6.3 PROXIMITY SENSORS

Limit Switches

A **proximity sensor** simply tells the controller whether a moving part is at a certain place. A **limit switch** is an example of a proximity sensor. A limit switch is a mechanical push-button switch that is mounted in such a way that it is actuated when a mechanical part gets to the end of its intended travel. For example, in an automatic garage-door opener, all the controller needs to know is if the door is all the way open or all the way closed. Limit switches can detect these two conditions. Switches are fine for many applications, but they have at least two drawbacks: (1) Being a mechanical device, they eventually wear out, and (2) they require a certain amount of physical force to actuate. (Chapter 4 has more on limit switches.) Two other types of proximity sensors, which use either optics or magnetics to determine if an object is near, do not have these problems. The price we pay for these improved characteristics is that they require some support electronics.

Optical Proximity Sensors

Optical proximity sensors, sometimes called *interrupters*, use a light source and a photo sensor that are mounted in such a way that the object to be detected cuts the light path. Figure 6.27 illustrates two applications of using photodetectors. In Figure 6.27(a), a photodetector counts the number of cans on an assembly line; in Figure 6.27(b), a photodetector determines if the read-only hole in a floppy disk is open or closed.

Figure 6.27

Two applications of a photodetector.

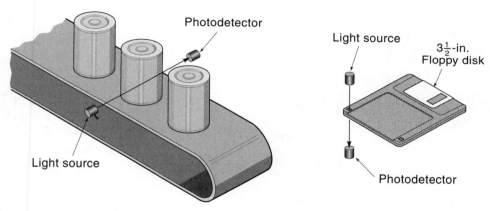

(a) Counting cans on a conveyor belt

(b) Detecting "read only" hole in floppy disk

Four types of photodetectors are in general use: photo resistors, photodiodes, photo transistors, and photovoltaic cells. A **photo resistor**, which is made out of a material such as cadmium sulfide (CdS), has the property that its resistance decreases when the light level increases. It is inexpensive and quite sensitive—that is, the resistance can change by a factor of 100 or more when exposed to light and dark. Figure 6.28(a) shows a typical interface circuit—as the light increases, R_{pd} decreases, and so V_{out} increases.

A **photodiode** is a light-sensitive diode. A little window allows light to fall directly on the PN junction where it has the effect of increasing the reverse-leakage current. Figure 6.28(b) shows the photodiode with its interface circuit. Notice that the photodiode is reversed-biased and that the small reverse-leakage current is converted into an amplified voltage by the op-amp.

A **photo transistor** [Figure 6.28(c)] has no base lead. Instead, the light effectively creates a base current by generating electron-hole pairs in the CB junction—the more light, the more the transistor turns on.

Figure 6.28
Photodetectors.

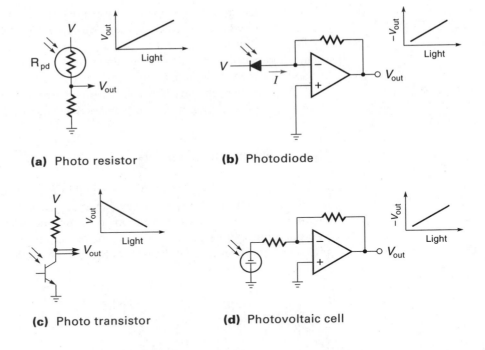

(a) Photo resistor

(b) Photodiode

(c) Photo transistor

(d) Photovoltaic cell

The **photovoltaic cell** is different from the photo sensors discussed so far because it actually creates electrical power from light—the more light, the higher the voltage. (A solar cell is a photovoltaic cell.) When used as a sensor, the small voltage output must usually be amplified, as shown in Figure 6.28(d).

Some applications make use of an optical proximity sensor called a **slotted coupler**, also called an *optointerrupter* (Figure 6.29). This device includes the light source and detector in a single package. When an object moves into the slot, the light path is broken. The unit comes in a wide variety of standard housings [Figure 6.29(a)]. To operate, power must be provided to the LED, and the output signal taken from the phototransistor. This is done in the circuit of Figure 6.29(b), which provides a TTL-level (5 V or 0 V) output. When the slot is open, the light beam strikes the transistor, turning it on, which grounds the collector. When the beam is interrupted, the transistor turns off, and the collector is pulled up to 5 V by the resistor.

Optical sensors enjoy the advantage that neither the light source, the object to be detected, nor the detector have to be near each other. An example of this is a burglar alarm system. The light source is on one side of the room, the burglar is in the middle, and the detector is on the other side of the room. This property can be important in a case where there are no convenient mounting surfaces near the part to be measured. On the other hand, keeping the lenses clean may be a problem in some industrial situations.

Hall-Effect Proximity Sensors

In 1879 E. H. Hall first noticed the effect that bears his name. He discovered a special property of copper, and later of semiconductors: They produce a voltage in the presence

Figure 6.29
An optical slotted coupler.

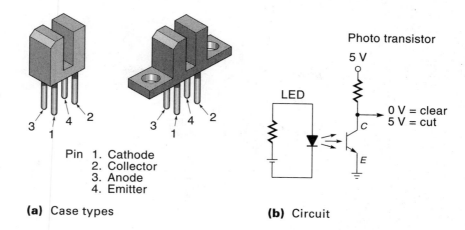

Pin 1. Cathode
2. Collector
3. Anode
4. Emitter

(a) Case types

(b) Circuit

of a magnetic field. This is especially true for germanium and indium. The **Hall effect**, as it is called, was originally used for wattmeters and gaussmeters; now it is used extensively for proximity sensors. Figure 6.30 shows some typical applications. In all cases, the Hall-effect sensor outputs a voltage when the magnetic field increases beyond a certain point. This is done by either moving a magnet or by changing the magnetic field path.

Figure 6.30
Typical applications of
Hall-effect sensors.

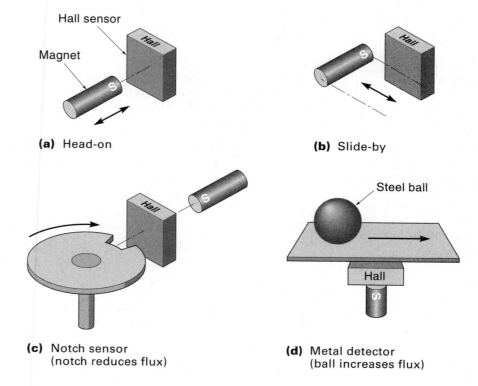

(a) Head-on

(b) Slide-by

(c) Notch sensor
(notch reduces flux)

(d) Metal detector
(ball increases flux)

Figure 6.31
The operation of a Hall-effect
sensor.

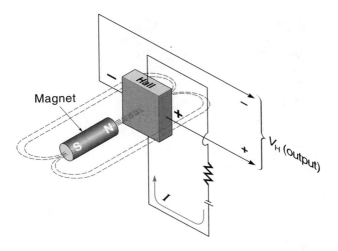

Figure 6.31
The operation of a Hall-effect
sensor.

Figure 6.31 shows how the Hall effect works. First, an external voltage source is used to establish a current (I) in the semiconductor crystal. The output voltage (V_H) is sensed across the sides of the crystal, perpendicular to the current direction. When a magnetic field is brought near, the negative charges are deflected to one side producing a voltage. The relationship can be described in the following equation:

$$V_H = \frac{KIB}{D} \qquad (6.4)$$

where

V_H = Hall-effect voltage
K = constant (dependent on material)
I = current from an external source
B = magnetic flux density
D = thickness constant

Equation 6.4 states that V_H is directly proportional to I and B. If I is held constant, then V_H is directly proportional to B (magnetic flux density). Therefore, the output is not really on/off but (over a short distance) somewhat linear. To get a switching action, the output must go through a **threshold detector** like that illustrated in Figure 6.32(a). This circuit uses two comparator amps to establish the high and low switching voltages. When V_H goes above 0.5 V, the top amp sets the R-S flip-flop. When V_H goes below 0.25 V, the bottom amp resets the flip-flop. For this circuit to work, we need to make sure that the magnet comes near enough to the sensor to make V_H go above 0.5 V and far enough away for V_H to drop below 0.25 V.

A complete Hall-effect switch can be purchased in IC form. One example is the Allegro 3175 [Figure 6.32(b)]; it includes the sensor (X), the cross-current drive, and the threshold detector. The transistor turns on when the magnetic field goes above + 100 gauss and turns off when the field drops below − 100 gauss. The transistor can sink 15 mA, which can drive a small relay directly or a TTL digital circuit.

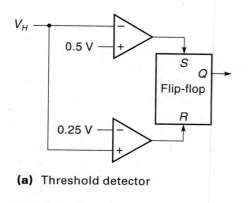

(a) Threshold detector

FUNCTIONAL BLOCK DIAGRAM

(b) Allegro UGN-3175

Figure 6.32 Hall-effect interface circuits.

Hall-effect sensors are used in many applications, for example, computer keyboard switches and proximity sensors in machines. They are also used in tachometers; for example, a small magnet is attached to the perimeter of a rotating part, and the stationary Hall-effect sensor detects each revolution.

6.4   LOAD SENSORS

Load sensors measure mechanical force. The forces can be large or small—for example, weighing heavy objects or detecting low-force tactile pressures. In most cases, it is the slight deformation caused by the force that the sensor measures, not the force directly. Typically, this deformation is quite small. Once the amount of tension or compression displacement has been measured, the force that must have caused it can be calculated using the mechanical parameters of the system. The ratio of the force to deformation is a constant for each material, as defined by **Hooke's law:**

$$F = KX \qquad (6.5)$$

where

K = spring constant of the material
F = applied force
X = extension or compression as result of force

For example, if a mechanical part has a spring constant of 1000 lb/in. and it compresses 0.5 in. under the load, then the load must be 500 lb.

Bonded-Wire Strain Gauges

The **bonded-wire strain gauge** can be used to measure a wide range of forces, from 10 lb to many tons. It consists of a thin wire (0.001 in.) looped back-and-forth a few times and cemented to a thin paper backing [Figure 6.33(a)]. More recent versions use printed-circuit technology to create the wire pattern. The entire strain gauge is securely bonded to some structural object and will detect any deformation that may take place. The gauge is oriented so the wires lie in the same direction as the expected deformation. The principle of operation is as follows: If the object is put under tension, the gauge will stretch and elongate the wires. The wires not only get slightly longer but also thinner. Both actions cause the total wire resistance to rise, as can be seen from the basic resistance equation:

$$R = \frac{\rho L}{A} \tag{6.6}$$

where

R = resistance of a length of wire (at 20°C)
ρ = resistivity (a constant dependent on the material)
L = length of wire
A = cross-sectional area of wire

The change in resistance of the strain-gauge wires can be used to calculate the elongation of the strain gauge (and the object to which it is cemented). If you know the elongation and the spring constant of the supporting member, then the principles of Hooke's law can be used to calculate the force being applied.

The resistance change in a strain gauge is small. Typically, it is only a few percent, which may be less than an ohm. Measuring such small resistances usually requires a bridge circuit [Figure 6.33(b)]. With this circuit, a small change in one resistor can cause a relatively large percentage change in the voltage across the bridge. Initially, the bridge is balanced by adjusting the resistances so that $V_1 = V_2$. Then, when the gauge resistance changes, the voltage difference $(V_1 - V_2)$ changes. The bridge also allows us to cancel out variations due to temperature, by connecting a **compensating gauge** (known as the *dummy*) as one of the bridge resistors. As shown in Figures 6.33 and 6.34, the actual compensation gauge is placed physically near the active gauge so as to receive

Figure 6.33
Strain gauges.

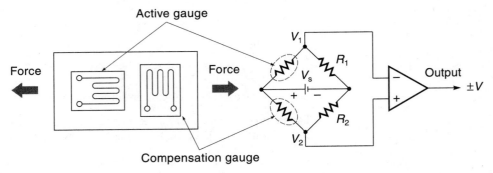

(a) Placement of gauges   **(b)** Interface circuit using a bridge

the same temperature, but it is oriented perpendicularly from the active gauge so the force will not elongate its wires.

If all four resistors in the bridge circuit [Figure 6.33(b)] are about the same value, then we can use Equation 6.7 to calculate the change in the strain-gauge resistance, based on the measured voltage change across the bridge:

$$\Delta R = \frac{4R(V_1 - V_2)}{V_s} \tag{6.7}$$

where

$$\Delta R = \text{change in the strain-gauge resistance}$$
$$R = \text{nominal value of all bridge resistors}$$
$$(V_1 - V_2) = \text{voltage across the bridge}$$
$$V_s = \text{source voltage applied to the bridge}$$

As the strain gauge is stretched, its resistance rises. The precise relationship between elongation and resistance can be computed using Equation 6.8 and is based on the **gauge factor** (GF), which is supplied by the strain-gauge manufacturer:

$$\epsilon = \frac{\Delta R/R}{GF} \tag{6.8}$$

where

$$\epsilon = \text{elongation of the object per unit of length } (\Delta L/L), \text{ called } strain$$
$$R = \text{strain-gauge resistance}$$
$$\Delta R = \text{change in strain-gauge resistance due to force}$$
$$GF = \text{gauge factor, a constant supplied by the manufacturer (GF is the ratio } (\Delta R/R)/(\Delta L/L)$$

One more equation is needed before we can solve a strain-gauge problem—an equation that relates *stress* and the resulting *strain* in an object. **Stress** is the force per

Figure 6.34
Strain-gauge configurations.

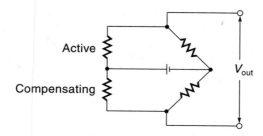

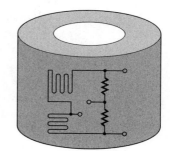

(a) Active and compensating gauges are placed together so they will be at the same temperature

(b) Load cell with strain gauge and bridge

cross-sectional area; for example, if a table leg has a cross-sectional area of 2 in^2 and is supporting a load of 100 lb, then the stress is 50 lb/in^2. **Strain** is the amount of length (per unit length) that the object stretches as a result of being subjected to a stress; for example, if an object 10 in. long stretches 1 in., then each inch of the object stretched 0.1 in., and so the strain would be 0.1 in./in. Stress and strain are related by a constant called **Young's modulus** (also called *modulus of elasticity*), as shown in Equation 6.9. Young's modulus (E) is a measure of how stiff a material is and could be thought of as a kind of spring constant:

$$E = \frac{\rho}{\epsilon} \qquad\qquad (6.9)$$

where
    E = Young's modulus (a constant for each material)
    ρ = stress (force per cross-sectional area)
    ϵ = strain (elongation per unit length)

Table 6.2 gives some values of E for common materials.

TABLE 6.2   Young's Modulus (E) for Common Materials

Substance	lb/in^2	N/cm^2
Steel	30×10^6	2.07×10^7
Copper	15×10^6	1.07×10^7
Aluminum	10×10^6	6.9×10^6
Rock	7.3×10^6	5.0×10^6
Hard wood	1.5×10^6	1.0×10^6

♦ **EXAMPLE 6.11**

A strain gauge and bridge circuit are used to measure the tension force in a steel bar (Figure 6.35). The steel bar has a cross-sectional area of 2 in^2. The strain gauge has a nominal resistance of 120 Ω and a GF of 2. The bridge is supplied with 10 V. When the bar is unloaded, the bridge is balanced so the output is 0 V. Then force is applied to the bar, and the bridge voltage goes to 0.0005 V. Find the force on the bar.

Solution   First, use Equation 6.7 to calculate the change in strain-gauge resistance due to the applied force:

$$\Delta R = \frac{4R(V_1 - V_2)}{V_s} = \frac{4 \times 120\ \Omega \times 0.0005\ V}{10\ V} = 0.024\ \Omega$$

Next, use Equation 6.8 to calculate the elongation (strain) of the strain gauge (how much it was stretched):

$$\epsilon = \frac{\Delta R/R}{GF} = \frac{0.024/120}{2} = 0.0001\ \text{in./in.}$$

Finally, use Equation 6.9

$$E = \frac{\rho}{\epsilon}$$

to calculate the force on the bar. This will require looking up the value of Young's modulus. From Table 6.2, we find it to be 30,000,000 lb/in^2 for steel. Rearranging Equation 6.9 gives

$$\rho = E\epsilon = 30{,}000{,}000\ \text{lb/in}^2 \times 0.0001\ \text{in./in.} = 3000\ \text{lb/in}^2$$

This result tells us that the tension force on the steel bar is 3000 lb/in^2, and because this bar has a cross-sectional area of 2 in^2, the total tension force in the bar is 6000 lb.

Figure 6.35

Strain-gauge measuring tension in a steel bar (Example 6.11).

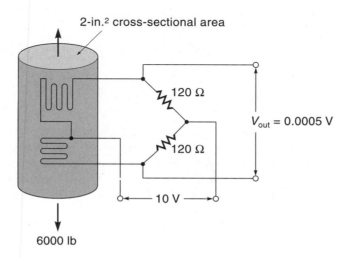

2-in.^{2} cross-sectional area

120 Ω

120 Ω

$V_{out} = 0.0005$ V

10 V

6000 lb

EXAMPLE 6.11 (Repeated with SI Units)

A strain gauge and bridge circuit are used to measure the tension force in a bar of steel that has a cross-sectional area of 13 cm^2. The strain gauge has a nominal resistance of 120 Ω and a GF of 2. The bridge is supplied with 10 V. When the bar is unloaded, the bridge is balanced so the output is 0 V. Then force is applied to the bar, and the bridge voltage goes to 0.0005 V. Find the force on the bar.

Solution   First, calculate the change in strain-gauge resistance due to the applied force:

$$\Delta R = \frac{4R(V_1 - V_2)}{V_s} = \frac{4 \times 120 \ \Omega \times 0.0005 \ V}{10 \ V} = 0.024 \ \Omega$$

Next, calculate the elongation (strain) of the strain gauge:

$$\epsilon = \frac{\Delta R/R}{GF} = \frac{0.024/120}{2} = 0.0001 \ cm/cm$$

Finally, use Equation 6.9

$$E = \frac{\rho}{\epsilon}$$

to calculate the force on the bar. This will require looking up the value of Young's modulus. From Table 6.2, we find it to be 2.07 × 10^7 N/cm for steel. Rearranging Equation 6.9 gives

$$\rho = E\epsilon = 20{,}700{,}000 \ N/cm^2 \times 0.0001 \ cm/cm = 2070 \ N/cm^2$$

This result tells us that the tension force on the steel bar is 2070 N/cm^2, and because this bar has a cross-sectional area of 13 cm^2, the total tension force in the bar is 26,910 N.   ♦

Strain-gauge force transducers (called *load cells*) are available as self-contained units that can be mounted anywhere in the system. A load cell may contain two strain gauges (active and compensating) and a bridge [Figure 6.34]. A typical application for load cells is monitoring the weight of a tank. The tank would be sitting on three or four load cells, so the weight of the tank is the sum of the outputs of the load cells [see Figure 6.60(c)].

Semiconductor Force Sensors

Another type of force sensor uses the **piezoresistive effect** of silicon. These units change resistance when force is applied and are 25–100 times more sensitive than the bonded-wire strain gauge. A semiconductor strain gauge is a single strip of silicon material that is bonded to the structure. When the structure stretches, the silicon is elongated, and the resistance from end to end increases (however, the resistance change is nonlinear).

Low-Force Sensors

Some applications call for low-force sensors. For example, imagine the sensitivity required for a robot gripper to hold a water glass without slipping and without crushing it. Strain gauges can measure low forces if they are mounted on an elastic substrate, like rubber—then a small force will cause a significant deflection and resistance change. Another solution would be to construct a low-force sensor with a spring and a linear-motion potentiometer (Figure 6.36). The spring compresses a distance proportional to the applied force, and this distance is measured with the pot.

Figure 6.36

A tactile force sensor using a spring-loaded linear pot.

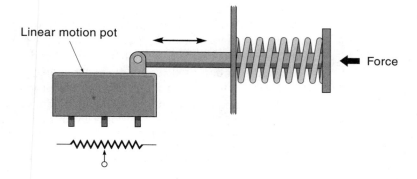

Linear motion pot

Force

♦ **EXAMPLE 6.12**

Construct a force sensor with the following characteristics;

    Range: 0–30 lb

    Deformation: 0.5 in. (maximum)

    Output: 0.1 V/lb

A 1 kΩ linear motion pot is available with a 1-in. stroke.

Solution

Using the concept of Figure 6.36, we first need to specify the spring. The specifications call for a spring that deforms 0.5 in. with 30 lb of force. Thus,

$$K \text{ (spring constant)} = \frac{30 \text{ lb}}{0.5 \text{ in.}} = 60 \text{ lb/in.}$$

Knowing we need a spring with a K of 60 lb/in., we could go to a spring catalog and select one.

The desired sensitivity of 0.1 V/lb dictates that the maximum output voltage will be 3 V when the force is 30 lb: voltage at maximum load = 30 lb × 0.1 V/lb = 3 V.

Finally, we must determine the supply voltage across the pot. The pot should output 3 V when it is moved 0.5 in. (one-half of its stroke). A ratio can be used to find the pot supply voltage for 1 in. of stroke:

$$\frac{3 \text{ V}}{0.5 \text{ in.}} = \frac{X}{1 \text{ in.}}$$

$$X = 6 \text{ V}$$

Therefore, the supply voltage should be 6 V. Figure 6.37 shows the final setup.

Figure 6.37

A tactile sensor setup
(Example 6.12).

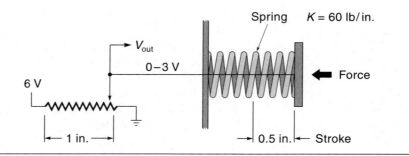

A very low-force tactile sensor can be made using conductive foam. This is the principle used in membrane keypads illustrated in Figure 6.38. The conductive foam is a soft foam rubber saturated with very small carbon particles. When the foam is squeezed, the carbon particles are pushed together, and the resistance of the material falls. Therefore, in some fashion, resistance is proportional to force. At present this concept has found limited application in such things as calculator keypads; because of its simplicity and low cost, however, it is a viable option for other applications such as robot tactile sensors.

Figure 6.38

Conductive foam tactile
sensor.

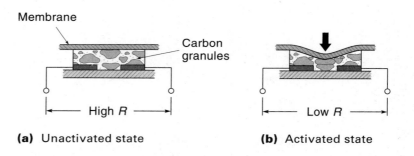

Finally, a very simple tactile sensor can be made with two or more limit switches mounted side-by-side with spring actuators that are set to switch at different pressures. As the pressure increases, the first switch closes, then with more pressure the next switch closes, and so on.

6.5 PRESSURE SENSORS

Pressure is defined as the force per unit area that one material exerts on another. For example, consider a 10-lb cube resting on a table. If the area of each face of the cube is 4 in^2, then 10 lb is distributed over an area of 4 in^2, so the cube exerts a pressure on the table of 2.5 lb/in^2 (10 lb/4 in^2 = 2.5 in^2, or 2.5 **psi**). In SI units, pressure is measured in Newtons per square meter (N/m^2), which is called a Pascal (Pa). For a liquid, pressure is exerted on the side walls of the container as well as the bottom.

Pressure sensors usually consist of two parts: The first converts pressure to a force or displacement, and the second converts the force or displacement to an electrical signal. Pressure measurements are made only for gases and liquids. The simplest pressure measurement yields a **gauge pressure**, which is the difference between the measured pressure and ambient pressure. At sea level, ambient pressure is equal to atmospheric pressure and is assumed to be 14.7 psi, or 101.3 kiloPascals (kPa). A slightly more complicated sensor can measure **differential pressure**, the difference in pressure between two places where neither pressure is necessarily atmospheric. A third type of pressure sensor measures **absolute pressure**, which is measured with a differential pressure sensor where one side is referenced at 0 psi (close to a total vacuum).

Bourdon Tubes

A **Bourdon tube** is a short bent tube, closed at one end. When the tube is pressurized, it tends to straighten out. This motion is proportional to the applied pressure. Figure 6.39 shows some Bourdon tube configurations. Notice that the displacement can be either linear or angular. A position sensor such as a pot or LVDT can convert the displacement into an electrical signal. Bourdon tube sensors are available in pressure ranges from 30 to 100,000 psi. Typical uses include steam- and water-pressure gauges.

Figure 6.39
Bourdon tube sensors.

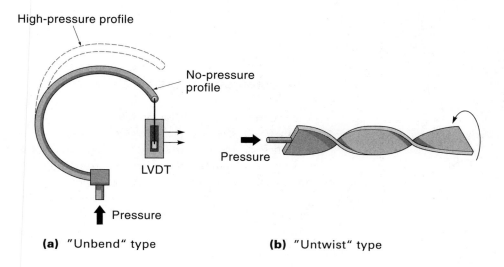

(a) "Unbend" type   (b) "Untwist" type

Bellows

This sensor uses a small metal bellows to convert pressure into linear motion [Figure 6.40(a)]. As the pressure inside increases, the bellows expand against the resistance of a spring (the spring is often the bellows itself). This motion is detected with a position sensor such as a pot. Figure 6.40(b) illustrates a differential pressure sensor, which can be made by enclosing the bellows in a canister. Here, the pressure from outside the bellows (pressure 2) tends to make the bellows compress, whereas pressure 1 tends to make the bellows expand. The position of the shaft is a function of the difference in pressure inside and outside the bellows. Bellows are capable of more sensitivity than the Bourdon tube in the lower-pressure range of 0–30 psi. Typical uses of bellows include automobile engine vacuum systems.

Figure 6.40
Bellows pressure sensors.

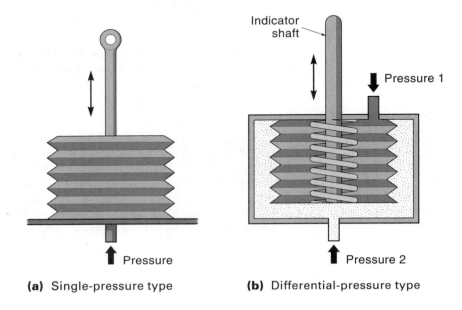

(a) Single-pressure type   **(b)** Differential-pressure type

Semiconductor Pressure Sensors

Some commercially available pressure sensors use the piezoresistive property of silicon (Figure 6.41). The piezoresistive element converts pressure directly into resistance, and resistance can be converted into voltage. These sensors have the advantage of "no moving parts" and are available in pressure ranges from 0–1.5 psi to 0–5000 psi. An example of a commercial semiconductor pressure sensor is the ST2000 series from Sen Sym Inc. (Figure 6.42). This unit can be used with fluids or gases, has an internal amplifier, and outputs a voltage that is directly proportional to absolute pressure.

6.6   TEMPERATURE SENSORS

Temperature sensors give an output proportional to temperature. Many control systems require these sensors, if only to know how much to compensate other sensors that are temperature-dependent. Some common types are discussed next.

Figure 6.41

A semiconductor pressure sensor.

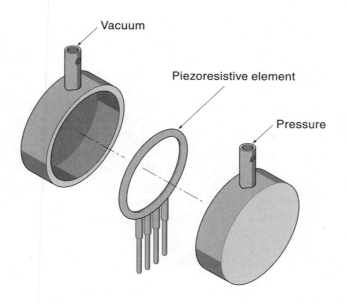

Vacuum

Piezoresistive element

Pressure

Figure 6.42

The ST2000 semiconductor pressure sensors. (Courtesy of Sen Sym Inc.)

Bimetallic Temperature Sensors

The **bimetallic temperature sensor** consists of a bimetallic strip wound into a spiral (Figure 6.43). The bimetallic strip is a laminate of two metals with different coefficients of thermal expansion. As the temperature rises, the metal on the inside expands more than the metal on the outside, and the spiral tends to straighten out. These sensors are typically used for on–off control as in a household thermostat where a mercury switch is rocked from on to off. In Figure 6.43, when the temperature increases, the tube containing liquid mercury rotates clockwise. When the tube rotates past the horizontal, the mercury runs down to the right and completes the electrical connection between the electrodes. One distinct advantage of this system is that the output from the switch can be used directly without further signal conditioning. Currently, mercury switches are being phased out because of environmental reasons, but contact-type switches are taking their place.

Figure 6.43

A bimetallic thermal sensor controlling a mercury switch (shown in "cold" state).

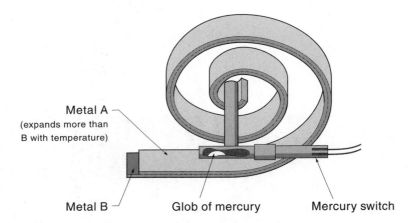

Metal A
(expands more than
B with temperature)

Metal B   Glob of mercury   Mercury switch

Thermocouples

The **thermocouple** was developed over 100 years ago and still enjoys wide use, particularly in high-temperature situations. The thermocouple is based on the **Seebeck effect**, a phenomenon whereby a voltage that is proportional to temperature can be produced from a circuit consisting of two dissimilar metal wires. For example, a thermocouple made from iron and constantan (an alloy) generates a voltage of approximately 35 μV/°F. Figure 6.44(a) illustrates this situation. We can think of the junctions at each end of the dissimilar metal wires as producing a voltage, so the net voltage (V_{net}) is actually the difference between the junction voltages. One junction is on the probe and is called the **hot junction**. The other junction is kept at some known reference temperature and is called the **cold junction**, or *reference junction*. The output voltage from this system can be expressed as follows:

$$V_{net} = V_{hot} - V_{cold} \qquad (6.10)$$

In practice, the thermocouple wires must connect to copper wires at some point as shown in Figure 6.44(b), so there are really three junctions. However, it turns out that

Figure 6.44

A thermocouple circuit
(iron–constantan type).

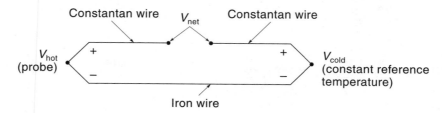

(a) Basic principle

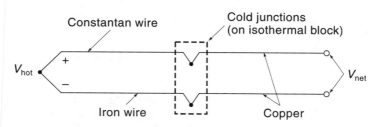

(b) Thermocouple connected to copper wires

the total voltage from the two copper junctions will be the same as the single cold-junction voltage (V_{cold}) of Figure 6.44(a) (assuming the copper junctions are at the same temperature), so the analysis is unchanged.

Traditionally, the cold junction was kept at 32°F in an **ice-water bath**, which is water with ice in it. Ice water was used because it is one way to produce a known temperature, and so V_{cold} becomes a constant in Equation 6.10, leaving a direct relationship between V_{net} and V_{hot}:

$$V_{hot} = V_{net} + \underbrace{\text{constant}}_{\substack{V_{cold} \text{ at } 32°F \\ \text{or some other} \\ \text{reference temperature}}} \tag{6.10a}$$

Modern systems eliminate the need for ice water. One method is to maintain the cold junction at constant temperature with a control system. This can be useful if there are many thermocouples in a system—they can all be referenced to the same temperature. Another method (used by a computer controller) is to simply look up in a table the value of V_{cold} for the ambient temperature and add this value to V_{net} to yield V_{hot}.

Still another way to eliminate the ice-water bath is to use a temperature-sensitive diode (in an interface circuit) that makes the thermocouple output behave as if the cold junction were still at freezing, even though it's not. Figure 6.45 shows such a circuit for an iron–constantan thermocouple. The cold junctions are maintained at the same temperature as the diode by mounting them all on an *isothermal* block. As the ambient temperature increases, the diode forward-voltage drop (about 0.6 V) decreases at a rate

Figure 6.45

A diode being used to compensate for cold-junction voltage.

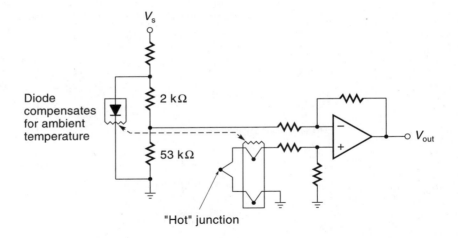

of about 1.1 mV/°F. This voltage is scaled down (with R_2 and R_3) to 28 μV/°F, *which is the same rate that the real cold-junction voltage increases with ambient temperature.* By subtracting (with the op-amp) the effect of the ambient temperature changes on the cold junction, we get a single thermocouple voltage that is directly proportional to temperature.

Commercial thermocouples are available with different temperature ranges and sensitivities (sensitivity being a measure of volts/degree). Figure 6.46 shows the volts versus temperature curves of the major classes of thermocouples. As you can see, type J (iron–constantan) has the highest sensitivity but the lowest temperature range, type K (chromel–alumel) has a higher temperature range but a lower sensitivity, and type R (platinum– rhodium) has an even lower sensitivity but can work at higher temperatures. Tables are available that give the precise thermocouple voltage–temperature relationship (for both °F and °C).

Figure 6.46

Thermocouple outputs (referenced at 32°F).

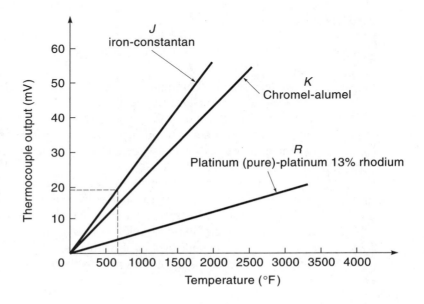

Thermocouples are simple and rugged but require extra electronics to deal with the inherent low-sensitivity and cold-junction problems. However, because they are linear (over a limited range), reliable, and stable, they enjoy wide use in measuring high temperatures in furnaces and ovens.

♦ **EXAMPLE 6.13**

An oven is supposed to be maintained at 1000°F by a control system, but you suspect that the temperature is much cooler. You have at your disposal a type-J thermocouple and a voltmeter. How would you use this equipment to check the oven temperature?

Solution

First, put the thermocouple in the oven and connect the thermocouple leads to the voltmeter (Figure 6.47). Try to make the cold-junction connections to the meter probes be at ambient temperature, which is about 90°F (as reported by a thermometer on the wall).

The voltmeter reads about 17 mV, which is V_{net} in Equation 6.10: $V_{net} = V_{hot} - V_{cold}$. The graph of Figure 6.46 (for type J) is based on the cold junction being at freezing (32°F), which it certainly is not in this case. From the graph, we can see that the 90°F would create about 2.5 mV by itself. You can see from Equation 6.10 that if V_{cold} increases, it will reduce V_{net}, so if we are going to use the graph of Figure 6.46, we must compensate by increasing our reading of 17 mV to 19.5 mV.

Now using the graph of Figure 6.46 for 19.5 mV, we read the temperature is 675°. This temperature is much lower than the desired 1000°F, so clearly there is something wrong with the temperature-control system.

Figure 6.47

Measuring oven temperature with a thermocouple (Example 6.13).

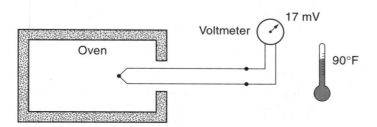

♦

Resistance Temperature Detectors

The **resistance temperature detector** (RTD) is a temperature sensor based on the fact that metals increase in resistance as temperature rises. Figure 6.48 shows a typical RTD. A wire, such as platinum, is wrapped around a ceramic or glass rod (sometimes the wire coil is supported between two ceramic rods). Platinum wire has a temperature coefficient of 0.0039 Ω/Ω/°C, which means that the resistance goes up 0.0039 Ω for each ohm of wire for each Celsius degree of temperature rise. RTDs are available in different resistances, a common value being 100 Ω. Thus, a 100-Ω platinum RTD has a resistance of 100 Ω at 0°C, and it has a positive temperature coefficient of 0.39 Ω/°C.

Figure 6.48

A resistance temperature detector.

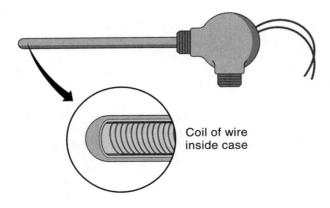

Coil of wire
inside case

◆ **EXAMPLE 6.14**

A 100-Ω platinum RTD is being used in a system. The present resistance reading is 110 Ω. Find the temperature.

Solution

A reading of 110 Ω means that the resistance has gone up 10 Ω from what it would be at 0°C. Therefore, knowing the temperature coefficient of RTD to be 0.39 Ω/°C, we can calculate the current temperature:

$$10 \ \Omega \times \frac{°C}{0.39 \ \Omega} = 25.6°C$$

◆

RTDs have the advantage of being very accurate and stable (characteristics do not change over time). The disadvantages are low sensitivity (small change in resistance per degree), relatively slow response time to temperature changes, and high cost.

Thermistors

A **thermistor** is a two-terminal device that changes resistance with temperature. Thermistors are made of oxide-based semiconductor materials and come in a variety of sizes and shapes. Thermistors are nonlinear; therefore, they are *not* usually used to get an accurate temperature reading but to indicate temperature changes, for example, overheating. Also, they have a negative temperature coefficient, which means the resistance decreases as temperature increases, as illustrated with the solid line in the graph of Figure 6.49(a). A very desirable feature of these devices is their high sensitivity. A relatively small change in temperature can produce a large change in resistance.

Figure 6.49
A thermistor.

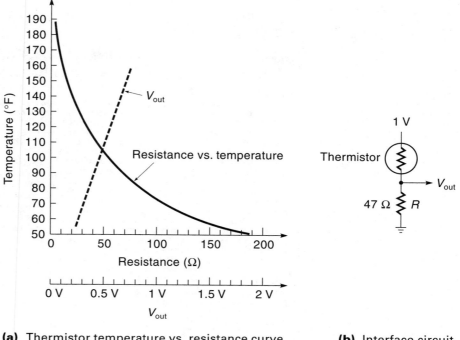

(a) Thermistor temperature vs. resistance curve   **(b)** Interface circuit

Figure 6.49(b) shows a simple thermistor interface circuit. By placing the thermistor in the top of a voltage divider, the resulting output voltage is relatively linear and has a positive slope [shown as a dashed line in Figure 6.49(a)]. The resistor (R) value selected should be close to the nominal value of the thermistor.

Thermistors come in a wide range of resistances, from a few ohms to 1 MΩ, selection of which depends on the temperature range of interest. Higher-resistance models are used for higher temperatures, to increase the sensitivity, and to keep the sensor from drawing too much current. Consider, for example, what would happen if we used the thermistor of Figure 6.49 in the temperature range of 150–200°F; the sensitivity is only 0.1 Ω/°F, and the nominal resistance is very low (15–20 Ω). If we were to operate the same thermistor in the 50–100°F temperature range, the sensitivity is much higher (2.6 Ω/°F), and the nominal resistance is higher (between 50 and 180 Ω).

Integrated Circuit Temperature Sensors

Integrated circuit temperature sensors come in various configurations. A common example is the LM34 and LM35 series. The LM34 produces an output voltage that is proportional to Fahrenheit temperature, and the LM35 produces an output that is proportional to Celsius temperature. Figure 6.50 gives the specification (spec) sheet for the LM35. Notice that it has three active terminals: supply voltage (V_s), ground, and V_{out}.

Figure 6.50

The LM35 temperature sensor.

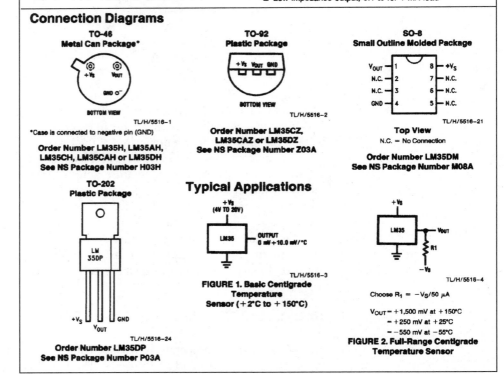

National Semiconductor

LM35/LM35A/LM35C/LM35CA/LM35D
Precision Centigrade Temperature Sensors

General Description

The LM35 series are precision integrated-circuit temperature sensors, whose output voltage is linearly proportional to the Celsius (Centigrade) temperature. The LM35 thus has an advantage over linear temperature sensors calibrated in ° Kelvin, as the user is not required to subtract a large constant voltage from its output to obtain convenient Centigrade scaling. The LM35 does not require any external calibration or trimming to provide typical accuracies of ± ¼°C at room temperature and ± ¾°C over a full −55 to +150°C temperature range. Low cost is assured by trimming and calibration at the wafer level. The LM35's low output impedance, linear output, and precise inherent calibration make interfacing to readout or control circuitry especially easy. It can be used with single power supplies, or with plus and minus supplies. As it draws only 60 μA from its supply, it has very low self-heating, less than 0.1°C in still air. The LM35 is rated to operate over a −55° to +150°C temperature range, while the LM35C is rated for a −40° to +110°C range (−10° with improved accuracy). The LM35 series is available packaged in hermetic TO-46 transistor packages, while the LM35C, LM35CA, and LM35D are also available in the plastic TO-92 transistor package. The LM35D is also available in an 8-lead surface mount small outline package and a plastic TO-202 package.

Features

■ Calibrated directly in ° Celsius (Centigrade)
■ Linear + 10.0 mV/°C scale factor
■ 0.5°C accuracy guaranteeable (at +25°C)
■ Rated for full −55° to +150°C range
■ Suitable for remote applications
■ Low cost due to wafer-level trimming
■ Operates from 4 to 30 volts
■ Less than 60 μA current drain
■ Low self-heating, 0.08°C in still air
■ Nonlinearity only ± ¼°C typical
■ Low impedance output, 0.1 Ω for 1 mA load

Connection Diagrams

TO-46
Metal Can Package*

TL/H/5516-1

*Case is connected to negative pin (GND)

Order Number LM35H, LM35AH, LM35CH, LM35CAH or LM35DH
See NS Package Number H03H

TO-92
Plastic Package

TL/H/5516-2

Order Number LM35CZ, LM35CAZ or LM35DZ
See NS Package Number Z03A

SO-8
Small Outline Molded Package

TL/H/5516-21

Top View
N.C. = No Connection

Order Number LM35DM
See NS Package Number M08A

TO-202
Plastic Package

TL/H/5516-24

Order Number LM35DP
See NS Package Number P03A

Typical Applications

+V$_S$
(4V TO 28V)

OUTPUT
0 mV +10.0 mV/°C

TL/H/5516-3

FIGURE 1. Basic Centigrade Temperature Sensor (+2°C to +150°C)

TL/H/5516-4

Choose R$_1$ = −V$_S$/50 μA

V$_{OUT}$ = +1,500 mV at +150°C
    = +250 mV at +25°C
    = −550 mV at −55°C

FIGURE 2. Full-Range Centigrade Temperature Sensor

The output voltage of the LM35 is directly proportional to °C, that is,

$$V_{out} = 10 \text{ mV/°C}$$

This equation states that for each 1° increase in temperature, the output voltage increases by 10 mV. If only positive temperatures need to be measured, then the simple circuit shown in the spec sheet (bottom middle of Figure 6.50) can be used. If positive and negative temperatures must be measured, then the circuit on the bottom right can be used, which requires a positive- and negative-supply voltage.

◆ **EXAMPLE 6.15**

Construct a temperature sensor using the LM35 that has the following specifications:

Range: 5–100°C

Supply voltage: 5 V

Output: 0.1 V/°C

Solution   The range requirement is no problem because the LM35 has an operating range of −55° to 150°C. The task comes down to specifying the circuit and amplifying the output to meet the specifications.

Because the temperature range is positive, we can use the simple circuit designated as "Figure 1" in the spec sheet (Figure 6.50), using 5 V for the supply voltage.

The specifications also call for 0.1 V = 1°C, which is ten times greater than the LM35 output. This requirement can be met with the op-amp illustrated in Figure 6.51. (Op-amps are discussed in Chapter 3.) The gain of the amplifier can be set to 10 by proper selection of resistors:

$$\text{Gain} = A = \frac{R_f}{R_a} + 1 = \frac{90 \text{ k}\Omega}{10 \text{ k}\Omega} + 1 = 10$$

Figure 6.51

An IC temperature sensor circuit (Example 6.15).

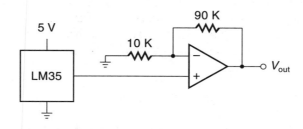

◆

The LM35 is a convenient IC to work with because the output voltage is in degrees Celsius. Some ICs, such as the LM135, provide an output that is in degrees kelvin. One degree of kelvin or Celsius represents the same interval of temperature, but the **Kelvin scale** starts at absolute zero temperature, which is 273°C below freezing. There is also an absolute zero temperature scale for Fahrenheit degrees, called the **Rankine scale**. These four temperature scales are compared in Figure 6.52.

Another device, the LM3911 temperature controller, is a complete temperature-measurement system on a single chip. It includes a temperature sensor, a stable

Figure 6.52

Comparison of the Rankine, Fahrenheit, Kelvin, and Celsius temperature scales.

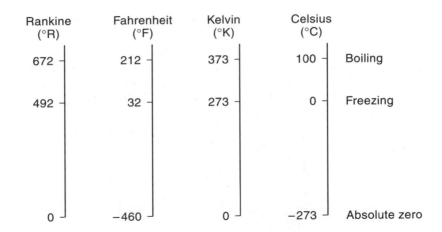

reference voltage, and an op-amplifier. With the addition of a few external resistors, any temperature scale can be obtained, and by connecting the op-amp as a comparator, the output can be made to switch on-and-off at any desired set-point temperature.

Still another device, the TMP01 (made by Analog Devices), was designed specifically to be a single-chip thermostat. Three external resistors establish the upper and lower set points. The TMP01 outputs can directly drive relays to turn either heating or cooling on as needed.

6.7   FLOW SENSORS

Flow sensors measure the quantity of fluid material passing by a point in a certain time. Usually, the material is a gas or a liquid and is flowing in a pipe or open channel. The flowing of solid material, such as gravel traveling on a conveyer belt, will not be considered here. Flow transducers come in several types—those that use differential pressure, those where the flow spins a mechanical device, and a smaller class of sensors that use more sophisticated technologies.

Pressure-Based Flow Sensors

This group of flow sensors is based on the fact that pressure in a moving fluid is proportional to the flow. The pressure is detected with a pressure sensor; based on the physical dimensions of the system, the flow can be calculated. The simplest flow sensor is called the **orifice plate** (Figure 6.53), a restriction in the pipe that causes a pressure drop in the flow, much like a resistor that causes a drop in voltage in a circuit. This sensor requires two pressure ports, one upstream and one downstream of the restriction. The flow is proportional to the pressure difference between these ports and is calculated as follows:

$$Q = CA \sqrt{\frac{2g}{d}(P_2 - P_1)} \qquad (6.11)$$

where

$$Q = \text{flow (in}^3\text{/s)}$$
$$C = \text{coefficient of discharge (approximately 0.63 for water if the orifice}$$
$$\text{hole is at least half the pipe size)}$$
$$A = \text{area of the orifice hole (in}^2\text{)}$$
$$d = \text{weight density of the fluid (lb/in}^3\text{)}$$
$$P_2 - P_1 = \text{pressure difference (psi)}$$
$$g = \text{gravity (384 in./s}^2\text{)}$$

The flow equation (6.11) is an approximation because, in addition to the pressure drop, the actual flow is dependent on velocity effects, the area ratio A_1/A_2, and the surface condition of the pipes. To get the correct constant (C in Equation 6.11) for a particular application, the flow sensor would have to be calibrated.

Figure 6.53

An orifice plate flow sensor.

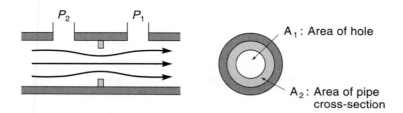

♦ **EXAMPLE 6.16**

A flowmeter for water is made from a 3-in. ID (inside diameter) pipe. The orifice plate has a 2-in. opening. What is the flow if the pressure drop across the orifice plate is 0.2 psi?

Solution   We will use Equation 6.11 to solve this problem, but we first need to figure out the values of the various terms:

$$C = 0.63 \text{ (we can start with this value because 2 in. is more than half of 3 in.)}$$
$$A = \pi r^2 = 3.14 \times (1)^2 = 3.14 \text{ in}^2 \qquad \text{(area of hole)}$$
$$g = 32 \text{ ft/s}^2 = 384 \text{ in./s}^2 \qquad \text{(gravity is a constant)}$$
$$P_2 - P_1 = 0.2 \text{ psi} \qquad \text{(a given)}$$
$$d = 64.4 \text{ lb/ft}^3 = 0.037 \text{ lb/in}^3 \qquad \text{(weight density of water)}$$

Now we can put these terms into Equation 6.11 and solve it:

$$Q = CA\sqrt{\frac{2g}{d}(P_2 - P_1)}$$

$$= (.63)(3.14 \text{ in}^2)\sqrt{\frac{2(384 \text{ in.})}{s^2} \frac{\text{in}^3}{0.037 \text{ lb}} \frac{(0.2 \text{ lb})}{\text{in}^2}} = 127 \text{ in}^3\text{/s}$$

Therefore, 127 in³/s of water flow in the pipe. To make this a more accurate flowmeter, we would have to measure the flow with some other reliable flowmeter and then adjust C in the flow equation to make the readings consistent with what we know them to be.   ♦

Another pressure-based flow sensor uses the venturi to create the pressure difference, as illustrated in Figure 6.54. A **venturi**, a gradual restriction in the pipe, causes the fluid velocity to increase in the restricted area. This area of higher velocity has a lower pressure. The flow is proportional to the difference in pressure between P_2 and P_1. The venturi flow sensor tends to keep the flow more laminar (smooth), but both the orifice plate and the venturi cause pressure drops in the pipe, which may be objectionable.

Figure 6.54

A venturi flow sensor.

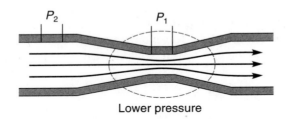

Lower pressure

A pressure-based flow sensor that causes minimum restriction is the pitot tube. The **pitot tube** is a small open tube that faces into the flow (Figure 6.55). The probe actually consists of two tubes: One faces into the flow and reports the **impact pressure** (often called the *velocity head*), and one opens perpendicularly to the flow and reports the **static pressure**. The impact pressure is always greater than the static pressure, and the difference between these two pressures is proportional to velocity and hence to flow. Common uses for the pitot tube are for aircraft and marine speed indicators.

Figure 6.55

A pitot tube flow sensor.

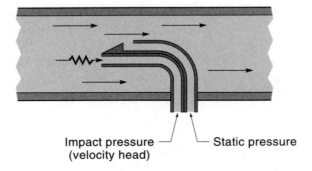

Impact pressure — └─ Static pressure
(velocity head)

Turbine Flow Sensors

Turbine, or spin-type flow sensors (also called flowmeters), employ a paddle wheel or propeller placed in the line of flow. The rotational velocity of the wheel is directly proportional to flow velocity. Figure 6.56 gives an illustration of this type of flow sensor. A small magnet is attached to one of the blades, and a Hall-effect sensor is mounted in the housing. The Hall sensor gives a pulse for each revolution of the blades. The fact that the bearings are in the flow medium may eliminate this type of sensor for some applications, especially high-temperature or abrasive-type fluids.

Figure 6.56

A turbine flow sensor.

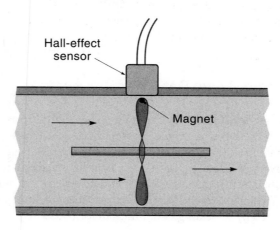

Hall-effect sensor

Magnet

Magnetic Flowmeters

If a liquid is even slightly conductive (and many are), a magnetic flowmeter can be used. Shown in Figure 6.57, the magnetic flowmeter has no moving parts and presents no obstruction to the flow. A nonconducting section of pipe is placed in a magnet field. The moving fluid in the pipe is like the moving conductor in a generator—it produces a voltage. The voltage, which is proportional to the fluid velocity, is detected from electrodes placed in the sides of the pipe.

Figure 6.57

A magnetic flowmeter.

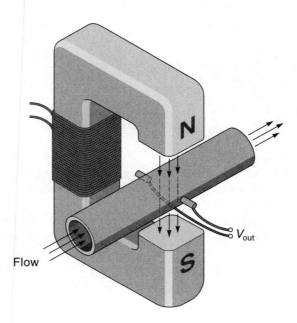

N

S

V_{out}

Flow

6.8 LIQUID-LEVEL SENSORS

Liquid-level sensors, which measure the height of a liquid in a container, have two classifications: discrete and continuous. Discrete-level detectors can only detect whether the liquid is at a certain level. The continuous-level detector provides an analog signal that is proportional to the liquid level.

Discrete-Level Detectors

Discrete-level detectors determine when a liquid has reached a certain level. An application of this type would be determining when to stop the fill cycle of a washing machine. The simplest type of level detector uses a float and a limit switch. There are many possible configurations of a float-based level detector—one is illustrated in Figure 6.58(a). In this case, the float is attached to a vertical rod. At a certain liquid level, the cam, which is attached to the rod, activates the limit switch. The activation level can be adjusted by relocating either the cam or the switch.

Figure 6.58

Discrete liquid-level detectors.

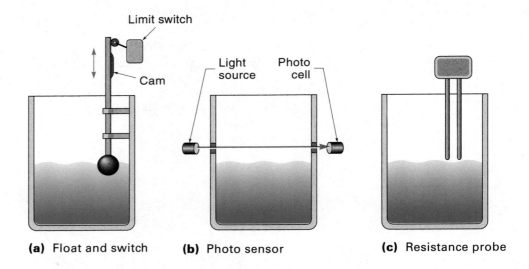

(a) Float and switch **(b)** Photo sensor **(c)** Resistance probe

   Another type of level detector is based on a photocell [Figure 6.58(b)]. When the liquid level submerges the light path, the photodetector signal changes, thus indicating the presence of the liquid.

   Many liquids—such as tap water, weak acids, beer, and coffee (to name a few)—are slightly conductive, which offers another means of detection. As illustrated in Figure 6.58(c), an electric probe is suspended over the liquid. When the liquid reaches the probe, the resistance in the circuit abruptly decreases. One way to detect the resistance change is to use a specially designed IC, such as the ULN-2429A fluid detector (Allegro).

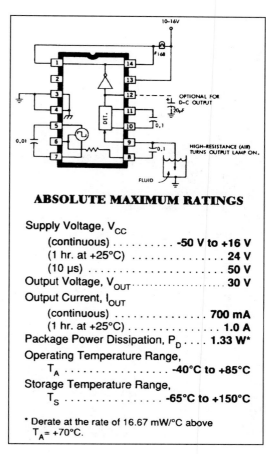

ABSOLUTE MAXIMUM RATINGS

Supply Voltage, V_{CC}

(continuous) . . . . -50 V to +16 V

(1 hr. at +25°C) . . . . 24 V

(10 µs) . . . . 50 V

Output Voltage, V_{OUT} . . . . 30 V

Output Current, I_{OUT}

(continuous) . . . . 700 mA

(1 hr. at +25°C) . . . . 1.0 A

Package Power Dissipation, P_D . . . . 1.33 W*

Operating Temperature Range,

T_A . . . . -40°C to +85°C

Storage Temperature Range,

T_S . . . . -65°C to +150°C

* Derate at the rate of 16.67 mW/°C above
  T_A = +70°C.

Primarily designed for use as an automotive low coolant detector, the ULN2429A monolithic bipolar integrated circuit is ideal for detecting the presence or absence of many different types of liquids in automotive, home, or industrial applications. Especially useful in harsh environments, reverse voltage protection, internal voltage regulation, temperature compensation, and high-frequency noise immunity are all incorporated in the design.

A simple probe, immersed in the conductive fluid being monitored, is driven with an ac signal to prevent plating problems. The presence, absence, or condition of the fluid is determined by comparing the loaded probe resistance with an internal (pin 8) or external (pin 6) resistance. Typical conductive fluids which can be sensed are tap water, sea water, weak acids and bases, wet soil, wine, beer, and coffee.

The high-current output is typically a square wave signal for use with an LED, incandescent lamp, or loudspeaker. A capacitor can be connected (pin 12) to provide a dc output for use with inductive loads such as relays and solenoids.

These devices are furnished in an improved 14-lead dual in-line plastic package with a copper alloy lead frame for superior thermal characteristics. However, in order to realize the maximum current-handling capability of these devices, both of the output pins (1 and 14) and both ground pins (3 and 4) should be used.

FEATURES

■ High Output Current
■ AC or DC Output
■ Single-Wire Probe
■ Low External Parts Count
■ Internal Voltage Regulator

Figure 6.59   The ULN-2429A fluid detector. (Courtesy of Allegro Microsystems)

Figure 6.59 shows the data sheet for this IC. Notice that the probe is connected directly to the IC. The output of the IC is either an AC signal, which can drive a loudspeaker directly, or a DC signal, which is capable of driving a relay.

Continuous-Level Detectors

Continuous-level detectors provide a signal that is proportional to the liquid level. There are a number of ways in which this can be done. One of the most direct methods (used in the gas tank of your car) is a float connected to a position sensor. Figure 6.60(a) illustrates one implementation of this method.

Figure 6.60

Continuous-level-detection methods.

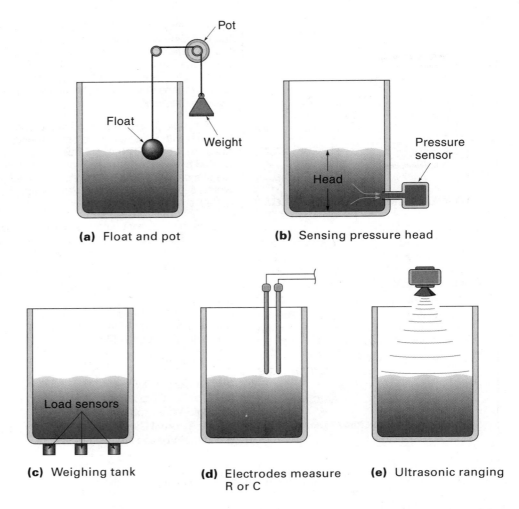

(a) Float and pot

(b) Sensing pressure head

(c) Weighing tank

(d) Electrodes measure R or C

(e) Ultrasonic ranging

Another way to measure liquid level is to measure the pressure at the bottom of the container [Figure 6.60(b)]. This method is based on the fact that the pressure at the bottom of the tank (called the **head**) is directly proportional to the level, as expressed in Equation 6.12:

$$\text{Pressure} = P = dH \tag{6.12}$$

where

P = gauge pressure at the bottom (head)
d = weight density (liquid weight per unit volume)
H = height of liquid in the tank

♦ **EXAMPLE 6.17**

Calculate what the pressure (head) would be at the bottom of a 10-ft deep-water tank.

Solution

To use Equation 6.12, we need to know the weight density of the fluid (which can usually be found in a data book). The density for water is about 64 lb/ft^3:

$$P = dH = 64 \text{ lb/ft}^3 \times 10 \text{ ft} = 640 \text{ lb/ft}^2$$

We may want to convert this result to psi (lb/in^2):

$$640 \text{ lb/ft}^2 \times \frac{1 \text{ ft}^2}{144 \text{ in}^2} = 4.44 \text{ lb/in}^2$$

Recall that 4.44 psi is the gauge pressure, which means the pressure at the bottom of the tank is 4.44 psi more than it is at the surface. The absolute pressure at the bottom would be

$$\underset{\text{Gauge}}{4.44 \text{ psi}} + \underset{\text{Ambient}}{14.7 \text{ psi}} = \underset{\text{Absolute}}{19.1 \text{ psi}}$$

EXAMPLE 6.17 (Repeated with SI Units)

Calculate what the pressure (head) would be at the bottom of a 3-m deep-water tank.

Solution

To use Equation 6.12, we need to know the weight density of the fluid (which can usually be found in a data book). The weight density for water is 9800 N/m^3:

$$P = dH = \frac{9800 \text{ N}}{\text{m}^3} \times 3 \text{ m} = 29,400 \text{ N/m}^2 = 29,400 \text{ Pa}$$

The gauge pressure at the bottom of the tank is 29,400 Pa.   ♦

Monitoring the weight of the liquid with load cells is another technique that can determine liquid level [Figure 6.60(c)]. The level can then be calculated knowing the diameter and weight of the tank (empty) and the density of the fluid. Note that the total weight of the tank is the *sum* of the weights reported by the three load cells.

Some devices can detect the liquid level directly. The device shown in Figure 6.60(d) is simply two vertical electrodes mounted inside the tank. The output of the device, which must be amplified or otherwise processed, is either resistance or capacitance and is proportional to the level. Figure 6.60(e) shows another direct level-sensing system. This system uses an ultrasonic-range detector mounted over the tank. The complete unit, which includes the transducer and electronics, can be purchased as a module and is rather inexpensive.

SUMMARY

Sensors, also called transducers, are devices that sense physical parameters such as position, temperature, or pressure. In most cases, the sensor outputs an analog voltage (or digital value) that is proportional to the parameter being measured.

Position sensors measure the physical position of an object. Potentiometers (variable resistors) measure angular position and give an analog output. The optical encoder, another type of position sensor, uses a slotted disk and a photo sensor. The output of the optical encoder is in digital form. Linear motion can be detected with a linear-motion potentiometer or a linear variable differential transformer (LVDT). The LVDT uses a movable slug inside a special transformer. The phase and magnitude of the AC output can be processed to provide position information.

Position sensors can determine velocity by processing the data from two sequential position samples. The more direct method to measure velocity is to use a tachometer. A DC tachometer is actually a small generator that gives a DC voltage proportional to velocity. Optical tachometers give a pulse for each revolution of a shaft.

Proximity sensors sense whether an object has arrived at a certain place. The simplest way to do this is with a mechanical limit switch. Other methods would include using a photo sensor or a Hall-effect switch. The Hall-effect switch gives an output if a magnet is brought near a specially configured semiconductor.

Load sensors can determine force by measuring the small deformation that the force causes. The traditional method for measuring large forces is with a bonded-wire strain gauge. This device incorporates a pattern of thin wires. When stretched, the resistance of the wires change. Another method for measuring force uses the piezoresistive effect of semiconductors—that is, the resistance changes when the material is compressed.

Pressure sensors measure the pressure of liquids and gases. One class of pressure sensors, such as a bellows, uses the pressure to cause mechanical motion. Semiconductor pressure sensors convert pressure directly into electrical resistance.

A wide variety of temperature sensors are in use. Simple bimetallic strips will bend when heated and can then activate switch contacts. The thermocouple is a traditional high-temperature-sensing device that makes use of the fact that the junction of two dissimilar metals will create a small voltage when heated. The resistance temperature detector (RTD) uses the fact that a wire will increase in resistance when heated. Numerous semiconductor devices are available that "convert" temperature directly into resistance or voltage.

Flow sensors measure the flow of a fluid in a pipe or open channel. Many flow sensors work by placing a restriction in the pipe and then measuring the pressure before and after the restriction. The pressure difference between the two places is proportional to fluid velocity. Turbine flow sensors use the moving fluid to spin a propeller. The rpm of the propeller is proportional to velocity.

Liquid-level sensors determine the level of liquid in a tank. The discrete type can sense only if the level is at or above a certain point. Examples of the discrete level detectors are a float activating a limit switch or a photo sensor. The continuous-level detector gives an analog output proportional to fluid level. There are many different ways this can be done—for example, connecting a float to a potentiometer, monitoring the pressure at the bottom of the tank, or monitoring the weight of the tank with load cells.

GLOSSARY

absolute optical encoder An optical rotary encoder that outputs a binary word representing the angular position.

absolute pressure The pressure difference between the measured value and an absolute vacuum.

bimetallic temperature sensor A strip made from two metals with different coefficients of thermal expansion; as the strip heats up, it bends.

bonded-wire strain gauge A force sensor consisting of a small pattern of thin wires attached to some structural member; when the member is stressed, the stretched wires slightly increase the resistance.

Bourdon tube A type of pressure sensor consisting of a short bent tube closed at one end; pressure inside the tube tends to straighten it.

cold junction   One of two temperature-sensitive junctions of the thermocouple temperature sensor. The cold junction is usually kept at a reference temperature.

compensating gauge   A nonactive strain gauge that is used solely for the purpose of canceling out temperature effects.

continuous-level detector   A sensor that can determine the fluid level in a container.

differential pressure   The difference between two pressures where neither may be ambient.

direct current tachometer   Essentially, a DC generator that gives an output voltage proportional to angular velocity.

discrete-level detector   A sensor that can determine if the fluid in a container has reached a certain level.

flow sensor   A sensor that measures the quantity of fluid flowing in a pipe or channel.

gauge pressure   The difference between measured and ambient pressure (ambient is 14.7 psi or 101.3 kPa).

Grey code   A sequence of digital states that has been designed so that only 1 bit changes between any two adjacent states.

Hall effect   The phenomena of a semiconductor material generating a voltage when in the presence of a magnetic field; used primarily as a proximity sensor.

head   The fluid pressure in a tank, which is caused by the weight of the fluid above and is therefore proportional to the level.

Hooke's law   The deformation in a spring is directly proportional to the force on the spring (spring force = constant × spring deformation).

hot junction   One of two temperature-sensitive junctions of the thermocouple temperature sensor. The hot junction is used on the probe.

ice-water bath   A traditional way to create a known reference temperature for the cold junction of a thermocouple.

impact pressure   The pressure in an open (pitot) tube pointed "upstream."

incremental optical encoder   An optical rotary encoder that has one track of equally spaced slots; position is determined by counting the number of slots that pass by a photo sensor.

Kelvin scale   The Kelvin temperature scale starts at absolute zero, but a Kelvin degree has the same temperature increment as a Celsius degree ($0°C = 273$ K).

least significant bit (LSB)   The right-most bit in a binary number, it represents the smallest quantity that can be changed—that is, the difference between two successive states.

limit switch   A switch used as a proximity sensor—that is, a switch mounted so that it is activated by some moving part.

linearity error   A measurement error induced into the system by the sensor itself, it is the difference between the actual quantity and what the sensor reports it to be.

linear variable differential transformer (LVDT)   A type of linear-motion position sensor. The motion of a magnetic core, which is allowed to slide inside a transformer, is proportional to the phase and magnitude of the output voltage.

loading error   An error that may occur in an analog voltage signal when the circuit being driven draws too much current, thus "loading down" the voltage.

LSB   *See* **least significant bit**.

LVDT   *See* **linear variable differential transformer**.

optical rotary encoder   A rotary position sensor that works by rotating a slotted disk past a photo sensor.

optical tachometer   Mounted next to a rotating shaft, a photo sensor that gives a pulse for each revolution (a stripe is painted or fixed on the shaft).

orifice plate   A type of flow sensor whereby a restriction is placed in a pipe causing a pressure difference (between either side of the restriction) that is proportional to flow.

photodiode   A type of optical sensor, it increases its reverse-leakage current when exposed to light.

photo resistor   A type of optical sensor; its resistance decreases when exposed to light.

photo transistor   A type of optical sensor; light acts as the base current and turns on the transistor.

photovoltaic cell   A device that converts light into electrical energy; used as an optical sensor, or as a solar cell.

piezoresistive effect   A property of semiconductors in which the resistance changes when subjected to a force.

potentiometer   A variable resistor that can be used as a position sensor.

proximity sensor   A sensor that detects the physical presence of an object.

psi   Pounds per square inch, a unit of pressure.

pitot tube   A velocity sensor for fluids whereby a small open tube is placed directly into the flow; the pressure in the tube is proportional to fluid velocity.

Rankine scale   The Rankine temperature scale starts at absolute zero, but a Rankine degree has the same temperature increment as a Fahrenheit degree ($0°F = 460°R$).

resistance temperature detector (RTD)   A temperature sensor based on the fact that the resistance of a metal wire will increase when the temperature rises.

resolution The smallest increment of data that can be detected or reported. For a digital system, the resolution is usually the value of the least significant bit.

RTD *See* **resistance temperature detector.**

Seebeck effect The property used by a thermocouple, a voltage proportional to temperature developed in a circuit consisting of junctions of dissimilar metal wires.

sensor A device that measures some physical parameter such as temperature, pressure, or position.

slider (wiper) The moving contact in a potentiometer, usually the center of three terminals.

slotted coupler An optical proximity sensor that is activated when an object "cuts" a light beam.

static pressure The pressure measured when the open tube is directed perpendicularly to the flow.

strain The deformation (per unit length) as a result of stress.

stress Subjecting an object to tension or compression forces. Stress is the force per unit area within the object.

thermistor A temperature sensor based on the fact the resistance of some semiconductors will decrease as the temperature increases.

thermocouple A temperature-measuring sensor made from the junction of two dissimilar metal wires; when the junction is heated, a small voltage is generated.

threshold detector A circuit that provides a definite off–on signal when an analog voltage rises above a certain level.

transducer A term used interchangeably with *sensor*. Literally means that energy is converted, which is what a sensor does.

turbine A flow sensor based on having the fluid rotate a propeller of some kind.

venturi A type of flow sensor whereby fluid is forced into a smaller channel, which increases its velocity; the higher velocity fluid has a lower pressure (than the fluid in the main channel), and the pressure difference is proportional to velocity.

wiper *See* **slider.**

wire-wound potentiometer A variable resistor that uses a coil of resistance wire for the resistive element; can be used as a position sensor.

Young's modulus A constant that relates stress and strain for a particular material (Young's modulus = stress/strain).

EXERCISES

Section 6.1

1. A potentiometer with a total range of 350° is supplied with a voltage of 12 Vdc. Find the output voltage for an angle of 135°.

2. A potentiometer with a total range of 350° is supplied with a voltage of 8 Vdc. The voltage at the wiper is 3.7 Vdc. What is the present angle of the pot?

3. A 10-kΩ pot is used as a position sensor with a 10-V supply. The input resistance to the interface circuit is 50 kΩ. Find the loading error when the pot is in the middle of its range.

4. A 5-kΩ pot is used as a position sensor with a 10-V supply. The input resistance to the interface circuit is 40 kΩ. Find the loading error when the pot is set to three-quarters of its range.

5. A potentiometer with a total range of 350° has a linearity of 0.25% and is connected to a 10-V source. If the pot is used as a position sensor, find the maximum possible error in degrees.

6. A 350° wire-wound potentiometer has 300 turns and a total resistance of 1 kΩ. What is the resolution in ohms? In degrees?

7. A 350° pot is connected to a shaft through a 3 : 1 gear ratio (the pot rotates three times further than the shaft). The pot has a linearity of 0.2%.
   a. What is the maximum angle the shaft can turn?
   b. What is the maximum difference in degrees from where the shaft is and where the pot reports it to be?

8. A pot with a total range of 350° is supplied with 5 Vdc. The output of the pot is converted to binary with an 8-bit ADC (which is also referenced at 5 Vdc). Find the 8-bit binary output of the ADC for a pot angle of 60°.

9. A 350° pot is connected through a 4 : 1 gear ratio to a shaft that rotates 80° (the pot rotates four times further

than the shaft). The pot has a supply voltage of 5 V and feeds to an 8-bit ADC converter. The ADC has a reference voltage of 5 V. The LSB of the ADC is to be 0.4° or smaller. Will this system meet the requirements?

10. An absolute optical rotary encoder has five tracks. How many bits is the output, and what is the resolution in degrees per state (that is, degrees per LSB)?

11. An absolute optical rotary encoder in a certain application must have a resolution of 3°. How many tracks must it have?

12. An incremental optical rotary encoder has 720 slots. Starting from the reference point, the disk turns 200 slots CW, then 80 slots CCW, then 400 slots CW. What is the final angle of the shaft?

13. The counter for a 500-slot incremental optical rotary encoder has a value of 101100011.
    a. What is the resolution in degrees—that is, what is the value of the LSB?
    b. What is the current angle of the encoder shaft?

14. Compare an LVDT with a potentiometer as a position sensor. What are some advantages and disadvantages of using an LVDT?

Section 6.2

15. Explain the basic principle of extracting velocity data from position sensors.

16. Position data from a pot is being processed to yield velocity data. The sample period is 0.5 s. The position is 68° at the first sample and 73° for the second sample. Calculate the velocity.

17. Data from an optical shaft encoder are providing velocity information. The sample period is 0.25 s. The present data are 10000111, and the previous data were 10000101. Find the present velocity (LSB = 1°).

18. The CK20-B tachometer is being used in a system. The output is 0.85 V. What is the velocity in rpm?

Section 6.3

19. Describe one application of using a limit switch as a proximity sensor.

20. Both optical and Hall-effect sensors can be used as proximity switches. Name one possible application for each.

21. Explain the operation of a slotted coupler optical sensor. Use an application in your explanation.

Section 6.4

22. When a 200-lb man gets in his car, the car body sinks 1 in. What is the spring constant for the suspension system of the car?

23. When a 180-lb man gets in his car, the car body sinks 1.25 in. What is the spring constant for the suspension system of the car?

24. When an 80-kg man gets in his car, the car body sinks 2 cm. What is the spring constant for the suspension system of the car? [*Hint:* Remember to convert the 80-kg mass into a weight using $W = mg$.)

25. How does a bonded-wire strain gauge work as a force sensor?

26. What is the purpose of using a bridge circuit with a strain gauge, and what is the purpose of the compensating gauge?

27. A strain gauge is used to measure the tension force in a 1-in. diameter bar of steel. The strain gauge has a nominal resistance of 140 Ω and a gauge factor (GF) of 4. The strain gauge is connected to a bridge (which is supplied with 10 Vdc). The bridge was initially balanced. After the bar is put under tension, the bridge output voltage goes to 0.0008 Vdc. Draw a schematic of the setup and calculate the force on the bar.

28. A strain gauge is used to measure the tension force in a 0.75-in. diameter bar of steel. The strain gauge has a nominal resistance of 180 Ω and a GF of 4. The strain gauge is connected to a bridge (which is supplied with 10 Vdc). The bridge was initially balanced. After the bar is put under tension, the bridge output voltage goes to 0.0006 Vdc. Draw a schematic of the setup and calculate the force on the bar.

29. A strain gauge is used to measure the tension force in a 2-cm diameter bar of steel. The strain gauge has a nominal resistance of 160 Ω and a GF of 5. The strain gauge is connected to a bridge (which is supplied with 10 Vdc). The bridge was initially balanced. After the bar is put under tension, the bridge output voltage goes to 0.0008 Vdc. Draw a schematic of the setup and calculate the force on the bar.

30. A low-force sensor uses a spring and a 1-kΩ linear pot. The pot has a 1-in. stroke and a supply voltage of 10 Vdc. If the sensor is to have an output of 1 V per 10 lb of load, what is the spring constant required?

Section 6.5

31. How does a Bourdon tube work as a pressure sensor?

32. What is the difference between *gauge pressure*, *differential pressure*, and *absolute pressure*?

33. A differential pressure bellows sensor is receiving 4 psi (gauge) on one side and 12 psi (absolute) on the other. What pressure does the sensor report?

34. A differential pressure bellows sensor is receiving 5 psi (gauge) on one side and 16.2 psi (absolute) on the other. What pressure does the sensor report?

35. A differential pressure bellows sensor is receiving 20 kPa (gauge) on one side and 90 kPa (absolute) on the other. What pressure does the sensor report?

Section 6.6

36. How does a bimetallic temperature sensor work?

37. List the three ways that the cold junction of a thermocouple is handled in modern systems.

38. An iron–constantan thermocouple is referenced at 32°F and has an output voltage of 45 mV. What is the temperature at the hot junction?

39. An iron–constantan thermocouple is used to measure the temperature in an oven. The cold junction is at ambient temperature of 80°F, and the thermocouple voltage is 12 mV. What is the approximate oven temperature?

40. A 100-Ω platinum RTD is being used to measure temperature in an oven. The present resistance is 122 Ω. Find the temperature.

41. A 100-Ω platinum RTD is being used to measure temperature in an oven. The present resistance is 130 Ω. Find the temperature.

42. Construct a temperature sensor using the LM35 with the following specifications: Supply voltage = 12 V, and output = 0.5 V/°C.

Section 6.7

43. Describe the operating principles of the following flow sensors:
    a. Orifice plate
    b. Venturi
    c. Pitot tube
    d. Turbine
    e. Magnetic

44. An orifice plate flow sensor is used to measure the flow of water in a 2-in. ID pipe. The orifice plate has a 1.5-in. diameter hole. The pressure difference across the plate is 0.3 psi. Find the approximate water flow in the pipe.

Section 6.8

45. How would you use two discrete-level detectors to maintain the liquid level in a tank at between 3 and 4 ft?

46. A pressure sensor is used to measure the water level in a tank. What pressure would you expect for 5 ft of water?

47. A pressure sensor is used to measure the water level in a tank. What pressure would you expect for 2 m of water?

48. A pressure sensor is used to measure the level of a liquid in a tank. The density of the liquid is 52 lb/ft^3. If the gauge pressure (from the sensor) is 5.7 psi, what is the height of the liquid?

49. A pressure sensor is used to measure the level of a liquid in a tank. The weight density of the liquid is 6000 N/m^3. If the gauge pressure (from the sensor) is 12,000 Pa, what is the height of the liquid?

7 Direct Current Motors

After studying this chapter, you should be able to:

- Explain the theory of operation of electric motors in general and DC motors in particular.
- Distinguish the characteristics of series-wound, shunt-wound, compound, and permanent magnet motors.
- Use the torque–speed curve of a motor to predict its performance.
- Select a DC motor based on mechanical requirements.

- Understand the operation of linear amplifier drivers for DC motors that incorporate power transistors, IC amplifiers, Darlington transistors, and power MOSFETS.
- Understand DC motor–speed control using pulse-width-modulation concepts.
- Understand operating a DC motor from rectified AC, using silicon-controlled-rectifier circuits.
- Understand the operating principles of brushless DC motors.

INTRODUCTION

An indispensable component of the control system is the actuator. The actuator is the first system component to actually move, converting electrical energy into mechanical motion. The most common type of actuator is the electric motor.

Motors are classified as either DC or AC, depending on the type of power they use. AC motors (covered in Chapter 9) have some advantages over DC motors: They tend to be smaller, more reliable, and less expensive. However, they generally run at a fixed speed that is determined by the line frequency. DC motors have speed-control capability, which means that speed, torque, and even direction of rotation can be changed at any time to meet new conditions. Also, smaller DC motors commonly operate at lower voltages (for example, a 12-V disk drive motor), which makes them easier to interface with control electronics.

233

7.1   THEORY OF OPERATION

The discovery that led to the invention of the electric motor was simply this: A current-carrying conductor will experience a force when placed in a magnetic field. The conductor can be any metal—iron, copper, aluminum, and so on. The direction of the force is perpendicular to both the magnetic field and the current (Figure 7.1). A demonstration of this principle is easy to perform with a strong magnet, flashlight battery, and a wire and is highly recommended! Place the wire between the magnet poles and alternately connect and disconnect the wire from the battery. Each time you complete the circuit, you should feel a little tug on the wire. The magnitude of the force on the wire can be calculated from the following equation:

$$F = IBL \sin \theta \tag{7.1}$$

where

F = force on the conductor (in Newtons)
I = current through the conductor (in amperes)
B = magnetic flux density (in gauss)
L = length of the wire (in meters)
θ = angle between the magnetic field and current

Figure 7.1

Action of force on a wire in a magnetic field.

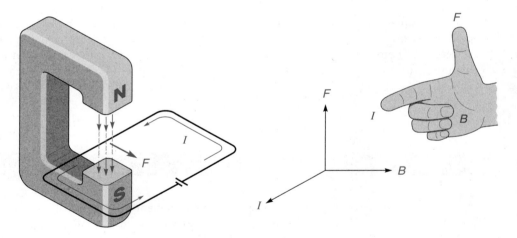

(a) Experimental setup

(b) Directions of I, F, B are mutually orthogonal (Conventional current flow)

An electric motor must harness this force in such a way as to cause a rotary motion. This can be done by forming the wire in a loop and placing it in the magnetic field (Figure 7.2). The loop (or *coil*) of wire is allowed to rotate about the axis shown and is called the **armature** winding. The armature is placed in a magnetic field called the **field**. The **commutator** and **brushes** supply current to the armature while allowing it to rotate.

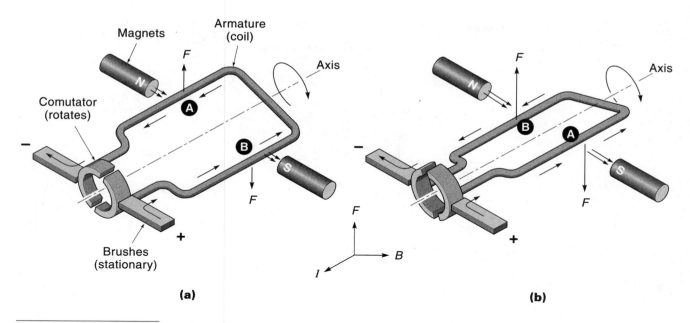

Figure 7.2   A simple DC motor action (conventional flow).

To understand how the motor works, look at Figure 7.2(a). Notice that wire segments A and B of the coil are in the same magnetic field, but the current in wire segment A is coming out of the page, whereas the current in wire segment B is going in. Applying the force diagram from Figure 7.1(b) (and repeated in Figure 7.2), we see that wire segment A of the coil would be forced up, whereas wire segment B would be forced down. These forces would cause the coil to rotate clockwise. Figure 7.2(b) shows the situation after the coil has rotated about 90°. The current has now reversed direction in the coil because the commutator contacts have rotated and are now making contact with the opposite brush. Now wire segment A of the coil will be forced down and wire segment B up, which causes the armature to continue rotating clockwise. **Torque**, as explained in Chapter 5, is the rotational force a motor can exert. Notice that the torque will be at a maximum when the coil is horizontal and will drop to zero when the coil is vertical (similar to peddling a bicycle).

To reverse the direction of the motor, polarity of the voltage to the commutator is reversed. This causes the forces on the armature coil to be reversed, and the motor would then run in the opposite direction.

Figure 7.3 shows the armature of a practical motor. Notice that there are multiple coils and each coil experiences the forces described in the preceding paragraph and so contributes to the overall torque of the motor. Each coil is connected to a separate pair of commutator segments, causing the current in each coil to switch directions at the proper time for that individual coil. The overall effect is to provide approximately the same torque for any armature position (like a multipiston engine).

One of the most important operating parameters of any motor is torque. Electric motor torque is directly proportional to the force on the armature wires. From Equation 7.1, we see that the force is proportional to the magnetic field and current (not voltage).

Figure 7.3

DC motor armature.

By gathering the mechanical parameters of the motor (such as number of poles) into a single constant K, the motor torque can be expressed as

$$T = K_T I_A \phi \qquad (7.2)$$

where

  T = motor torque
  K_T = a constant based on the motor construction
  I_A = armature current
  ϕ = magnetic flux

So far we have been looking at how the motor converts electrical energy to mechanical energy. It turns out that the very same device (motor) is also capable of converting mechanical energy to electrical energy, in which case it is called a generator. For example, if the armature coil of Figure 7.2 were rotated in the magnetic field by some external force, a voltage [called the *electromotive force* (EMF)] would appear on the commutator segments. The magnitude of the EMF is given in Equation 7.3:

$$\text{EMF} = K_E \phi S \qquad (7.3)$$

where

  EMF = voltage generated by the turning motor
  K_E = a constant based on motor construction
  ϕ = magnetic flux
  S = speed of motor (rpm)

Although it may seem strange, this EMF voltage is being generated even when the motor is running on its own power, but it has the opposite polarity of the line voltage; hence, it is called the **counter-EMF** (CEMF). *Its effect is to cancel out some of the line voltage.* In other words, the actual voltage available to the armature is the line voltage minus the CEMF:

$$V_A = V_{ln} - \text{CEMF} \qquad (7.4)$$

where

  V_A = actual voltage available to the armature
  V_{ln} = line voltage supplied to the motor
  CEMF = voltage generated within the motor

You can never actually measure V_A with a voltmeter because it is an effective voltage inside the armature. However, there *is* physical evidence that the CEMF exists because the armature current is also reduced, as indicated in Equation 7.5:

$$I_A = \frac{V_{\text{ln}} - \text{CEMF}}{R_A} \tag{7.5}$$

where

$$
\begin{aligned}
I_A &= \text{armature current} \\
V_{\text{ln}} &= \text{line voltage to the motor} \\
R_A &= \text{armature resistance} \\
\text{CEMF} &= \text{voltage generated within the motor}
\end{aligned}
$$

Equation 7.5 (which is in the form of Ohm's law) tells us that the armature current is a function of the applied voltage minus the CEMF. Because CEMF increases with motor speed, the faster the motor runs, the less current the motor will draw, and consequently its torque will diminish. This explains why most DC motors have a finite maximum speed; eventually, if the motor keeps going faster, the CEMF will nearly cancel out the line voltage, and the armature current will approach zero.

The actual relationship between motor speed and CEMF follows and is derived from Equation 7.3:

$$S = \frac{\text{CEMF}}{K_E \phi} \tag{7.6}$$

where

$$
\begin{aligned}
S &= \text{speed of the motor (rpm)} \\
\text{CEMF} &= \text{voltage generated within the motor} \\
K_E &= \text{a motor constant} \\
\phi &= \text{magnetic flux}
\end{aligned}
$$

Looking at Equation 7.6, we see that the motor speed is directly proportional to the CEMF (voltage) and (surprisingly) inversely proportional to the field flux.

♦ **EXAMPLE 7.1**

A 12 Vdc motor has an armature resistance of 10 Ω and generates a CEMF at the rate of 0.3 V/100 rpm. Find the actual armature current at 0 rpm and at 1000 rpm.

Solution   We can find the armature current with Equation 7.5. For the first case, when the motor isn't turning at all (0 rpm), the CEMF will be 0 V:

$$I_A = \frac{V_{\text{ln}} - \text{CEMF}}{R} = \frac{12\text{ V} - 0\text{ V}}{10} = 1.2\text{ A}$$

For the second case (1000 rpm), determine the CEMF before applying Equation 7.5. Given the CEMF rate of 0.3 V/100 rpm,

$$\text{CEMF} = \frac{0.3\text{ V}}{100\text{ rpm}} \times 1000\text{ rpm} = 3\text{ V}$$

Then

$$I_A = \frac{V_{ln} - CEMF}{R_a} = \frac{12\,V - 3\,V}{10} = \frac{9\,V}{10} = 0.9\,A$$

Thus, we see that the armature current is reduced in the running motor.   ◆

7.2   WOUND-FIELD DC MOTORS

Wound-field motors use an electromagnet called the **field winding** to generate the magnetic field. The only other way to generate a magnetic field is with permanent magnets, which will be discussed later. The speed of wound-field motors is controlled by varying the voltage to the armature or field windings. Figure 7.4 shows an example of a wound-field motor. This particular series of motors is available from ¼ to 1 hp and at several standard rated speeds. The **rated speed** of a motor is the speed when it is supplying the rated horsepower—when the motor is unloaded, it will go faster than the rated speed. These motors are designed to run at 90 Vdc because 90 V is about what a practical rectifier circuit can produce from 120 Vac. The speed can be controlled by adjusting the rectified voltage, as will be discussed later in this chapter. There are three basic types of wound field motors: series wound, shunt wound, and compound.

Series-Wound Motors

In a **series-wound motor**, the armature and field windings are connected in series [Figure 7.5(a)]. This configuration gives the motor a large starting torque, which is useful in many situations—for example, car starter motors. The explanation for this large initial torque is as follows: When the motor is stopped, there is no CEMF, and the full-line voltage is available to the windings. Therefore, the initial armature current is large; and being in series with the field windings, this same current creates a large magnetic field as well. The combination of a large armature current *and* a large field flux is what produces the large start-up torque (see Equation 7.2). Once the motor starts turning, the increasing CEMF reduces the motor currents and hence the torque. Because the field coil carries the full armature current, it must have a low resistance; thus, it consists of a few turns of heavy-gauge wire.

Another characteristic of the series-wound motor is that it tends to "run away" (go faster and faster) under no-load conditions. As the field current diminishes from the increasing CEMF, the magnetic field flux also decreases, which according to Equation 7.6 tends to speed up the motor, which increases the CEMF even more. The overall effect of this seemingly circular logic is that the motor will continue to accelerate until the torque is balanced by friction forces. Most smaller motors can handle the high speeds without causing any damage, but larger motors may literally fly apart if operated with no load.

WOUND FIELD

EXPLOSION-PROOF
$^1/_4$–1 HP
TOTALLY ENCLOSED ● FAN-COOLED

FEATURES

- ADJUSTABLE SPEED
- 90 V ARMATURE 50/100 V FIELD(3)
- CLASS F INSULATION
- 40°C AMBIENT
- EXPLOSION-PROOF(1)
- FOR OPERATING FROM FULL WAVE SINGLE-PHASE RECTIFIED POWER (TYPE K)
- RIGID MOUNT C-FACE

Wound Field – Explosion-Proof

HP	RPM	Enclosure	Frame ■
$^1/_4$	3500	TEFC-XP	HM56HC
	2500	TEFC-XP	HB56HC
	1750	TEFC-XP	HB56HC ◆
	1150	TEFC-XP	HU56HC
$^1/_3$	3500	TEFC-XP	HG56HC
	2500	TEFC-XP	HU56HC
	1750	TEFC-XP	HB56HC ◆
	1150	TEFC-XP	HJ56HC
$^1/_2$	3500	TEFC-XP	HU56HC
	2500	TEFC-XP	HG56HC
	1750	TEFC-XP	HJ56HC ◆
	1150	TEFC-XP	HE56HC
$^3/_4$	3500	TEFC-XP	HU56HC
	2500	TEFC-XP	HJ56HC
	1750	TEFC-XP	HJ56HC ◆
1	3500(2)	TEFC-XP	HJ56HC
	2500(2)	TEFC-XP	HE56HC
	1750(2)	TEFC-XP	HE56HC ◆

Figure 7.4 Wound-field DC motor. (Courtesy of Reliance Electric)

Figure 7.5(b) shows a typical **torque–speed curve** for a series-wound motor. Understanding this graph is important because it is the most useful tool in describing a particular motor's operating characteristics. The vertical axis has two scales: speed and current. The horizontal axis represents torque. As described earlier, torque is a measure of the motor's strength in turning its shaft and is the direct result of the forces on the armature conductors. The graph shows the torque–speed relationship for a particular constant applied voltage. A lower motor voltage would give a curve of the same shape but lying to the left of the one shown.

Looking at the curve [Figure 7.5(b)], notice that *speed decreases as torque increases.* This is true of all electric motors, and it makes intuitive sense; when the load is increased, the motor slows down. As an example, consider an electric circular saw. You can hear the motor slowing down under the load of each board being cut. *The maximum torque the motor can deliver occurs when the motor is loaded so much it comes to a stop.* This is called the **stall torque**. Some motors are designed to operate right down to the stall torque; in fact, many control system motors operate in this mode all the time. For example, a motor driving a robot arm must move the loaded arm from one rest position to another. A useful coincidence is that the stall torque is also the maximum torque because most mechanical systems require more force to get things moving from a resting position than they do at other times.

The top end of the curve [Figure 7.5(b)] reveals the **no-load speed**. This is the speed the motor would attain if it were allowed to run with absolutely no external load on it. In series motors, the no-load speed is established by a number of factors, including bearing friction and wind resistance, and so is somewhat unpredictable. In other types of motors, the effect of CEMF is the primary factor in determining the no-load speed. Notice that at the no-load speed the motor is capable of doing absolutely no useful work because it is supplying zero torque to the outside world. Therefore, *any motor that is doing something useful must be going at less than its no-load speed.* As pointed out before, larger series motors could be damaged if run with no external load and, for that reason, are never connected to a load with a belt—belts break!

The other parameter graphed in Figure 7.5(b) is the current (shown as a dashed line). The current increases as the torque increases because the force on the armature conductors is proportional to armature current. It may seem backward that when the motor turns faster it draws less current, but a faster turning motor means it has less of a load on it.

The speed–torque curve of the series-wound motor is highly nonlinear. The unusually high-stall torque is a desirable quality for many workhorse applications such as cranes, portable power tools, and automobile starter motors. Series motors are less desirable in control applications, however, because the nonlinear characteristics complicate the mathematics done by the controller. One last comment on the series motor: Like all wound field motors, reversing the polarity of the line voltage reverses both the armature and field windings, causing the motor to continue to rotate in the same direction. To reverse the direction of rotation, the polarity of *only* the armature (or field) would have to be reversed.

Shunt-Wound Motors

In the **shunt-wound motor**, the armature and field windings are connected in parallel [Figure 7.5(c)]. With this configuration, the current in the field is dependent only on the supply voltage. In other words, the field flux is not affected by variations in current due to the CEMF. This results in a motor with a more natural speed regulation. **Speed regulation** is the ability of a motor to maintain its speed when the load is applied. The basis of this self-regulation is the CEMF. When the motor's load is increased, the speed tends to decrease, but the lower speed reduces the CEMF, which allows more current into the armature. The increased current results in increased torque, which prevents the motor from slowing further. Speed regulation is usually expressed as a percentage, as shown in Equation 7.7:

$$\% \text{ speed regulation} = \frac{S_{NL} - S_{FL}}{S_{FL}} \times 100 \qquad (7.7)$$

Figure 7.5
Series and shunt DC motors.

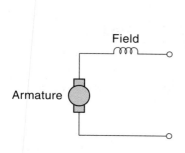

(a) Series motor circuit

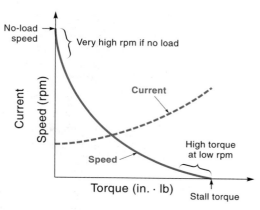

(b) Torque-speed curve for series motor

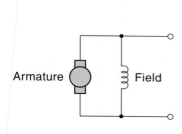

(c) Shunt motor circuit

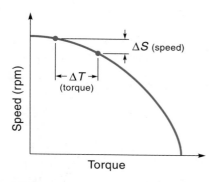

(d) Torque-speed curve for shunt motor

where
S_{NL} = no-load speed
S_{FL} = full-load speed

♦ **EXAMPLE 7.2**

A motor has a no-load speed of 1150 rpm. When the maximum load for a certain application is applied to the motor, the speed drops to 1000 rpm. Find the speed regulation for this application.

Solution Applying Equation 7.7, we get

$$\frac{S_{NL} - S_{FL}}{S_{FL}} \times 100 = \frac{1150 \text{ rpm} - 1000 \text{ rpm}}{1000 \text{ rpm}} \times 100 = 15.0\%$$ ♦

The torque–speed curve of Figure 7.5(d) graphically shows the shunt-wound motor's characteristics. Notice that both the stall torque and no-load speed values are relatively low when compared to the series motor. Also, the top portion of the curve tends to be more horizontal. This (more) horizontal region is the area of good speed regulation. An increase in torque [ΔT in Figure 7.5(d)] reduces the motor's speed by only ΔS. In practice, this means a shunt-wound motor will slow down somewhat when the load increases and speed up somewhat when the load is reduced, but on average, a ¼-hp shunt motor will change speed only 15% from a no-load to a rated full-load operation.

Because the shunt motor tends to run at a relatively constant speed, it has tradition-ally been used in such applications as fans, blowers, conveyer belts, and machine tools. The shunt motor can be designed to operate at almost any torque–speed combination at a given line voltage by specifying the number of turns in the field windings. For a fixed line voltage, the speed can be controlled (within limits) by inserting a rheostat in series with the field winding. As the resistance is increased, the current to the field is reduced, which reduces the field flux and [according to the motor-speed equation (7.6)] speeds up the motor. Speed control can also be achieved by adjusting the line voltage directly. As is the case with the series motor, the shunt motor is reversed by reversing the polarity of only the armature or field coil, but not both (usually it's the armature).

◆ **EXAMPLE 7.3**

An old 90-Vdc shunt motor on a conveyer belt needs to be replaced. The identifica-tion plate on the old motor is unreadable, but you know it was turning at about 1750 rpm. Using your ingenuity and common measuring instruments, determine the specifications for a new motor and select one from the list of Figure 7.4.

Solution

The problem comes down to specifying the horsepower, and to know that you need to know how much torque the old motor was supplying. Using a torque wrench, you turn the conveyer belt shaft and determine it requires about 2 ft · lb to keep it moving.

To convert this data into horsepower requires two formulas;

$$P = TS \tag{7.8}$$

where

P = power
T = torque
S = motor speed

$$1 \text{ hp} = 33,000 \ \frac{\text{ft} \cdot \text{lb}}{\text{min}} \tag{7.9}$$

First, use Equation 7.8 to calculate the power required to run the conveyer belt (*revolutions* must be converted to *radians*):

$$P = TS = 2 \text{ ft} \cdot \text{lb} \times \frac{1750 \text{ rev}}{\text{min}} \times \frac{2\pi \text{ rad}}{\text{rev}} = 21,990 \ \frac{\text{ft} \cdot \text{lb}}{\text{min}}$$

Now use Equation 7.9 to convert this result into horsepower:

$$\frac{21,990 \text{ ft} \cdot \text{lb}}{\text{min}} \times \frac{1 \text{ hp} \cdot \text{min}}{33,000 \text{ ft} \cdot \text{lb}} = 0.666 \text{ hp}$$

So, based on our calculations, we need a motor that rotates at 1750 rpm with at least 0.666 hp. Looking down the list of Figure 7.4, we see a ¾-hp model with a rated speed of 1750, which should work fine.   ◆

Compound Motors

The **compound motor** has both shunt and series field windings, although they are not necessarily the same size. Figure 7.6(a) illustrates a compound motor circuit. Typically, the series and shunt coils are wound in the same direction so that the field fluxes add. The main purpose of the series winding is to give the motor a higher starting torque. Once the motor is running, the CEMF reduces the strength of the series field, leaving the shunt winding to be the primary source of field flux and thus providing some speed regulation. Also, the combination of both fields acting together tends to straighten out (linearize) a portion of the torque–speed curve [Figure 7.6(b)].

Figure 7.6

A compound DC motor.

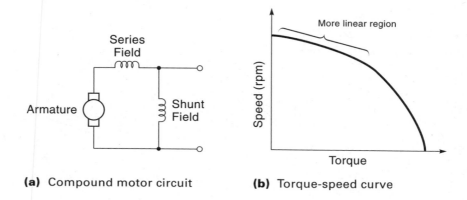

(a) Compound motor circuit   **(b)** Torque-speed curve

The motor discussed so far, where the fields add, is called a **cumulative compound motor**. Less common is the **differential compound motor**, where the field coils are wound in opposite directions. The differential compound motor has very low starting torque but excellent speed regulation. However, because it can be unstable at higher loads, it is rarely used. The compound motor direction of rotation is reversed by reversing the polarity of the armature windings.

7.3   PERMANENT MAGNET MOTORS

Permanent magnet (PM) motors use permanent magnets to provide the magnetic flux for the field. The armature is similar to those in the wound-field motors discussed earlier. Three types of magnets are used: (1) The Alnico magnet (iron based alloy) has a high-flux density but loses its magnetization under some conditions such as a strong armature field during stalled operation; (2) ferrite (ceramic) magnets have a low-flux density, so they have to be larger, but they are not easily demagnetized; (3) the newer, so-called rare-earth magnets, made from samarium-cobalt or neodymium-cobalt, have the combined desirable properties of high-flux density and high resistance to demagnetization. At the present time, the size of PM motors is limited to a few horsepower or less. Small PM motors are used extensively in office machines such as printers and disk drives, toys, equipment such as VCRs and cameras (for zoom and autofocus), and many places in industry. Larger PM motors are used in control systems such as industrial robots.

Torque and Speed Relationship

The fact that the field flux of a PM motor remains constant regardless of speed makes for a very linear torque–speed curve. This is very desirable for control applications because it simplifies the control equations. Figure 7.7 shows a typical PM motor symbol and torque–speed curve. Notice that the "curves" are straight lines for both speed and current. The absence of field coils is apparent in the schematic [Figure 7.7(a)], which shows the applied voltage feeding only the armature. The PM motor is easily reversed by reversing the polarity of the applied voltage. Example 7.4 illustrates how the torque–speed curve can be used to predict the performance of a motor under any load condition.

Figure 7.7

A permanent magnet motor.

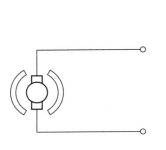

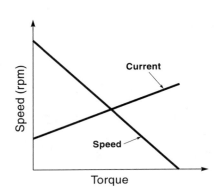

(a) PM motor symbol   **(b)** PM motor torque-speed curve

♦ **EXAMPLE 7.4**

Figure 7.8(a) shows the torque–speed curve of a PM motor. Find the speed and motor current for the following:

    a. No-load and stall conditions

    b. Lifting a 10-oz load with a 2-in. radius pulley

    c. A motor driving a robot arm with a weight

Solution

    a. If voltage is applied to the motor with no load attached to the shaft, the motor would turn at its no-load speed of 1000 rpm. On the other hand, if the shaft was clamped so it could not turn, the motor would exert the stall torque of 100 in. · oz on the clamp and draw 260 mA of current.

    b. A 10-oz weight is hung from a 2-in. radius pulley. The pulley is on the motor shaft [Figure 7.8(b)]. Torque equals force times distance, so the motor torque required to lift the weight is

$$2 \text{ in.} \times 10 \text{ oz} = 20 \text{ in.} \cdot \text{oz}$$

From the graph, we can see that, at a torque of 20 in. · oz, the speed has declined to 800 rpm, and the current is up to 125 mA.

Figure 7.8

Torque–speed curve and hardware setups for Example 7.4.

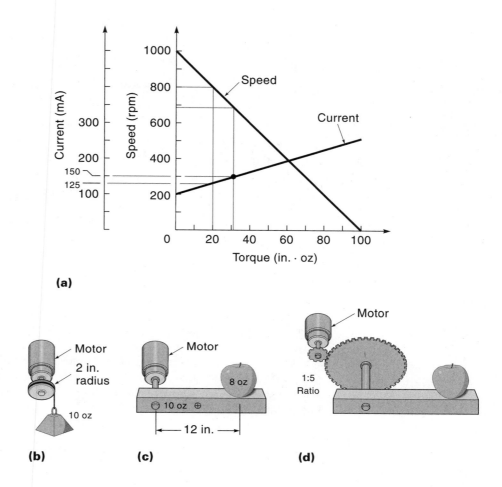

(a)

(b)   (c)   (d)

c. The motor is attached to a 12-in. robot arm (weighing 10 oz). On the end of the arm rests an 8-oz apple [Figure 7.8(c)]. The total load on the motor is calculated as shown below. Note the arm is considered to be a point mass of 10 oz at its center of gravity, which is 6 in. from the motor shaft.

$$\underbrace{(6 \text{ in.} \times 10 \text{ oz})}_{\text{Arm}} + \underbrace{(12 \text{ in.} \times 8 \text{ oz})}_{\text{Apple}} = 156 \text{ in.} \cdot \text{oz}$$

From the graph, we see that 156 in. · oz exceeds the stall torque of 100 in. · oz, so the motor will not be able to lift this load at all. One solution to the problem would be to insert a gear train of, say, 5 : 1 between the motor and load, [Figure 7.8(d)]. Now the torque required of the motor is only one-fifth of what it was, or

$$\frac{156 \text{ in.} \cdot \text{oz}}{5} = 31.2 \text{ in.} \cdot \text{oz}$$

Which is within the torque range of the motor, so it will rotate at 690 rpm and require a current of 150 mA. However, because of the gear train, the load will rotate at only one-fifth of the motor shaft, or

$$\frac{690 \text{ rpm}}{5} = 138 \text{ rpm}$$

Real-world loads are not always constant. In this case, the action of gravity causes the load to be greatest when the arm is horizontal and zero when the arm is vertical, so we would expect the motor speed to increase and decrease with each revolution.   ◆

A single torque–speed curve represents the motor performance for a particular voltage. Control motors are not usually operated at a single voltage because varying the voltage is a way of controlling the power (and therefore speed). Figure 7.9 shows a family of torque–speed curves, where each curve represents the motor response for a different voltage. As the voltage increases, the stall torque and no-load speed also rise, but saying that speed increases with voltage is too simple. What is really increasing is the torque, and in most cases, increased torque results in increased speed.

Figure 7.9

A family of torque–speed curves (Example 7.5).

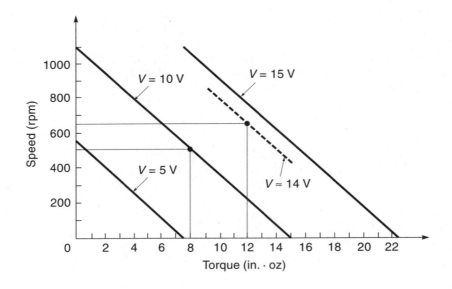

◆ **EXAMPLE 7.5**

A small PM motor is used on a tape deck. During rewind the motor runs at 500 rpm at 10 V. To shorten the rewind time, we want the motor to run at 650 rpm. It has been estimated that the increased speed will increase the load torque by 50%. Based on that estimate, determine what voltage must be applied to the motor to make it run at 650 rpm (use the torque–speed curve of Figure 7.9).

Solution

On the torque–speed curve of Figure 7.9, identify the existing operating point of the motor on the 10-V curve at 500 rpm. The motor is supplying 8 in. · oz at that speed and voltage; for the faster rewind, the motor needs to run at 650 rpm supplying 12 in. · oz (12 in. · oz is 50% greater than 8 in. · oz). The new operating point does not lie directly on one of the voltage curves, so we must sketch in some parallel lines and estimate the new voltage to be approximately 14 V. ◆

Figure 7.10(a) shows an example of an "instrument-type" PM motor. This small motor (about 1 in. in diameter) is available in seven models from 3 to 36 V. Consider the 3-V model (22-45-30), which has a no-load speed of 7500 rpm and a stall torque of 1.89 in. · oz [Figure 7.10(a)]. The torque–speed curve could be constructed from these figures (see Example 7.6). This motor is available with small **gearheads**, which are gear trains that are designed to connect directly to the motor. In this case, gear ratios from 20 : 1 to 2000 : 1 are available.

◆ **EXAMPLE 7.6**

For motor 22-45-30 [Figure 7.10(a)],

    a. Draw the torque–speed curve.

    b. Determine the speed for a load torque of 1 in. · oz.

Solution

    a. From Figure 7.10(a), we see that motor 22-45-30 has a no-load speed of 7500 rpm and a stall torque of 1.89 in. · oz. To draw the torque–speed curve, draw the axes and establish an appropriate scale so that 7500 is on the y-axis and 1.89 is on the x-axis. Connect these two points with a straight line as shown in Figure 7.10(b) (remember, only a PM motor has a straight line).

    b. Using the torque–speed curve, we can now determine the speed when the motor is delivering 1 in. · oz of torque. From Figure 7.10(b), we see that the speed would be 3400 rpm. ◆

Circuit Model of the PM Motor (Optional)

[*Note*: This section can be omitted without affecting the comprehension of the rest of this chapter.]

BERTSCH U.S. MICROMOTORS

Series 22-45 · 6 Watts · 10,000 RPM   Dimensions – mm   1 mm = 0.0394 in

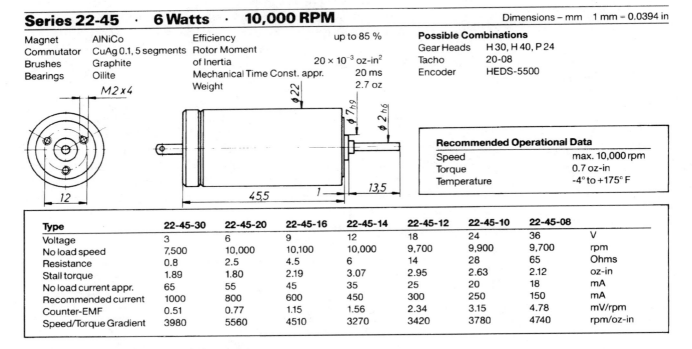

Magnet	AlNiCo	Efficiency	up to 85 %
Commutator	CuAg 0.1, 5 segments	Rotor Moment	
Brushes	Graphite	of Inertia	20×10^{-3} oz-in^2
Bearings	Oilite	Mechanical Time Const. appr.	20 ms
		Weight	2.7 oz

Possible Combinations

Gear Heads	H 30, H 40, P 24
Tacho	20-08
Encoder	HEDS-5500

Recommended Operational Data

Speed	max. 10,000 rpm
Torque	0.7 oz-in
Temperature	-4° to +175° F

Type	22-45-30	22-45-20	22-45-16	22-45-14	22-45-12	22-45-10	22-45-08	
Voltage	3	6	9	12	18	24	36	V
No load speed	7,500	10,000	10,100	10,000	9,700	9,900	9,700	rpm
Resistance	0.8	2.5	4.5	6	14	28	65	Ohms
Stall torque	1.89	1.80	2.19	3.07	2.95	2.63	2.12	oz-in
No load current appr.	65	55	45	35	25	20	18	mA
Recommended current	1000	800	600	450	300	250	150	mA
Counter-EMF	0.51	0.77	1.15	1.56	2.34	3.15	4.78	mV/rpm
Speed/Torque Gradient	3980	5560	4510	3270	3420	3780	4740	rpm/oz-in

Gear Heads H 30 and H 40 with Oilite Bearings

Standard Gear Ratios

| H 30: | 20 : 1 | 60 : 1 | 200 : 1 | 600 : 1 | 2500 : 1 | |
| H 40: | 40 : 1 | 100 : 1 | 200 : 1 | 500 : 1 | 1000 : 1 | 2000 : 1 |

Gear Head	H 30	H 40	
Max. torque	30	42	oz-in
Max. Motor Speed	5000	7500	rpm
Max. Power	1	2	W
Max. Axial Load	36	54	oz
Max. Radial Load	36	36	oz
Weight with motor	4.2	5	oz

(a)

Figure 7.10   Small PM motor specifications. (Courtesy of Bertsch U.S.)

So far we have been examining the performance of the PM motor and found that it is dependent on both the electrical and mechanical conditions present. To understand *all* factors that affect the motor's performance in a particular situation, examining a comprehensive motor–load mathematical model is helpful. This model is simply an equation, where each term represents some electrical or mechanical condition. In this section, we will look at the basic equations that describe motor-system performance.

Figure 7.10
(*continued*)

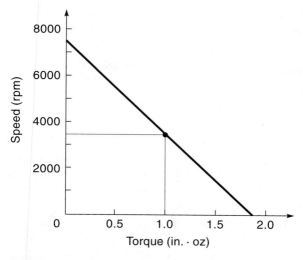

(b) Torque-speed curve for motor 22–45–30

Figure 7.11 shows a simple model of the PM motor driving a mechanical load. The equivalent circuit of the motor is a series circuit consisting of an inductance (L_A) from the armature windings, a resistance (R_A) to account for wire and commutator resistance, and a voltage source CEMF. Recall that the CEMF is an internally generated voltage that increases with motor speed. The polarity of the CEMF is such that it subtracts from the line voltage (V_{ln}), creating an effective armature voltage. The inductance (L_A) limits how fast the armature current can change but doesn't affect the steady-state operation. Writing Kirchhoff's equation around the loop of Figure 7.11 (for steady state), we get

$$V_{ln} = (R_A \times I_A) + \text{CEMF}$$

Solving for I_A, we get

$$I_A = \frac{V_{ln} - \text{CEMF}}{R_A} \tag{7.10}$$

Therefore, for steady-state operation, the motor current is directly proportional to the difference between the applied voltage and the CEMF. *The value of motor current is particularly important because it is the electrical parameter that most directly relates to*

Figure 7.11

Equivalent circuit of a PM motor with a mechanical load.

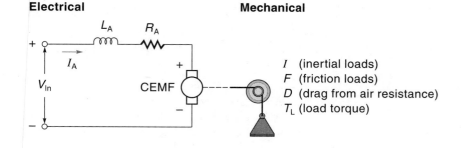

mechanical torque. For the PM motor, the relationship between torque and current is very simple, as expressed in the following equation:

$$T = K'I_A \qquad (7.11)$$

where

T = mechanical torque
K' = motor constant
I_A = armature current

The motor, together with its load, is a mechanical as well as an electrical system. Figure 7.11 indicates some of the mechanical factors that will effect the motor's performance. These include the **moment of inertia** (*I*), the bearing friction (*F*), and the drag (*D*). The drag, called *windage*, comes from air resistance on moving parts. All these factors cause resisting torques that the motor must overcome. The moment-of-inertia factor opposes any change in speed, the bearing friction is a constant, and the air drag increases with speed. Equation 7.12 describes the mechanical torque on a *free-running motor*, that is, a motor not connected to any external load:

$$T = (I_m \times \Delta S) + F_m + (DS) \qquad (7.12)$$

where

T = torque need to spin a motor with no load
I_m = moment of inertia of the armature and shaft
S = motor speed
ΔS = change in the motor speed per unit time (acceleration)
F_m = friction in the motor from bearings
D = drag factor from wind resistance

The effect of the moment of inertia is only present when the motor speed is increasing or decreasing. The larger the *I* factor, the more the speed resists change. Consequently, the system response time is slowed by the moment of inertia, whereas the steady-state speed is unaffected.

When the motor is connected to an external load, an additional set of I_L, F_L, and T_L influences the motor speed. The external load will certainly have its own moment of inertia and friction, and there is now an external **load torque** (T_L), which is the only useful work being done by the motor and the reason for it being there in the first place. The complete mathematical picture of what the motor has to drive is shown in Equation 7.13:

$$T = \underbrace{(I_M \times \Delta S) + F_M + (DS)}_{\text{Motor}} + \underbrace{(I_L \times \Delta S) + F_L + (D_LS) + T_L}_{\text{Load}} \qquad (7.13)$$

where

I_L = load moment of inertia
F_L = load friction
D_L = wind-resistance drag of the load
T_L = load torque (useful work)

If there is a gear train between the motor and load, the moment of inertia and friction will be affected. In most cases, the motor is geared down to the load, so from the motor's point of view, the load values of I_L, F_L, D_L, and T_L are reduced. In summary, the motor's actual performance is determined by its internal characteristics *and* the nature of the load.

7.4   DC MOTOR–CONTROL CIRCUITS

In this section, we discuss the two basic approaches for controlling the speed of a DC motor. The term *speed control* is somewhat inaccurate because, technically speaking, the motor converts electrical energy into torque, not speed (the precise speed is determined by *both* the motor torque and the mechanical load). Still, in general, it is safe to say that increased voltage will result in increased speed.

To drive the motor, an interface circuit is required to convert the low-level motor-control signal from the controller into a signal strong enough to run the motor. The classical way to do this is with an **analog drive**. In this method, a linear power amplifier amplifies the drive signal from the controller and gives the motor a "strengthened" analog voltage [Figure 7.12(a)]. A DAC (digital-to-analog converter) would be required if the controller is digital.

Figure 7.12

Methods of speed control for DC motors.

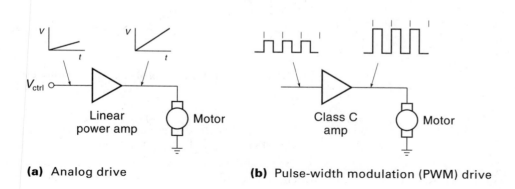

(a) Analog drive   **(b)** Pulse-width modulation (PWM) drive

The other technique for controlling a DC motor is **pulse-width modulation** (PWM). In this system, power is supplied to the motor in the form of DC pulses of a fixed voltage [Figure 7.12(b)]. The width of the pulses is varied to control the motor speed. The wider the pulses, the higher the average DC voltage, so more energy is available to the motor. The frequency of the pulses is high enough that the motor's inductance averages them, and it runs smoothly. This system has two advantages over the analog drive: (1) The power amplifier can be of the efficient class C type, and (2) the DAC is not needed because the amplifier is either on or off and can be driven directly with a digital signal.

DC Motor Control Using an Analog Drive

The analog-drive system of DC motor control uses a linear-power amplifier to drive the motor. The amplifier is the interface between the controller and the motor. Typically, it is a current amplifier, meaning its primary function is to boost the current; the output voltage may or may not be larger than the input voltage.

The simplest analog-drive circuit is a class A amplifier using a single power transistor. The circuit could be either the *common emitter* (CE) configuration, which gives current and voltage gain, or the *common collector* (CC) configuration, which gives only current gain; these circuits are shown in Figure 7.13. The operation of both circuits is similar; when the base voltage (V_B) is increased (beyond the forward-bias voltage), the transistor begins to turn on and let the collector current (I_C) flow. The collector current is 30–100 times greater than the base current, depending on the gain (h_{fe}) of the transistor. Once the transistor starts to conduct, I_C increases with V_B more or less linearly. Note that all of I_C goes through the motor, providing the drive current.

Figure 7.13

Analog-drive output configurations for a DC motor (conventional current flow).

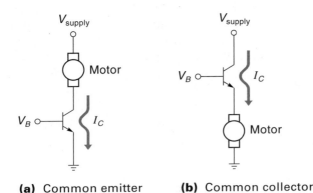

(a) Common emitter   **(b)** Common collector

The problem with this arrangement is that class A amplifiers are very inefficient. Consider the case of a 12-V, 2-A motor connected as shown in Figure 7.14(a). The DC power available is 12 V. When the transistor is all the way on, it is acting like a closed switch, so the entire 12 V is applied across the motor. When the transistor is half-on, it is acting like a resistor in series with the motor, reducing I_c by one-half, in this case to 1 A. Now, only 6 V is dropped across the motor, and the remaining 6 V is dropped across the transistor. The power being drawn from the supply is

$$P = IV$$
$$= 1\,A \times 12\,V = 12 \text{ watts (W)}$$

whereas the power being dissipated by the transistor is

$$P = VI$$
$$= 6\,V \times 1\,A = 6\,W$$

Half the power (6 W) of this system is wasted in the form of heat by the transistor when the motor is running at half-speed. In many control systems, the motor may be running

Figure 7.14
Power transistors.

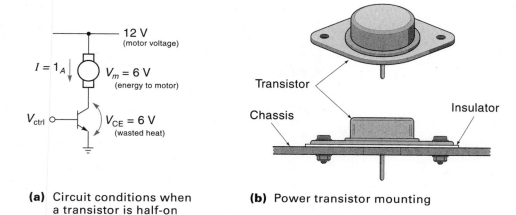

(a) Circuit conditions when a transistor is half-on

(b) Power transistor mounting

with an average speed of even less than half. For smaller motors, these losses may be acceptable, but for larger motors or battery operation a more efficient arrangement is needed.

The circuit just described uses a power transistor as the motor driver amp. As shown in Figure 7.14(b), power transistors and other power-driver circuits are made so that they can be mounted securely on the chassis. The purpose of the mounting is to provide a path for the heat to escape. A power transistor that is not mounted (or mounted incorrectly) cannot handle anywhere near its rated power. Typical power transistors can carry up to 60 A and dissipate as much as 300 W, and special models can go much higher. (Power transistors and heat considerations are covered in some detail in Chapter 4.) Other motor drivers besides the power transistor include the power IC, the Darlington power transistor, and the power MOSFET; these are described next.

The power IC driver is a single-package DC amplifier with a relatively high current output. An example is the LM12 (National Semiconductor) shown in Figure 7.15. This high-power operational amplifier can supply up to 13 A with a maximum voltage of ± 30 V. As in any op-amp circuit, feedback resistors are added to adjust the gain to any desired value. The voltage gain for the circuit of Figure 7.15 is 21 $[A_v = (R_f/R_i) + 1]$.

Figure 7.15
LM12 power operational amplifier (National Semiconductor).

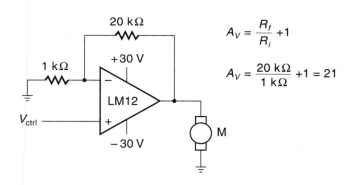

$$A_v = \frac{R_f}{R_i} + 1$$

$$A_v = \frac{20 \text{ k}\Omega}{1 \text{ k}\Omega} + 1 = 21$$

Figure 7.16 shows a motor-driver circuit using a power Darlington transistor. The Darlington configuration consists of two CC amplifiers connected in such a way that the first transistor directly drives the second. Although the voltage gain is only 1, the current gain can be very high. The transistor shown in Figure 7.16 is a TIP 120, which has a current gain (h_{fe}) of 1000 and a maximum output current of 5 A. The motor must be placed in the emitter path of the output transistor. A separate small-signal amplifier, probably an op-amp, would be needed to provide any voltage gain required.

Figure 7.16

A DC motor drive using a Darlington power transistor.

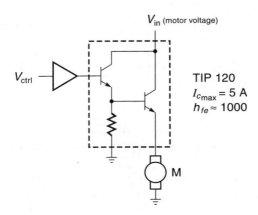

Another device capable of providing analog drive for a motor is the power MOSFET, known by such names as VFET, TMOS, and HEXFET. The shape of its construction allows for large currents [Figure 7.17(a)]. These FETs are usually designed to operate in the enhancement mode, which means the biasing voltage is always positive. Figure 7.17(b) shows a typical set of curves for the power MOSFET. Notice the output current (I_D) is 0 A when the input voltage (V_{GS}) is in the 0–5-V range but then climbs to 12 A when V_{GS} rises to 13 V. Figure 7.17(c) shows the basic motor-driver circuit using a power MOSFET. In this case, the motor is in series with the drain, which means the FET will

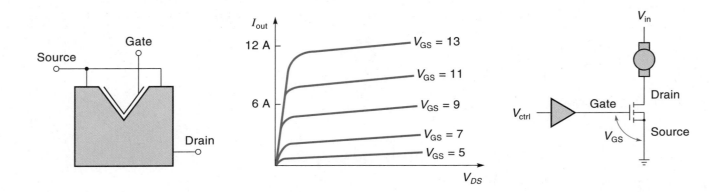

(a) Power MOSFET Construction **(b)** Output curves **(c)** Motor driver circuit

Figure 7.17 Power MOSFETS.

provide both voltage and current gain. The gate voltage is supplied from an op-amp circuit that is designed to interface the controller with the FET.

Reversing the PM Motor

To reverse the rotation direction of the PM motor, the polarity of the applied voltage must be reversed. One way to accomplish this is to have a motor-driver amp capable of outputting a positive and negative voltage (Figure 7.18). When the drive voltage is positive with respect to ground, the motor turns clockwise (CW). When the drive voltage is negative with respect to ground, the voltage polarity at the motor terminals reverses, and the motor rotates counterclockwise (CCW). The LM12 power op-amp (Figure 7.15) is capable of providing positive and negative output voltages.

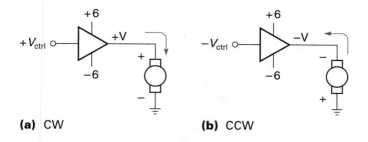

(a) CW   **(b)** CCW

Figure 7.18

Reversing motor rotation with negative voltage (conventional current).

In many applications, the drive amplifier cannot output both positive and negative voltages, in which case a switching circuit must be added to reverse the motor. One approach is to use a double-pole relay (Figure 7.19). When the relay contacts are up, the positive voltage is connected to terminal A of the motor, and terminal B is connected to the negative voltage. When the relay contacts are down, the positive voltage is connected to terminal B, and terminal A goes to the negative voltage, thus effectively reversing the polarity.

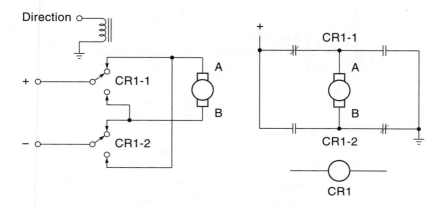

Figure 7.19

Reversing a motor with relay switching (two schematic representations).

Forward–reverse switching can also be done with solid-state devices. Figure 7.20 shows a motor-reversing circuit that uses four FETs. When Q_1 and Q_4 are on, the current $I_{1,4}$ causes the motor to turn clockwise. When Q_2 and Q_3 are on, the current $I_{3,2}$ flows in the opposite direction and causes the motor to turn counterclockwise. The entire switching operation can be performed by a single IC, such as the Allegro A3952 shown in Figure 7.21. This IC contains four separate driver transistors that are controlled by internal logic to operate in pairs (in the manner of Figure 7.20). The A3952 controls a motor-supply voltage of up to 50 V with up to 2 A of output current.

Figure 7.20

Reversing a DC motor with solid-state switching (conventional current).

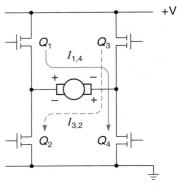

Figure 7.21

A full-bridge PWM motor driver. (Courtesy of Allegro Micro Systems Inc.)

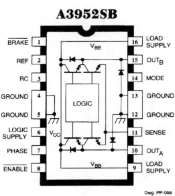

The A3952 is very simple to use, as illustrated in Figure 7.22. Basically, the IC connects the supply voltage (V_{BB}) and ground to the motor through pins Out$_A$ and Out$_B$. The polarity of the motor voltage (that is, motor direction) is controlled by the Phase input, and the Enable input can be used to turn the motor on and off. Both Phase and Enable inputs are digital TTL-compatible signals.

DC Motor Control Using Pulse-Width Modulation

Pulse-width modulation is an entirely different approach to controlling the torque and speed of a DC motor. Power is supplied to the motor in a square wave of constant voltage but varying pulse width or duty cycle. **Duty cycle** refers to the percentage of time the pulse is high (per cycle). Figure 7.23 shows the waveforms for four different speeds. For the slowest speed, the power is supplied for only one-quarter of the cycle time (duty

Figure 7.22

Application of the Allegro
A3952.

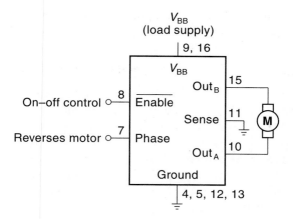

cycle of 25%). The frequency of the pulses is set high enough to ensure that the mechanical inertia of the armature will smooth out the power bursts, and the motor simply turns at a constant velocity of about one-quarter speed. For a 50% duty cycle (power on one-half the time), the motor would turn at about half speed, and so on. In real life, nonlinear factors cause the motor to go slower than the straight proportions suggest, but the principle still holds—that is, the speed of a motor can be regulated by pulsing the power.

Figure 7.23

PWM waveforms.

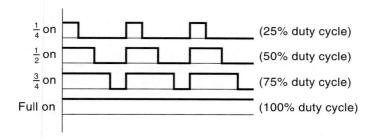

PWM offers two distinct advantages over analog drive. First, it is digital in nature—the power is either on or off, so it can be controlled directly from a computer with a single bit, eliminating the need for a DAC. Second, the drive amplifier can use efficient class C operation. The class C amplifier is efficient because the conditions that cause power dissipation are minimized. Consider the circuit shown in Figure 7.24. A control voltage (V_{ctrl}) of 2 V will turn on Q_1 all the way (saturation), and I_C flows freely through the motor. That is, Q_1 acts like a closed switch, making V_{CE} near 0 V and so dropping the entire line voltage (V_{ln}) across the motor. We can calculate the power dissipated by the transistor using the power equation: $P = IV$. Applying this to Q_1,

$$P = I_C \times V_{CE}$$

If V_{CE} is near 0 V, then

$$P = I_C \times 0\,V = 0\,W$$

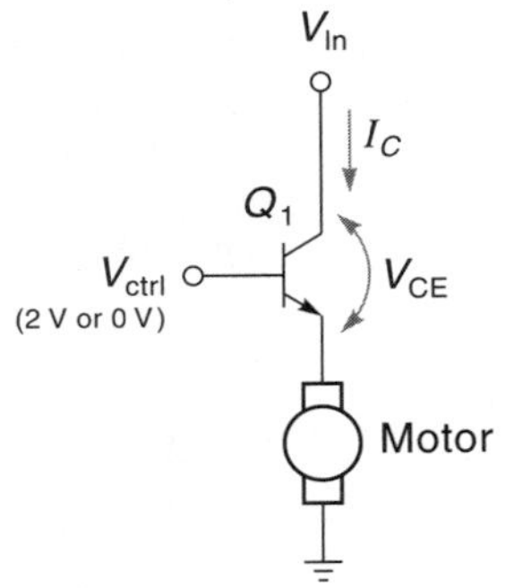

Figure 7.24

Class C amp operation using a power transistor (conventional current).

This tells us that almost no power is lost by the transistor when I_C is flowing. On the other hand, when V_{ctrl} is 0 V, Q_1 is turned completely off, thus reducing I_C to near 0 A. Recalculating the power under these conditions,

$$P = 0\,\text{A} \times V_{CE} = 0\,\text{W}$$

Again, theoretically no power is dissipated by the transistor. Therefore, if the transistor is operated so that it is either all the way on or all the way off, we have a very efficient amplifier. This is class C operation, and it can be used for PWM. In practice, switching transistors do have losses because of leakage and because they can't turn on (or off) instantaneously. Therefore, a little dissipation occurs with each transition, which means that the higher the frequency, the hotter the transistor gets!

The driver amplifier circuit used for PWM is essentially the same as for an analog drive. The power transistor, power Darlington, power MOSFET, and power IC amp can all be used. One important difference from analog drive is that PWM amplifiers do not have to be linear; this tends to make the PWM amp less complicated. Biasing is designed so that the amplifier is normally off. The input signal will look like that shown in Figure 7.23. The magnitude of the input must be high enough to ensure that the amp will turn on all the way during the pulses. If not, the resulting inefficiency may cause the amplifier to fail from overheating.

The Allegro A3952 (Figures 7.21 and 7.22) is ideally suited to provide a PWM drive. The load supply terminal (V_{BB}) is connected to a DC source and provides the motor voltage. A TTL-level PWM waveform (such as in Figure 7.23) is connected directly to the Enable input. Recall that the Enable input turns on and off the motor voltage.

Some problems are associated with pulsing power to a DC motor. In fact, we usually try to avoid rapidly turning on and off an inductive load like a motor because of the large voltage spikes generated when the interrupted armature current has no place to go. Recognizing this problem, the PWM circuit includes *free-wheeling* or *flyback diodes* to provide a nondestructive return path for the current. Figure 7.21 shows four flyback diodes—one across each transistor. Each time the voltage is withdrawn from the motor, a brief voltage spike will appear at the motor terminals. The diodes redirect this energy back into the power source.

Because the PWM drive current is not a steady DC value, a unique kind of inefficiency arises, which the analog drive does not have. Figure 7.25 shows the PWM input voltage and current for two different duty cycles. Notice that the current waveform looks more like a sawtooth than a square wave. The current is trying to follow the voltage waveform, but the inductance of the motor restricts the current from changing very fast. As the pulses get shorter, the relative peak-to-peak value of the current increases. Like any periodic waveform, this current waveform has both an average value (I_{av}) and an rms value (I_{rms}), but note that the difference between the two values gets larger as the pulses get shorter. Here is where the inefficiency occurs because the armature heat losses are proportional to the (higher) I_{rms} value and the mechanical torque of the motor is proportional to the (lower) I_{av} value. These two relationships are mathematically described below:

$$T = I_{av} \times K_t \tag{7.14}$$

where

T = motor torque
I_{av} = average armature current
K_t = motor constant

$$P_d = I^2_{rms} \times R_a \qquad (7.15)$$

where

P_d = power dissipated through the armature
I_{rms} = value of the rms component of the current
R_a = armature resistance

Figure 7.25

Difference of I_{ave} and I_{rms} in PWM waves.

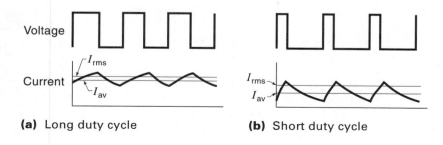

(a) Long duty cycle   **(b)** Short duty cycle

In summary, as the PWM pulses get shorter, the I_{rms} component rises, as does the armature losses. The implication here is that slow motor speeds require the duty cycle to be more than a simple proportion; for example, for 25% speed, the duty cycle might have to be 35%.

PWM Control Circuits

There are many ways to create the PWM waveform, using either hard-wired circuits or software. However, so not to burden the main processor with such a repetitive task, PWM is usually generated with dedicated cicuits, or with special programmable timing circuits built-in to the microcontroller. In this section, we examine some different methods that can be used to produce PWM. In all designs, it is the duty cycle, not the frequency, that determines motor speed. The frequency stays constant and is usually in the range of 40 Hz–10 kHz. Lower frequencies can sometimes cause objectional vibrations, and power transfer to the motor declines at higher frequencies due to the inductive reactance of the armature windings.

Figure 7.26 shows a 4-speed PWM control circuit. Operation of the circuit is as follows: Two *one-shots*, O/S$_1$ and O/S$_3$, are connected together to make an oscillator. (A one-shot generates a single pulse when triggered.) The 3-ms one-shot triggers the 1-ms one-shot, which feeds back and triggers the 3-ms one-shot, and so on. The output of this circuit is a series of pulses that are high (5 V) for 1 ms and low for 3 ms. This is the ¼-speed waveform shown in Figure 7.26(b). Note that the basic period of the waveform has been established at 4 ms. A third one-shot (O/S$_2$) creates a 2-ms pulse and is triggered by the "falling edge" of the O/S$_1$ output. The output of O/S$_2$ is high for one-half of the 4-ms period and is shown as the ½-speed waveform in Figure 7.26(b). By taking the output of the 3-ms O/S$_3$, we create a ¾-speed waveform as shown. Control gates A, B, and C admit one of the three signals into the OR gate. A fourth signal into the OR gate

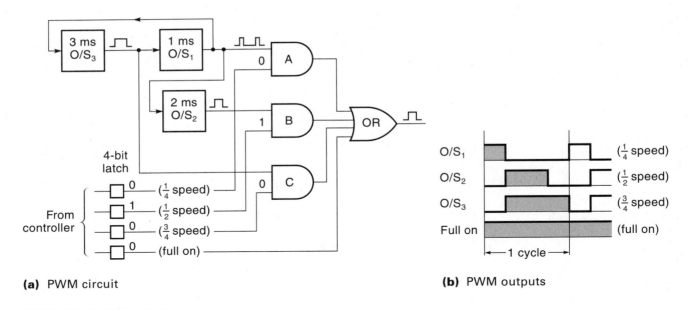

(a) PWM circuit

(b) PWM outputs

Figure 7.26   Circuit for a 4-speed PWM motor control (shown with ½-speed selection).

called "full on" (duty cycle = 100%) provides for straight DC to the motor. The control signals could come directly from the computer. The computer would merely send a 4-bit data word to a latching output port, with a 1 in the bit position of the desired speed. The motor would operate at that speed until a new command was sent.

Some ICs can generate the PWM waveform directly. One such device is the LM3524 (National Semiconductor) shown in the block diagram of Figure 7.27(a). The duty cycle can be varied from 0% to almost 100% by applying a DC voltage called "compensation" on pin 9. An on-board oscillator establishes the PWM frequency in the form of a sawtooth waveform (frequency can be adjusted with an external R and C). This sawtooth waveform is compared with the compensation voltage in a comparator amp. At the start of each cycle, the output transistor is turned on; when the sawtooth voltage rises to the value of the compensation voltage, the output transistor turns off and stays off for the rest of that cycle [Figure 7.27(b)]. As the compensation voltage is increased, the output transistor stays on longer, increasing the duty cycle.

One of the most common ways to produce PWM is with programmable hardware built into the microcontroller. For example, PWM can be generated from the MC68HC11 microcontroller by loading two numbers into registers corresponding to the high time and the low time of each cycle. The PWM output is set low while a high-speed counter counts up to the value in the low-time register, then the output is set high while the counter counts up to the value in the high-time register. Interrupt routines automatically take care of the whole PWM process, so the microcontroller is basically free to do other things. Other microcontrollers, such as the MC68HC711K4, come with built-in PWM channels. Each channel can be initialized for a specific period and duty

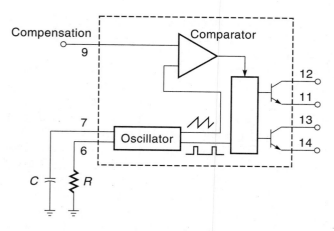

(a) Block diagram of a PWM IC (LM3524)

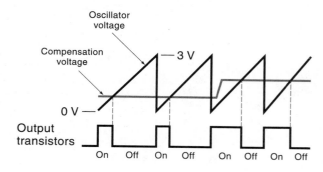

(b) Output transistors turn off when the oscillator voltage reaches compensation

Figure 7.27   Pulse width modulation IC (LM3254). (Courtesy of National Semiconductor)

cycle, then the hardware takes over and produces the PWM wave. If the application program needs to change the duty cycle, a new value is simply written to the duty-cycle register.

Figure 7.28 shows an MC68HC711K4 microcontroller (Motorola) being used to provide PWM drive to a DC motor. The output of the microcontroller is a 0–5-V logic signal that must be amplified to provide the necessary motor current. In this case, an A3952 (discussed earlier) will be used. Notice that the PWM signal goes to the enable input and the signal that specifies motor direction goes to the phase input.

Figure 7.28

Controlling a DC motor with PWM (from a micro-controller).

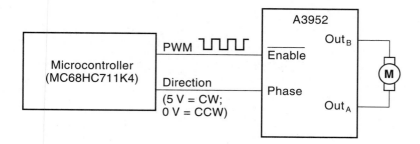

DC Motor Control for Larger Motors

DC motors come in all sizes, from small instrument motors requiring a fraction of an ampere to large industrial motors requiring hundreds of amperes. The discussion thus far has assumed that a DC supply voltage was available and that all the control circuit had to do, broadly speaking, was to connect this voltage to the motor. For larger motors— say, 20 A or more—the hardware needed to supply pure DC becomes bulky and expensive. An alternative solution is to drive the DC motor with rectified AC, where no attempt is made to smooth the waveform. A device that is frequently used in this application to provide both rectification and some measure of control is the **silicon-controlled rectifier** (SCR), which is covered in some detail in Chapter 4. The following paragraph is only a summary of SCR operation.

The SCR is a member of the thyristor family of semiconductor switching devices. Figure 7.29(a) shows the symbol of the SCR, which has three terminals: anode, gate, and cathode. The SCR is used as an electronic switch and can handle currents up to over 100 A. This power current (I_A) flows from the anode to the cathode as indicated in Figure 7.29(a). The SCR is switched on by applying a low voltage to the gate (V_{GT}), which is typically 0.6–3 V. Once turned on, it stays on, even if the gate voltage is removed. In fact, the only way to turn it off is to reduce the power current (I_A) below a low threshold called the **holding current**, which is typically a few milliamps. Notice that the SCR symbol is similar to that of a diode, which reminds us that the SCR is a rectifier as well as a switch—that is, current can flow through it in only one direction.

Figure 7.29

A silicon-controlled rectifier.

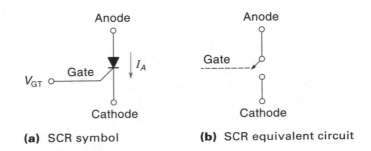

(a) SCR symbol   (b) SCR equivalent circuit

Figure 7.30(a) shows the basic SCR motor control circuit. Notice that the power source is single-phase AC and that the DC motor is connected in series with the SCR. The gate of the SCR is driven by a trigger circuit that provides one pulse for each cycle of the AC. The free-wheeling diode (D) across the motor provides an escape path for the energy stored in the motor windings when the SCR switches off.

Figures 7.30(b) and (c) show how the circuit can control motor speed. The top waveform of Figure 7.30(b) is the AC power. Notice that the motor voltage (V_m) stays at

Figure 7.30

DC motor control from an AC source using an SCR.

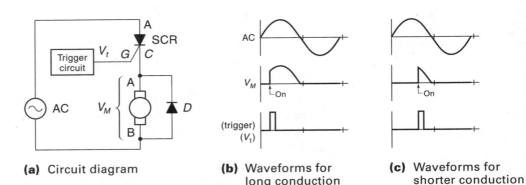

(a) Circuit diagram

(b) Waveforms for long conduction period

(c) Waveforms for shorter conduction period

0 V until the trigger pulse (V_t) turns on the SCR. Once turned on, the motor voltage equals the AC voltage for the remainder of the positive half of the cycle. During the negative half of the AC cycle, the SCR remains off, and there is no power to the motor. Figure 7.30(b) shows the motor getting almost all the power this circuit can deliver, which is still only about half of the AC power available.

In Figure 7.30(c), the trigger pulse is delayed from what it was in Figure 7.30(b). Therefore, the motor is not connected to the AC power until later in the cycle, and consequently it receives even less power—that is, it runs slower. So, by controlling the delay time of the trigger pulse, we can control the speed of the motor.

A circuit with one SCR is a half-wave rectifier, so the load gets a maximum of only half the available power. Circuits using multiple SCRs to create a full-wave rectifier overcome this problem. Figure 7.31 shows such a circuit. Here, SCR_1 is triggered during the positive half of the AC cycle, and SCR_2 is triggered during the negative half-cycle. The result? The motor receives two power pulses per cycle. Figure 7.32 shows another full-wave motor-control circuit. In this case, four diodes are used for the full-wave rectifier, and a single SCR controls the delay of each half cycle.

Figure 7.31

A full-wave SCR DC motor–control circuit using multiple SCRs.

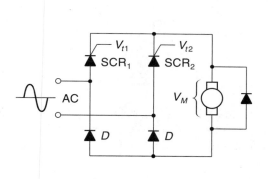

(a) Circuit diagram

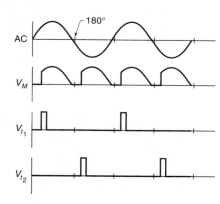

(b) Waveforms

Figure 7.32

A full-wave SCR motor control using a full-wave bridge.

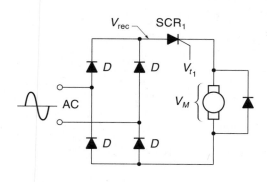

(a) Circuit diagram

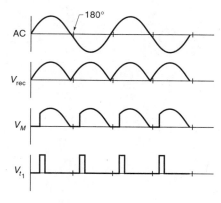

(b) Waveforms

Electric motors have a large starting current that is many times more than the running current. For smaller motors, this may not present a problem; for larger motors (over the range of 1–2 hp), however, special **reduced voltage–starting circuits** are used. A reduced voltage–starting circuit will limit the armature current to some acceptable value when the motor starts. One way to do this is to have a resistor in series with the armature. After the motor comes up to speed, a relay is used to bypass the resistor, allowing the full line voltage to the motor.

7.5 A COMPREHENSIVE APPLICATION USING A SMALL DC MOTOR

Example 7.7 illustrates a practical motor-application problem. The process requires you to understand the characteristics of the entire system. This example will give you an idea of the real-life factors involved in motor selection. (This example integrates some material from Chapter 5.)

♦ **EXAMPLE 7.7**

A PM motor turns a large 24-in. diameter, 10-lb turntable through a 20 : 1 gear train [Figure 7.33(a)]. A particular requirement is that the turntable must be able to accelerate from a rest position to 90° in 0.2 s. Determine the necessary motor voltage. The torque–speed curves of the motor are given in Figure 7.33(b).

Solution

To solve this real-life problem, we must know how to calculate the forces required to get parts moving and understand the torque–speed characteristics of DC motors. Accelerating from 0 to 90° in 0.2 s is a response-time requirement and is determined by the system's moment of inertia (I). Because the turntable is massive compared to the motor, we will only consider the turntable's moment of inertia in the problem. First, calculate the turntable's I. For a disk,

$$I = \frac{1}{2}mr^2 \tag{7.16}$$

where

I = moment of inertia
m = mass (in this case, $m = w/g = 10 \text{ lb}/32 \text{ ft/s}^2$)
r = radius (in this case, 12 in., or 1 ft)

After plugging in the numbers, we get

$$I = \frac{1}{2} \frac{10 \text{ lb s}^2}{32 \text{ ft}} 1 \text{ ft}^2$$

$$= 0.156 \text{ lb} \cdot \text{s}^2 \cdot \text{ft}$$

Figure 7.33

Motor driving a turntable (Example 7.7).

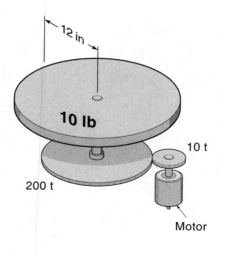

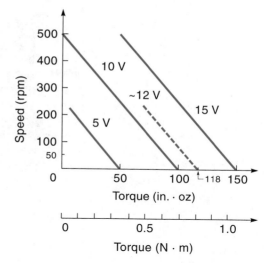

(a) Hardware setup

(b) Motor torque-speed curves

Converting I to in. $\cdot$ oz $\cdot$ s^2 gives us

$$I = 0.156 \text{ lb} \cdot \text{s}^2 \cdot \text{ft} \times \frac{16 \text{ oz}}{\text{lb}} \times \frac{12 \text{ in.}}{\text{ft}} = 30 \text{ in.} \cdot \text{oz} \cdot \text{s}^2$$

which is the moment of inertia for the turntable.

Second, we need to know the acceleration required. From Chapter 5, recall the equation that relates position to acceleration:

$$\theta = \frac{1}{2}\alpha t^2 \tag{7.17}$$

where

θ = angle in radians
α = angular acceleration
t = elapsed time since the object was at rest

Solving for acceleration, we find that

$$\alpha = \frac{2\theta}{t^2} = \frac{2(90°)}{(0.2 \text{ s})^2} \times \frac{\pi \text{ rad}}{180°} = 79 \text{ rad/s}^2$$

Now we need to determine the motor torque required to cause this acceleration. The basic equation of angular motion is

$$T = I\alpha \tag{7.18}$$

where T is the motor torque. But remember that the motor is connected to the turntable through a 20 : 1 gear ratio. Thus, the motor itself must accelerate 20 times faster than the turntable, and the turntable moment of inertia reflected back to the motor is only $\frac{1}{20}$ squared, as shown in the following equation:

$$T = I\alpha = \frac{(30 \text{ in.} \cdot \text{oz} \cdot \text{s}^2)}{20^2} \times \frac{(79 \times 20)}{\text{s}^2} = 118 \text{ in.} \cdot \text{oz}$$

This result shows that 118 in. · oz of motor torque are required to accelerate the turntable. Looking along the x-axis of the torque–speed curves [Figure 7.33(b)] we see that for a stall torque of 118 in. · oz, the motor must get approximately 12 V. The answer to the original question seems to be as follows: The motor voltage to spin the turntable 90° in 0.2 s equals 12 V.

Final Comment

There is a problem, however, with this conclusion because it is based on the assumption that the torque is a *constant* 118 in. · oz. It is true that when the motor is at a dead stop, 12 V will cause a torque of 118 in. · oz, but by the time the turntable has moved 90°, the motor is turning at 50 rps and the torque is down to 110 in. · oz. In this case, the difference between 118 and 110 in. · oz is not that much and may (with minor adjustments) be considered an acceptable solution. Getting the precise answer to this problem would require either mathematics beyond the scope of this text or empirical testing.

EXAMPLE 7.7   (Repeated with SI Units)

A PM motor turns a large 60-cm diameter, 4.5-kg turntable through a 20 : 1 gear train [Figure 7.33(a)]. A particular requirement is that the turntable must be able to accelerate from a rest position to 90° in 0.2 s. Determine the necessary motor voltage. The torque–speed curves of the motor are given in Figure 7.33(b).

Solution   Accelerating from 0 to 90° in 0.2 s is a response-time requirement and is determined by the system's moment of inertia (I). Because the turntable is massive compared to the motor, we will only consider the turntable's moment of inertia in the problem. First, calculate the turntable's I. For a disk,

$$I = \frac{1}{2}mr^2 \tag{7.16}$$

where

I = moment of inertia
m = mass (in this case, 4.5 kg)
r = radius (in this case, 30 cm)

After plugging in the numbers, we get

$$I = \frac{1}{2}\ 4.5\ \text{kg}\ 30\ \text{cm}^2$$

$$= 2025\ \text{kg} \cdot \text{cm}^2 = 0.2025\ \text{kg} \cdot \text{m}^2$$

which is the moment of inertia for the turntable.

Second, we need to know the acceleration required:

$$\theta = \frac{1}{2}\alpha t^2 \tag{7.17}$$

where

θ = angle in radians
α = angular acceleration
t = elapsed time since the object was at rest

Solving for acceleration, we find that

$$\alpha = \frac{2\theta}{t^2} = \frac{2(90°)}{(0.2s)^2} \times \frac{\pi \text{ rad}}{180°} = 79 \text{ rad/s}^2$$

Now we need to determine the motor torque required to cause this acceleration. The basic equation of angular motion is

$$T = I\alpha \tag{7.18}$$

where T is the motor torque. But remember that the motor is connected to the turntable through a 20 : 1 gear ratio. Thus, the motor itself must accelerate 20 times faster than the turntable, and the turntable moment of inertia reflected back to the motor is only $\frac{1}{20}$ squared, as shown in the following equation:

$$T = I\alpha = \frac{(0.2025 \text{ kg} \cdot \text{m}^2)}{20^2} \times \frac{(79 \times 20)}{s^2} = 0.8 \text{ N} \cdot \text{m}$$

[*Note:* The relationship $1 \text{ kg} = 1\text{N}/(1 \text{ m/s}^2)$ was used to get units to cancel.]

This result shows that 0.8 N · m of motor torque is required to accelerate the turntable. Looking along the x-axis of the torque–speed curves [Figure 7.33(b)] we see that for a stall torque of 0.8 N · m, the motor must get approximately 12 V. The answer to the original question seems to be as follows: The motor voltage to spin the turntable 90° in 0.2 s equals 12 V. (However, see "Final Comment" on page 266.) ◆

7.6   BRUSHLESS DC MOTORS

The weak point in the mechanical design of the DC motor is the brushes rubbing against the rotating commutator (to get current into the armature). Brushes wear out, get dirty, cause dust, and are electrically noisy. The **brushless DC motor** (BDCM) operates without brushes by taking advantage of modern electronic switching techniques. Although this adds some complexity, the result is a motor that is extremely reliable, very efficient, and easily controlled—all very desirable qualities. The BDCM is becoming increasingly popular, particularly in those cases where the motor must be operated from a DC source such as a battery.

Figure 7.34(a) shows a diagram of a three-phase BDCM. The armature (called the **rotor**) is a permanent magnet, and it is surrounded by three field coils. Each field coil can be switched on and off independently. When a coil is on, such as coil A in Figure 7.34(a), the north pole of the rotor magnet is attracted to that coil. By switching the coils on and off in sequence (A, B, C), the rotor is "dragged" around clockwise—that is, the field has rotated electronically.

BDCMs have much in common with stepper motors, which are discussed in detail in Chapter 8. The major difference between these two types of motors is that the BDCM is

Figure 7.34

A three-phase brushless DC motor.

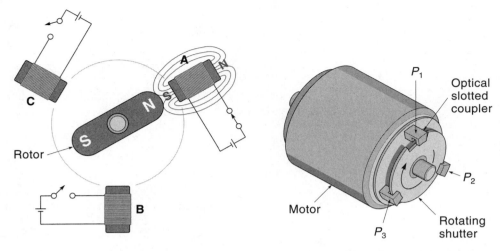

(a) Field coil arrangement   **(b)** Rotating shutter and sensors

used as a source of rotary power, like a regular electric motor, whereas the stepper motor is used when it is necessary to step to precise positions and then stop. Unlike the stepper motor, the BDCM has a built-in sensor system to direct the switching from one field coil to the next. Figure 7.34(b) shows the three-phase BDCM with three optical slotted couplers and a rotating shutter (Hall-effect sensors can also be used for this application). These position sensors control the field windings. When the shutter is open for sensor P_1 as shown, field coil A [Figure 7.34(a)] is energized. When the rotor actually gets to field coil A, sensor P_1 is turned off and P_2 is turned on, energizing field coil B and pulling the rotor on around to coil B, and so on. In this manner, the rotor is made to rotate with no electrical connection between the rotor and the field housing.

Figure 7.35 shows a schematic of a generalized three-phase BDCM. The three position sensors connect to the control circuitry. In the simplest case, such as described in the preceding paragraph, these signals are passed directly on to solid-state switches that drive the motor coils. A more sophisticated motor-control system would provide for

Figure 7.35

Schematic of a three-phase BDCM.

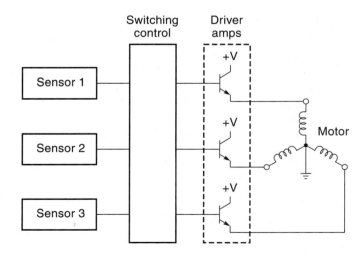

the motor to reverse direction (by reversing the sequencing) and would control the speed by using PWM techniques.

BDCMs range in size from small instrument motors to 10 hp and larger. They seem to be the wave of the future for DC motors because they combine the mechanical simplicity of a brushless motor with the ultimate in control possibilities.

SUMMARY

The electric motor, the most common type of actuator, is based on the principle that a current-carrying conductor will experience a physical force when in the presence of a magnetic field. The DC motor consists of a rotating armature and a stationary magnetic field. The current in the armature, which must come through brushes, causes the rotating forces. The stationary magnetic field is provided by either electromagnets (in which case, it is called a wound-field motor) or by permanent magnets.

There are three types of wound field motors. The series motor has the armature and field windings connected in series. This type of motor is characterized by a high starting torque and a high no-load speed but poor speed regulation (the speed changes considerably if the load changes). The shunt motor has the armature and field windings connected in parallel. This type of motor has much better speed regulation than does the series motor. The compound motor has both series and shunt-type field windings and combines the good characteristics of both the series and shunt motors.

Permanent magnet (PM) motors use permanent magnets to provide the stationary magnetic field. This results in a very linear torque–speed curve, which makes it easy to calculate the motor speed for various load conditions and thus attractive for control system applications.

There are two ways to control the speed of a DC motor: (1) Analog drive uses linear amplifiers to provide a varying DC voltage to the motor. Although simple and direct, this method is very power-inefficient and is usually used only with smaller motors. (2) Pulse-width modulation (PWM), a more efficient method, provides the motor with constant voltage pulses of varying widths: the wider the pulse, the more energy transferred to the motor. For larger DC motors, SCRs power the motor with pulses taken directly from the AC waveform. This is a form of PWM, and it eliminates the need for a large DC-power supply.

The newest type of DC motor is the brushless DC motor (BDCM). This motor uses permanent magnets instead of coils in the armature (called the rotor) and so does not need brushes. The field coils are switched on and off in a rotating sequence that pulls the rotor around. BDCMs have built-in sensors that direct when the individual field coils are to be switched on and off.

GLOSSARY

analog drive   A method of controlling an electric motor's speed by varying the DC supply voltage.

armature   The part of a motor that responds to the magnetic field; typically, the armature is the rotating assembly of an electric motor.

BDCM   *See* **brushless DC motor**.

brush   A stationary conductive rod that rubs on the rotating commutator and conducts current into the armature windings.

brushless DC motor (BDCM)   The newest type of DC motor that does not use a commutator or brushes; instead, the field windings are turned on and off in sequence, and the permanent magnet rotor (armature) is pulled around by the apparently rotating magnetic field.

CEMF   *See* **counter-EMF**.

commutator   The part of an armature that makes contact with the brushes.

compound motor   A motor that has both shunt and series field windings.

counter-EMF (CEMF)   A voltage that is generated inside an electric motor when running under its own power; the polarity of the CEMF is always opposite to the applied voltage.

cumulative compound motor A compound motor where the magnetic fields of the shunt and series fields aid each other.

differential compound motor A compound motor where the magnetic fields of the shunt and series windings oppose each other.

duty cycle The percentage of cycle time that the PWM pulse is high—that is, a true square wave has a duty cycle of 50%.

field The part of an electric motor that provides a magnetic field; typically, it is the stationary part.

field winding A stationary electromagnet used to provide the magnetic field needed by the armature.

gearhead In smaller motors, a gear train attached or built-in to the motor assembly, which effectively gives the motor more torque at less rpm.

holding current Once the SCR has been turned on, the small current (to the SCR) necessary to keep the SCR in the conduction state.

I *See* **moment of inertia**.

load torque The torque required of the motor to rotate the load.

moment of inertia (I) A quantity based on mass and shape of a rotating part; the larger *I* is, the more torque it takes to change the rpm of the part.

no-load speed The speed of a motor when there is no external load on it; it will always be the maximum speed (for a particular voltage).

permanent magnet (PM) motors A motor that uses permanent magnets to provide the magnetic field. The PM motor has a linear torque–speed relationship, making it desirable for control applications.

PM motor *See* **permanent magnet motor**.

pulse-width modulation (PWM) A method of controlling an electric motor's speed by providing pulses that are at a constant DC voltage. The width of the pulses is varied to control the speed.

PWM *See* **pulse-width modulation**.

rated speed The speed of a motor when producing its rated horsepower.

reduced voltage–starting circuit A special circuit that limits the motor current during start-up.

rotor The rotating part of a motor; if the field poles are stationary (as they usually are), then the rotor is known as the armature.

SCR *See* **silicon-controlled rectifier**.

series-wound motor A motor that has the field windings connected in series with the armature; the series-wound motor has a high starting torque and a high no-load speed.

shunt-wound motor A motor that has the field windings connected in parallel with the armature windings; the shunt-wound motor has a measure of natural speed regulation—that is, it tends to maintain a certain speed despite load changes.

silicon-controlled rectifier (SCR) A semiconductor device that provides speed control to a DC motor from an AC-power source (without the need of a power supply).

speed regulation In general, a motor's ability to maintain its speed under different loads; specifically, a percentage based on no-load speed and full-load speed.

stall torque The torque of the motor when the shaft is prevented from rotating; it will always be the maximum torque (for a particular voltage).

torque In general, the measure of the motor's strength in providing a twisting force; specifically, the product of a tangential force times the radius.

torque–speed curve A graph of a motor's torque versus speed; can be used to predict the motor's speed under various load conditions.

EXERCISES

Section 7.1

1. What is CEMF? How and why does it affect electric motor performance?

2. A DC motor is running with a line voltage of 90 V, a CEMF of 80 V, and an armature resistance of 3 Ω. Find the armature current.

3. A motor generates a CEMF at a rate of 0.4 V/100 rpm and has an armature resistance of 20 Ω. Ten volts are applied to the (stopped) motor. Find the armature current just after the power is applied and when the speed is 1500 rpm.

4. A motor generates a CEMF at a rate of 0.5 V/100 rpm and has an armature resistance of 15 Ω. Twelve volts

are applied to the (stopped) motor. Find the armature current just after the power is applied and when the speed is 1500 rpm.

Section 7.2

5. What are the distinguishing characteristics of the series-wound DC motor?

6. Explain *stall torque* and *no-load speed*.

7. What are the distinguishing characteristics of the shunt-wound motor?

8. When 24 V are applied to a motor with no load, the speed levels off at 2300 rpm. When the load is connected, the speed drops to 2050 rpm. Calculate the speed regulation of this motor.

9. When 90 V are applied to a motor with no load, the speed levels off at 2000 rpm. When the load is connected, the speed drops to 1750 rpm. Calculate the speed regulation of this motor.

10. Select a motor from the list of Figure 7.4 that meets the following requirements: Load torque = 3 ft · lb, and speed = 1750 rpm.

Section 7.3

11. Why is the PM motor particularly suited to control system applications?

12. A PM motor supplied with 10 V has the torque–speed curves of Figure 7.9.
    a. Find the stall torque.
    b. Find the no-load speed.
    c. Find the rpm of the motor if it is lifting a 4-oz load by a string over a 3-in. radius pulley.
    d. A 3 : 1 step-down gear pass is inserted between the motor and pulley shaft (of part c). Sketch the setup and determine the motor rpm.

13. A PM motor has the torque–speed curves of Figure 7.9. You want to use the motor in an application that requires a speed of 450 rpm and a load of 11 in. · oz. What voltage must be supplied to the motor?

14. A motor has the torque–speed curves of Figure 7.9. If the supply voltage is 12 V, what is the largest torque the motor can deliver and still maintain a speed of 600 rpm?

15. For the motor in Figure 7.10(a), draw the torque–speed curve for model 22-45-16.

16. For model 22-45-12 in Figure 7.10(a), find the approximate motor torque if the applied voltage is 15 V and speed is 5000 rpm.

Section 7.4

17. Comment on the power efficiency of an analog-drive motor-control system that uses a class A power amplifier.

18. An analog-drive circuit uses a power Darlington transistor. The supply voltage is 10 V; at 300 rpm, the motor is drawing 500 mA with a voltage of 3 V. How much power is the transistor dissipating?

19. An analog-drive circuit uses a power Darlington transistor. The supply voltage is 12 V; at 400 rpm, the motor is drawing 1 A with a voltage of 7 V. How much power is the transistor dissipating?

20. Explain the principle of PWM for motor-speed control.

21. What duty cycle would you specify to make a 2000-rpm, 12-V motor run at 1500 rpm? (Assume 12-V pulses and no PWM losses.)

22. What duty cycle would you specify to make a 1500-rpm, 12-V motor run at 1000 rpm? (Assume 12-V pulses and no PWM losses.)

23. What are the advantages of PWM over analog drive for motor control?

24. Explain how an SCR is turned on and off.

25. Sketch the voltage waveform of a full-wave SCR-driven motor for the following conditions:
    a. SCR starts conducting at 30°.
    b. SCR starts conducting at 120°.

Section 7.5

26. A rotating antenna has a moment of inertia (I) of 0.8 lb · ft · s^2. It is driven through a 10 : 1 gearbox by a DC motor (motor turns 10 × faster). The antenna must be able to rotate 180° from rest in 1 s. The motor characteristics are in Figure 7.33. Find the motor voltage required.

27. A rotating antenna has a moment of inertia (I) of 1 kg · m^2. It is driven through a 10 : 1 gearbox by a DC motor. The antenna must be able to rotate 180° from rest in 1 s. The motor characteristics are in Figure 7.33. Find the motor voltage required.

Section 7.6

28. Explain how a brushless DC motor works without brushes.

29. What are the advantages of a BDCM over a regular DC motor?

Stepper Motors

After studying this chapter, you should be able to:

- ❑ Explain what a stepper motor is, how it is different from a "regular" motor, and the applications it is used in.
- ❑ Understand the basic parts and operation of the three kinds of stepper motors: permanent magnet, variable reluctance, and hybrid.
- ❑ Differentiate between two-phase, three-phase, and four-phase stepper motors.

- ❑ Understand the different operational modes—single-step versus slew, single- and dual-phase excitation, half-step, and microstepping.
- ❑ Calculate the final position of a stepper motor, given the sequence of drive pulses.
- ❑ Explain the operation of stepper motor driver circuits.

INTRODUCTION

A stepper motor is a unique type of DC motor that rotates in fixed steps of a certain number of degrees. Step size can range from 0.9 to 90°. Figure 8.1 illustrates a basic stepper motor, which consists of a **rotor** and **stator**. In this case, the rotor is a permanent magnet, and the stator is made up of electromagnets (field poles). The rotor will move (or step) to align itself with an energized field magnet. If the field magnets are energized one after the other around the circle, the rotor can be made to move in a complete circle.

Stepper motors are particularly useful in control applications because the controller can know the exact position of the motor shaft without the need of position sensors. This is done by simply counting the number of steps taken from a known reference position. Step size is determined by the number of rotor and stator poles, and there is no **cumulative error** (the angle error does not increase, regardless of the number of steps taken). In fact, most stepper motor systems operate open-loop—that is, the controller sends the motor a determined number of step commands and assumes the motor goes to the right place. A common example is the positioning of the read/write head in a disk drive.

Figure 8.1
A PM 90° stepper motor.

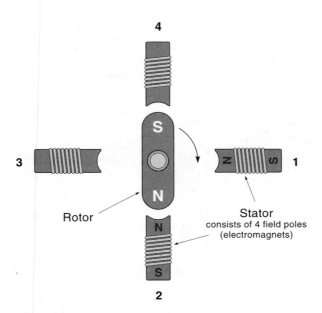

Steppers have inherently low velocity and therefore are frequently used without gear reductions. A typical unit driven at 500 pulses/second rotates at only 150 rpm. Stepper motors can easily be controlled to turn at 1 rpm or less with complete accuracy.

There are three types of stepper motors: permanent magnet, variable reluctance, and hybrid. All types perform the same basic function, but some differences among them may be important in some applications.

8.1 PERMANENT MAGNET STEPPER MOTORS

The **permanent magnet (PM) stepper motor** uses a permanent magnet for the rotor. Figure 8.1 shows a simple PM stepper motor. The field consists of four poles (electromagnets). The motor works in the following manner: Assume the rotor is in the position shown with the south end up. When field coil 1 is energized, the south end of the rotor is attracted to coil 1 and moves toward it. Then field coil 1 is deenergized, and coil 2 is energized. The rotor pulls itself into alignment with coil 2. Thus, the rotor turns in 90° steps for each successive excitation of the field coils. The motor can be made to reverse by inverting the sequence.

One desirable property of the PM stepper motor is that the rotor will tend to align up with a field pole even when no power is applied because the PM rotor will be attracted to the closest iron pole. You can feel this "magnetic tug" if you rotate the motor by hand; it is called the detent torque, or residual torque. The detent torque is a desirable property in many applications because it tends to hold the motor in the last position it was stepped to, even when all power is removed.

As mentioned earlier, one big advantage of the stepper motor is that it can be used open-loop—that is, by keeping track of the number of steps taken from a known point, the exact shaft position is always known. Example 8.1 demonstrates this.

◆ **EXAMPLE 8.1**

A 15°/step stepper motor is given 64 steps CW (clockwise) and 12 steps CCW (counterclockwise). Assuming it started at 0°, find the final position.

Solution

After completing 64 steps CW and 12 steps CCW, the motor has ended up 52 steps CW (64 − 12 = 52). Because there are 24 15°-steps per revolution (360°/15° = 24),

$$\frac{52}{24} = 2\frac{1}{6} = 2 \text{ rev} + \frac{1}{6} \text{ rev}$$

$$= 2 \text{ rev} + \frac{360°}{6}$$

$$= 2 \text{ rev} + 60°$$

Therefore, the motor has made two complete revolutions and is now sitting at 60° CW from where it started.   ◆

Effect of Load on Stepper Motors

For the open-loop concept to work, the motor must actually step once each time it's commanded to. If the load is too great, the motor may not have enough torque to make the step. In such a case, the rotor would probably rotate a little when the step pulse was applied but then fall back to its original position. This is called **stalling**. If feedback is not used, the controller has no way of knowing a step was missed.

Within each step, the torque developed by the stepper motor is dependent on the shaft angle. In fact, the torque on the rotor is actually zero when it is exactly aligned with an energized field coil. Figure 8.2 illustrates how the motor can only provide torque when the rotor is *not* aligned. The first frame of Figure 8.2 shows a rotor pole approaching an energized field pole. The actual force of attraction is between the south (S) end of the rotor and the north (N) end of the field pole. As the rotor pole approaches the field pole, the attraction force (F) gets stronger but the torque component (T) gets weaker. When the rotor is pointing directly at the field pole (last frame in Figure 8.2), the torque component is zero. In practice, this means that the rotor may come to a stop before it is completely aligned with the energized field pole, at the point where the diminishing step torque just equals the load torque.

For the simple motor under discussion (Figure 8.1), the maximum torque occurs when the rotor is 90° (one step) away from the field pole (first frame of Figure 8.2.) It might seem that we should just plan to let the rotor lag one step behind the energized field pole to take advantage of the maximum torque, but this approach might cause the

Figure 8.2
Torque goes to zero as the rotor aligns with the field pole.

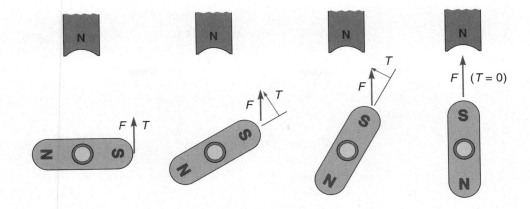

motor to take a step backward instead of forward. Consider the situation in Figure 8.3(a) where the rotor has been stepping CCW and we have allowed it to lag a full step behind the energized pole (currently, pole 1). The next pole to be energized in the CCW sequence is pole 2 [Figure 8.3(b)]. The first problem here is that the rotor is pointing directly away from pole 2, so there will be little or no torque exerted. The second problem is that, in this balanced condition, the rotor will be equally attracted in either direction and we cannot reliably predict if it will turn CW or CCW.

For proper operation, *the rotor lag must not be to allowed exceed one-half the step size*, which would be 45° for the motor illustrated in Figure 8.3. This solves the preceding problems—namely, the motor will always turn in the direction it's supposed to, and it

Figure 8.3
Illustrating what would happen if the rotor was allowed to lag a full step behind the field poles.

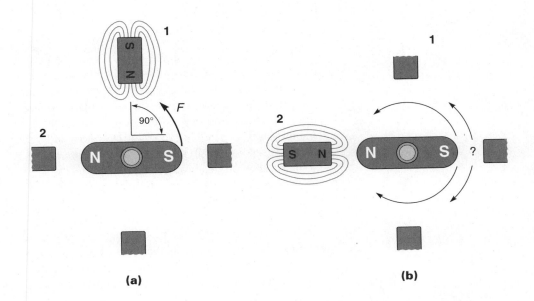

(a)   (b)

will not stall. (Recall that stalling occurs when the motor is too weak to take a step.) In practical terms, the dynamic torque, which is the power available when the motor is running, may only be about half of the maximum *static torque* (the torque required to displace the rotor when stopped). There is an exception to this rule: When the rotor is stepping rapidly (called *slewing*), the inertia can be counted on to keep the rotor going in the right direction. Slewing is discussed in the next section.

Modes of Operation

The stepper motor has two modes of operation: single step and slew. In the **single-step mode** or bidirectional mode, the frequency of the steps is slow enough to allow the rotor to (almost) come to a stop between steps. Figure 8.4 shows a graph of position versus time for single-step operation. For each step, the motor advances a certain angle and then stops. If the motor is only lightly loaded, overshoot and oscillations may occur at the end of each step as shown in the figure.

Figure 8.4

Position vs time for single-step mode.

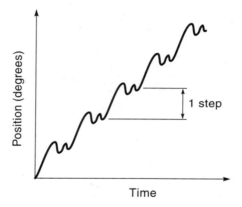

The big advantage of single-step operation is that each step is completely independent from every other step—that is, the motor can come to a dead stop or even reverse direction at any time. Therefore, the controller has complete and instantaneous control of the motor's operation. Also, there is a high certainty that the controller will not lose count (and hence motor position) because each step is so well defined. The disadvantage of single-step mode is that the motion is slow and "choppy." A typical single-step rate is 5 steps/second which translates to 12.5 rpm for a 15°/step motor.

In the **slew mode**, or unidirectional mode, the frequency of the steps is high enough that the rotor does not have time to come to a stop. This mode approximates the operation of a regular electric motor—that is, the rotor is always experiencing a torque and rotates in a smoother, continuous fashion. Figure 8.5 shows a graph of position versus time for the slew mode. Although the individual steps can still be discerned, the motion is much less choppy than in single-step mode.

A stepper motor in the slew mode cannot stop or reverse direction instantaneously. If attempted, the rotational inertia of the motor would most likely carry the rotor ahead

Figure 8.5

Position vs time for the slew mode.

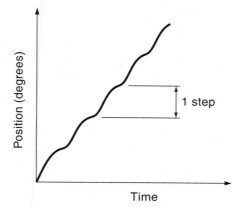

a few steps before it came to rest. The step-count integrity would be lost. It *is* possible to maintain the step count in the slew mode by slowly ramping up the velocity from the single-step mode and then ramping down at the end of the slew. This means the controller must know ahead of time how far the motor must go. Typically, the slew mode is used to get the motor position in the "ballpark," and then the fine adjustments can be made with single steps. Slewing moves the motor faster but increases the chances of losing the step count.

Figure 8.6 shows the torque–speed curves for both the single-step and slew modes. The first observation is that available load torque diminishes as the stepping rate rises (this is true of all DC motors). Also, for the single-step mode, the price paid for the ability to stop or reverse instantaneously is less torque and speed. Looking along the *x*-axis, notice three different kinds of torques. The **detent torque** is the torque required to overcome the force of the permanent magnets (when the power is off). It is the little tugs

Figure 8.6

Torque–speed curves for single-step and slew modes.

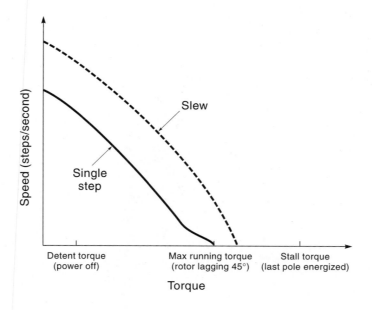

you feel if you manually rotate the unpowered motor. The **dynamic torque**, which is the maximum running torque, is obtained when the rotor is lagging behind the field poles by half a step. The highest stall torque shown in Figure 8.6 is called the **holding torque** and results when the motor is completely stopped but with the last pole still energized. This is really a detent type of torque because it represents the amount of external torque needed to rotate the motor "against its wishes."

♦ **EXAMPLE 8.2**

A stepper motor has the following properties:

Holding torque:	50 in. · oz
Dynamic torque:	30 in. · oz
Detent torque:	5 in. · oz

The stepper motor will be used to rotate a 1-in. diameter printer platen (Figure 8.7). The force required to pull the paper through the printer is estimated to not exceed 40 oz. The static weight of the paper on the platen (when the printer is off) is 12 oz. Will this stepper motor do the job?

Solution

The torque required to rotate the platen during printing can be calculated as follows:

$$\text{Force} \times \text{radius} = 40 \text{ oz} \times 0.5 \text{ in.} = 20 \text{ in.} \cdot \text{oz}$$

Therefore, the motor, with 30 in. · oz of dynamic torque, will be strong enough to advance the paper.

The torque on the platen from just the weight of the paper is calculated as follows:

$$\text{Force} \times \text{radius} = 12 \text{ oz} \times 0.5 \text{ in.} = 6 \text{ in} \cdot \text{oz}$$

Figure 8.7

A stepper motor driving a printer platen (Example 8.2).

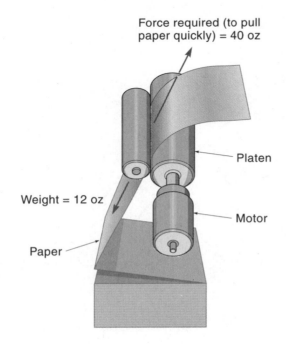

Force required (to pull paper quickly) = 40 oz

Platen

Weight = 12 oz

Motor

Paper

When the printer is on, the powered holding torque of 50 in. · oz is more than enough to support the paper. However, when the printer is off, the weight of the paper exceeds the detent torque of 5 in. · oz, and the platen (and motor) would spin backward. Therefore, we conclude that this motor is not acceptable for the job (unless some provision such as a ratchet or brake is used to prevent back spinning).   ♦

Excitation Modes for PM Stepper Motors

Stepper motors come with a variety of winding and rotor combinations. In addition, there are different ways to sequence energy to the field coils. All these factors determine the size of each step. **Phase** refers to the number of separate winding circuits. There are two-, three-, and four-phase steppers, which are discussed next.

Two-Phase (Bipolar) Stepper Motors

The **two-phase** (bipolar) **stepper motor** has only two circuits but actually consists of four field poles. Figure 8.8(a) shows the motor symbol, and Figure 8.8(b) shows how it is wired internally. In Figure 8.8(b), circuit AB consists of two opposing poles such that when voltage is applied ($+A$ $-B$), the top pole will present a north end to the rotor and the bottom pole will present a south end. The rotor would tend to align itself vertically (position 1) with its south pole up (because, of course, opposite magnetic poles attract).

The simplest way to step this motor is to alternately energize either AB or CD in such a way as to pull the rotor from pole to pole. If the rotor is to turn CCW from position 1, then circuit CD must be energized with polarity $C+$ $D-$. This would pull the rotor to position 2. Next, circuit AB is energized again, but this time the polarity is reversed ($-A$ $+B$), causing the bottom pole to present a north end to the rotor, thereby pulling

Figure 8.8

A two-phase (bipolar) stepper motor.

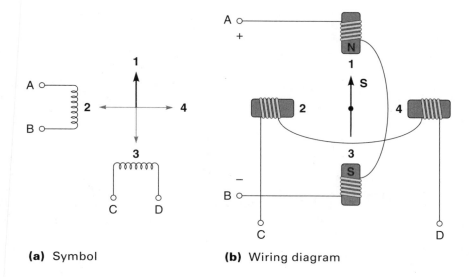

(a) Symbol   (b) Wiring diagram

it to position 3. The term **bipolar** applies to this motor because the current is sometimes reversed. The voltage sequence needed to rotate the motor one full turn is shown below. Reading from top to bottom gives the sequence for turning CCW, reading from bottom up gives the CW sequence:

Circuit	Position
A+ B−	1
C+ D−	2
A− B+	3
C− D+	4

Another way to operate the two-phase stepper is to energize both circuits at the same time. In this mode, the rotor is attracted to two adjacent poles and assumes a position in between. Figure 8.9(a) shows the four possible rotor positions. The excitation sequence for stepping in this dual mode is as follows:

Circuits			Position
A+ B−	and	C+ D−	1
A− B+	and	C+ D−	2
A− B+	and	C− D+	3
A+ B−	and	C− D+	4

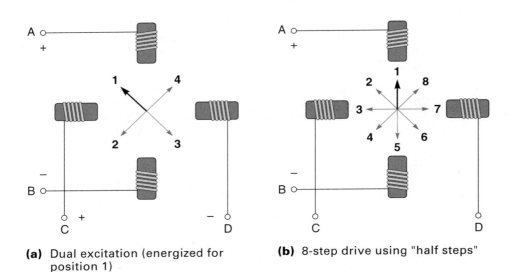

Figure 8.9

Additional operating modes for stepper motors.

(a) Dual excitation (energized for position 1)

(b) 8-step drive using "half steps"

Having two circuits on at the same time produces considerably more torque than the single-excitation mode; however, more current is consumed, and the controller is more complex.

Both methods produce **four-step drives**, that is, four steps per cycle. By alternating the single- and dual-excitation modes, the motor can be directed to take **half-steps**, as shown in Figure 8.9(b). Positions 1, 3, 5, and 7 are from the single-excitation mode, and positions 2, 4, 6, and 8 are from the dual-excitation mode. When driven this way, the motor takes eight steps per revolution and is called an **eight-step drive**. This is desirable for some applications because it allows the motor to have twice the position resolution. Even smaller steps are possible with a process called *microstepping,* which is discussed later in the chapter.

The motor described thus far in this section steps 90° in the four-step mode (and 45° in the eight-step mode). PM stepper motors commonly have a smaller step, as low as 30°. This is done by increasing the poles in the rotor. Figure 8.10 shows a 30° stepper motor; the rotor has six rotor poles. Assume that field poles AB have been energized, pulling the rotor into the position shown. Next, circuit CD is energized. Rotor poles *yz* will be attracted to poles CD and will have to rotate only 30° to come into alignment.

Figure 8.10

A 30° stepper motor with a six-pole rotor.

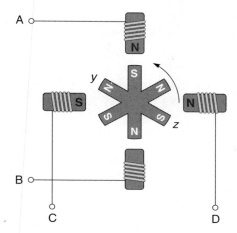

Four-Phase (Unipolar) Stepper Motors

The **four-phase** (unipolar) is the most common type of stepper motor (Figure 8.11). The term *four-phase* is used because the motor has four field coils that can be energized independently, and the term **unipolar** is applied because the current always travels in the same direction through the coils. The simplest way to operate the four-phase stepper motor is to energize one phase at a time in sequence (known as *wave drive*). To rotate CW, the following sequence is used:

    A B
    C D
    E F
    G H

Compared with the two-phase bipolar motor, the four-phase motor has the advantage of simplicity. The control circuit of the four-phase motor simply switches the poles on and off in sequence; it does not have to reverse the polarity of the field coils. (However, the two-phase motor produces more torque because it is pushing and pulling at the same time.)

Figure 8.11

A four-phase (unipolar) stepper motor.

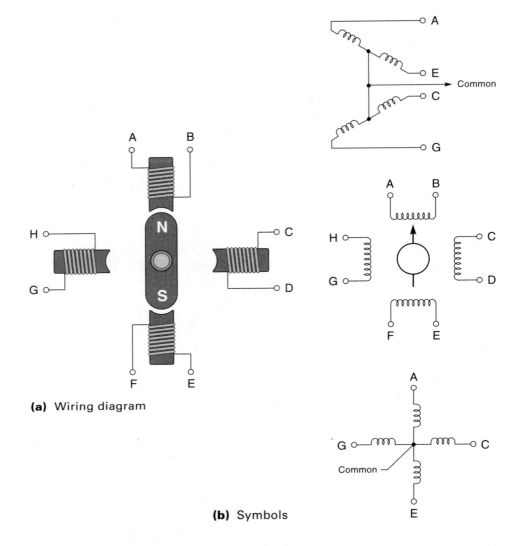

(a) Wiring diagram

(b) Symbols

The torque of a four-phase stepper motor can be increased if two adjacent coils are energized at the same time, causing the rotor to align itself between the field poles [similar to that shown in Figure 8.9(a)]. Although twice the input energy is required, the motor torque is increased by about 40%, and the response rate is increased. By alternating single- and dual-excitation modes, the motor steps in half-steps, as shown in Figure 8.9(b).

Constructing motors so they can be used in either a two- or four-phase mode is common practice. This is done by bringing out two additional wires (from the two-phase motor) that are internally connected to points between the opposing field coils. Figure 8.12(a) shows the symbol for this type of motor, and Figure 8.12(b) shows the interior motor wiring. When such a motor is used in the two-phase mode, the center taps (terminals 2 and 5) are not used. When the motor is operated in the four-phase mode, the center taps become a *common return*, and power is applied to terminals 1, 4, 3, and 6 as required.

Figure 8.12
A four-phase stepper with center tap windings.

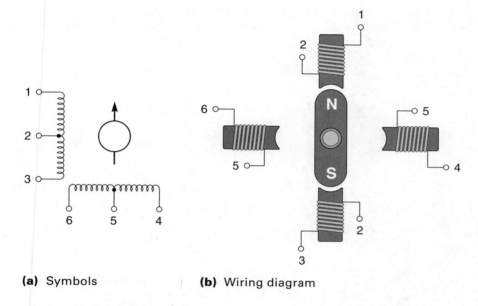

(a) Symbols   **(b)** Wiring diagram

Available PM Stepper Motors

Almost all PM stepper motors available today have smaller step sizes than the simplified motors discussed so far. These motors are made by stacking two multipolled rotors (offset by one-half step), as illustrated in Figure 8.13. Typical step sizes for four-phase PM stepper motors are 30°, 15°, and 7.5°. Figure 8.14 is a specification sheet for a family of PM stepper motors. For example, the PF35–48C (continuous duty) motor has 48 steps/revolution (which is 7.5°), requires 133 mA at 12 V, and has a holding torque of 2.78 in. · oz.

Figure 8.13
Stacked-rotor design allows smaller steps.

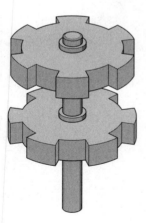

8.2   VARIABLE RELUCTANCE STEPPER MOTORS

The **variable reluctance (VR) stepper motor** does not use a magnet for the rotor; instead, it uses a toothed iron wheel [see Figure 8.15(b)]. The advantage of not requiring the rotor to be magnetized is that it can be made in any shape. Being iron, each rotor tooth is

PF35 Series		Models					
		PF35-48				**PF35-24**	
Excitation Mode		2-2					
Step Angle	(°)	7.5				15	
Step Angle Tolerance	(%)	±5					
Steps per Revolution		48				24	
Rating		Continuous			Intermittent	Continuous	
Letter Designator		C	D	Q	C	C	D
Winding Type		Unipolar	Unipolar	Bipolar	Unipolar	Unipolar	Unipolar
DC Operating Voltage	(V)	12	5	5	24	12	5
Operating Current	(mA/ø)	133	313	310	266	133	313
Winding Resistance	(Ω/ø)	90	16	17	90	90	16
Winding Inductance	(mH/ø)	48	8.9	12	48	48	8.9
Holding Torque	(oz-in)	2.78		3.25	3.88	2.08	
Rotor Inertia	(oz-in²)	24.1x10⁻³					
Starting Pulse Rate, Max.	(pps)	500		400	680	310	
Slewing Pulse Rate, Max.	(pps)	530		500	770	410	
Ambient Temp Range, Operating	(°C)	-10~+50					
Temperature Rise	(°C)	55			-	55	
Weight	(oz)	2.8					

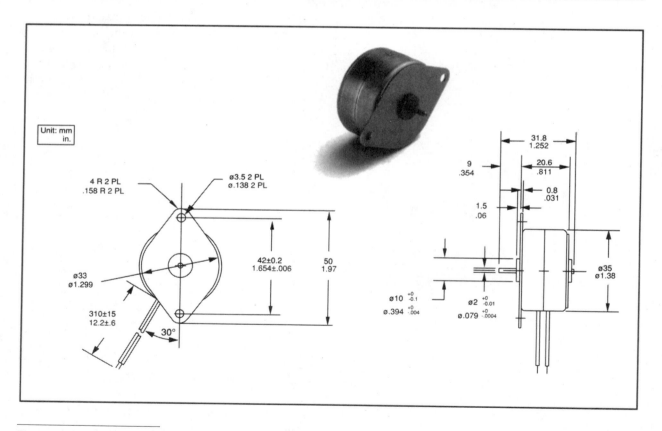

Figure 8.14   PM stepper motors. (Courtesy of Kollmorgen Motion Technologies Group)

Figure 8.15

A three-phase VR stepper motor (15° step). (Wires for Ø2 and Ø3 left out for clarity.)

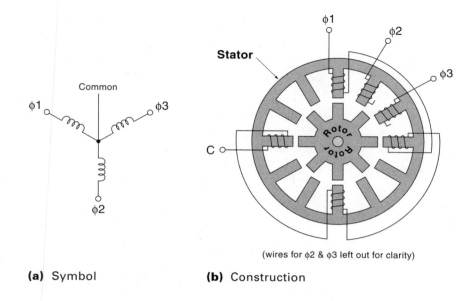

(wires for φ2 & φ3 left out for clarity)

(a) Symbol   **(b)** Construction

attracted to the closest energized field pole in the stator, but not with the same force as in the PM motor. This gives the VR motor less torque than the PM motor.

A VR motor usually has three or four phases. Figure 8.15(a) shows a typical **three-phase stepper motor**. The stator has three field pole circuits: Ø1, Ø2, and Ø3. Figure 8.15(b) shows that the actual motor has 12 field poles, where each circuit energizes four windings; you can see this by closely observing the Ø1 wire in Figure 8.15(b). Notice that the rotor has only 8 teeth even though there are 12 teeth in the stator. Therefore, the rotor teeth can never line up "one for one" with the stator teeth, a fact that plays an important part in the motor's operation.

Figure 8.16 illustrates the operation of the VR stepper motor. When circuit Ø1 is energized, the rotor will move to the position shown in Figure 8.16(a)—that is, a rotor tooth (A) is lined up with the Ø1 field pole. Next, circuit Ø2 is energized. Rotor tooth B,

Figure 8.16

A 15° three-phase VR stepper motor (only four field poles shown for clarity.)

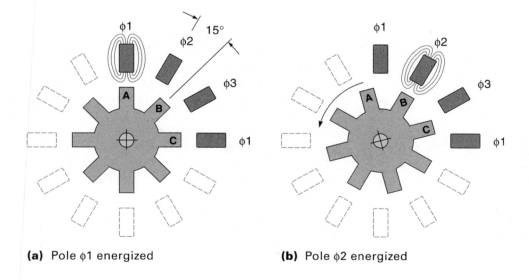

(a) Pole φ1 energized   **(b)** Pole φ2 energized

being the closest, is drawn toward it [Figure 8.16(b)]. Notice that the rotor has to move only 15° for this alignment. If circuit Ø3 is energized next, the rotor would continue CCW another 15° by pulling tooth C into alignment.

The step angle of the VR motor is the difference between the rotor and stator angles. For the motor of Figure 8.16, the angle between the field poles is 30°, and the angle between the rotor poles is 45°. Therefore, the step is 15° (45° − 30° = 15°). By using this design, the VR stepper motor can achieve very small steps (less than 1°). Small step size is often considered to be an advantage because it allows for more precise positioning.

The VR stepper motor has a number of functional differences when compared with the PM type. Because the rotor is not magnetized, it is weaker than a similar-sized PM stepper motor. Also, it has no detent torque when the power is off, which can be an advantage or disadvantage depending on the application. Finally, because of the small step size and reduced detent torque, the VR stepper motor has more of a tendency to overshoot and skip a step. This is a serious matter if the motor is being operated open-loop, where position is maintained by keeping track of the number of steps taken. To solve the problem, some sort of damping may be required. This can be done mechanically by adding friction or electrically by providing a slight braking torque with adjacent field poles.

8.3   HYBRID STEPPER MOTORS

The **hybrid stepper motor** combines the features of the PM and VR stepper motors and is the type in most common use today. The rotor is toothed, which allows for very small step angles (typically 1.8°), and it has a permanent magnet providing a small detent torque even when the power is off.

Recall that the step size of a PM motor is limited by the difficulty in making a multipole magnetized rotor. There is simply a limit to the number of different magnetizations that can be imposed on a single iron rotor. The VR stepper motor gets around this by substituting iron teeth (of which there can be many) for magnetized poles on the rotor. This approach allows for a small step angle, but it sacrifices the strength and detent torque qualities of the PM motor. The hybrid motor can effectively magnetize a multitoothed rotor and thus has the desirable properties of both the PM and VR motors.

Figure 8.17 illustrates the internal workings of the hybrid motor, which is considerably more complicated than the simple PM motor. The rotor consists of two toothed wheels with a magnet in between—one wheel being completely north in magnetization and the other being completely south. For each step, two opposing teeth on the north wheel are attracted to two south field poles, and two opposing teeth on the south wheel are attracted to two north field poles. The internal wiring is more complicated than it is for the PM or VR motors, but to the outside world this motor is just as simple to control.

Figure 8.17

Internal construction of the hybrid stepper motor (only 2 poles per stator shown for clarity).

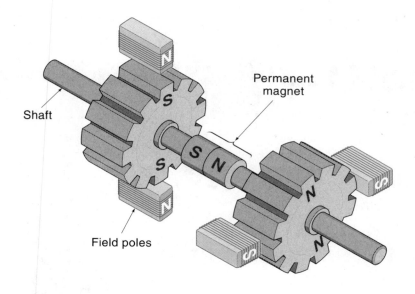

The theory of operation of the hybrid motor is similar to the VR motor in that the rotor and stator have a different number of teeth, and for each step, the closest energized teeth are pulled into alignment. However, the principles of magnetics require that, at any one time, half the poles be north and the other half be south. To maintain the magnetic balance, each pole must be able to switch polarity so that it can present the correct pole at the correct time. This is accomplished in one of two ways: For a bipolar motor, the applied voltage must be reversed by the driver circuit (similar to the two-phase PM motor). On the other hand, a unipolar motor has two separate windings of opposite direction on each field pole (called a *bifilar winding*), so each pole can be a north or a south. Therefore, the unipolar hybrid motor does not require a polarity-reversing driver circuit.

Figure 8.18 shows a specification sheet for a family of 1.8° hybrid stepping motors. For example, the 11-SHBD-45AB draws 0.3 A at 13.8 V and has a holding torque of 9.5 in. · oz, a running (dynamic) torque of 5.9 in. · oz, and a detent torque (unpowered) of 0.36 in. · oz. Unloaded, it can step at a rate of 1385 steps/minute.

8.4   STEPPER MOTOR CONTROL CIRCUITS

Figure 8.19 shows the block diagram for a stepper motor driving circuit. The *controller* decides on the number and direction of steps to be taken (based on the application). The *pulse sequence generator* translates the controller's requests into specific stepper motor coil voltages. The *driver amplifiers* boost the power of the coil drive signals. It should be

Bulletin SM 11/1.8

Size 11
1.8°

Compact, light weight, high resolution stepper motors with high torque-to-size ratio

Features and Benefits

- Small size: 1.067" diameter
  1.500" long (excl. shaft)
- ±5% accuracy
- Half or full step (200 or 400 steps per revolution)
- High torque-to-size ratio
- Ball bearing construction
- Direct drive - no gearing
- Light weight
- Low cost

Typical applications

- Medical equipment
- Avionic instruments
- Robotics and automation
- Scanners
- Office equipment
- Battery operated equipment
- Chart recorders
- Test equipment
- Laser optics

STEPPER MOTOR SPECIFICATIONS	11–SHBD–45AB	11–SHBD–47AB	11–SHBD–49AB
INDEX ANGLE	1.8°±5%	1.8°±5%	1.8°±5%
FUNCTION	2ø HYBRID	4ø HYBRID	4ø HYBRID
STEPS PER REVOLUTION	200	200	200
INPUT VOLTAGE (DC)	13.8 REF.	9.6 REF.	14
INPUT CURRENT PER PHASE (AMPS) ± 10%	0.300	0.300	0.311
DC RESISTANCE PER PHASE (OHMS) ± 10%	46	32	45
INDUCTANCE PER PHASE (MH) REF.	51.3	22.4	28.4
NO LOAD RESPONSE RATE (PPS) MIN.	1140 (2)	875 (2)	810 (2)
NO LOAD SLEW RATE (PPS) MIN.	1385 (2)	1130 (2)	1225 (2)
HOLDING TORQUE (OZ-IN) MIN.	9.5 (2)	6.2 (2)	7.8 (2)
DYNAMIC TORQUE (OZ-IN) MIN.	5.9 (2)	3.9 (2)	4.3 (2)
DETENT TORQUE (OZ-IN) REF.	0.36	0.30	0.30
SHAFT RADIAL PLAY MAX.	0.0006 (0.015)	0.0006 (0.015)	0.0006 (0.015)
SHAFT END PLAY MAX. (1)	0.005 (0.13)	0.005 (0.13)	0.005 (0.13)
ROTOR INERTIA (GM-CM^2) REF.	3.3	3.8	3.6

(1) Shaft end play is spring loaded toward front of unit.   (2) Measured with two phases on, L/R.

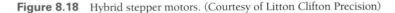

Figure 8.18 Hybrid stepper motors. (Courtesy of Litton Clifton Precision)

clear that the stepper motor is particularly well suited for digital control; it requires no digital-to-analog conversion, and because the field poles are either on or off, efficient class C driver amplifiers can be used.

Figure 8.19

Block diagram of stepper motor control circuit.

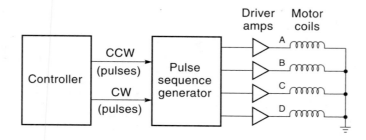

Controlling the Two-Phase Stepper Motor

Controlling the two-phase bipolar stepper motor requires polarity reversals, making it more complicated than four-phase motor controllers. Figure 8.20 shows a two-phase stepper motor. The two circuits are designated AB and CD. The timing diagram shows the required waveforms for A, B, C, and D (CCW rotation). Looking down the position 1 column in Figure 8.20(b), we see A is positive and B is negative, so current will flow from A to B in circuit AB. Meanwhile C and D are both negative, effectively turning off circuit CD. For position 2 in the timing diagram, C is positive, and D is negative; causing current to flow from C to D in circuit CD while coil AB is completely off, and so on, for positions 3 and 4.

Figure 8.20

Two-phase (bipolar) stepper motor operation.

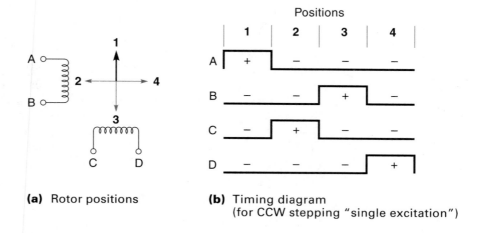

(a) Rotor positions

(b) Timing diagram
(for CCW stepping "single excitation")

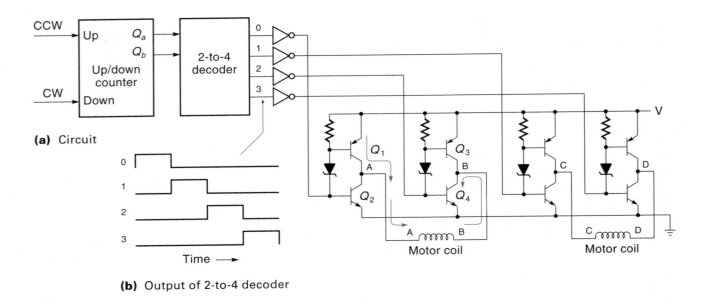

(a) Circuit

(b) Output of 2-to-4 decoder

Figure 8.21 Complete interface circuit for a two-phase (bipolar) stepper motor.

A digital circuit such as the one shown in Figure 8.21(a) can be used to generate the timing waveforms. The up/down counter is a 2-bit counter that increments for each pulse received on its up input and decrements for each pulse received on the down input. Q_a and Q_b of the up/down counter are decoded in a 2-to-4 decoder. Because the counter is always in one of four states (00, 01, 10, 11), one (and only one) of the four decoder outputs is "high" at any particular time. Figure 8.21(b) shows the output of the decoder when the counter counts up (a result of CCW pulses from the controller).

The next task is to connect the timing signals from the decoder in such a way as to drive the motor coils. This can be accomplished with the power amplifier circuit shown on the right side of Figure 8.21(a). Notice there are four complementary–symmetry drivers, one for each end of each motor coil. When Q_1 and Q_4 are on, the current can flow through the motor in the direction shown. On the other hand, when Q_3 and Q_2 are on, the polarity is reversed, and the current flows the opposite direction through the motor. Finally, if Q_1 and Q_3 are off, no current flows through the motor coil.

The four outputs of the decoder (which must be inverted in this case) control the four complementary–symmetry transistor circuits. The resistor and diode in each circuit cause the upper transistor to be on when the lower transistor is off, and vice versa. Trace through the circuit for each step of the decoder, and you will see that the timing diagram of Figure 8.20(b) is reproduced. This arrangement will provide for the motor to step CCW when the counter counts up. When the counter counts down, the sequence will be backward, and the motor will step CW.

Controlling the Four-Phase Stepper Motor

The electronics needed to drive the four-phase stepper motor is simpler than for a two-phase motor because polarity reversals are not required. Figure 8.22 identifies the coils in a four-phase motor and shows the timing diagram for simple single-excitation

Figure 8.22
Four-phase (unipolar) stepper motor operation.

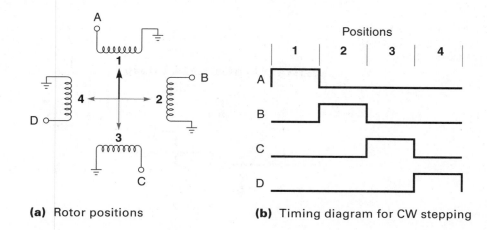

(a) Rotor positions

(b) Timing diagram for CW stepping

Figure 8.23
Complete interface for a four-phase stepper motor (simplified for clarity).

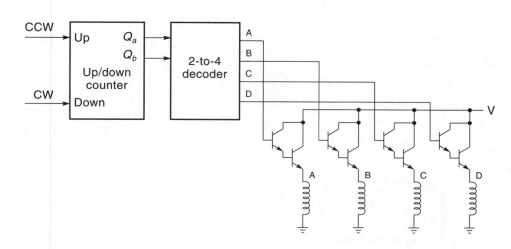

CW stepping. The timing is very straightforward and can easily be generated with the counter and decoder circuit shown in Figure 8.23.

The outputs of the decoder (Figure 8.23) are connected to four Darlington driver transistors. As timing signals A, B, C, and D go high in sequence, the corresponding transistors are turned on, energizing coils in the motor. The necessary four Darlingtons are available on a single IC such as the Allegro ULN-2064B (Figure 8.24). These amplifiers can supply up to 1.5 A and can be driven directly from TTL 5-V logic. (Note the diode to allow an escape path for current in the coil when the transistor is turned off.)

ICs designed specifically to drive stepper motors contain both the timing logic and power drivers in one package. One example is the Allegro UCN-5804B (Figure 8.25). The basic inputs are the step input (pin 11) and direction (pin 14). The motor will advance one step for each pulse applied to the step input pin, and the logic level on the direction pin determines if rotation will be CW or CCW. Notice that the output transistors are in the common emitter configuration (called *open collector*). The motor coils should be connected between the output pins and the supply voltage (as shown). When an output transistor turns on, it completes the circuit by providing a path to

Figure 8.24
The Allegro ULN-2064B with four Darlington 1.5-A switches. (Courtesy of Allegro MicroSystems)

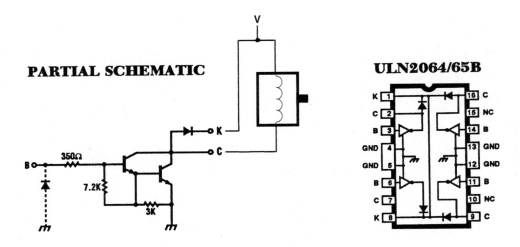

Figure 8.25 A unipolar stepper motor translator/driver (Allegro UCN-5804B). (Courtesy of Allegro MicroSystems)

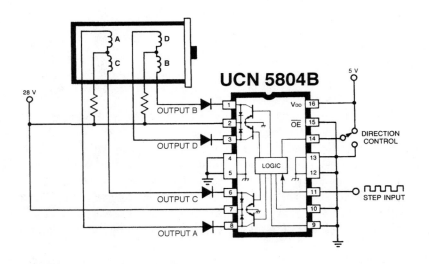

(a) Driver circuit

(b) Modes of operation

WAVE-DRIVE SEQUENCE

	Half Step = L, One Phase = H			
Step	A	B	C	D
POR	ON	OFF	OFF	OFF
1	ON	OFF	OFF	OFF
2	OFF	ON	OFF	OFF
3	OFF	OFF	ON	OFF
4	OFF	OFF	OFF	ON

DIRECTION = L ↓   ↑ DIRECTION = H

TWO-PHASE DRIVE SEQUENCE

	Half Step = L, One Phase = L			
Step	A	B	C	D
POR	ON	OFF	OFF	ON
1	ON	OFF	OFF	ON
2	ON	ON	OFF	OFF
3	OFF	ON	ON	OFF
4	OFF	OFF	ON	ON

DIRECTION = L ↓   ↑ DIRECTION = H

HALF-STEP DRIVE SEQUENCE

	Half Step = H, One Phase = L			
Step	A	B	C	D
POR	ON	OFF	OFF	OFF
1	ON	OFF	OFF	OFF
2	ON	ON	OFF	OFF
3	OFF	ON	OFF	OFF
4	OFF	ON	ON	OFF
5	OFF	OFF	ON	OFF
6	OFF	OFF	ON	ON
7	OFF	OFF	OFF	ON
8	ON	OFF	OFF	ON

DIRECTION = L ↓   ↑ DIRECTION = H

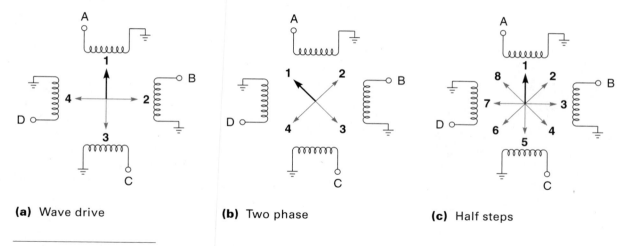

(a) Wave drive   **(b)** Two phase   **(c)** Half steps

Figure 8.26   Three modes of operation for the Allegro UCN-5804B.

ground for the motor coil current. The three operating modes of the UCN-5804B are given in tabular form in Figure 8.25(b), and will be explained using Figure 8.26:

1. Figure 8.26(a) shows how the motor responds to the wave-drive sequence, a single-excitation mode where the coils A, B, C, and D are energized one at a time in sequence.

2. Figure 8.26(b) shows how the motor responds to the two-phase drive sequence, a dual-excitation mode where two adjacent phases are energized at the same time to give more torque.

3. Figure 8.26(c) shows how the motor responds to the half-step drive sequence, where the operation alternates between single- and dual-excitation modes, yielding eight half-steps per cycle.

Microstepping

Microstepping, a technique that allows a stepper motor to take fractional steps, works by having two adjacent field poles energized at the same time, similar to half-steps described earlier. In microstepping the adjacent poles are driven with different voltage levels, as demonstrated in Figure 8.27(a). In this case, pole 1 is supplied with 3 V and pole 2 with 2 V, which causes the rotor to be aligned as shown—that is, three-fifths of the way to pole 1. Figure 8.27(b) shows the voltages (for poles 1 and 2) to get five microsteps between each "regular" step. The different voltages could be synthesized with pulse-width modulation (PWM). The most commonly used microstep increments are 1/5, 1/10, 1/16, 1/32, 1/125, and 1/250 of a full step. Another benefit of microstepping (for delicate systems) is that it reduces the vibrational "shock" of taking a full step—that is, taking multiple microsteps creates a more "fluid" motion.

Two other points on microstepping: It does not require a special stepper motor, only special control circuitry, and the actual position of the rotor (in a microstepping system) is very dependent on the load torque.

Figure 8.27
Microstepping.

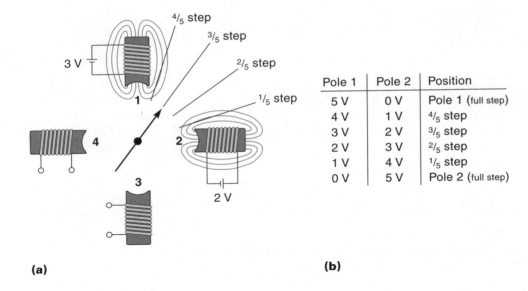

Pole 1	Pole 2	Position
5 V	0 V	Pole 1 (full step)
4 V	1 V	4/5 step
3 V	2 V	3/5 step
2 V	3 V	2/5 step
1 V	4 V	1/5 step
0 V	5 V	Pole 2 (full step)

(a)   **(b)**

Improving Torque at Higher Stepping Rates

It is important that the stepper motor develop enough torque with each step to drive the load. If it doesn't, the motor will stall (not step). When steps are missed, the controller no longer knows the exact position of the load, which may render the system useless.

At higher stepping rates, two problems occur. First, if the load is accelerating, extra torque is needed to overcome inertia; second, the available motor torque actually diminishes at higher speeds. Recall that motor torque is directly proportional to motor current and that the average current decreases as the stepping rate increases. This is illustrated in Figure 8.28, which shows the motor current for three stepping rates. The problem is that the rate of change of current is limited by the circuit-time constant τ.

$$\tau = \frac{L}{R} \tag{8.1}$$

where

τ = time constant
L = motor inductance
R = motor coil resistance

Figure 8.28
Coil current as a function of stepping speed.

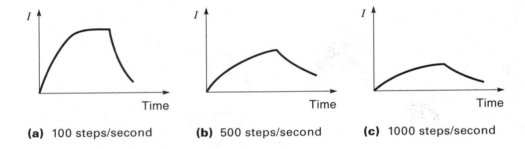

(a) 100 steps/second   **(b)** 500 steps/second   **(c)** 1000 steps/second

You can see that as the stepping rate increases the current cannot build up in the field coils to as great a value. If we could reduce the value of the motor-time constant, the current could build up faster. One way to do this would be to increase the value of R in Equation 8.1. This can easily be done by adding external resistors (R) in series with the motor coils as shown in Figure 8.29(a). Such resistors are called **ballast resistors**, and their purpose is to improve the torque output at higher stepping rates (it also limits the current, which may be important in some cases). Stepper motor driver circuits that use ballast resistors are called **L/R drives**. Figure 8.29(b) shows the motor current with the ballast resistor added (for a rate of 1000 steps/second). Compare this with the last graph in Figure 8.28 to see the improvement.

Figure 8.29

Effect of adding an external ballast resistor.

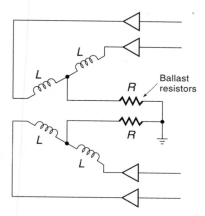

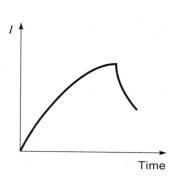

(a) Motor coils and ballast resistors

(b) Coil current at 1000 steps/second with ballast resistor

Another way to improve motor torque at higher stepping rates is to use **bilevel drive**. In this approach, a high voltage is momentarily applied to the motor at the beginning of the step to force a fast in-rush of current. Then a lower voltage level is switched on to maintain that current. Figure 8.30(a) shows a simplified circuit to provide bilevel drive.

Figure 8.30

Principle of a bilevel drive.

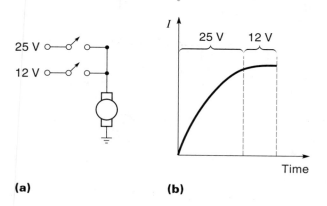

The 25-V circuit is switched on, and the current rises rapidly [Figure 8.30(b)]. When the desired current level is reached, the 25-V circuit is switched off, and the 12-V circuit is switched on, which keeps the current at the desired level for the rest of the step time.

Figure 8.31 shows a bilevel-drive interface circuit. In this case, the higher voltage is 12 V, and the lower voltage is 5 V. The 12-V is switched through either Q_5 or Q_6 in response to a pulse from a timing circuit (not shown). The 5-V is applied through D_1 and D_2. These diodes keep the 12-V pulses from backing up into the 5-V power supply. The bilevel drive is more complex but allows the stepper motor to have more torque at higher stepping rates.

Figure 8.31

A bilevel-drive circuit.

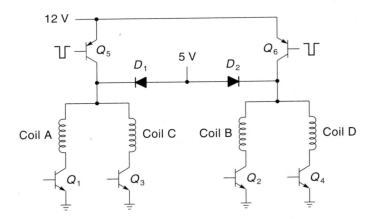

Another approach for providing higher torque at faster stepping rates is the **constant current chopper drive**. Using PWM techniques, this driver circuit can deliver almost the same current to the motor at all speeds. A chopper-drive waveform is shown in Figure 8.32 and works in the following manner: A relatively high voltage is switched to the motor coil, and the current is monitored. When the current reaches a specified level, the voltage is cut off. After a short time, the voltage is reapplied, and the current again

Figure 8.32

A PWM (chopper drive) used to regulate stepper motor coil current.

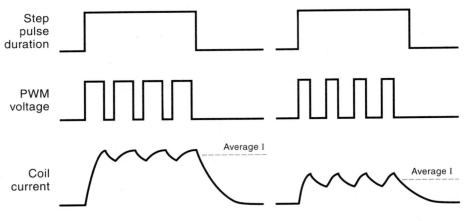

(a) Higher average coil current **(b)** Lower average coil current

increases, only to be cut off, and so on. Thus, in the same way that a thermostat can maintain a constant temperature by switching the furnace on and off, the chopper drive maintains a constant average current (within each drive pulse) by rapidly switching the voltage on and off. In summary, the chopper drive is another technique for providing good torque at high stepping rates. Stepper motor driver ICs are available, such as the Allegro A2919SB, with built-in PWM constant current capability.

8.5   STEPPER MOTOR APPLICATION: POSITIONING A DISK DRIVE HEAD

Example 8.3 illustrates many of the principles presented in this chapter and extends the discussion to show how software can control a stepper motor.

◆ **EXAMPLE 8.3**

A 30° four-phase stepper motor drives the read/write head on a floppy disk drive (Figure 8.33). The in-and-out linear motion is achieved with a leadscrew connected directly to the motor shaft. Each magnetic track on the disk is 0.025 in. apart (40 tracks/in.). The leadscrew has 20 threads/in.

The motor is driven by the UCN-5804B stepper motor interface IC. This IC requires only two inputs: step input and direction. A computer will supply these signals in response to toggle switch settings. A front panel contains eight toggle switches; seven are used to input (in binary) the number of tracks to move, and the eighth switch specifies direction—in or out. Write a program in BASIC that will cause the motor to step the number of tracks and direction specified by the switch settings.

Solution

First we need to find the number of steps required to advance one track on the disk. If the leadscrew has 20 threads per inch, then rotating it one revolution (360°) will advance it 1 thread, which is $\frac{1}{20}$ in. The following equation was set up by multiplying all the component transfer functions, including conversion factors as necessary (and oriented so that, if possible, the units would cancel):

$$\frac{0.025 \text{ in.}}{\text{track}} \times \frac{20 \text{ threads}}{\text{in.}} \times \frac{360°}{\text{thread}} \times \frac{1 \text{ step}}{30°} = \frac{6 \text{ steps}}{\text{track}}$$

Thus, the stepper motor must take six steps to advance one track on the disk.

The program must first read the switches, then calculate the required number of steps, and, finally, output the step command pulses, one by one, to the UCN-5804B. The direction bit must also be read and passed along to the UCN-5804B. Figure 8.34 shows a simplified flowchart of the program.

The next step is to translate the flowchart into a BASIC program. The complete program, along with line-by-line explanations, is given in Table 8.1. With BASIC we can only input or output 8 bits at a time. The input is the 8 bits from the switches. For output, only 2 of the 8 bits are used: the least significant bit (LSB) (D_0) for the step input pulse and the most significant bit (MSB) (D_7) for the direction command. The

Figure 8.33
The hardware setup for the stepper motor example.

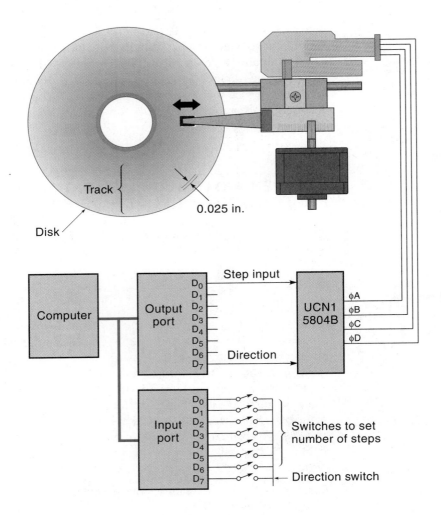

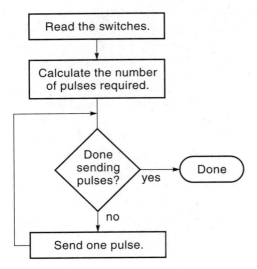

Figure 8.34
Flowchart for a program to drive a stepper motor.

TABLE 8.1 BASIC Program to Control a Disk Drive Head

	Instruction	Explanation
10	SW = INP(209)	Input 8 bits of switch data.
20	IF SW < 128 THEN	If MSB = 0, then the direction bit = 0.
	DIR = 0	
	ELSE	
	DIR = 128	Otherwise, set direction bit = 1 (10000000 = 128)
	SW = SW − 128	and remove the MSB to leave the number of tracks.
	END IF	
30	PUL = SW*6	Calculate the number of pulses needed at 6 pulses per track.
40	REM***** SEND PULSES****	
50	FOR I = 1 TO PUL	Prepare to send PUL (number of pulses).
60	HIGH = 1 + DIR	Make the LSB a 1 and add a direction bit.
70	OUT 208, 1	Send out a "high" for the step pulse.
80	FOR J=1 TO 100	This is the time-delay loop for the pulse width.
90	NEXT J	
100	LOW = 0 + DIR	Make the LSB a 0 and add a direction bit.
110	OUT 208, 0	Send out a "low" to define the end of the pulse.
120	FOR J = 1 TO 100	This is the time-delay loop for the pulse being "low."
130	NEXT J	
140	NEXT I	Go back and send the next pulse.
END		

step input pulse is created by making the output (D_0) go "high" for a period of time and then bringing it "low" for a period of time. The direction bit, brought in as the MSB from the switches, is simply passed on through by the program and sent out as the MSB (actually it is stripped off the input data and added back later to the output word). The parallel I/O port addresses would, of course, depend on the system being used; in this case, the input switch port is 209 (decimal), and the output port to the motor has an address of 208 (decimal). ◆

SUMMARY

A stepper motor, a unique type of DC motor, rotates in fixed steps of a certain number of degrees: 30°, 15°, 1.8°, and so on. The rotor (the part that moves) is made of permanent magnets or iron and contains no coils and therefore has no brushes. Surrounding the rotor is the stator, which contains a series of field pole electromagnets. As the electromagnets are energized one after the other, the rotor is pulled around in a circle. Stepper motors used in position systems are usually operated open-loop, that is, without feedback sensors. The controller will step the motor so many times and expect the motor to be there.

The permanent magnet (PM) stepper motor uses permanent magnets in the rotor. This type of motor has a detent torque—a small magnetic tug that tends to keep it in the last position stepped to, even when the power is removed. PM motors have good torque capability but cannot take very small steps. The field coils of the PM motor can be driven in the two-phase or four-phase mode. The two-phase mode requires polarity reversals. Four-phase operation does not require polarity reversals and the timing is more straightforward.

The variable reluctance (VR) stepper motor uses a toothed iron wheel for the rotor (instead of permanent magnets). This allows VR motors to take smaller steps, but they are weaker and have no detent torque.

The hybrid stepper motor combines the features of both the PM and VR stepper motors and is the type in most common use today. Hybrid motors can take small steps (typically 1.8°) and have a detent torque. The internal construction of the hybrid motor is more complicated than the PM or VR motors, but the electrical operation is just as simple.

Stepper motors are driven from digital circuits that provide the desired number of stepping pulses (in the correct order) to the field coils. These pulses must usually be amplified with driver transistors (operating as switches) before being applied to the motor coils. ICs are available that can provide the proper sequencing and amplification in one chip.

GLOSSARY

ballast resistor A resistor placed in series with the motor coils to improve the torque at higher stepping rates; it works by reducing the motor-time constant.

bilevel drive A technique that uses two voltages to improve torque at higher stepping rates. A higher voltage is applied to the motor at the beginning of the step, and then a lower voltage is switched in.

bipolar motor A motor that requires polarity reversals for some of the steps; a two-phase motor is bipolar.

constant current chopper drive A drive circuit for stepper motors that uses PWM techniques to maintain a constant average current at all speeds.

cumulative error Error that accumulates; for example, a cumulative error of 1° per revolution means that the measurement error would be 5° after five revolutions.

detent torque A magnetic tug that keeps the rotor from turning even when the power is off; also called *residual torque*.

dynamic torque The motor torque available to rotate the load under normal conditions.

eight-step drive A two- or four-phase motor being driven in half-steps; the sequencing pattern has eight steps.

four-phase stepper motor A motor with four separate field circuits; this motor does not require polarity reversals to operate and hence is unipolar.

four-step drive The standard operating mode for two- or four-phase stepper motors taking full steps; the sequencing pattern has four states.

half-steps By alternating the standard mode with dual excitation mode, the angle of step will be half of what it normally is.

holding torque The motor torque available to keep the shaft from rotating when the motor is stopped but with the last field coil still energized.

hybrid stepper motor A motor that combines the features of the PM and VR stepper motors—that is, it can take small steps and has a detent torque.

L/R drive A stepper motor driver circuit that uses ballast resistors in series with the motor coils to increase torque at higher stepping rates.

microstepping A technique that allows a regular stepper motor to take fractional steps; it works by energizing two adjacent poles at different voltages and by balancing the rotor between.

permanent magnet (PM) stepper motor A motor that uses one or more permanent magnets for the rotor; this motor has a detent torque.

phase The number of separate field winding circuits.

rotor The internal part of the stepper motor that rotates.

single-step mode Operating the motor at a slow enough rate so that it can be stopped after any step without overshooting.

slew mode Stepping the motor at a faster rate than the single-step mode; used to move to a new position quickly. The motor will overshoot if the speed is not ramped up or down slowly.

stalling A situation wherein the motor cannot rotate because the load torque is too great.

stator The part of a stepper motor that surrounds the rotor and consists of field poles (electromagnets).

stepper motor A motor that rotates in steps of a fixed number of degrees each time it is activated.

three-phase stepper motor   A motor with three separate sets of field coils; usually found with VR motors.

two-phase stepper motor   A motor with two field circuits. This motor requires polarity reversals to operate and hence is bipolar.

unipolar motor   A motor that does not require polarity reversals. A four-phase motor is unipolar.

variable reluctance (VR) stepper motor   A motor that uses a toothed iron wheel for the rotor and consequently can take smaller steps but has no detent torque.

EXERCISES

Section 8.1

1. A 15° stepper motor is commanded to go 100 steps CW and 30 steps CCW from a reference point. What is its final angle?

2. A 7.5° stepper motor is commanded to go 50 steps CCW, 27 steps CW, and 35 steps CCW again. What is its final angle (referenced from its original position)?

3. Why can a stepper motor be operated open-loop in a control system?

4. What is the detent (or residual) torque in a PM stepper motor, and what causes it?

5. A stepper motor is being used as a crane motor in an expensive toy. The motor has the following properties: Holding torque = 35 in. · oz, dynamic torque = 20 in. · oz, and detent torque = 5 in. · oz. The motor turns a 1.5-in. diameter pulley around which a string is wound. How much weight can the crane lift? How much weight can the crane continue to support with the power on; with the power off?

6. List the stepping sequence you would use to make the two-phase motor of Figure 8.8 rotate CW.

7. List the stepping sequence you would use to make the two-phase motor of Figure 8.9 operate as an eight-step drive (CCW).

8. List the stepping sequence you would use to make the four-phase motor of Figure 8.11(a) operate as an eight-step drive (CCW).

Section 8.2

9. Explain the principle of operation of a VR stepper motor.

10. Does a VR stepper motor have a detent torque? Explain.

Section 8.3

11. Explain the principle of operation of a hybrid stepper motor.

12. What are the advantages of the hybrid stepper motor?

Section 8.4

13. A 5-V stepper motor is to be microstepped with one-tenth steps. List the voltage table required for this [similar to Figure 8.27(b)].

14. What is the purpose of a ballast resistor on a stepper motor drive, and how does it work?

15. What is the purpose of bilevel drive, and how does it work?

16. How does a chopper drive improve torque at higher stepping rates?

Section 8.5

17. A 1.8° stepper motor turns a leadscrew that has 24 threads per inch.
    a. How many steps will it take to advance the leadscrew 1.25 in.?
    b. What is the linear distance the leadscrew advances for each step?

18. A 7.5° stepper motor (four phase), controlled by a computer, is used to position a telescope through a gear train. The telescope must be positioned to within 0.01°. A front panel has toggle switches that are used to specify how far the telescope is to move (LSB = 0.01°). The total range of the telescope is 0–60°.
    a. How many toggle switches would be required?
    b. What gear ratio would you specify?
    c. Draw a block diagram of the system, showing all parts of the system and specifying the gear ratio.

9 Alternating Current Motors

After studying this chapter, you should be able to:

❏ Describe the characteristics of single-phase AC and three-phase AC motors.

❏ Understand the principles of operation of the AC induction motor.

❏ Explain how single-phase induction motors are started, including the concept of run and start windings and start capacitors.

❏ Explain how to reverse the two-phase (split-phase) motor.

❏ Understand the principles of operation of the synchronous motor.

❏ Describe the concept of power factor and power-factor correction.

❏ Understand the principles of AC motor control, including start–stop control, jogging, reduced voltage starting, and variable-speed control with a DC link converter.

INTRODUCTION

In terms of sheer numbers, the AC induction motor (Figure 9.1) is the most widely used type of electric motor in the modern world. AC motors are primarily used as a source of constant-speed mechanical power but are increasingly being used in variable speed–control applications. They are popular because they can provide rotary power with high efficiency, low maintenance, and exceptional reliability—all at relatively low cost. These desirable qualities are the result of two factors: (1) AC motors are usually connected directly to power lines—DC motors require the added expense of a rectifier circuit; (2) most AC motors do not need brushes as DC motors do. In most cases, the AC power is connected only to the motor's stationary field windings. The rotor gets its power by electromagnetic induction, a process that does not require physical electrical contact. Maintenance is reduced because brushes do not have to be periodically replaced. Also, the motor tends to be more reliable and last longer because there are fewer parts to go wrong and there is no "brush dust" to contaminate the bearings or windings.

Figure 9.1
An AC induction motor.

Despite these advantages, there is a problem with using AC motors in control systems: These high-efficiency AC motors are by nature constant speed, and control systems usually require the motor speed to be controllable. As you recall, the speed of a DC motor can be controlled by simply adjusting the applied voltage. For complete speed control of an AC motor, both voltage and frequency must be adjusted, which requires using special electronic speed-control circuitry.

Still, the most common use of AC motors is for the many applications where speed control is not necessary. Included here are fans, pumps, mixers, machine tools, hydraulic power supplies, and household appliances, to name but a few.

Like DC motors, AC motors come in all sizes. **Integral horsepower (HP) motors** are those with a power rating of 1 hp to over 1000 hp, and **fractional horsepower (HP) motors** are those rated less than 1 hp. Typically, the larger motors use three-phase AC power with voltages from 208 to 600 Vac, and the smaller motors use single-phase AC power, with either 120 Vac or 240 Vac.

There are a number of different types of AC motors. The most common by far is the general-purpose induction motor, which is the familiar type found in major appliances and in machine shops. Other types of AC motors include the synchronous motor, the AC servomotor, and the universal motor.

9.1   AC POWER

Background

Thomas Edison proposed the first power-distribution system, which was to be a DC system. A major drawback of this proposed system was that the power consumer had to be within 10 miles or so of the generating station because of the significant power losses along the wires. Then in the early 1900s, Nikola Tesla invented the AC motor and generator, and with George Westinghouse built a large AC-generating station at Niagara Falls, New York. This system demonstrated the tremendous advantage of using AC for power distribution, namely, much reduced line losses. This is achieved by using a transformer at the generating station to increase the voltage and reduce the current. Thus modified, the power can be transmitted through smaller wires with much less power loss. At the destination site, another transformer reduces the voltage back to usable levels (and increases the current). Because transformers work only with AC, the AC system was destined to become the standard commercial form for transmitting electric power.

Single-Phase AC

Before discussing how AC motors work, let's briefly review exactly what AC power is. The simplest form of AC, and the kind most people are familiar with, is **single-phase AC**, the standard diagram of which is shown in Figure 9.2(a). How does this sine-wave diagram relate to the actual instantaneous voltages that would be found on prongs A and B of a standard power plug [Figure 9.2(b)]? To really understand AC, we must go back to the original definition of voltage, which is the difference in potential between two points (notice it says nothing about *ground*). Therefore, the sine wave in Figure 9.2(a) is

Figure 9.2 120/240 Vac single-phase power.

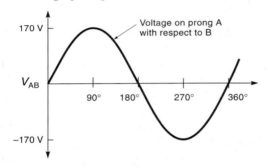

(a) 120 Vac waveform

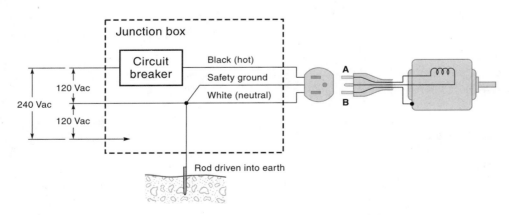

(b) 120 Vac power distribution circuit

showing only the voltage difference between prongs A and B for one complete cycle. For example, in the first half of the cycle, prong A is more positive than B by up to 170 V. At exactly midcycle (180°), there is no voltage difference between A and B. During the second half of the cycle, prong A has a negative voltage with respect to B. The power company maintains the voltage of prong B at ground, which means it is not only the return path for the current but also is actually connected to a metal rod driven into the earth. Therefore, the wire that goes to the wide prong (B) is called the **neutral** or *cold side* (usually white in color) and is theoretically safe to touch. (However, this is not recommended because it could be wired wrong!) The wire going to prong A is called the *hot side* because it carries the voltage (usually black in color and should definitely *not* be touched!). The third pin on the plug [Figure 9.2(b)] is the *safety ground* and is kept separate from the neutral wire until the actual grounding point. This ground wire is typically connected to the chassis or motor frame, which prevents you from getting a shock from touching the equipment because it is at the same potential as the earth.

The highest instantaneous voltage on (prong A) is 170 V (called V_{PEAK}) and is considerably higher than the AC designation of 120 Vac (called V_{RMS}) because the number 120 represents the effective voltage if the power content were constant over the whole cycle. Single-phase AC is supplied as 120 Vac and 240 Vac [usually combined together on three wires as shown in Figure 9.2(b)]. Fractional hp motors are commonly 120 Vac, and larger motors usually use 240 Vac. There are three common line frequencies in use—60 Hz in the United States and Canada, 50 Hz in most of the rest of the world, and 400 hz on many aircraft and shipboard systems. (The higher frequency reduces the amount of iron required in transformers and motors and thus reduces the weight.)

Ground-Fault Interrupters

The principle of a **ground-fault interrupter** (GFI) is to disconnect the power if a current leakage to ground is detected. There are two types of GFIs: those designed for *life protection* and those designed for *equipment protection*.

A GFI used for *life protection* is to prevent shock hazard, particularly in bathrooms and around water. Without a GFI, a person who touches a hot wire while standing in water or on a wet floor will get shocked because his or her body provides an electrical path to ground. The GFI circuit opens a circuit breaker quickly if it senses that more current is in the hot wire than in the neutral wire, indicating that the current has found an alternate path to ground—that is, through someone's body. Figure 9.3 shows a GFI used for life protection. Notice that power-line conductors pass through a toroidal coil. If this coil senses any (current) unbalance, a solid-state circuit directs a circuit breaker to open.

Figure 9.3

A ground-fault interrupter for life protection.

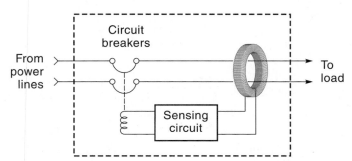

A GFI used for *equipment protection* is to prevent equipment damage and possibly fire due to internal arcing. Under normal circumstances, a current should not flow in the safety ground wire. For example, in Figure 9.2(b), the motor current should flow only through the hot and neutral wires, not in the ground wire. Only if there is a short circuit or fault between the hot wire and the motor chassis will there be a current in the ground wire. The fault might not draw enough current to trigger the overload device, but nonetheless could be a fire hazard. The GFI monitors the current in the ground wire and opens a circuit breaker if the current exceeds a preset level.

Three-Phase AC

Single-phase AC is the most common form of power in homes and businesses, but it is not the best type of AC for transmitting power or running AC motors. That distinction belongs to **three-phase AC**, which is more complicated; however, because of its advantages, it is used extensively by heavy power users.

Three-phase power is created by a three-phase generator that has three separate field windings spaced 120° apart. Two configurations are possible: the **wye-connection** (or Y-connection), shown in Figure 9.4(a), and the **delta-connection** (or Δ-connection), shown in Figure 9.4(b). A schematic of a wye-connection generator is shown in the left side of Figure 9.5. One end of each coil is connected to a common wire labeled N (neutral), and the other ends of the coils are labeled A, B, and C, which are the three-phase outputs. The relationship of **phase voltages** A, B, and C is presented in Figure 9.6, which shows each phase voltage with respect to neutral. Each phase makes a complete cycle in one period and is 120° apart; at any particular time, all phases add to 0 V (for example, add the magnitudes of the three voltages when $t = 0$). Notice in Figure 9.6 that the **phase sequence** is ABC (first A, then B, then C). The other possible sequence would be ACB.

Figure 9.4

Two types of connections for three-phase AC.

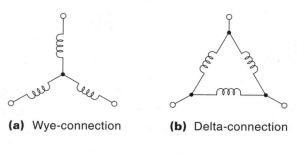

(a) Wye-connection   **(b)** Delta-connection

Figure 9.5

A three-phase power system (wye-connection).

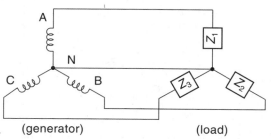

(generator)   (load)

Figure 9.6

Three-phase AC waveforms.

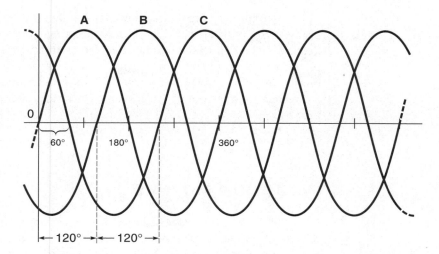

Typically, three-phase power is distributed within a facility as wye-connected. Figure 9.5 shows a complete wye-connected system where a three-phase generator is connected to a three-phase load (such as a three-phase motor). If each of the load impedances (Z_1, Z_2, Z_3) are the same, then the system is said to be balanced, and there would actually be no current in the neutral wire. Under these conditions, the neutral wire could be removed, and the motor would continue to run just fine. In real life, it is unlikely that any particular facility would present a purely balanced load to the power lines, because each phase voltage, if considered separately, can be used as a standard single-phase voltage. Therefore, a single-phase load such as a light bulb could be (and frequently is) connected between one of the phase voltages and neutral. The system would now be unbalanced, and some current will flow in the neutral wire. In practice, effort is made to keep the load on each phase equal. Finally, the specified **line voltage** of a three-phase wye-connected system is the vector sum of two individual phase voltages in series. This relationship is given in Equation 9.1:

$$\text{For wye-connected:} \qquad E = \sqrt{3}\ V_P \qquad\qquad (9.1)$$

where

E = three-phase line voltage (between any two lines)
V_P = individual phase voltage (between a line and neutral)

For a delta-connected system, the line and phase voltages are the same.

♦ **EXAMPLE 9.1**

A three-phase wye-connected system has individual phase voltages of 120 Vac. Find the three-phase line voltage.

Solution   Using Equation 9.1,

$$E = \sqrt{3}V_P = 1.73 \times 120 \text{ Vac} = 208 \text{ Vac} \qquad\qquad ♦$$

Wye-connected three-phase power is mainly used for distribution within a facility because the neutral wire allows separate single-phase loads to be pulled off as needed. Delta-connected systems have no neutral wire and are inherently balanced, so the power company typically uses the delta-connection to transmit the three-phase power over the power lines. Once at the site, delta-connected power is easily converted to wye-connected power with special transformers designed for that purpose.

9.2 INDUCTION MOTORS

By far the most commonly used type of AC motor is the induction motor, the simple, reliable, "workhorse" that powers most domestic and industrial machines. The basic parts of the induction motor are the **frame**, **stator**, and **rotor** [Figure 9.7(a)]. The stator consists of the stationary field coil windings. The rotor is positioned inside the stator and rotates as a result of electromagnetic interaction with the stator. The frame supports the stator and rotor in the proper position shown in [Figure 9.7(b)].

Figure 9.7 An induction motor.

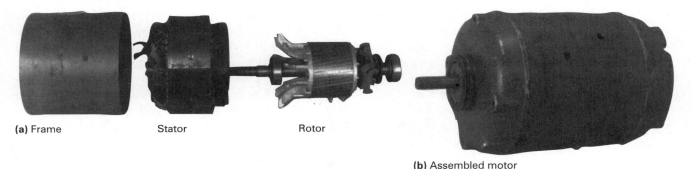

(a) Frame   Stator   Rotor

(b) Assembled motor

Theory of Operation

The theory of operation of the AC induction motor has some similarities to that of the stepper motor or BDCM (brushless DC motor). Recall that those motors work by having their field poles energized in sequence around the stator. The rotor is pulled around because it is attracted to the sequentially energized poles. With stepper motors and BDCMs, special switching circuits are required to turn the field windings on and off. The AC motor also works by rotating the stator field, but it makes use of the natural alternating nature of the AC wave to turn the field coils on and off sequentially. The AC induction motor does not need brushes because the rotor is essentially a passive device that is continuously being pulled in one direction. To use an old analogy, the rotor is the "horse," and the rotating stator field is the "carrot."

To explain the principles of how the AC wave can be used to sequentially energize the field coils, we will examine the operation of a two-phase motor.* Two-phase AC consists of two individual phase voltages (Figure 9.8). Notice that phase B is lagging behind phase A by 90°—that is, phase A peaks at 0°, and phase B peaks 90° later. The two-phase motor (Figure 9.9) is connected so that phase A energizes the top and bottom poles and phase B energizes the left and right poles. The action of two-phase AC on the motor is to cause the stator magnetic field to effectively rotate clockwise (called a **rotating field**), even though the coils themselves are stationary.

Figure 9.8

How two-phase AC causes a rotating field.

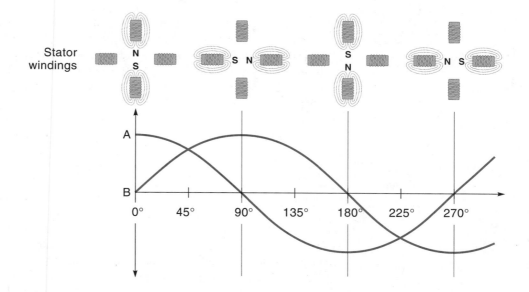

In Figure 9.8, at 0° phase A is at peak voltage while phase B is 0 V. At this point, phase A has all the voltage, and phase B has none; therefore, the windings connected to phase A (top and bottom) will be energized, and the windings connected to phase B (left and right) will be off. This situation is depicted in the coil drawing (upper left) of Figure 9.8. The polarity of the applied voltage causes the top winding to present a north (N) magnetic pole to the rotor and the bottom winding to present a south (S) magnetic pole to the rotor.

At 90° later in the power cycle (Figure 9.8), phase A voltage has gone to 0 V (deenergizing the top and bottom windings), and phase B has risen to peak voltage, energizing the left and right windings. Specifically, the positive phase B voltage will cause the right-side winding to present a north magnetic pole to the rotor and the left winding to present a south magnetic pole (as indicated in the coil drawing of Figure 9.8).

At 180°, phase B voltage has gone back to 0 V (deenergizing the left and right windings), and phase A has descended to a negative peak voltage. Once again, the top and bottom windings are energized but this time with the opposite polarity from what they were at 0°, causing the magnetic poles to be reversed. Now the bottom winding presents a north magnetic pole to the rotor, and the top winding presents a south magnetic pole.

---

*Power companies do not supply two-phase power, only single-phase and three-phase. However, two-phase power *is* used in some AC control motors (being "created" from single-phase power). This discussion is based on two-phase power because I feel it is the best way to explain the concept of a rotating field.

At 270°, phase A has ascended to 0 V (deenergizing the top and bottom windings), and phase B has gone to a negative peak. Once again, the left and right windings are energized but this time with the left winding presenting a north magnetic pole to the rotor and the right winding a south magnetic pole.

This analysis explains how two-phase AC causes the magnetic field to act as if it were rotating in a clockwise (CW) direction. (You can see this in the coil drawings of Figure 9.8, where the north pole apparently rotates CW.) What was not apparent from the discussion is that the rotation of the field is smooth and continuous—it doesn't jump from pole to pole as might be inferred from the discussion. For example, consider the situation at 45°. From Figure 9.8, you can see that both sets of poles are partially energized, causing the resultant N–S magnetic field to be halfway between the two poles.

Mechanical rotation of the rotor is the result of the rotor being pulled around in a CW direction, "chasing" the rotating field. (The rotor is magnetized by an "induced" current from the field—a concept that will be explained later.) The field makes one complete revolution per cycle; thus, for a line frequency of 60 Hz, the field would rotate at 3600 rpm—(60 cycle/s) × (60 s/min) = 3600 rpm. The speed of the rotating field (3600 rpm in this case) is called the **synchronous speed**. For an induction motor, the rotor speed does not exactly match the synchronous speed, but it's usually close.

The motor under discussion (Figure 9.9) would be called a *two-pole motor*, even though it has four individual poles, because the "pole" designation of an AC motor refers to the number of individual *poles per phase*. The reason? Motor speed is proportional to poles per phase, not simply the number of poles. In this case, 4 poles/ 2 phase = two-pole. An AC motor can be built to go slower than 3600 rpm by increasing the number of pole sets. For example, if the motor in Figure 9.9 had two complete sets of poles, it would turn one-half revolution for each cycle, giving it a speed of 1800 rpm. (This would be called a *four-pole motor*, because 8 poles/2 phase = four-pole.) If the motor had three sets of poles (making it a *six-pole motor*), it would turn at 1200 rpm (which is one third of 3600 rpm). This idea can be expressed in a synchronous equation:

$$S_S = \frac{120f}{P} \tag{9.2}$$

where

S_S = synchronous speed in rpm (motor speed is actually somewhat less, as will be explained later)
f = frequency of the AC line
P = number of field poles per phase

♦ **EXAMPLE 9.2**

a. Find the synchronous speed of a 60-Hz, four-pole, single-phase motor.
b. How many individual poles does this motor have?

Solution   a. For a four-pole motor, $P = 4$ in Equation 9.2:

$$S_S = \frac{120 \times 60}{4} = 1800 \text{ rpm}$$

which is the synchronous speed.

b. To find the number of individual poles, recognize that this motor has four poles per phase and one phase; therefore, it has four poles. ◆

Figure 9.9
A two-phase motor.

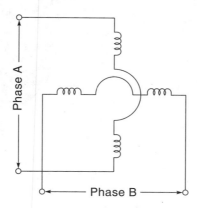

Phase A

Phase B

◆ **EXAMPLE 9.3**

a. Find the synchronous speed of a 60-Hz, four-pole, three-phase motor.
b. How many individual poles does this motor have?

Solution

a. For a four-pole motor, $P = 4$ in Equation 9.2:

$$S_S = \frac{120 \times 60}{4} = 1800 \text{ rpm}$$

b. To find the number of individual poles, recognize that this motor has four poles per phase and three phases; therefore, it has 12 poles. ◆

Some motors are designed so that the number of poles can be changed by rewiring. This way the same motor can be set up to run at different speeds as the application requires; for example, when it's operating as a two-pole motor, the speed is 3600 rpm, and when it's operating as a four-pole motor, the speed is 1800 rpm.

The rotor of the AC motor must act like a magnet in order for it to be pulled around by the rotating stator field. A diagram of a rotor is presented in Figure 9.10(a). It consists of a number of aluminum or copper bars connected with two end rings. Because this configuration reminded someone of a squirrel cage, it is called a **squirrel cage rotor**.

The squirrel cage rotor has no magnetic properties when the power is off. However, when AC power is applied to the stator windings and the stator field starts rotating, the rotor becomes magnetized by induction. It works like this: As the stator field rotates past an individual bar, the field strength in the bar rises and falls. This changing magnetic field induces a voltage in the bar, and the voltage causes a current to flow. The current flows through the bar, through the end rings, and back through other bars. This current causes the bar to have a magnetic field, and it is this field, interacting with the rotating stator field, that produces the mechanical torque. Figure 9.10 (b) shows a rotor with bars that are actually embedded in slots of a ferromagnetic core. No insulation is between the

Figure 9.10

An induction motor "squirrel cage" rotor.

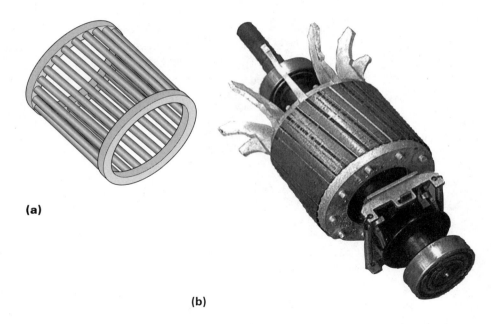

(a)

(b)

bars and the iron core because the bars present a much lower resistance than the laminated iron sheets that make up the core.

For this motor to work, a voltage must be induced from the stator to the rotor, but this can only happen when the magnetic field is changing. As far as the rotor is concerned, the changing field occurs as a result of the stator field rotating past it. This means that, even though the rotor is chasing the rotating stator field, it can never quite catch up to the synchronous speed; if it did, no relative motion would be between the rotor and stator and consequently no induced voltage in the rotor. Thus, the rotor is always going slower than the rotating field, and this difference is referred to as **slip**. Slip is frequently given as a percentage (usually less than 10%) of the stator speed as described in the following equation:

$$\text{Slip} = \frac{S_S - S_R}{S_S} \tag{9.3}$$

where

S_S = synchronous (stator) speed (in rpm)
S_R = rotor speed (rpm)

♦ **EXAMPLE 9.4**

A typical washing machine motor has a rated speed of 1725 rpm. What is the slip in percentage at this speed?

Solution

Because it is a 60-Hz motor, we know that the stator speed must be a simple fraction (½, ⅓, ¼, and so on) of 3600 rpm. Quick calculations reveal that ½ of 3600 is 1800, which is slightly above 1725 rpm. Therefore, 1800 must be the synchronous speed. Now we use Equation 9.3 to find the slip:

$$\frac{S_S - S_R}{S_S} = \frac{1800 - 1725}{1800} = 0.042 = 4.2\% \qquad ♦$$

Some speed control of the AC motor is possible by controlling the amount of slip by varying the voltage. For a constant load, the slip will increase as the voltage is lowered because, as the field weakens, the rotor must slip more to get its required induced current. This type of control can be used to change the speed 10–15% or more, depending on the rotor design.

Induction motors are classified by the National Electrical Manufacturers Association (NEMA). The most common type, NEMA class B, has a low-resistance rotor, which allows it to operate with low slip and high efficiency. Another type, NEMA class D, has a high-resistance rotor, which allows it to operate with more slip and consequently more torque.

A torque–speed curve for the motor of Example 9.4 is shown in Figure 9.11, and it is very useful for understanding motor behavior. This curve shows the torque available at various motor speeds. The maximum possible speed (1800 rpm in this case) is the speed of the rotating stator field, that is, the synchronous speed. However, there is no torque at this speed because the motor needs some slip to operate. To get full-load torque, a point defined by the manufacturer, the rotor must be slipping some—in this case, reducing the speed to 1725 rpm. These motors have good **speed regulation** in the operating range—that is, the speed changes little over a wide range of load torque demands. Also in Figure 9.11, notice that the AC motor behaves something like the DC motor in the operating range—that is, as the load increases the speed decreases. However, at a certain point called the **pull-out torque**, the torque drops off sharply, even though the speed is still decreasing. In real life, this means that if the load on a motor was increasing, the speed would decrease until the motor load reached the pull-out torque, at which point the motor would **stall** (stop turning). The **starting torque** will depend on the type of motor, but it is typically 2.5 times greater than the rated load, and unlike the DC motor, it is not the maximum torque available.

Figure 9.11

Torque–speed curve of an induction motor.

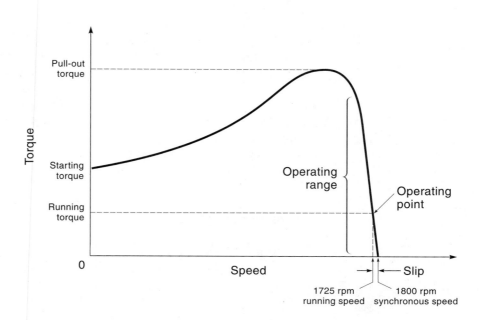

Three-Phase Motors

AC induction motors are grouped into classifications depending on the type of power they use: three-phase or single-phase (two-phase motors actually use single-phase power). Each type is based on the same operating principles presented previously in this chapter, and each type has its advantages and disadvantages. The three-phase motor is simpler and smaller than its single-phase counterpart, but it can be used only where three-phase power is available. Usually, this is a heavy industrial site; therefore, these motors tend to be large, from 2 hp up. Figure 9.12 shows a diagram of the three-phase motor. It has three sets of stator windings, with each set of windings being powered by one of the phase voltages. The natural timing sequence of the three individual phase voltages produces the rotating stator field that pulls the rotor around. The rotor is the squirrel cage type described earlier. The reason this motor is so simple (and hence reliable) is that it is self-starting—just apply the power, and it starts.

Figure 9.12

A three-phase induction motor.

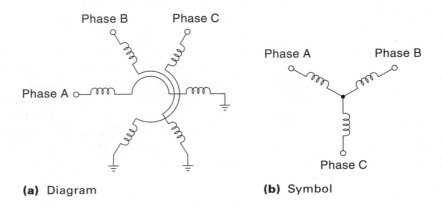

(a) Diagram   (b) Symbol

The three-phase motor can be wired to run in either direction by simply reversing any two of the three leads. For example, if the three motor terminals are connected to phases ABC and it turns CW, then if wires B and C are reversed (to ACB), the motor will rotate CCW. (Reversing C and A then brings it back to CW.)

A three-phase motor, once started, will continue to run even when one of the phases is disconnected, because two-thirds of the rotating field is still working and the mechanical inertia of the spinning rotor will carry it over the "dead spot" caused by the missing wire. However, vibration and noise will increase, torque will decrease, and the motor may overheat due to greater current in the active field windings.

Single-Phase Motors

Single-phase AC is the most common form of power and is available almost everywhere. For this reason, the single-phase motor is in wide use for domestic and industrial purposes, typically for applications requiring up to 1 or 2 hp. The basic operational theory is the same as for the motors discussed before, but this motor has one major problem: It can't start itself. To understand this, look at the single-phase motor diagram shown in Figure 9.13. This motor has only one set of stator windings (called the **run windings**), which are positioned across from each other. As the AC voltage cycles from positive to negative, the magnetic field reverses itself [Figures 9.13(a) and (b)]. These field reversals cannot truly be said to be a rotating field, so they cannot provide a starting torque to the rotor. They can, however, keep the rotor turning if it's started in some other way. The trick, then, is to provide a way to start the single-phase motor.

The most common way to start a single-phase motor is to use a second set of windings, called the **start windings**, which are only energized during the start-up period. In effect, the motor temporarily becomes a two-phase motor, which *is* self-starting. This type of motor is known by the name **split-phase**. The AC in the start winding should ideally be 90° out of phase with the run winding. The simple split-phase motor shown in Figure 9.14 relies on the fact that the start winding, which consists of a few turns of thin wire, has much less inductance than the run windings and so creates its own phase shift. When the motor is up to about 80% speed, a switch opens automatically by centrifugal force (Figure 9.15), disconnecting the start windings. The motor then continues running on only the run windings. This starting method provides about 40–50° of phase shift, enough to start the motor but not enough to provide a lot of start-up torque.

Figure 9.13

A single-phase motor showing how the field reverses with each half-cycle of AC.

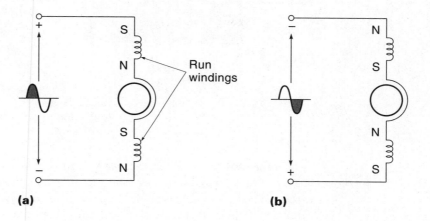

(a)   **(b)**

Figure 9.14

A single-phase motor (split-phase).

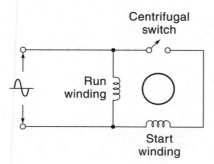

Figure 9.15

A centrifugal switch disconnects the start winding when the rotor is up to speed.

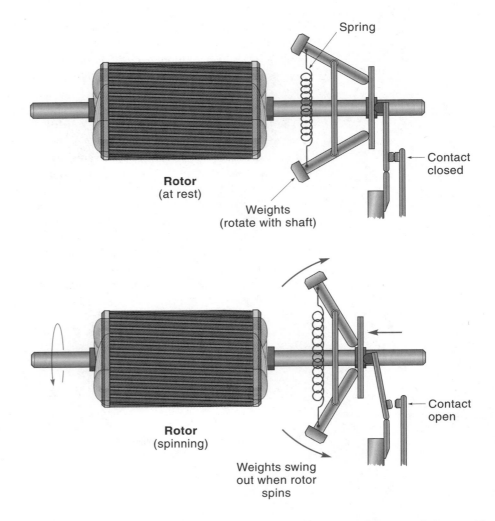

Another way to create the necessary phase shift between the run and start windings is with a nonpolarized capacitor. This motor, shown in Figure 9.16, is a **capacitor start motor**. When the motor is started, the centrifugal switch is closed, and the capacitor causes the current in the start winding to lead the current in the run winding. Once the motor is running, the switch opens, and the motor continues to operate in single-phase

Figure 9.16

A single-phase motor (split-phase) using a capacitor with a centrifugal switch.

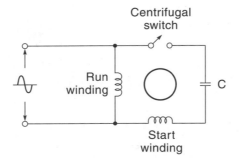

mode on the run windings. The capacitor start motor can achieve a full 90° phase shift between run and start windings, with the consequence of a high starting torque.

There are a number of other variations of the split-phase type of the single-phase motor. In one design, called the **permanent-split capacitor motor**, the start windings (with capacitor) are permanently connected, eliminating the need for the centrifugal switch. This design decreases the efficiency of the motor somewhat but simplifies the hardware and increases the reliability. Another motor design, the **two-capacitor motor**, is shown in Figure 9.17. This motor uses a larger value capacitor for starting and then switches to a lower value for running. Two capacitor motors run with less vibration because the run capacitor introduces some phase shift, which (as shown in Figure 9.8) more closely approximates a rotating field.

Figure 9.17
A two-capacitor motor.

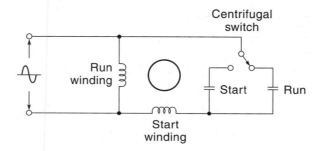

All single-phase motors discussed so far will start off turning in the same direction with each power up. By reversing the polarity of one of the windings, the motor will start and run in the opposite direction.

Fractional hp motors can be **shaded-pole motors**, which do not require a separate start winding for them to start. Instead, a small portion of each pole is separated from the rest and encircled by a copper band, as shown in Figure 9.18. When power is applied to the motor, a current is induced in this copper band, which results in the magnetic field being delayed in its vicinity. This unbalance creates a weak rotating component to the stator field, which is enough to start the motor turning. This motor design is simple and easy to construct but is inefficient and has low torque. Consequently, it is usually used where only low power and a nearly constant load is required, such as small blowers.

Figure 9.18
A shaded-pole motor.

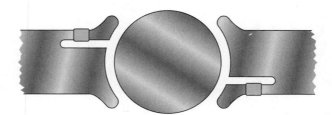

Split-Phase Control Motors (Two-Phase Motors)

The **split-phase control motor** is technically a two-phase motor because it has two sets of windings (Figure 9.19). Not as common as the three-phase or single-phase types, these motors do find application in control systems. The operating parameters that make them desirable are (1) they are self-starting and (2) they can easily be controlled to turn in either direction. Typically, they are small (less than 1 hp) and are used to move something back-and-forth, such as opening and closing a valve or raising and lowering a garage door.

Figure 9.19
A two-phase motor.

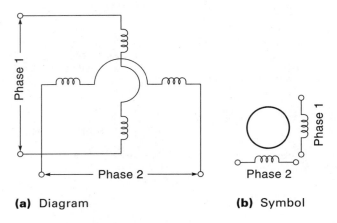

(a) Diagram   **(b)** Symbol

The problem is that two-phase AC is not available directly from the power company; it must be created, usually from single-phase AC. This adds some complexity to the system, but it does provide the opportunity to control the direction of rotation of the motor and hence its value as a back-and-forth type of actuator in a control system.

The required two-phase power is created from single-phase AC by placing a capacitor in series with one of the windings (Figure 9.20). The capacitor causes the current in winding 2 to lead the line current in winding 1 by almost 90° (it doesn't have to be exact). To change the direction of rotation, the capacitor must be able to switch so that it is in series with the other winding. Such a circuit is shown in Figure 9.21(a). Notice that the switch is in the down position, which causes the line voltage to be applied directly to winding 1, while winding 2 is fed through the capacitor. In Figure 9.21(b), the switch is in the up position, which allows winding 2 to get the line voltage, while winding 1 is fed through the capacitor. This causes the motor to rotate in the opposite direction.

Figure 9.20
A split-phase motor
(two-phase) driven with
single-phase AC.

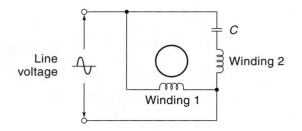

Figure 9.21

Controlling the direction of a split-phase control motor (two-phase motor).

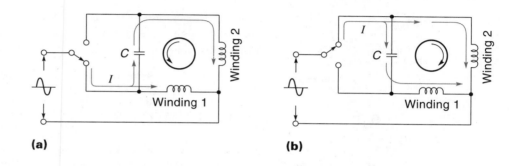

(a)   **(b)**

AC Servomotors

A special case of the two-phase motor is the **AC servomotor**. This is a high-slip, high-torque motor, designed specifically for control systems, and it has a relatively linear torque–speed curve (Figure 9.22). As you can see, the maximum torque occurs when the speed is zero. When the motor is running, the speed is inversely proportional to the load torque; put another way: the lighter the load, the faster the motor runs. This is very similar to the way a DC motor behaves. Figure 9.23 shows a diagram of the AC servomotor. The two windings are called the **main winding** and the **control winding**. The main winding is connected to an AC source, usually 120 Vac. The control winding is driven by an electronic circuit that (1) causes the phase to be either leading or lagging the main winding (thereby controlling the motor direction) and (2) sets the magnitude of the control-winding voltage, which determines the speed. Typically, the maximum

Figure 9.22

AC servomotor torque–speed curve.

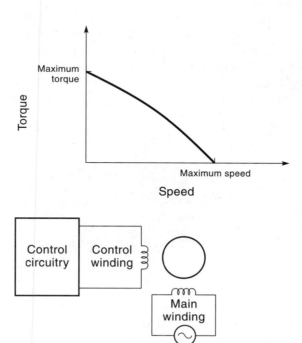

Figure 9.23

Diagram of an AC servomotor.

control winding voltage is about 35 Vac. If the control winding has 0 V, the motor will coast to a stop, even though the main winding is still connected to the line voltage. This is different from a normal induction motor that *will* continue to run on a single phase.

9.3   SYNCHRONOUS MOTORS

The **synchronous motor** is similar to the induction motor with one important difference: The rotor in the synchronous motor rotates at exactly the speed of the rotating field—there is no slip. Put another way, the speed of the synchronous motor is always an exact multiple of the line frequency. This feature is extremely desirable in industrial applications, for example, when several motors along a conveyer belt must all be going exactly the same speed. Although many synchronous motors are large, the concept is also used extensively in small clock or timing motors where an exact relationship must exist between frequency and speed.

Theory of Operation

Recall that the rotor in an induction motor receives its power through induction, which requires a difference (slip) between the speed of the rotor and the rotating field. To make a synchronous motor work, the power to form a magnetic field in the rotor must come from another source. Traditionally, this is done by supplying DC power into the rotor via **slip rings** and brushes (Figure 9.24). Slip rings and brushes on the synchronous motor are similar to the commutator assembly used in DC motors, with one important difference; here the electrical contact from stator to rotor is made through a smooth ring, not the multiple contacts of the DC motor's commutator. The action is smoother, the components last far longer, and less electrical noise is generated. The rotor of the synchronous motor uses DC power (called *excitation*) to energize electromagnets around its perimeter. These magnets tend to lock on to the rotating magnetic field in the stator and cause the rotor to rotate at the exact speed of the rotating field, that is, the synchronous speed.

Figure 9.24

The rotor of a synchronous motor showing slip rings.

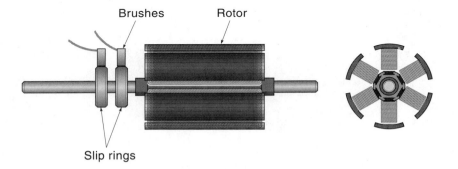

Unfortunately, the synchronous motor is not self-starting. When the power is switched on, the field starts rotating full speed, causing the magnetized rotor to be first tugged in one direction and then the other. Experiencing no sustained torque, the rotor just sits there. To start the motor, additional hardware is required. Any one of the following methods can be used:

1. Provide for the frequency to start slow and then speed up. As long as the rotating field does not get too far ahead of the rotor, the rotor will continue to be tugged in one direction. The problem is that it takes extra electronics to provide the variable frequency. However, the advantage of this system is that if you can control the frequency, you can control the speed. This approach is being used more and more, with the continuing advancement of high-power solid-state electronics.

2. Use another motor (called a **pony motor**) to bring the synchronous motor up to speed. This approach has the obvious drawback of requiring significant additional hardware.

3. Insert some squirrel cage rotor bars (as used in induction motors) in the rotor of the synchronous motor. This allows the motor to get started as if it were an induction motor.

4. Use hysteresis-synchronous motors (covered later in this chapter).

Power-Factor Correction and Synchronous Motors

The AC power from the power company arrives with the voltage and current in phase with each other. This is the most efficient way to transmit power because, by definition, power is the product of voltage times current. Therefore, for maximum power transfer, the voltage and current must be occurring at the same time, that is, in phase.

The windings of induction motors present an inductive-type load to the power line; if there are many motors in a facility, this inductive load can be significant. The problem is that the inductive load causes the current to lag behind the voltage, which means the motor will draw more current to deliver the same power. This situation is described in terms of the power factor. The **power factor** is the cosine of the angle between the voltage and current, so when the voltage and current are exactly in phase, the power factor is 1. For a purely reactive load of a 90° phase shift between voltage and current, the power factor is 0. We say that an inductive load creates a *lagging power factor*, and a capacitive load creates a *leading power factor*.

Plant engineers would like to keep the power factor as close to 1 as possible, but how can this be done with a plant full of inductive motors? One way is to have the induction motors running with a full load—a lightly loaded induction motor is much more inductive than a fully loaded one. Another way to improve a lagging power factor is to add a capacitive load to cancel out all or part of the inductive load. Synchronous motors are sometimes used for this purpose. With proper adjustment of the rotor excitation current, the synchronous motor will present a capacitive load, which will improve (increase) the power-factor of the plant. Synchronous motors installed specifically for power-factor correction are called **synchronous condensers**; they do not drive a load—they simply spin.

Small Synchronous Motors

Small synchronous motors are used as timing or clock motors or simply as a source of rotary power in instruments and small machines. They are available to be used as a direct drive or with an attached gear train. Figure 9.25 shows a selection of these motors. Many of these motors are split-phase motors designed to operate on single-phase AC with the use of a capacitor. The simple switching arrangement shown in Figure 9.26 allows for the motor to rotate in either direction. When the switch is in the up position, winding 1 gets the line voltage and winding 2 is phase-shifted through the capacitor. This arrangement causes the motor to turn in one direction. When the switch is in the middle position, the motor is off; when the switch is in the down position, winding 2 gets the line voltage, and winding 1 is driven through the capacitor. This causes the motor to turn in the other direction.

Small synchronous motors do not use electromagnets in their rotors and so do not need slip rings. Instead, the rotor is magnetized in other ways. There are three designs: the reluctance motor, the hysteresis motor, and the permanent magnet motor. All three types can be made to be self-starting. The hysteresis motor is particularly interesting because the rotor consists of a smooth, round piece of steel. Its operation is very smooth and quiet, which makes it an ideal choice for such applications as tape drives.

Figure 9.25

A selection of small synchronous motors. (Courtesy of Kollmorgen Motion Technologies Group)

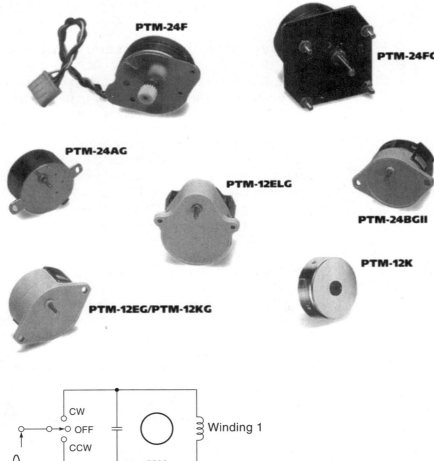

Figure 9.26

A small synchronous motor being driven with single-phase AC (switch controls direction of rotation).

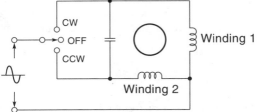

9.4   UNIVERSAL MOTORS

The **universal motor** is so named because it can be powered with either AC or DC. Basically, it is a series-wound DC motor that has been specifically designed to operate on AC as well. Like its DC counterpart, it is reversible by changing the polarity of either the field or the rotor windings, but not both. Physically, the universal motor is similar to a DC motor except that more attention is paid to using laminations (thin sheets of lacquered metal) for the metal parts (to reduce the AC eddy currents) and the inductance of the windings is minimized as much as possible.

The operating characteristics of the universal motor are similar to those of the DC motor. For a fixed voltage, the speed is inversely proportional to the load torque—as the load increases, the speed decreases. For a constant load, as the applied voltage increases, the speed will increase. Typically, universal motors are designed to operate at high speeds—from 3600 to 20,000 rpm—but, because they use a commutator and brushes (which wear out), they have a limited lifetime. Being a series-wound motor, they have high starting torque, and for this reason are widely used for handheld power tools (for example, a hand drill motor).

9.5   AC MOTOR CONTROL

When the AC motor is first switched on, there is an initial rush of current that may be five times higher than the average running current. In fractional hp motors, this momentary high current is usually acceptable, but for larger motors special provision is usually made to reduce the initial current surge to prevent electrical and mechanical components from being overstrained. Once the motor is running, its speed will be determined by the frequency, which is why the AC motor is usually thought of as a constant-speed motor. At one time, "controlling" an AC motor meant "turning it on and off." Currently, this notion of the AC motor being constant speed is being revised because efficient electronic power circuits can vary the frequency, thus allowing the speed to change. This and other topics having to do with motor control are the subjects of the subsequent sections.

Start–Stop Control

The simplest way to start a fractional hp AC motor is to connect it to the line voltage through a switch, and this is how it's done in many cases. Figure 9.27 shows two different schematics for this approach. Both schematics are for the same circuit, but Figure 9.27(b) is drawn in a form called a **ladder diagram** and is typical of an industrial motor-wiring diagram. In the ladder diagram, Figure 9.27(b), L_1 and L_2 are the AC line voltage wires, and each "rung" is a circuit. (Ladder diagrams are covered in Chapter 12.).

Figure 9.27

A simple on–off motor control circuit.

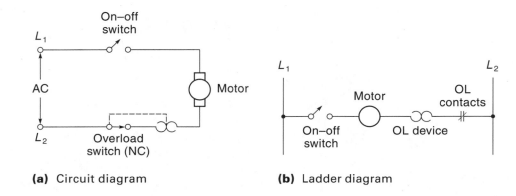

(a) Circuit diagram   **(b)** Ladder diagram

The symbol labeled "OL" in Figure 9.27(b) represents the normally closed contacts of the **overload device** (normally open contacts would not have the diagonal line). The overload device is designed to open its contacts if the motor gets too hot. This device may be a thermal sensor built into the motor that will trigger when a certain temperature is reached, or it may be separate from the motor—in which case it is actually triggering on a sustained current-overload condition.

The start–stop circuit for motors used in industrial processes is typically a little more complicated than an on–off toggle switch. The motor might well be controlled from a control panel some distance from the motor. In such a case, routing the motor power through a relay located near the motor and runing the low-current wires (which only need to energize the relay coil) to the control panel make sense. This approach is particularly appropriate for larger motors. An everyday example of this system is the starter motor in a car. When you turn the key, a low-current ignition switch provides power to energize a high-current capacity relay (starter solenoid) that, in turn, connects the starter motor to the battery.

Another problem with using a simple toggle switch for motor control has to do with safety. Imagine the following situation: A motor driving a potentially hazardous machine overheats, and the overload protection circuit shuts it down. Somebody comes along, sees the motor is stopped, and investigates. Meanwhile, the motor cools down, the thermal protection device resets, and—because the power switch is still on—the motor suddenly restarts. The unsuspecting investigator may get hurt.

The standard industrial motor-control circuit has been designed to address both of these situations. The motor is connected to the power lines through a relay, and the relay is energized with separate start and stop buttons on a control panel. To put it simply, the motor is started by pushing the start button and stopped by pushing the stop button. If the motor stops because of overheating, the start button *must* be pushed again to restart it. Figure 9.28(a) shows the circuit diagram for the start–stop circuit. The stop button is normally closed, and the start button is normally open. When the start button is pushed, power is applied to relay coil M, which closes contacts M-1 and M-2. Contact M-2 connects the motor to the AC power, and the motor starts. Contact M-1 connects the relay coil to a point in the circuit ahead of the start button; consequently, when the start button is released, the relay coil will remain energized. This process is called **sealing**, or **latching**, the relay. When the stop button is pushed, power is removed from the relay coil, which opens contacts M-1 and M-2. The opening of the M-2 contact stops the motor, and the opening of the M-1 contact "breaks the seal" that was keeping the relay coil energized. Consequently, when the stop button is released, the relay coil remains unenergized, and the motor stays off. If the overload contact should open at any time, the result is the same as when the stop button is pushed: The motor stops and will not restart until the start button is pushed. A ladder diagram of the start–stop circuit is shown in Figure 9.28(b). Notice that the relay coil is represented by a circle.

Figure 9.28

A start–stop circuit for a single-phase AC motor.

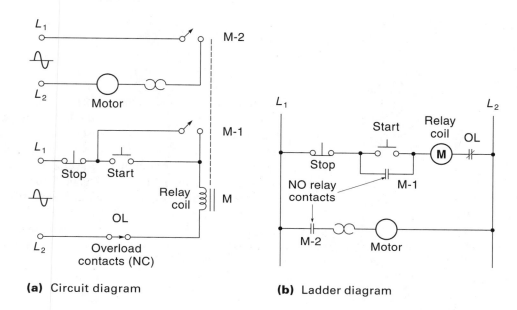

(a) Circuit diagram

(b) Ladder diagram

Besides safety reasons, sealing circuits are useful when the entire plant loses power. When the power is restored, you don't want all motors starting up at the same time, causing a tremendous current surge in the main power lines. Instead, the motors can be restarted individually or in groups, in some orderly fashion. Also, starting and stopping motors from multiple locations is common; for example, a cooling pump in a nuclear power plant might be started from three places: for example, the main control room, the pump site, and an auxiliary control room.

Figure 9.29 shows the start–stop circuit for a three-phase motor. This is basically the same as for the single-phase motor except that the relay now provides three sets of contacts for the motor (M-2, M-3, and M-4). Also, the relay coil circuit, which requires only single-phase AC, is connected across L_1 and L_2.

Jogging

Jogging refers to applying power to the motor in short bursts for the purpose of "inching" it into some desired position. For example, a commercial mixer might need

Figure 9.29

A start–stop circuit for a
three-phase AC motor.

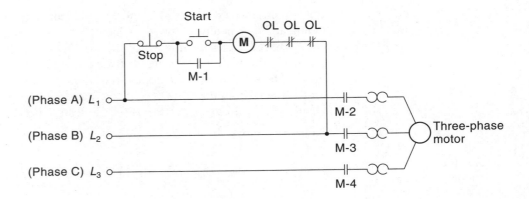

to be jogged into a certain position so the blades can be replaced, or a motor driving a large shaft might need to be jogged to get the keyway lined up with the mating part.

The traditional motor-control circuit discussed in the previous section (which uses a sealing relay) does not lend itself to jogging because it's not responsive enough—that is, it can't cycle on and off fast enough. Jogging requires that you have direct push-button-power control to the motor, which usually means bypassing the regular on–off relay circuit. Figure 9.30 shows a simple motor-control circuit that has provision for jogging. When switch A is set to start, this circuit behaves exactly like the circuit of Figure 9.28; that is, the relay is sealed every time the start button is pushed. However, when switch A is set to jog, the sealing circuit is disabled, and power is supplied to the motor only when the start button is pushed.

Figure 9.30

A start–stop circuit with a
provision for jogging.

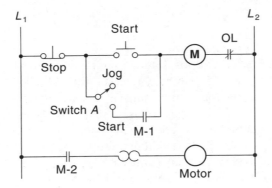

Reduced-Voltage Starting

The nature of any motor is to have a large initial current when first connected to a voltage source. The power-distribution system can usually cope with the increased start-up current of fractional hp motors (although the lights may dim perceptibly in the immediate area), but larger motors usually need special starting circuits that limit the starting current to some acceptable value. One of the simplest ways to limit start-up current is to connect resistors in series with the motor during the start-up period. After the motor has come up to speed and is drawing a reduced amount of current, a switching device removes the resistors from the circuit, and the motor is connected to the full line voltage.

Figure 9.31 shows a reduced-voltage starting circuit for a three-phase motor. Notice the presence of resistors R_1, R_2, and R_3 in series with the motor power lines. Also notice that each resistor is bypassed with a relay contact. When the start button is pushed, the MR relay coil is energized, and contacts MR-1, MR-2, and MR-3 close. This action connects the motor to the power lines, but the current must go through the resistors. Another MR contact (MR-5) applies power to the coil of a **timer relay** (TR), a relay with a built-in delay mechanism (based on pneumatics or electronics) that causes its contacts to close a certain time *after* the TR coil is energized. The delay is set to give the motor enough time to come up to near-operating speed. When the relay "times out," contact TR-1 closes and allows relay coil RB to be energized. Relay RB has three sets of contacts (RB-1, RB-2, and RB-3) that connect the motor directly to the power lines, bypassing the resistors.

Figure 9.31

A reduced-voltage starting circuit.

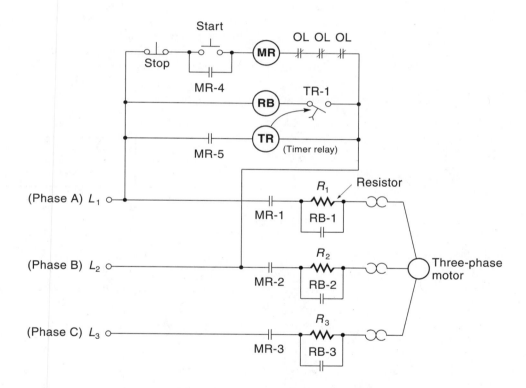

The reduced-voltage starting circuit is simple and relatively inexpensive, but it does have the drawback of reducing the motor start-up torque. Remember that torque is proportional to current, and if current is limited, so is the torque. One way to improve the situation is to use transformers instead of resistors in the start-up circuit. The advantage of using transformers (actually autotransformers) is that they can provide the motor increased current, while not straining the power-distribution system. The transformers are only engaged during the start-up period. When the motor gets up to speed, the transformers are switched out, and the motor is powered directly from the power lines.

Variable-Speed Control of AC Motors

AC motors enjoy many advantages over DC motors: They are lighter in weight, more reliable, less expensive, and require less routine maintenance. And although small speed changes can be made by changing the voltage, AC motors are essentially limited to running at a speed determined by the line frequency. This fact has traditionally kept them from being used in control systems where the speed needs to be varied. The introduction of high-power solid-state switching circuits has changed this picture. Clearly, if you want to fully control the speed of an AC motor, you must be able to change the frequency. This can be done with off-the-shelf power-conversion circuits that are capable of converting the line voltage at 60 Hz into a wide range of voltages and frequencies.

The basic steps for adapting 60-Hz line power for variable-speed AC motors are shown in the block diagram of Figure 9.32, and the circuit is known as a **DC link converter**. The first step is to convert 60-Hz AC into DC power. The second step is to convert this DC power back into AC at the desired frequency. Figure 9.33 shows a typical generalized circuit to perform these functions for a three-phase motor. In this circuit, 60-Hz AC line power is converted to DC with a silicon-controlled rectifier (SCR) network. SCRs are used so the magnitude of the DC voltage can be controlled. Recall that

Figure 9.32
A block diagram of a variable-speed-control system.

Figure 9.33
A variable-speed-control circuit (trigger and timing circuits not shown).

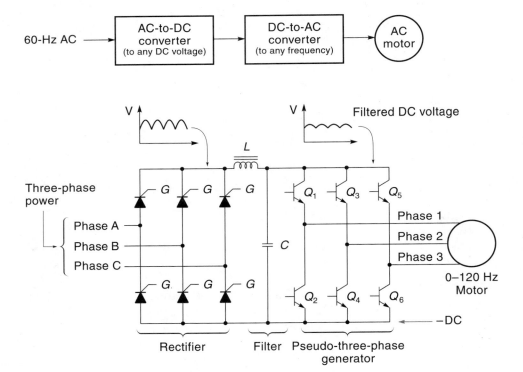

the SCR is essentially a diode, but in order for it to conduct, two conditions must be met. First, it must be forwarded-biased; second, it must receive a trigger pulse on its gate input. Referring to the circuit of Figure 9.33, if the gate (G) of each SCR is pulsed at the beginning of each positive AC cycle, the conduction time will be maximum, and the maximum DC voltage will result. If the trigger pulse to the gate is delayed, then the SCR is conducting for a shorter time, and a lower DC voltage is established (note that the L and C in Figure 9.33 help smooth out the power pulses from the SCRs). To sum up, the average value of the DC voltage can be determined by controlling the delay of the trigger pulses.

The next job of this circuit (Figure 9.33) is to create a sort of artificial three-phase AC power at any desired frequency. This is accomplished with the six transistors on the right side of the circuit. Each transistor is turned on and off in sequence by a controller circuit (not shown) in such a way as to cause three pseudo-sine waves. Figure 9.34 shows the three-phase output voltage waveforms of this circuit. In particular, notice the phase 1 voltage waveform. During time period A, both transistors (Q_1 and Q_2) are off, so the output, which is taken from between the transistors, is neither positive or negative. Then during time period B, transistor Q_1 is on, connecting the phase 1 output to the plus DC voltage. During time period C, both transistors are again off; finally, during time period D, transistor Q_2 is on, connecting the phase 1 output to the minus side of the power supply. This same-shape waveform is generated by transistor pair Q_3–Q_4 and again by Q_5–Q_6, with each phase lagging the one ahead of it by 120°. Clearly, the apparent frequency of the output is determined by how fast transistors Q_1–Q_6 are sequenced. The typical frequency range for these systems is 2–120 Hz.

Figure 9.34

Output-voltage waveforms of a speed-control circuit.

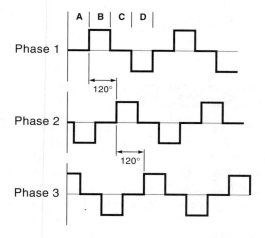

One further consideration concerning AC motor-speed control must be taken into account. For the motor to work well at various speeds, the voltage to the motor must be modified each time the frequency is changed. Specifically, the voltage and frequency should be held proportional—that is, when the frequency is increased, the voltage should be increased, and vice versa. The reason for this requirement is that the current in the stator windings must be maintained at a certain design value for the magnetic induction process (to the rotor) to work. Most motors are designed to operate at 60 Hz and 120 V (or 240 V), so the stator is wound to create the proper magnetic field with

Figure 9.35

Voltage and frequency are controlled in a variable-speed-control circuit.

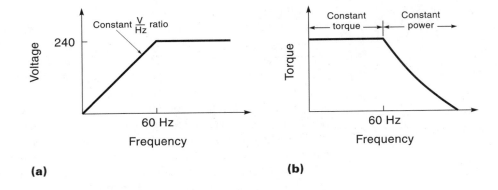

(a)

(b)

those conditions. If the frequency drops below 60 Hz, the inductive impedance of the windings also drops, which would allow in more current. Consequently, the voltage must be lowered as the frequency is lowered in order to maintain the proper stator current. Figure 9.35(a) shows how the voltage should increase linearly with frequency in the range 0–60 Hz. The voltage is usually not allowed to increase beyond the motor's rated voltage (for its own health). So, in practice, there are really two distinct operating ranges. The first range (0–60 Hz) is called the *constant-torque region* because the motor produces a constant torque in this speed range, as shown in Figure 9.35(b). This is the same torque that the motor has at normal (60 Hz) operating speed. The region above 60 Hz is known as the *constant-power region* because, even though the torque is falling off, the speed is increasing, so the actual mechanical power stays the same (power is the product of speed times torque). Commercial motor-control circuits are capable of meeting these qualifications.

SUMMARY

The AC motor is a common source of electric rotary power. Most AC motors do not use commutator/brushes, which makes them more reliable and less expensive than DC motors. The primary use of AC motors is for constant-speed applications because the speed of an AC motor is dependent on the line frequency.

AC power is supplied in two forms: single phase and three phase. Single-phase AC is the common form supplied to homes and businesses and requires two wires. Three-phase power is usually only available in industrial locations and consists of three separate 60-Hz sine waves, 120° apart, on three wires (four including the neutral).

The most common type of AC motor is the induction motor. In the induction motor, power is supplied only to the stator windings, where it creates a rotating magnetic field around the rotor. The rotor receives some electric power

through induction from the stator, which it uses to generate a magnetic field. This field is attracted to the stator's rotating field, which causes the rotor to rotate (but slightly slower than the rotating field).

Motors are designed to run on either three-phase power or single-phase power. Three-phase motors are the simplest because they are self-starting, but they can only be used where three-phase power is available. Single-phase motors are not inherently self-starting. There are a number of ways used to start single-phase motors, and most involve adding a second set of stator windings (called a split-phase motor), which are only energized during start-up.

The synchronous motor is designed to rotate at an exact multiple of the line frequency. For larger motors, this means that the rotor receives its power through slip rings and brushes, which complicate the design (induction will not work in this application). Smaller synchronous motors (timing motors) do not use slip rings.

The universal motor is so named because it can run on AC or DC. It is essentially a DC motor (with brushes) designed to minimize the losses that occur when running on AC.

AC motors have a large initial current when first switched on. For larger motors, a special starting circuit may be required to prevent damage to the power system and/or motor. It works by reducing the voltage to the motor during start-up, either by resistors or transformers.

The speed of an AC motor can be controlled in two ways. A small amount of speed control can be had from simply reducing the voltage, which causes the rotor speed to slip more. For complete speed control, the frequency must be controlled with control circuits that convert 60-Hz AC to other frequencies.

GLOSSARY

AC servomotor   A two-phase motor used in control applications; designed to have a near-linear torque–speed curve.

capacitor start motor   A single-phase AC induction motor that uses a capacitor to phase-shift the AC for the start windings.

control winding   One of two windings in an AC servomotor.

DC link converter   A circuit that converts line voltage at 60 Hz to AC power at different voltage and frequencies, for the purpose of controlling the speed of AC motors.

delta-connection   One of two ways in which to connect three-phase motor or generator coils. With the delta-connection, each of the three coils is connected between two phase wires; there is no neutral.

fractional horsepower (hp) motor   A motor with less than 1 hp.

frame   A basic part of an AC motor; an overall case that supports the stator and the rotor bearings.

GFI   *See* **ground-fault interrupter.**

ground-fault interrupter (GFI)   A saftey device that will shut off the power if it senses a current leakage to the safety ground.

integral horsepower (hp) motor   A motor with 1 hp or more, that is, a large motor.

jogging   The practice of "inching" a motor into position by repeated short bursts of power.

ladder diagram   A type of wiring diagram typically used for industrial motor-control circuits.

latching   *See* **sealing.**

line voltage   A single number that represents the voltage of a power system; for three-phase power, the line voltage is the vector sum of two of the three-phase voltages.

main winding   One of two windings in an AC servomotor.

neutral   The return wire or cold side in an AC power-distribution system; the neutral wire is at ground potential.

overload device   A device that can stop the motor if an overload condition is detected. Overload is detected by sensing the temperature of the motor or the current that the motor is drawing.

permanent-split capacitor motor   A capacitor start motor where the start winding remains engaged during operation—that is, the motor does not have a centrifugal switch.

phase sequence   The sequence of phase voltages possible with three-phase power: ABC and ACB.

phase voltage   The individual voltage in a multiphase power system.

pony motor   A separate motor used to start a synchronous motor.

power factor   The cosine of the angle between the current and voltage. A power factor of 1 means the current and voltage are in phase, which is desirable. A lagging power factor means the current is lagging the voltage, probably due to the inductive load of motors.

pull-out torque   The maximum torque that an AC motor can provide just before it stalls. However, unlike the DC motor, this maximum torque condition occurs at about 75% of the unloaded speed.

rotating field   The phenomenon of what is apparently happening in the stator coils of an AC motor. Even though the coils themselves are stationary, the magnetic field is transferred from coil to coil sequentially, producing the effect that the field is rotating.

rotor   A basic part of an AC motor, the part in the center that actually rotates.

run windings The main stator windings in a single-phase motor.

sealing A process whereby a relay is initially energized by a start button and then stays energized from current coming through one of its own closed contacts.

shaded-pole motor A single-phase induction motor, usually small, that does not use a start winding. Instead, a modification to the single-phase poles causes the motor to have a small starting torque.

single-phase AC The type of AC supplied to residences (and industry). It consists of a single alternating waveform that makes 60 complete cycles per second.

slip The difference between the rotor speed of an induction motor and the synchronous speed of the stator field; the amount of slip is typically less than 10%.

slip rings Sliding electrical contacts that allow power to be fed to the rotor of the synchronous motor.

speed regulation The ability of a motor to maintain its speed under different loads.

split-phase A general term applied to an induction motor that has two sets of windings but operates on single-phase AC with some provision to phase-shift the AC for the second set of windings.

split-phase control motor A two-phase motor, usually small, powered by single-phase AC using a capacitor for the phase shift. The direction is reversible, making these motors useful in controlling back-and-forth motion.

squirrel cage rotor The type of rotor used in an induction motor—named as such because someone thought it looked like a squirrel cage.

stall When the load on the AC motor is increased (about 75% of unloaded speed), the motor torque drops back abruptly, and the motor stalls (stops).

starting torque The motor torque available when the motor is first started; the starting torque of most AC motors is less than its maximum torque but more than its rated load.

start windings A second set of windings (besides the run windings) in the single-phase induction motor; the start windings are usually only engaged during the starting process.

stator A basic part of an AC motor, a collection of coils that surrounds the rotor and does not move.

synchronous condenser A term applied to a synchronous motor that is being used solely for power-factor correction.

synchronous motor An AC motor that runs at the synchronous speed, which means an exact multiple of the line frequency. A synchronous motor has no slip.

synchronous speed The speed of the rotating field in the stator, which is always an exact multiple of the line frequency. A synchronous motor spins at the synchronous speed, whereas an induction motor turns somewhat slower than the synchronous speed.

three-phase AC The type of AC available to industry. It consists of three AC waveforms (called phases) on three wires, where each phase is delayed 120° from the previous phase and all phases operate at 60 Hz.

timer relay (TR) A relay with a built-in time delay mechanism so that the contacts close or open a specified time after the relay is energized.

two-capacitor motor A capacitor start motor that uses one value of capacitance for starting and switches to a lower value of capacitance for running.

TR *See* **timer relay**.

universal motor An electric motor that can run on AC or DC power; it is essentially a series DC motor.

wye-connection One of two ways in which to connect three-phase motor or generator coils; with the wye-connection, each of the three coils is connected between a phase wire and neutral.

EXERCISES

Section 9.1

1. A single-phase voltage waveform is shown in Figure 9.2.
   a. What is the voltage between prong A and ground at 270°?
   b. What is the voltage between prong B and ground at 90°?
   c. What is the voltage between prong A and prong B at 360°?

2. Explain the distinguishing features of the AC waveform shown in Figure 9.6.

3. Redraw Figure 9.5 with a delta-connected load.

4. In a balanced three-phase wye-connected system, what percentage of the current flows in the neutral wire? Explain.

5. A three-phase wye-connected system has individual phase voltages of 240 Vac. Find the three-phase line voltage.

Section 9.2

6. Explain what a *rotating field* in an AC motor is and briefly how it is created.

7. Referring to Figure 9.8, draw a diagram showing how the four coils would be energized at 45° (indicate the magnetic orientation).

8. Define the term *slip*. Why is slip essential for an induction motor to work?

9. An induction motor is powered by 120 Vac at 60 Hz and runs at approximately 1200 rpm. What pole motor must this be?

10. What is the synchronous speed of a 60-Hz, six-pole, three-phase motor? How many individual poles does it have?

11. What is the synchronous speed of a 400-Hz eight-pole, single-phase motor? How many individual poles does it have?

12. A single-phase 60-Hz AC motor has a rated speed of 1695 rpm. Find the slip.

13. A certain six-pole 60-Hz AC motor has a slip of 5% (under normal-load conditions). How fast is it rotating?

14. Explain the important parts of the torque–speed curve shown in Figure 9.11. Include in your explanation the following terms: synchronous speed, operating range, pull-out torque, and starting torque.

15. Is a three-phase motor reversible, and if so, how?

16. Draw a diagram of a capacitor start motor (which incorporates a centrifugal switch). Explain what happens during the start-up period. (What is the purpose of the capacitor?)

17. Draw a diagram of a permanent-split phase motor and explain how it operates.

18. Explain how a shaded-pole motor operates.

19. Draw a diagram and explain the operation of a split-phase control motor being powered by single-phase AC that can go in either direction.

20. What are the basic characteristics of an AC servomotor?

Section 9.3

21. List the differences between the *synchronous motor* and the *induction motor*.

22. What is the purpose of the slip rings used in synchronous motors?

23. List three methods used to start a synchronous motor.

Section 9.5

24. Describe the operation of the start–stop circuit of Figure 9.28.

25. Redraw Figure 9.29 and then include the additional circuitry necessary to have the motor start and stop in the reverse direction.

26. In the reduced-voltage-starting circuit of Figure 9.31,
   a. What is the purpose of relay MR?
   b. What is the purpose of relay RB?
   c. What is the purpose of relay TR ?

27. Name the two methods used to vary the speed of an AC motor.

28. Explain the operation of the variable-speed-control circuit shown in Figure 9.33.

10 Actuators: Electric, Hydraulic, and Pneumatic

OBJECTIVES

After studying this chapter, you should be able to:

- ❏ Understand what linear actuators are and how they are used.
- ❏ Explain how an electric leadscrew linear actuator works and calculate extension times.
- ❏ Understand solenoid operation and characteristics.
- ❏ Describe the components of a basic hydraulic system and understand how they work.

- ❏ Calculate the force generated by a hydraulic cylinder, given the system parameters.
- ❏ Describe the components of a basic pneumatic system and understand how they work.
- ❏ Calculate the force generated by a pneumatic cylinder, given the system parameters.
- ❏ Understand the basic control-valve operation.

INTRODUCTION

The **prime mover** or **actuator** provides the source of mechanical power in a control system. It is easy to identify the prime mover in a system because it will be the source of the first physical movement. Typical prime movers are motors, hydraulic cylinders, and control valves. Technically speaking, a prime mover is a transducer as well because it converts one kind of energy into another, usually electrical energy into mechanical energy. By far the most common type of prime mover is the electric motor, which is an efficient and versatile converter of electrical to mechanical energy. However, many real-life control applications require a slow, powerful, back-and-forth straight-line motion instead of the higher speed rotary motion produced by the electric motor. An actuator that can provide a back-and-forth straight-line motion is called a **linear actuator**.* Many examples of linear actuators can be found in control systems, particularly in the field of robotics—for example, an object is picked up, moved in a linear fashion to another spot, and set down.

---

*This text uses the term *linear actuator* in a general sense to mean any straight-line actuator. A more particular definition for a linear actuator is "a device where there is a linear relationship between the applied control signal and the resulting displacement."

Some devices (such as the solenoid) can convert electrical energy directly into linear motion, but more often it is done in two stages. First, a motor is used to create rotary mechanical motion; second, the rotary motion is converted into linear motion. An example would be a motor driving a rack-and-pinion gear train (Figure 5.23), which moves the print head across the paper on your printer. Hydraulic systems are more complicated—that is, a motor drives a pump that pushes fluid into a hydraulic cylinder that causes the piston to move.

The three general types of linear actuators are electric, hydraulic, and pneumatic. Each type has its particular advantages and disadvantages, and all are in wide use. This chapter will introduce you to these systems and to flow-control valves.

10.1 ELECTRIC LINEAR ACTUATORS

Of the three types of linear actuators (electric, hydraulic, and pneumatic), electric actuators require the least hardware and maintenance. However, they tend to be relatively slow-moving, which make them suitable in only a limited number of applications. For example, electric actuators work well for lowering the landing gear in an airplane but could not be used for a spray-painting robot, which must make quick, long sweeps.

Leadscrew Linear Actuators

A **leadscrew linear actuator** has a threaded shaft (called the *leadscrew* or *drive screw*), which is rotated by an electric motor. A "nut" on the shaft moves in a linear motion as the shaft rotates. Figure 10.1 shows a diagram of the electric linear actuator. Notice that

Figure 10.1

Diagram of an electric linear actuator (Courtesy of Warner Electric).

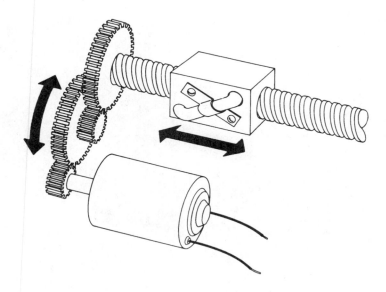

the rotary torque of the motor is transferred through a gear train to the drive screw. The output motion results as the nut advances or retreats along the threads, depending on which direction the shaft is rotating. Note that the nut must be constrained from rotating.

To reduce the friction and increase the efficiency, many units use ball bearings to transfer the force from the drive screw to the nut. Figure 10.2 shows a cutaway diagram of this design, called a **ball bearing screw**. When the shaft rotates, the balls roll in the grooves. When the balls get to one end of the nut, they are recirculated to the other end through small pipes. Ball bearing screws practically eliminate friction when compared with the traditional nut on thread design.

Figure 10.3 shows a cutaway diagram of a practical electric linear actuator. The unit resembles and acts like a hydraulic cylinder, the main visual difference being the attached electric motor. Many units come with a built-in electrically activated brake that prevents the shaft from rotating when the motor is not energized. This feature ensures that a heavy load cannot back drive the motor after it is turned off. Also, limit switches should be included in the application design, to cut off power to the motor if it attempts to drive beyond its intended travel.

Electric linear actuators come in all sizes, with rated loads from under 100 lb to over 1000 lb and strokes from 4 to 36 in. and beyond. An important parameter for electric actuators is **extension rate**, that is, how fast the unit can extend and contract. A typical extension rate is around 1 in./s, but this depends on the size of the motor and the load. However, even the larger units are no match for hydraulic cylinders (covered in Section 10.2) when it comes to speed and power. The electric actuator is characterized by slower, steady, linear movement. Figure 10.4 shows the characteristics of a 12-Vdc linear actuator. Notice that as the load increases, the extension rate decreases, and the required current increases.

Figure 10.2

Cutaway drawing of a ball bearing screw (Courtesy of Warner Electric).

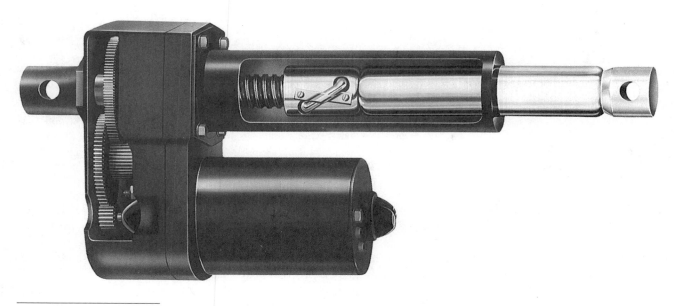

Figure 10.3   An electric linear actuator (Courtesy of Warner Electric).

Figure 10.4

Performance characteristics of
a typical electric linear
actuator.

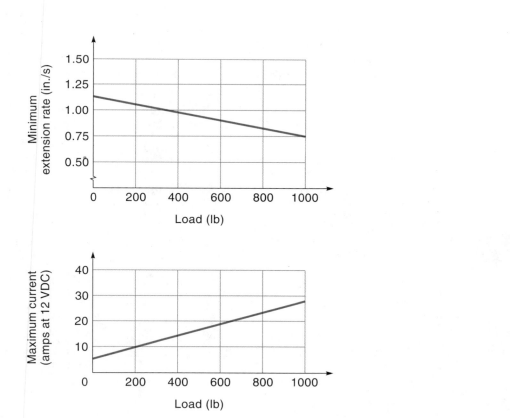

◆ **EXAMPLE 10.1**

An electric linear actuator is used to lift 600 lb castings for a vertical distance of 8 in. Using the data from Figure 10.4, how long will the lifting operation take?

Solution   Using the top graph of Figure 10.4, we see that for a load of 600 lb the extension rate is about 0.9 in./s. The time required to lift a distance of 8 in. is calculated as follows:

$$\text{Time to lift} = 8 \text{ in.} \times \frac{s}{0.9 \text{ in.}} = 8.9 \text{ s} \qquad ◆$$

For applications that can tolerate the slower extension rates, the electric linear actuator has some distinct advantages over hydraulic and pneumatic systems. Specifically, electric actuators (1) require only wires to be connected—not tubing, (2) do not require extra supporting hardware such as pumps and tanks, (3) use power only when they are actually moving, and (4) when compared with hydraulic systems, are less "messy." On the other hand, compressed air for pneumatic systems may already be available, and for working around explosive gasses, a nonelectric actuator would be safer (no sparks!).

Solenoids

A **solenoid** is a simple electromagnetic device that converts electrical energy directly into linear mechanical motion, but it has a very short **stroke** (length of movement), which limits its applications. The solenoid consists of a coil of wire with an iron plunger that is allowed to move through the center of the coil. Figure 10.5(a) shows the solenoid in the unenergized state. Notice that the plunger is being held about halfway out of the coil by a spring. When the coil is energized [Figure 10.5(b)], the resulting magnetic field pulls the plunger to the middle of the coil. The magnetic force is unidirectional—a spring is required to return the plunger to its unenergized position. Figure 10.6 shows two practical solenoids.

The main limitation of the solenoid is its short stroke, which is usually under an inch. Still, there are many applications for short-stroke linear motion; examples are activating electric car-door locks, opening and closing valves, and triggering mechanical latches.

Figure 10.5

A solenoid.

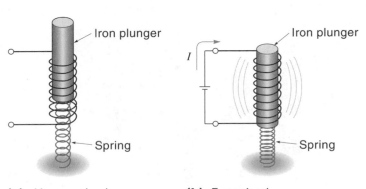

(a) Unenergized   **(b)** Energized

Figure 10.6
Solenoids.

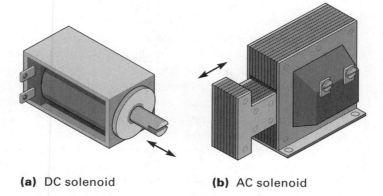

(a) DC solenoid   **(b)** AC solenoid

Most applications use the solenoid as a on or off device—that is, the coil is either completely energized or switched off. However, variable-position control is possible by varying the input voltage.

Both AC and DC solenoids are used, the major difference being that AC solenoids use a plunger and frame made from laminations instead of solid iron. *Laminations* are

TABLE 10.1   A Selection of Solenoids

| Type | Volts | Duty | Ohms | Lifts and Strokes (ounces at inches) | | Body |
				(minimum)	(maximum)	
870	120 AC	Inter.	300	8 @ ⅛	2 @ ½	1
871	120 AC	Cont.	675	3 @ ⅛	0.5 @ ⅞	1
078	120 AC	Inter.	60	33 @ ⅛	9 @ ⅞	1
079	120 AC	Cont.	166	10.5 @ ⅛	3 @ ⅞	1
1897	120 AC	Inter.	36	83 @ ⅛	16 @ ¾	1
1898	120 AC	Cont.	113	17 @ ⅛	4 @ ¾	1
1899	24 DC	Inter.	20	100 @ ⅛	5 @ ¾	1
1900	24 DC	Cont.	64	50 @ ⅛	1 @ ⅝	1
872	120 AC	Inter.	37	24 @ ⅛	22 @ 1	2
873	120 AC	Cont.	133	8 @ ⅛	6 @ 1	2
885	24 DC	Inter.	15	110 @ ⅛	4 @ 1	2
886	24 DC	Cont.	54	55 @ ⅛	2.5 @ ⅞	2
887	115 DC	Inter.	346	90 @ ⅛	4 @ 1	2
888	115 DC	Cont.	1300	55 @ ⅛	2.5 @ ⅞	2
1901	24 DC	Inter.	15	156 @ ⅛	21 @ ¾	2
1902	24 DC	Cont.	50	70 @ ⅛	4 @ ¾	3
1903	115 DC	Inter.	350	156 @ ⅛	21 @ ¾	3
1904	115 DC	Cont.	1200	70 @ ⅛	4 @ ¾	3
5600	120 AC	Inter.	220	17 @ ⅛	8 @ ½	4
5601	120 AC	Cont.	350	11 @ ⅛	4 @ ½	4
5602	24 DC	Inter.	52	16 @ ⅛	1 @ ½	4

thin sheets of lacquered iron that are riveted together to form the frame and plunger. Laminations prevent power-consuming *eddy currents* (induced by the AC) from circulating in the metal parts of the solenoid. The laminations can easily be seen in Figure 10.6(b).

Table 10.1 lists various models of solenoids. Note the various coil voltages and whether the unit can be operated intermittently or continuously. An *intermittent-duty solenoid* is designed to operate for a short time and then take time to cool. The problem is that, like relays, it takes more voltage to pull in the solenoid than it takes to hold it in the retracted position, Thus, if a solenoid is left on, it is drawing more current than it really needs, and it tends to heat up. A *continuous-duty solenoid* is designed to deal with the heat, so it can operate all the time. The "Lifts and Strokes" data in Table 10.1 tell us the maximum pull force (which occurs when the unit is first energized). The minimum pull force occurs at the end of the stroke. For example, the unit on the first line of Table 10.1 exerts (at least) 8 oz of force for the first ⅛ inch of stroke but falls to only 2 oz of force when the solenoid is extended to its maximum of ½ inch.

10.2 HYDRAULIC SYSTEMS

Hydraulic systems are ideally suited to provide a strong, fast, or slow linear motion. The system uses a fluid (light-grade oil) to transfer energy from a pump to an actuator. A simplified system, illustrated in Figure 10.7, consists of the following components: a tank of hydraulic fluid, a pump, a control valve, and a cylinder. The pump pushes the fluid through a tube to the control valve. The control valve directs the fluid to the cylinder, causing the piston to move down in response to the fluid pressure. The pump is the actual source of mechanical power, and it is physically separate from the cylinder actuator. The practical significance of this fact is that *only the cylinder needs to be mounted at the place where the motion is needed*; the pump can be elsewhere. This allows a relatively small component such as a hydraulic cylinder to provide far more power than a similarly sized electric actuator, which must have the motor attached.

Figure 10.7
A simplified hydraulic system.

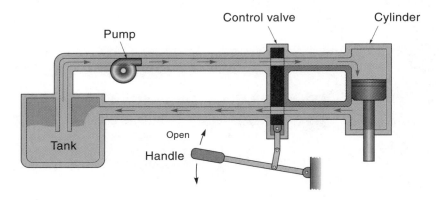

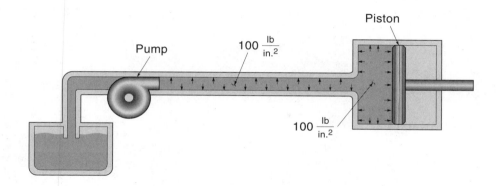

Figure 10.8
Hydrostatic pressure pushes out in all directions.

Basic Principles of Hydraulics

One major difference between a liquid and a gas is that a gas will squeeze down into a smaller volume when pressurized, whereas a liquid will not. We say that liquids (including hydraulic fluid) are incompressible. This is an important concept to remember when studying hydraulic systems. It means that *each cubic inch of fluid pumped into the system must appear someplace else downstream as 1 in^3 of fluid volume.*

Another important concept is that of **hydrostatic pressure**, which results when a fluid is under pressure and it is not moving (hence, static). Pascal's principle states that *a fluid under hydrostatic pressure will exert the same pressure uniformly on the inside walls of all the interconnected components.* Recall that pressure is defined as a "force per unit area." For example, a pressure* of 100 lb/in^2 (psi) means that the container must withstand a force of 100 lb for every square inch of surface area. The unit of pressure for SI units is the Pascal (Pa), where 1 Pa = 1 Newton/meter^2 (N/m^2). This pressure is so small that a more common unit is the kiloPascal (1000 Pa = 1 kPa). The relationship between psi and kPa is 1 psi = 6.89 kPa.

The principle of hydrostatic pressure is illustrated in Figure 10.8, which shows a tank, pump, piston, and hydraulic cylinder. The fluid on the high-pressure side of the pump is pressurized to 100 psi. This means that the tubing, cylinder walls, and piston face are all receiving 100 lb of force for each square inch of area, as indicated by the arrows in Figure 10.8. All internal surfaces are under pressure, but only the piston is movable, so it moves to the right. The force on the piston from the hydrostatic pressure is the product of the pressure times the area of the piston face, as given in Equation 10.1:

$$F = PA \tag{10.1}$$

where

F = total force exerted on the piston
P = hydrostatic pressure
A = area of the piston face

*In this chapter, the pressure values referred to are technically called *gauge pressure*, meaning "that pressure above the ambient atmospheric pressure of 14.7 psi or 10^5 Pa."

♦ EXAMPLE 10.2

The hydraulic cylinder in Figure 10.8 is 3 in. in diameter. Find the force exerted on the piston if the pressure is 100 psi.

Solution   We can use Equation 10.1 to find the force, but first we need to find the area of the piston face:

$$\text{Area} = \pi r^2 = 3.14 \times (1.5 \text{ in.})^2 = 7.07 \text{ in}^2$$

Now we can calculate the piston force, using Equation 10.1, knowing that the pressure is 100 psi:

$$\text{Force} = PA = \frac{100 \text{ lb}}{\text{in}^2} \times 7.07 \text{ in}^2 = 707 \text{ lb}$$

EXAMPLE 10.2 (Repeated with SI Units)

The hydraulic cylinder in Figure 10.8 is 8 cm in diameter. Find the force exerted on the piston if the pressure is 700 kPa.

Solution   We can use Equation 10.1 to find the force, but first we need to find the area of the piston face:

$$\text{Area} = \pi r^2 = 3.14 \times (4 \text{ cm})^2 = 50.24 \text{ cm}^2$$

Now we can calculate the piston force, using Equation 10.1, knowing that the pressure is 700 kPa:

$$700 \text{ kPa} = 700{,}000 \text{ Pa} = 700{,}000 \text{ N/m}^2$$

$$\text{Force} = PA = \frac{700{,}000 \text{ N}}{\text{m}^2} \times \frac{1 \text{ m}}{100 \text{ cm}} \times \frac{1 \text{ m}}{100 \text{ cm}} \times 50.24 \text{ cm}^2 = 3517 \text{ N}$$

EXAMPLE 10.3 (Mixed English and SI Units)

A 2-in. diameter hydraulic cylinder is being supplied with a pressure of 1000 kPa. Find the force exerted on the piston in pounds.

Solution   We can use Equation 10.1 to find the force, but first we need to find the area of the piston face:

$$\text{Area} = \pi r^2 = 3.14 \times (1 \text{ in.})^2 = 3.14 \text{ in}^2$$

Now we can calculate the piston force, using Equation 10.1, but first we must convert the pressure to psi:

$$1000 \text{ kPa} \times \frac{1 \text{ psi}}{6.89 \text{ kPa}} = 145.1 \text{ psi}$$

$$\text{Force} = PA = \frac{145.1 \text{ lb}}{\text{in}^2} \times 3.14 \text{ in}^2 = 455.6 \text{ lb} \quad ♦$$

Figure 10.9

A hydraulic system that "amplifies" force.

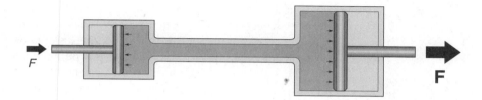

Static hydraulic systems can be used to "amplify" a force. This is the principle used in hydraulic jacks or hydraulic brake systems and is illustrated in Figure 10.9. The system contains two cylinders of different diameters that are connected by a tube. When a force is applied to the small piston, a pressure develops. This same pressure is transmitted through the tube and is applied to the large piston. Because the large piston has more surface area, it receives a larger net force. Of course, it is also true that the large piston will move a shorter distance than the small piston because the work done by each piston must be the same.

♦ **EXAMPLE 10.4**

For the system shown in Figure 10.9, the small piston has a face area of 2 in^2 and receives an external force of 10 lb. The large piston has a face area of 20 in^2. Calculate the force exerted by the large piston.

Solution

The small piston has a surface area of 2 in^2 and is being pushed with a force of 10 lb. Rearranging Equation 10.1 ($F = PA$) and solving for P, we can calculate the fluid pressure this will create:

$$P = \frac{F}{A} = \frac{10 \text{ lb}}{2 \text{ in}^2} = \frac{5 \text{ lb}}{\text{in}^2}$$

This pressure of 5 psi is conveyed to the large cylinder where it pushes on the 20-in^2 piston. Rearranging Equation 10.1 again, we can calculate the force on the large piston:

$$F = PA = \frac{5 \text{ lb}}{\text{in}^2} \times 20 \text{ in}^2 = 100 \text{ lb}$$

Thus, the large piston will exert a force ten times larger but will move only one-tenth the distance of the smaller piston because the volume change must be the same for both sides. ♦

Hydraulic Pumps

In an active hydraulic system, a pump is used to create the hydrostatic pressure. Figure 10.10 shows a common pump design called a **gear pump**, which consists of two meshed gears in a housing. As the gears rotate, fluid is trapped in the little spaces between the teeth and the housing (both top and bottom) and is conveyed from the inlet to the outlet. The mesh between the gears in the center is tight enough so that no fluid moves through either way at that point. This type of pump is also known as a **positive-displacement pump** because a constant volume of fluid is pumped for every revolution of the gears.

The **vane pump** is another type of hydraulic pump. Shown in Figure 10.11, it consists of an offset rotor in a housing with retractable vanes. The spring-loaded vanes push out and seal against the housing wall. Because there is more fluid between the vanes in the top half of the housing than in the bottom, there is a net transfer of fluid from the inlet to the outlet. In some designs, the position of the rotor axis is adjustable. The more offset the rotor axis, the more fluid is pumped. Such a pump is called a **variable-displacement pump**.

Figure 10.10

A hydraulic gear pump.

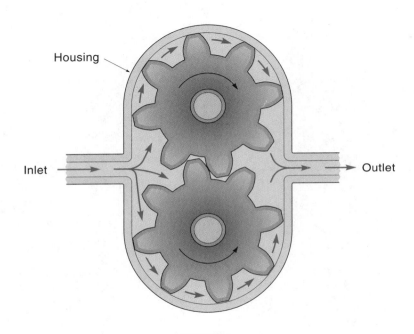

Figure 10.11

A hydraulic vane pump.

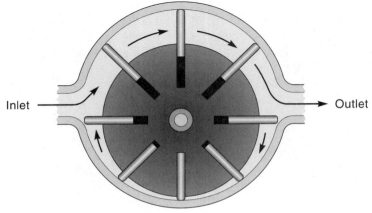

A third kind of pump, an **axial piston pump**, uses small pistons reciprocating back-and-forth to pump the fluid. As shown in Figure 10.12, the pump consists of a rotating cylinder and a metal ring called a **swash plate** (which does not rotate). The cylinder contains a number of small pistons that do the actual pumping. One end of each piston rides against the swash plate. Because the swash plate is at an angle to the cylinder, each piston is forced to move in-and-out with each rotation of the cylinder. By changing the angle of the swash plate, the quantity of fluid pumped per revolution is changed.

Figure 10.12

An axial piston pump.

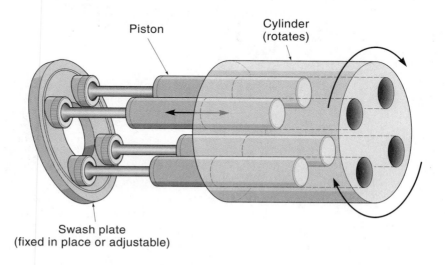

Hydraulic Actuators

The most common type of hydraulic actuator is the **hydraulic cylinder**. A typical cylinder is shown in Figure 10.13(a) and is known as a **double-acting cylinder** because it can provide force in either direction. It consists of a piston and a cylinder body. The piston has a rod that extends out one end of the cylinder. Fluid can enter and leave the cylinder on either side of the piston through ports. Under normal operating conditions, both ends of the cylinder are filled with fluid. If additional fluid enters port A, the piston will move toward the right, but the fluid must be able to escape through port B. A selection of hydraulic cylinders is shown in Figure 10.13(b).

Some hydraulic actuators can create rotary motion and are very similar to the pump designs discussed previously. For example, the **gear motor** rotary actuator shown in

Figure 10.13

A double-acting hydraulic cylinder.

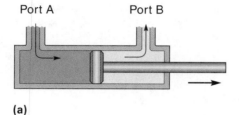

(a)

Figure 10.13 *continued*

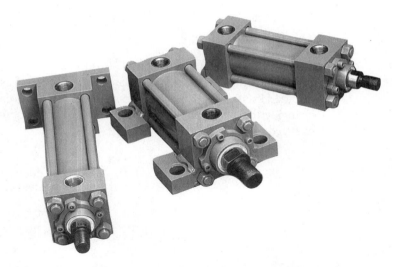

(b) Hydraulic cylinders (Courtesy of Miller Fluid Power)

Figure 10.14 is almost identical to the gear pump. For the motor, fluid is pumped in the left side of the case, putting that area under pressure (indicated by p). Within the pressurized area, all surfaces receive a force, but only those three surfaces indicated with arrows will affect rotation (the other surfaces are balanced out). The pressure on the teeth next to the case (top and bottom arrows) will cause the gears to rotate as shown. The pressure on the meshing teeth in the center would cause the gears to turn in the opposite direction, but this torque is overpowered because two teeth (top and bottom) are pushing the other way.

Figure 10.14

A hydraulic gear motor.

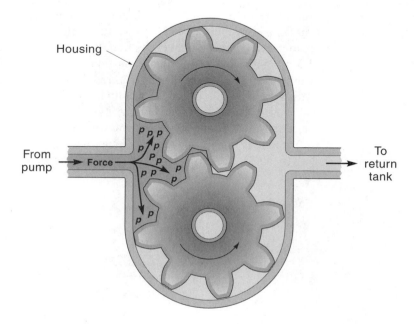

Pressure-Control Valves

A **pressure-control valve** is a spring-loaded valve that is capable of maintaining a constant pressure in a system, regardless of the flow rate. This is important because most pumps (such as the gear pump) are constant-displacement types—a constant volume of fluid is pumped for each revolution of the pump shaft. If the pump were connected directly to the cylinder, it would have to start and stop each time the piston moved to a new position. Such on–off cycling reduces the lifetime of machinery and is therefore undesirable. When a pressure-control valve is put into the system (as shown in Figure 10.15), the pump can remain on the whole time—when the fluid pressure exceeds the preset limit, the valve opens, and the surplus fluid is simply returned to the tank. In other words, when the piston is *not* moving, the fluid is simply circulating from the tank, through the pump, through the valve, and back to the tank. If the pressure-control valve opens at 1000 psi, then the pressure in the lines will never get much above 1000 psi.

Figure 10.15

A relief valve circulates surplus fluid to the tank.

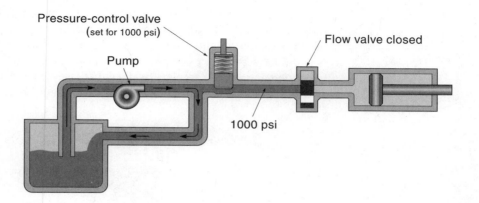

Figure 10.16 shows the inside of a simple pressure-control valve. Notice that the passage from the pump to the load is always open. If the hydraulic pressure in the main passage overcomes the spring pressure, the valve opens, and fluid can escape out a

Figure 10.16

A pressure-control valve (partially open position).

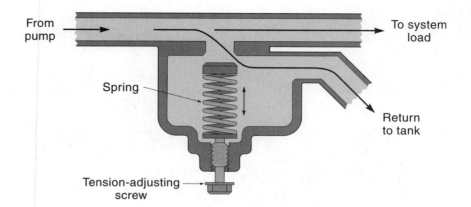

separate port that is connected back to the tank. The pressure at which the valve opens can be set by adjusting the tension on the spring.

Accumulators

An **accumulator**, which is connected into the system, is a special kind of spring-loaded storage tank for hydraulic fluid (Figure 10.17). The accumulator serves two functions. First, it acts as a low-pass filter to remove pressure pulsations from the pump; second, it stores extra fluid for those high-demand times when the actuator requires fluid at a faster rate than the pump can supply. This latter situation is illustrated in Figure 10.17, which shows a pump (capacity: 1 gal/min) supplying fluid to a cylinder. Notice that the accumulator is located between the pump and the cylinder. In Figure 10.17(a) the valve to the cylinder is closed; consequently, all fluid from the pump goes into the accumulator, compressing the accumulator spring. In Figure 10.17(b), the valve opens enough to allow fluid to the cylinder at a rate of 2 gal/min. Because the pump can supply only 1 gal/min, the other 1 gal/min comes from the stored fluid in the accumulator, which is pushed out by the spring. Naturally, this concept will work only if the cylinder's demand for fluid is intermittent, but this is often the case. In fact, this is one of the advantages of a hydraulic system; it can store energy within the system, to be used during periods of peak demand. For example, consider the case where a robot performs a 10s lifting operation (which requires 1 hp) once each minute. Because the average load is only

Figure 10.17
An accumulator in a system.

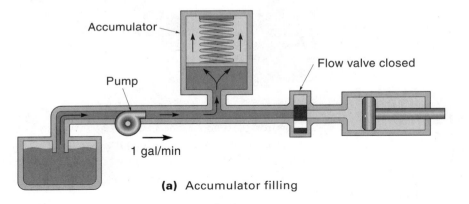

(a) Accumulator filling

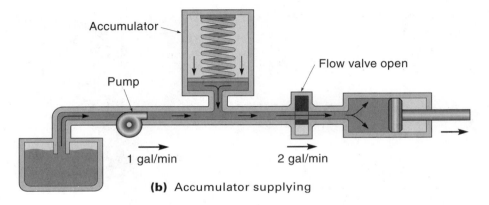

(b) Accumulator supplying

⅙ hp [1 hp × 10 s/60 s], a hydraulic system could get by with a ⅙-hp pump. For 50 s out of each minute, the pump would be "charging up" the accumulator; for 10 s, both the pump and the accumulator would be supplying the cylinder. Now suppose, instead of the hydraulic system, the robot uses an electric linear actuator. Because the electric motor cannot store up energy, it would have to be the full 1 hp, even though it is only on one-sixth of the time.

The three types of accumulators are: spring-loaded (already mentioned), weighted, and gas-pressurized. The weighted type uses a weight to provide the pressure on the fluid. This has the advantage of providing a constant pressure but the disadvantage of needing to be mounted in an upright and relatively stable position. The gas-pressurized type uses a gas (dry nitrogen) under pressure to provide the push on the fluid. This is the most popular type because the pressure it exerts is easily adjustable and it can be mounted in any position.

Flow-Control Valves

The flow-control valve is used to regulate the flow rate of hydraulic fluid. For this discussion, the most important type of flow valve is the **bidirectional control valve**, which regulates the movement of the piston in the cylinder. Figure 10.18 shows a diagram of this type of valve. A sophisticated device, it must be capable of admitting pressurized fluid to either end of the cylinder while providing a return path for fluid being squeezed out of the other end of the cylinder.

The control valve consists of two major parts: the valve body and the spool. Referring to Figure 10.18, you can see that the valve body has four fluid-connection ports: (1) high-pressure fluid supply (from the pump), (2) low-pressure return to the tank, and (3 and 4) connections to both ends of the cylinder. The **spool** is a solid-metal machined shaft that can slide back-and-forth in the valve body. The spool in Figure 10.18 has two deep grooves that allow fluid to pass from one port in the valve body to another. The position of the spool within the valve body determines how fast, and in what direction, the piston will move. The spool is moved by a hand lever, an electric actuator such as a solenoid, or a small hydraulic or pneumatic actuator.

Figure 10.18

A control valve for a double-acting cylinder.

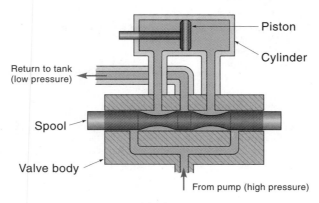

(a) Spool centered

Figure 10.18 *continued*

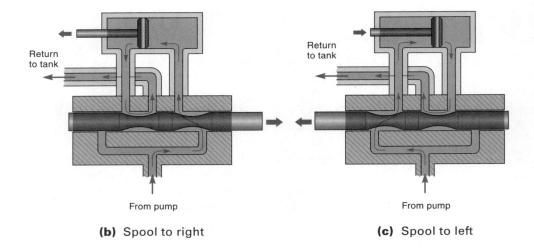

(b) Spool to right   **(c)** Spool to left

The following is a description of the operation of the spool-type valve of Figure 10.18. (To really understand this clever design, take the time as you read to trace the flow of fluid in each diagram.) Figure 10.18(a) shows the valve with the spool centered. In this position, the fluid from the pump is completely blocked by the spool. The lines connected to the cylinder are also blocked; with the cylinder fluid thus trapped, the piston is locked in place. This explains why a backhoe hydraulic shovel will remain in the upright position even with a heavy load. (However, don't trust your safety to this feature—hoses can leak!)

Figure 10.18(b) shows the same valve with the spool moved to the right. Notice that the right-hand groove is allowing the fluid to pass from the pump to the right end of the cylinder, causing the piston to move to the left. The left spool groove has moved into position to allow an escape route for the fluid being pushed out of the left end of the cylinder. This low-pressure fluid is allowed to return to the tank.

Figure 10.18(c) shows the same valve with the spool moved to the left of center. Now the high-pressure fluid from the pump is allowed to pass, via the left groove, to the left end of the cylinder. The return fluid from the cylinder passes through the right groove and then back to the tank.

Figure 10.19 shows a diagram (using standard symbols) of a complete simple hydraulic system, which includes the tank, filter, pump, accumulator, pressure-control valve, flow-control valve, and cylinder. The constant-displacement pump would be

Figure 10.19

A basic hydraulic system (standard symbols).

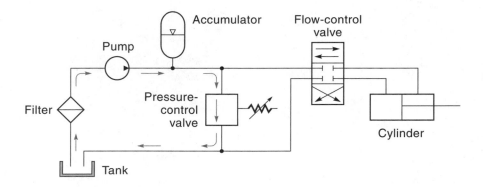

running all the time the system is on. During those times when the cylinder is not moving, the fluid from the pump, after filling the accumulator, returns to the tank through the pressure-control valve. Of special importance is the filter that removes small contaminants, which can get into the fluid. These contaminants can cause abrasive wear on the system components and reduce their lifetime considerably.

10.3 PNEUMATIC SYSTEMS

Pneumatic systems use air pressure to create mechanical motion. As shown in Figure 10.20, the basic system includes an intake filter that traps dirt before it enters the system, an air compressor that provides a source of compressed air, a dryer that removes the moisture in the air, a pressure tank that is a reservoir of compressed air, a pressure regulator that maintains air pressure, a valve that controls the air flow, and a pneumatic cylinder that creates the mechanical motion.

Figure 10.20

A basic pneumatic system.

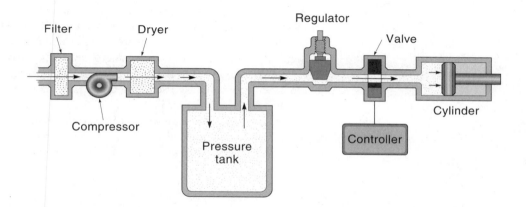

Pneumatic systems are very similar to hydraulic systems, but there are several important differences. The major functional difference is that air is compressible, whereas hydraulic fluid is not. In fact, air is so compressible that the air in the average car tire would occupy three times its volume at atmospheric pressure. To explore this concept further, consider a bicycle pump being used to inflate a tire. When you push down on the plunger, the air in the pump is compressed until the air pressure in the pump is greater than the pressure in the tire, at which point some portion of the air flows out of the pump (through a one-way check valve) into the tire. If you were to push the plunger down only halfway and let go, the compressed air in the pump cylinder would expand and push the plunger part way back out. If for some strange reason you were pumping an incompressible fluid (such as water) into your tire, you would notice a *big* difference. Pushing the plunger halfway down would not compress the water volume, so exactly half of the water in the cylinder would be expelled into the tire. And if you were to let go of the plunger in midstroke, the plunger would not rebound because water under pressure does not expand. The point of this analogy is that a pneumatic system tends to be springy, whereas a hydraulic system is rock solid. That is, the position of a hydraulic cylinder is strictly a function of the volume of fluid in the cylinder, whereas the position of a pneumatic cylinder is a function of the air pressure, which in turn depends on the resistance of the mechanical load, and the temperature. For example, if a hydraulic actuator is moved to position and stopped, the trapped fluid on each side of the piston will lock it in place; if a pneumatic actuator is moved into position and stopped, the piston *can* be displaced by simply pushing or pulling on it—the trapped air in the cylinder simply compresses somewhat. For this reason, pneumatic systems are not usually used when accurate feedback positioning is required; instead, positioning is achieved by having the piston travel back-and-forth between two mechanical stops, and enough air pressure is provided to guarantee that the piston can move any expected load.

Another difference between hydraulic and pneumatic systems is that pneumatic systems do not have to return the low-pressure air to the compressor: They simply exhaust it to the atmosphere—clearly an advantage. Another advantage is that pneumatic systems are not "messy." Hydraulic systems can and do leak; pneumatic systems also leak, but there are no spills to clean up. Also, pneumatic systems tend to be smaller and less expensive. All these qualities make them very attractive as a source of mechanical motion in cleaner, high-tech industrial environments.

A control system that uses pneumatic actuators would include a controller along with the pneumatic components shown in Figure 10.20. The controller specifies when the individual valves in the system are to be turned on and off. Usually, this controller is a digital electronic device such as a microcomputer or a programmable logic controller (PLC). However, some controllers are strictly pneumatic devices and use no electronics at all. Pneumatic controllers make use of pneumatic logic devices—actual AND, OR, and NOT gates that operate only on air pressure. The output signals from these controllers are, of course, pneumatic and are conveyed to the control valves through small tubes. Although pneumatic controllers are still used extensively in some applications (such as simple sequencers or around flammable gases), in general, industry is backing away from this concept.

Compressors, Dryers, and Tanks

The **air compressor** is a machine that pumps air from the atmosphere into a tank. There are a number of types of compressors, but one of the most common is the **reciprocating piston compressor** shown in Figure 10.21. The compressor crankshaft is driven by an external power source, typically an electric motor. As the crankshaft rotates, the piston is forced up-and-down in the cylinder. The air is drawn into the cylinder through a valve during the intake stroke [Figure 10.21(a)]; then during the exhaust stroke, the air is pushed out through another valve into the pressure tank.

Water vapor in the compressed air must be removed, or it will eventually damage the pneumatic components. Removing the moisture in the air is done by the **dryer**, of which there are several types. One type, an **aftercooler**, chills the air, causing the moisture to condense into drops, which can then be drained off. The **desiccant dryer** circulates the air through a moisture-absorbing chemical called a desiccant. When the desiccant becomes saturated, it must be changed.

The **pressure tank** receives air from the compressor and becomes the high-pressure air reservoir for the system. Figure 10.22 shows the compressor and tank system. Power to the compressor motor is controlled by a pressure switch on the tank. When the tank

Figure 10.21

A reciprocating piston compressor.

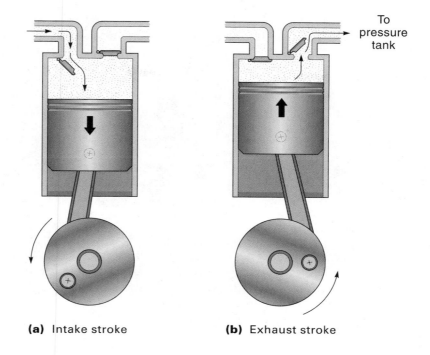

(a) Intake stroke   **(b)** Exhaust stroke

pressure falls below a set value, the switch closes, and the compressor motor starts. When the tank pressure increases to a specified value, the switch opens, turning off the compressor. One benefit of using the tank is that it tends to smooth out the pulsations of air pressure that result from a reciprocating piston compressor. The typical pressure range for pneumatic systems is 30–100 psi (200–700 kPa), although some systems go as high as 750 psi (5000 kPa).

Figure 10.22

A pressure tank system showing a pressure switch, which controls the compressor.

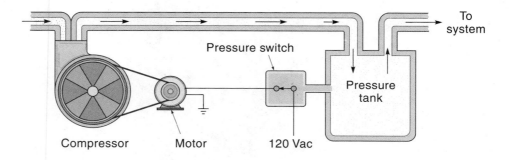

Pressure Regulators

The tank pressure can range anywhere between the high and low limits of the pressure switch. Some systems cannot tolerate this variation and hence require a pressure regulator to be installed between the tank and the system components. The **pressure regulator** can supply air at a constant pressure regardless of the source pressure *as long as the source pressure stays above the desired regulated pressure*. Figure 10.23 shows a

Figure 10.23

A pneumatic pressure regulator.

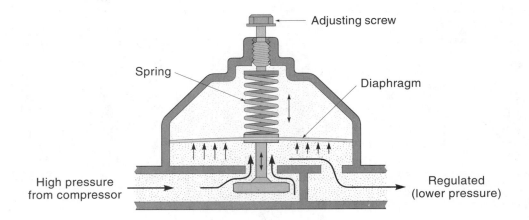

simple pressure regulator. A spring-loaded diaphragm is pushed on by regulated air pressure. If the regulated air pressure starts to fall, the reduced pressure on the diaphragm causes it to move downward, thus opening a valve and allowing more high-pressure air in.

Figure 10.24

Cutaway view of a pneumatic control valve (Courtesy of Festo Corp.).

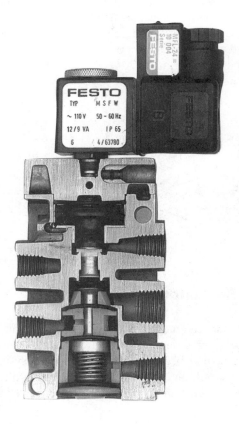

Pneumatic Control Valves

Pneumatic control valves regulate the air flow, which in most cases means on or off. Many configurations are possible, but of particular interest is the bidirectional control valve that causes a piston to move in either direction. A pneumatic bidirectional control valve is shown in Figure 10.24. It is functionally similar to the hydraulic control valve shown in Figure 10.18, but there are some differences. For instance, pneumatic control valves use *O rings* (rubber seals) to minimize internal leaks. Pneumatic valves are usually designed to be full on or completely off and are driven by either an electric solenoid or a pneumatic control signal. Finally, these valves require only one tube coming from the air supply; the "used" air from the cylinder is simply vented to the atmosphere.

Pneumatic Actuators

Pneumatic actuators convert air pressure into mechanical motion. There are two basic types: linear actuators (cylinder/piston or diaphragm types) and rotary actuators. Piston and rotary actuators are functionally similar to their hydraulic counterparts.

   Pneumatic cylinders are available in a variety of shapes and sizes (Figure 10.25). As shown in Figure 10.26, there are two basic internal configurations. The double-acting cylinder connects to the valve with two tubes and can be driven in either direction. The **single-acting cylinder** can only be driven in one direction with air pressure and is returned by a spring.

♦ **EXAMPLE 10.5**

A spring-loaded single-acting cylinder has a diameter of 2 in. and a stroke of 2 in. The return spring has a spring constant of 3 lb/in. The available air pressure is 30 psi. What force can this cylinder supply to a load at the end of its stroke?

Solution

The total force that the piston can supply can be calculated from Equation 10.1 in exactly the same manner as it was done for the hydraulic cylinder. First, we calculate the piston surface area:

$$A = \pi r^2 = 3.14 \times 1 \text{ in}^2 = 3.14 \text{ in}^2$$

Now find the force from Equation 10.1:

$$F = PA = \frac{30 \text{ lb}}{\text{in}^2} \times 3.14 \text{ in}^2 = 94.2 \text{ lb}$$

Some of this force is used to compress the spring and is *not* available to the external load. The force needed to compress the spring 2 in. is calculated as follows:

$$F = \text{spring constant} \times \text{length} = \frac{3 \text{ lb}}{\text{in.}} \times 2 \text{ in.} = 6 \text{ lb}$$

Therefore, the total available force from the cylinder is

$$F = \text{piston force} - \text{spring force} = 94.2 \text{ lb} - 6 \text{ lb} = 88.2 \text{ lb}$$

♦

Figure 10.25 Pneumatic cylinders (Courtesy of Festo Corp.).

Figure 10.26

Two types of pneumatic cylinders.

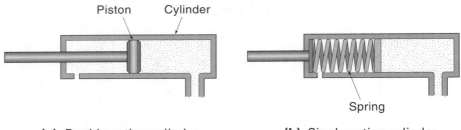

(a) Double-acting cylinder   **(b)** Single-acting cylinder

Rotary actuators convert air pressure into rotary mechanical motion. One common design is the **vane motor** (Figure 10.27). The motor consists of a rotor that is offset in a housing. Protruding from the rotor are spring-loaded vanes that seal against the housing and slide in-and-out of the rotor as it turns. Motion is achieved because the vanes on the top have more exposed surface area than those on the bottom and hence receive more force, causing the rotor to turn clockwise.

Figure 10.27
A pneumatic vane motor.

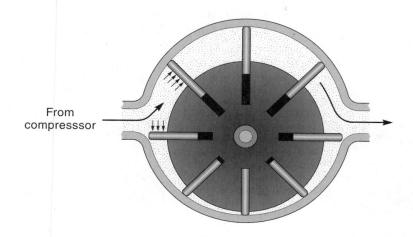

From compresssor

10.4   FLOW VALVES

One common type of actuator used in process control systems is the **control valve**, which regulates the flow of fluids. The control valve has a built-in valve-operating mechanism, allowing it to be controlled remotely by a signal from the controller. Usually, this signal is either electric or pneumatic.

Figure 10.28 shows three types of control valves. Figure 10.28(a) shows a solenoid-actuated, on–off valve. When the solenoid is energized, the valve is pulled open, and the fluid flows. When the solenoid is deenergized, a spring returns the valve to the

Figure 10.28
Flow valves.

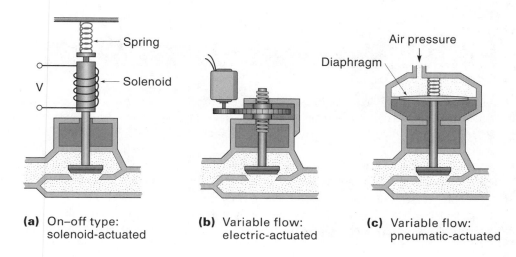

(a) On–off type: solenoid-actuated

(b) Variable flow: electric-actuated

(c) Variable flow: pneumatic-actuated

closed position. On–off valves are used in batch processes (for example, a washing machine where the tank is filled to a specified level as quickly as possible, agitated for a while, then emptied).

Many processes require the ability to vary the flow of a fluid in a pipe on a continuous basis. To do this, the valve stem must be controlled with a linear actuator of some type. Figure 10.28(b) shows an electrically operated valve. In this case, an electric motor drives a leadscrew-type valve stem, so it can be put in any position.

Pneumatically operated valves use air pressure as the control signal. Shown in Figure 10.28(c), you can see that as the air pressure is increased, the diaphragm will move down (against a spring) and close the valve. This type of valve could be used in an on–off or a variable-flow application.

SUMMARY

A linear actuator generates straight-line motion. There are three basic types of linear actuators: electric, hydraulic, and pneumatic.

Electric linear actuators use the principle of a nut unscrewing along a threaded shaft. The threaded shaft (the lead screw) is rotated by an electric motor, and the nut moves linearly. Compared with hydraulic systems, electric linear actuators are clean and self-contained but, generally speaking, are not as strong and cannot move as fast.

Hydraulic systems are used to create linear motion, but it takes more hardware. A typical hydraulic system includes a pump, a pressure-control valve, a bidirectional flow-control valve, and a piston/cylinder-type actuator. The pump takes fluid at atmospheric pressure from a tank and causes it to flow under pressure. The pressure-control valve

bleeds off some of the high-pressure fluid (back into the tank) in order to maintain a constant system pressure. The bidirectional flow-control valve directs the fluid into one end or the other of the cylinder, causing the piston to move.

Pneumatic systems are similar to hydraulic systems except that air is used instead of fluid. Because fluid is incompressible and air is not, an air cylinder is more springy, and it is difficult to control its position precisely. For this reason, most pneumatic systems use mechanical stops to determine the length of travel.

Control valves are used to regulate the flow of fluids. The control valve has a built-in valve-operating mechanism, allowing it to be controlled remotely by a signal from the controller.

GLOSSARY

accumulator Used in a hydraulic system, a spring-loaded tank connected into the high-pressure side of the system; the accumulator can store some fluid to be used at peak-demand periods.

actuator The source of energy or motion in a system; typical actuators would be electric motors or hydraulic cylinders.

aftercooler Used in pneumatic systems, a device that removes moisture from the air supply by chilling and thereby condensing any water vapor.

air compressor Used in pneumatic systems, a device that creates the supply of high-pressure air.

axial piston pump A type of hydraulic pump that uses a number of small pistons to pump the fluid, this is a variable-displacement pump because the stroke of the pistons is determined by the adjustable angle of the swash plate.

ball bearing screw A ball bearing type of a leadscrew, recirculating balls roll between the threaded shaft and the nut to remove almost all friction.

bidirectional control valve   Used in hydraulic and pneumatic systems, it can direct the fluid into either end of the actuator cylinder, allowing the piston to be moved in either direction.

control valve   Used to control the flow of fluids. A valve that has a built-in valve-operating mechanism, allowing it to be controlled remotely by a signal from the controller.

desiccant dryer   Used in hydraulic systems, a device that dries the air supply by passing the air through a water-absorbing chemical.

double-acting cylinder   A hydraulic or pneumatic cylinder and piston assembly in which the piston can be driven in either direction.

dryer   Used in pneumatic systems, a dryer that removes the moisture from the air supply.

extension rate   The rate at which a linear actuator can extend or retract; the term is usually applied to electric linear actuators.

gear motor   A simple hydraulic motor in which fluid pushes on the teeth of a set of constantly meshed gears (in a housing), causing them to rotate.

gear pump   A simple positive-displacement hydraulic fluid pump that uses two constantly meshed gears (in a housing).

hydraulic cylinder   A linear actuator powered by fluid under pressure, it consists of a cylinder, enclosed at both ends with a piston inside. The piston is moved by admitting the pressurized fluid into either end of the cylinder.

hydraulic system   A system that uses pressurized fluid (oil) to power actuators such as hydraulic cylinders.

hydrostatic pressure   The pressure that exists inside a closed hydraulic system; the significance is that the pressure is the same everywhere within the system (assuming no velocity effects).

leadscrew linear actuator   A type of linear actuator that uses an electric motor to rotate a threaded shaft; the linear motion is created by a nut advancing or retreating along the threads.

linear actuator   A prime mover that causes movement in a straight line, such as a hydraulic cylinder.

pneumatic control valve   Used in pneumatic systems, a valve that controls air flow to the actuators.

pneumatic cylinder   A linear actuator powered by compressed air, it consists of a cylinder, enclosed at both ends with a piston inside. The piston is moved by admitting the compressed air into either end of the cylinder.

pneumatic system   A system that uses compressed air to power actuators such as pneumatic cylinders.

positive-displacement pump   A hydraulic pump that expels a fixed volume of fluid for each revolution of the pump shaft.

pressure-control valve   Used in a hydraulic system, an adjustable spring-loaded valve that can maintain a set pressure in a system by bleeding off excess fluid.

pressure tank   Used in pneumatic systems, a reservoir for compressed air.

pressure regulator   Used in pneumatic systems, a regulator that maintains a specified pressure.

prime mover   A component that creates mechanical motion from some other energy source; the source of the first motion in the system.

reciprocating piston compressor   An air compressor that uses pistons and one-way valves to compress the air.

single-acting cylinder   A spring-loaded pneumatic cylinder; air pressure drives the piston in one direction, and spring pressure returns it.

solenoid   An electromechanical device that uses an electromagnet to produce short-stroke linear motion.

spool   The moving part within the bidirectional control valve.

stroke   The overall linear travel distance of a linear actuator.

swash plate   Part of the axial piston pump; when the angle of the swash plate is changed, the flow from the pump is changed.

vane pump   A hydraulic pump that uses an offset rotor with retractable vanes.

variable-displacement pump   A hydraulic pump that can be adjusted to change the flow while keeping the speed of rotation constant.

vane motor   A device that creates rotary motion from pneumatic or hydraulic pressure. The construction is similar to a vane pump.

EXERCISES

Section 10.1

1. Describe the operating principles of a leadscrew linear actuator and give the advantages/disadvantages of an electric linear actuator compared with a hydraulic linear actuator.

2. The characteristics of an electric linear actuator are given in Figure 10.4. If this actuator were lifting a 200-lb load, how long would it take to lift the load 6 in.?

3. The characteristics of an electric linear actuator are given in Figure 10.4. If this actuator were lifting a 800-lb load, how long would it take to lift the load 12 in.?

4. A 400-lb load needs to be lifted 7 in. in 5 s. Can the actuator described by Figure 10.4 do the job?

5. A solenoid application requires a stroke of 1 in. with at least 4 oz of force through the entire stroke. Select a solenoid that will do this job from the list given in Table 10.1. Coil voltage is to be 24 Vdc, intermittent duty.

6. A solenoid application requires a stroke of ¾ inch with a starting force of at least 15 oz. Select a solenoid that will do this job from the list given in Table 10.1. Coil voltage is to be 120 Vac, continuous duty.

Section 10.2

7. Calculate the force delivered by a 4-in. diameter hydraulic cylinder with a fluid pressure of 500 psi.

8. Calculate the force delivered by a 2-in. diameter hydraulic cylinder with a fluid pressure of 1000 psi.

9. Calculate the force delivered by a 10-cm diameter hydraulic cylinder with a fluid pressure of 3500 kPa.

10. Calculate the force delivered by a 6-cm diameter hydraulic cylinder with a fluid pressure of 7000 kPa.

11. A hydraulic system is operating with a pressure of 200 psi and is required to lift a load of 1400 lb. Find the cylinder diameter required (to the nearest ¼ inch).

12. A hydraulic jack has the same diagram as Figure 10.9, where the small piston has a diameter of 0.5 in. and the large piston has a diameter of 3 in. Find the output force (from the large piston) when the input force is 10 lb.

13. A hydraulic jack has the same diagram as Figure 10.9, where the small piston has a diameter of 0.35 in. and the large piston has a diameter of 4 in. Find the output force (from the large piston) when the input force is 10 lb.

14. A hydraulic jack has the same diagram as Figure 10.9, where the small piston has a diameter of 1 cm and the large piston has a diameter of 8 cm. Find the output force (from the large piston) when the input force is 45 N.

15. Draw a diagram of a complete hydraulic system, from tank to actuator, and identify each system component.

Section 10.3

16. Explain why a pneumatic system can be described as springy when compared with a hydraulic system.

17. Calculate the force delivered by a 1-in. diameter pneumatic cylinder with an air pressure of 100 psi.

18. A pneumatic system uses air pressure at 50 psi. What diameter cylinder would be required to provide a force of 75 lb?

19. A pneumatic system uses air pressure at 75 psi. What diameter cylinder would be required to provide a force of 50 lb?

20. A pneumatic system uses air pressure at 500 kPa. What diameter cylinder would be required to provide a force of 200 N?

21. A spring-loaded, single-action pneumatic cylinder has a diameter of 1.5 in. and a stroke of 4 in. The return spring has a spring constant of 3 lb/in. The available air pressure is 20 psi. What force will this cylinder exert at the end of its stroke?

22. Draw a diagram of a complete pneumatic system, from the intake filter to the actuator, and identify each of the system components.

11 Feedback Control Principles

After studying this chapter, you should be able to:

- Understand the terms and operation of a closed-loop control-system block diagram.
- Describe the basic operation of on–off control systems.
- Understand the concept and operation of a proportional control system (including bias) and calculate the error and controller output, given the system gain and inputs.
- Understand the concept of dead band and calculate the dead-band range for a proportional control system.
- Understand the concept and characteristics of integral control.
- Understand the concept and characteristics of derivative control.

- Understand the concepts and characteristics of PID control.
- Explain the circuit operation of an analog controller.
- Explain the principles of operation of a digital controller, including programming concepts and sample rate.
- Understand the concept of stability and interpret a Bode plot.
- Implement two methods of tuning a process control system.
- Explain the principles of operation and applications of fuzzy logic controllers.

INTRODUCTION

Broadly speaking, control systems can be classified into two groups: open-loop and closed-loop. In an **open-loop system** [Figure 11.1(a)], no feedback is used, so the controller must independently determine what signal to send to the actuator. The trouble with this approach is that *the controller never actually knows if the actuator did what it was supposed to do.*

In a **closed-loop system**, also known as a *feedback control system*, the output of the process is constantly monitored by a sensor [Figure 11.1(b)]. The sensor samples the system output and passes this information back to the controller. Because the controller knows what the system is actually doing, it can make any adjustments necessary to keep

Figure 11.1
Open- and closed-loop
systems.

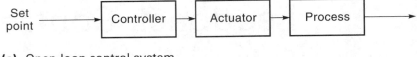

(a) Open-loop control system

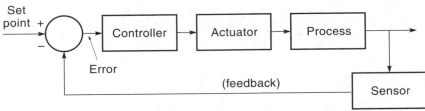

(b) Closed-loop control system

the output where it belongs. This self-correcting feature of closed-loop control makes it preferable over open-loop control in many applications. In this chapter, we deal with the principles and hardware of closed-loop controllers.

Clearly, the "heart" of the control system is the **controller**, an analog or digital circuit that accepts data from the sensors, makes a decision, and sends the appropriate commands to the actuator. In general, the controller is trying to keep the **controlled variable**—such as temperature, liquid level, position, or velocity—at a certain value called the **set point** (*SP*). A feedback control system does this by looking at the **error** (*E*) signal, which is the difference between where the controlled variable is and where it should be. Based on the error signal, the controller decides the magnitude and the direction of the signal to the actuator.

Figure 11.2 shows block diagrams of the two major classifications of feedback control systems: process control and servomechanisms.* For a **process control system** [Figure 11.2(a)], the job of the controller is to maintain a stationary set point despite disturbances—for example, maintaining a constant temperature in an oven whether the door is opened or closed. Often, in a process control system, the set point is established by another controller acting in a supervisory role. In a **servomechanism** [Figure 11.2(b)], the job of the controller is to have the controlled variable track a moving set point—for example, moving a robot arm from one position to another.

Engineers approach the problem of designing a controller for a servomechanism differently than they would for an industrial process control system. For a servomechanism such as a robot, the transfer functions of the individual system components are usually known or can be calculated. For example, the transfer functions of motors and sensors are usually provided by the manufacturer. The transfer function of mechanical components can be calculated based on the laws of physics (for example, moment of inertia) and empirical data. The response of the servomechanism can be modeled using higher mathematics, and the precise characteristics of the controller are then determined. On the other hand, for a large process control system, so many factors affect the system performance

*A word about terminology: Many different terms to describe the blocks and signals in Figure 11.2 are found in the literature. The terms selected for this text are a combination of the traditional and the modern.

Figure 11.2

Feedback-control block diagrams.

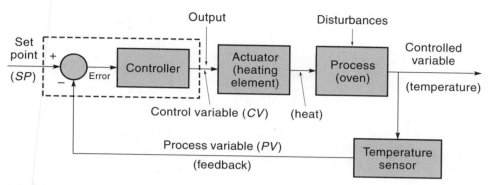

(a) Process control system (heating system)

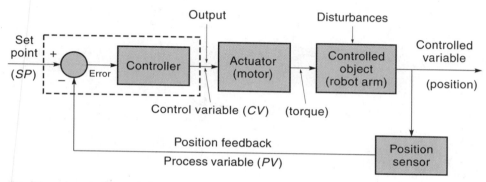

(b) Servomechanism (robot arm)

that engineers take a more empirical approach. This means that they use a general-purpose controller, and then tune it to meet the specifications of a particular system.

A new and increasingly important type of control is called fuzzy logic control. The fuzzy logic controller (Section 11.8) does not use mathematical models but mimics the skill and experience of a human operator.

11.1   PERFORMANCE CRITERIA

Performance criteria are various measurable parameters that indicate how good (or bad) the control system is. These are divided into transient (moving) and steady-state (not changing) parameters.

The exact path the controlled variable takes when going from one position to the next is called its **transient response**. Consider the behavior of the robot arm whose response is shown in Figure 11.3; it is directed to move from 0 to 30°, as shown by the dashed line. This type of command (changing instantaneously from one position to another) is called a **step change**. The actual response of the system is shown as a solid line. As you can see, there is a difference between the ideal path of the arm and the one it took. One major consideration is how fast the system picks up speed (called **rise time**). The real arm simply cannot move fast enough to follow the ideal path. Rise time (T) is usually defined as the time it takes for the controlled variable to go from 10 to 90% of the way to its new position. Another transient parameter is **overshoot**. Once the arm starts moving, its momentum will keep it going right on past where it was supposed to stop. Overshoot can be reduced by the controller but usually at the expense of a longer rise time. **Settling time** (T_s) refers to the time it takes for the response to settle down to within some small percentage (typically 2–5%) of its final value. In this case, it is the time it takes for the oscillations to die out. Rise time, settling time, and overshoot are all related; a change in one will cause a change in the others.

Figure 11.3

Transient response.

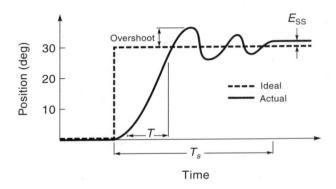

The **steady-state error** (E_{SS}) of the system is simply the final position error, which is the difference between where the controlled variable is and where it should be. In Figure 11.3, E_{SS} is shown as the position error after the oscillations have died out. This error is the result of friction, loading, and feedback-sensor accuracy. A sophisticated controller can reduce steady-state error to practically zero.

11.2   ON–OFF CONTROLLERS

Two-Point Control

Two-point control (also called **on–off control**) is the simplest type of closed-loop control strategy. The actuator can push the controlled variable with only full force or no force. When the actuator is off, the controlled variable settles back to some rest state. A good example of two-point control is a thermostatically controlled heating system. Consider a house sitting for a long time with the heat turned off and an outside temperature of 50°F. Eventually, the inside temperature would drop to 50°. This is its rest-state temperature. Now suppose the heat is turned on and the thermostat is set for an average temperature of 70°. As Figure 11.4(a) shows, the inside temperature begins to climb, rapidly at first, and then more slowly (as the heat losses increase). When the temperature reaches the 72° cutoff point, the furnace shuts down. The house temperature immediately starts to decline toward its rest state of 50°; but long before it gets there, it reaches the cut-on point of 68°, and the furnace comes back on. Notice that the temperature curve is like a charging and discharging capacitor. A larger furnace would steepen the "charging" curve, and a larger house (or poorer insulation) would steepen the "discharging" curve (because the inside temperature would fall more quickly). Notice there is a cycle time (T_{cyc}) associated with two-point control. This cycle time is affected by the capacity of the furnace and the house, as well as the temperature difference between the cut-on and cut-off points. If the limits were moved closer together—say, 69° and 71°, the temperature would be maintained closer to 70°, but the cycle frequency would increase, as illustrated in Figure 11.4(b). Generally, a high cycle rate is undesirable because of wear on motors and switches. Consequently, two-point control has only limited applications, mostly on slow-moving systems where it is acceptable for the controlled variable to move back-and-forth between the two limit points.

Figure 11.4

Temperature curve of a two-point heating system.

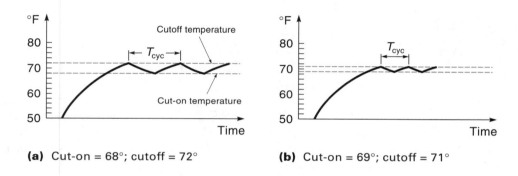

(a) Cut-on = 68°; cutoff = 72°

(b) Cut-on = 69°; cutoff = 71°

Three-Position Control

Three-position control is similar to two-point control, except in this case the controller has three states: forward-off-reverse, up-off-down, hot-off-cold, and so on. This strategy would be used in a system that has no particular self-seeking rest state. For example, consider the case of a floating oil-drilling platform that needs to stay over the wellhead on the ocean bottom, as illustrated in Figure 11.5(a). The platform must not drift more than 5 ft away from the center, or the pipe may break. Two motors, A and B, are used to keep the platform centered on the east–west axis (the north–south axis would be handled with other motors). If the platform drifts more than 5 ft east, motor A comes on and drives it back toward the center. Motor B will come on if the platform drifts more than 5 ft west. Figure 11.5(b) shows an example of the platform's east–west movement. Notice that the controlled variable (platform position) will tend to oscillate back-and-forth across the center because the motor size selected was based on worst-case conditions (strong winds). In calm weather, a brief on-period of motor A may give the platform enough momentum to drift completely across the dead zone, only to be pushed back by motor B. In fact, if the system is not designed properly, the back-and-forth oscillations could get bigger and bigger, in which case the system has gone unstable [Figure 11.5(c)]. On the other hand, if the wind is too strong for the motor, the platform will be driven outside the 5 ft limit, and the pipe will break [Figure 11.5(d)]. Three-position control is a simple control strategy that is appropriate for a limited number of applications, such as the one given.

Figure 11.5
Example of a three-position
control.

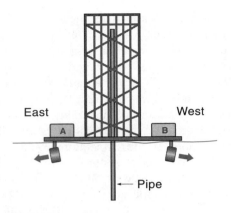

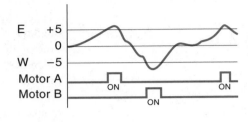

(a) Floating platform

(b) Typical operation

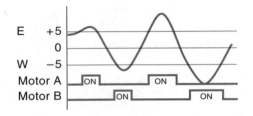

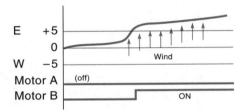

(c) Dead calm (motors too strong)

(d) High wind (motor too weak)

11.3   PROPORTIONAL CONTROL

We now consider more sophisticated control strategies that require "smart" controllers such as a microprocessor. The first and most basic of these strategies is called **proportional control**. With proportional control, the actuator *applies a corrective force that is proportional to the amount of error*, as expressed in Equation 11.1:

$$\text{Output}_P = K_P E \tag{11.1}$$

where

Output_P = system output due to proportional control
K_P = proportional constant for the system called **gain**
E = error, the difference between where the controlled variable should be and where it is

Consider the position control system shown in Figure 11.6. A robot arm is powered by a motor/gearhead. A potentiometer provides position information, which is fed back to a comparator. This feedback signal is called the **process variable** (*PV*). The comparator subtracts *PV* from the set point (*SP*) to determine the error (*E*) as expressed in Equation 11.2.

Figure 11.6
A proportional control
position system.

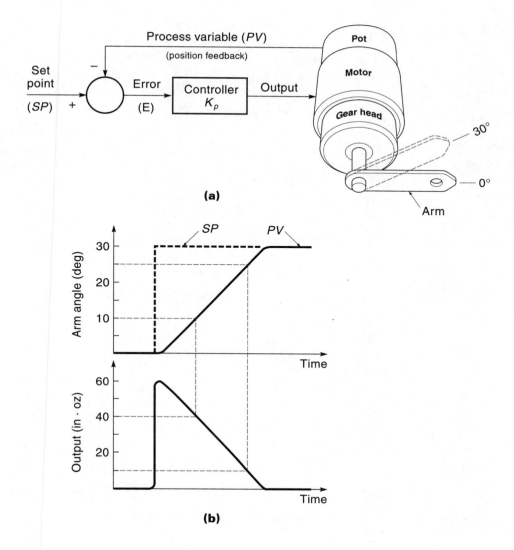

(a)

(b)

$$E = SP - PV \qquad (11.2)$$

where

E = error
SP = set point, a desired value of the controlled variable
PV = process variable, an actual value of the controlled variable

Referring to Equation 11.1, we see that the controller output is proportional to the error. This output directs the motor to move in a direction to reduce the error. As the position of the arm gets closer to the set point, the error diminishes, which causes the motor current to diminish. At some point, the error (and current) will get so small that the arm comes to a stop.

♦ **EXAMPLE 11.1**

Assume that the arm in Figure 11.6(a) was originally at 0° and then was directed to move to a new position at 30°. The gain of the system is $K_P = 2$ in. · oz/deg. Describe how the controller responds to this situation.

Solution

Originally, the arm is at rest at 0°. When the set point is first changed to 30°, an error signal of 30° results (because the arm is 30° away from where it should be):

$$\text{Error} = SP - PV = 30° - 0° = 30°$$

Using Equation 11.1, we can calculate the initial restoring torque the system would generate:

$$\text{Output}_P = K_P E = 2 \text{ in.} \cdot \text{oz/deg} \times 30° = 60 \text{ in.} \cdot \text{oz}$$

This means that the motor would initially create a torque of 60 in. · oz, causing the arm to rise rapidly. As the arm continues to rise, the output falls [Figure 11.6(b)].
    When the arm gets up to 10°,

$$\text{Error} = SP - PV = 30° - 10° = 20°$$

$$\text{Output}_P = K_P E = 2 \text{ in.} \cdot \text{oz/deg} \times 20° = 40 \text{ in.} \cdot \text{oz}$$

The torque has now reduced to 40 in. · oz, so the arm will slow down.
    When the arm gets up to 25°,

$$\text{Error} = SP - PV = 30° - 25° = 5°$$

$$\text{Output}_P = K_P E = 2 \text{ in.} \cdot \text{oz/deg} \times 5° = 10 \text{ in.} \cdot \text{oz}$$

With a torque of only 10 in. · oz, the arm is slowing way down.
    Finally, when the arm reaches the new set point of 30°,

$$\text{Error} = SP - PV = 30° - 30° = 0°$$

$$\text{Output}_P = K_P E = 2 \text{ in.} \cdot \text{oz/deg} \times 0° = 0 \text{ in.} \cdot \text{oz}$$

With the torque at 0 in. · oz, the arm stops.   ♦

Having the correcting force be proportional to the error makes sense when you consider the following argument. A large error implies there is a long way to go, and so you want some speed to get there (which requires a large torque from the motor). However, when the error is small, the arm should slow down (small torque) so as not to overshoot. Also, note that the system is bidirectional in the sense that the torque will always be applied in whatever direction is needed to reduce the error. For example, consider again the system shown in Figure 11.6 (Example 11.1). When the arm was directed to go to 30°, the controller outputted to the motor a signal corresponding to 60 in. · oz of torque. Later, if the motor was directed to return to 0°, a new negative error and torque would be generated:

$$\text{Error} = SP - PV = 0° - 30° = -30°$$

$$\text{Output}_P = K_P E = 2 \text{ in.} \cdot \text{oz/deg} \times -30° = -60 \text{ in.} \cdot \text{oz}$$

The negative sign of the output would result in a negative voltage to the motor, which would cause it to run backward. Thus, proportional control is capable of driving the arm in either direction.

If you had the opportunity to experiment with the proportional control system shown in Figure 11.6, you would notice that the arm has a springy feel. For example, if you were to push down on the arm, you would feel a restoring force, as if you were pressing on a spring. The more you displaced it, the more it would resist. What you are actually feeling is the control system's correcting force, where the resistance to movement is proportional to the error (displacement). Figure 11.7 shows how we might model this property for the motor of Example 11.1. Notice in the model that the arm is a leaf spring attached to a locked shaft. If we try to push the arm down, the leaf spring resists. The farther the arm is deflected, the greater the resistance force [Figure 11.7 (b)]. When the external force is removed, the arm is returned by the spring to the center position. When we command the controller to move the arm to a new position (say, 30°), it is like causing the entire spring to rotate, so it is now at rest in a new position and will resist being deflected from that point [Figure 11.7(c)].

Figure 11.7

Proportional control as modeled with a spring.

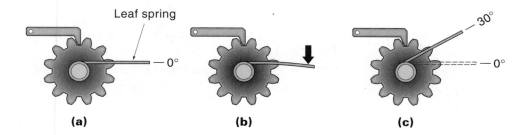

(a)   (b)   (c)

The Steady-State-Error Problem

Proportional control is simple, makes sense, and is the basis of most control systems, but it has one fundamental problem—steady-state error. In practical systems, proportional control cannot drive the controlled variable to zero error because as the load gets close to the desired position, the correcting force drops to near zero. This small force may not be enough to overcome friction, and the load comes to a stop just short of the mark.

Friction, *always present in mechanical systems*, is a nonlinear force that opposes the applied force. In fact, the load will not move at all until the friction force has been overcome. Figure 11.8(a) shows another spring model of a proportional control system. This time the load is sitting on a friction surface in the center of a frame to which it is connected with two springs. Figure 11.8(b) shows what happens when the frame quickly moves to the right. At first, the resulting push and pull from the springs drags the load to the right. As the load approaches the new center position, the declining force from the springs is overcome by friction, and the load stops *before* it gets to the new set point [Figure 11.7(c)]. The steady-state error is the distance between where the load stopped and the new set point. This region on either side of the set point, where the restoring force is incapable of precisely locating the controlled variable, is called the **dead**

band or **dead zone.** Other factors (besides friction) also contribute to the dead band, that is, backlash, flexing of mechanical parts, and poor controller design.

Figure 11.8

Friction causing steady-state error in a proportional system.

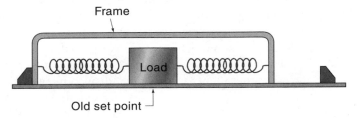

Frame

Load

Old set point

(a) Load is centered

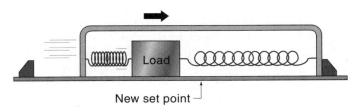

Load

New set point

(b) Load is commanded to new position by moving frame

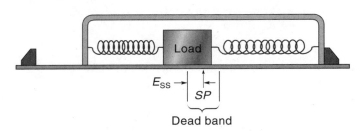

Load

E_{ss} — SP

Dead band

(c) Load is at rest in new position; friction keeps load from being centered

---

♦ **EXAMPLE 11.2**

A position control system has a gain K_P of 2 in. · oz/deg and works against a constant friction torque of 6 in. · oz. What is the size of the dead band?

Solution

To overcome the friction, the system must cause the motor to output 6 in. · oz. Because the input to the controller is the error signal, we need to find the value of error that results in an output of 6 in. · oz. Starting with the basic proportional equation (11.1),

$$\text{Output}_P = K_P E$$

Rearranging to solve for error,

$$E = \frac{\text{output}_P}{K_P} = \frac{6 \text{ in} \cdot \text{oz}}{2 \text{ in} \cdot \text{oz/deg}} = 3°$$

With 3° error on each side of the set point, dead band = 6°. ♦

Figure 11.9
The higher gain system has less steady-state error.

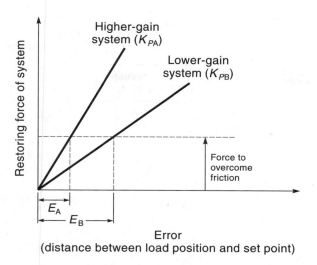

One way to decrease the steady-state error due to friction is to increase the system gain (K_P in Equation 11.1), which could be done in the model of Figure 11.8 by using stiffer springs. Consider the diagram in Figure 11.9. It shows the restoring force on the load for two different system gains, K_{PA} and K_{PB}. Notice in both cases that the restoring force is proportional to the error. The force necessary to overcome friction is shown as a dashed line. For the lower gain system, the error due to friction is shown as E_B. The higher gain system, acting like a stiffer spring, does not have to deflect as far from the set point to produce the force necessary to overcome friction. This system has less steady-state error, shown as E_A. It might seem reasonable to make the gain of every system very high; however, high K_P can lead to instability problems (oscillations). Increasing K_P independently without limit is *not* a sound control strategy.

The Gravity Problem

Another source of steady-state error is the *gravity problem*, which occurs when a constant external force is pushing on the controlled variable. Consider the robot arm lifting a weight in Figure 11.10. The set point for the arm is 0°, but when the weight is added, the arm sags. Its new position is where the restoring force of the system just balances the

Figure 11.10
How gravity causes an arm to sag.

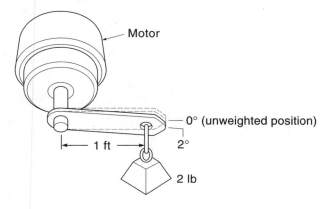

Figure 11.11

The control system for Example 11.3.

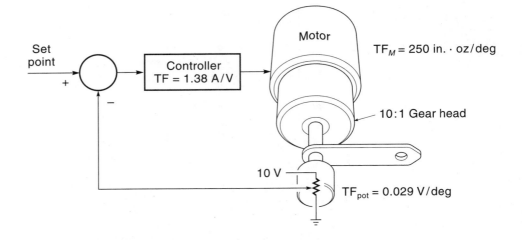

weight, in this case 2 ft. · lb. Herein is the paradox of the proportional control concept. *For the system to support the weight, there must be an error.* The reason? The proportional system only produces a restoring force when there *is* an error and the weight requires a constant force to support it. Increasing the system gain can reduce the error, but proportional control alone can never completely eliminate it.

◆ **EXAMPLE 11.3**

Specify the system gain (K_P) required to position the robot arm to within 5° of the set point (Figure 11.11). Total resistance from friction and gravity will be less than 50 in. · oz. The DC motor has a torque constant of 25 in. · oz/A with a 10 : 1 ratio gearbox built in. A 350° feedback pot is connected directly to the arm shaft.

Solution

To keep the error down to 5°, the system needs to supply at least 50 in. · oz of restoring torque when the arm is 5° from the set point. This specifies the stiffness, or gain (K_P), of the system and is shown in graph form in Figure 11.12. To calculate the required value of K_P, start with Equation 11.1:

$$\text{Output}_P = K_P E$$

$$K_P = \frac{\text{output}_P}{E} = \frac{50 \text{ in.} \cdot \text{oz}}{5°} = 10 \text{ in.} \cdot \text{oz/deg}$$

Thus, the controller must direct the motor to provide 10 in. · oz of torque for each degree that the arm is away from the set point. However, a controller cannot input degrees or output torque directly—the controller inputs and outputs voltages and currents.

Figure 11.12

The stiffness of the control system in Example 11.3.

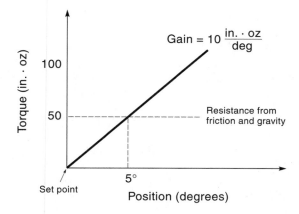

The motor by itself provides 25 in. · oz/A of torque, but because it is geared down by a factor of 10, the output shaft torque is

$$10 \times 25 \text{ in.} \cdot \text{oz/A} = 250 \text{ in.} \cdot \text{oz/A}$$

which is the steady-state motor/gearhead transfer function. Finally, the pot is connected so that 10 V = 350°; thus,

$$\frac{10 \text{ V}}{350°} = 0.029 \text{ V/deg}$$

which is the pot transfer function. To arrive at the gain required of the controller, multiply the system gain by the transfer functions:

$$K_P = \underbrace{\frac{10 \text{ in.} \cdot \text{oz}}{\text{deg}}}_{\substack{\text{Controller} \\ K_P}} \times \underbrace{\frac{1 \text{ A}}{250 \text{ in.} \cdot \text{oz}}}_{\text{Motor/gearhead}} \times \underbrace{\frac{1 \text{ deg}}{0.029 \text{ V}}}_{\text{Pot}} = 1.38 \text{ A/V}$$

Thus, the controller must act like a power amplifier that supplies 1.38 A to the motor for each volt of input. For example, if we want to move the arm to the 30° position, we would input a voltage of

$$30° \times 0.029 \text{ V/deg} = 0.87 \text{ V}$$

Assuming the arm had been at 0°, this command would cause an error signal of

$$0.87 \text{ V} - 0 \text{ V} = 0.87 \text{ V}$$

The 0.87-V error signal would be converted by the amplifier to

$$0.87 \text{ V} \times 1.38 \text{ A/V} = 1.2 \text{ A}$$

This initial value of current would drive the motor/gearhead to produce a torque of

$$250 \text{ in.} \cdot \text{oz/A} \times 1.2 \text{ A} = 300 \text{ in.} \cdot \text{oz}$$

However, as the error decreased, so would the torque. Finally, when the arm gets to 25° (5° short of 30°), the torque has dropped to 50 in. · oz, which (by design) is just enough to overcome friction.   ◆

Bias

One way to deal with the gravity problem is to have the controller add in a constant value (to its output) that is just sufficient to support the weight. This value is called the **bias** and is the same value that an open-loop system would give to the actuator. The equation for a proportional control system with bias is

$$\text{Output}_P = K_P E + \text{bias} \qquad (11.3)$$

The bias value is considered to be another input to the controller, as shown in the block diagram of Figure 11.13. A 2-lb weight places a constant torque of 2 ft · lb on the motor. A bias signal (corresponding to 2 ft · lb) is added to the output of the controller and provides the drive signal to support the weight. This system allows the error to go to zero because the static load is being supported by the bias signal, not the proportional part of the controller. However, the bias must be changed if the static load is changed.

Figure 11.13

Block diagram of control system with bias.

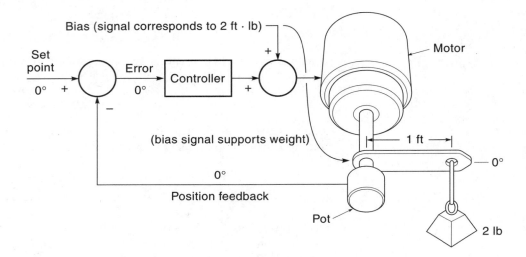

Analog Proportional Controllers

An analog controller typically uses op-amps to provide the necessary gain and signal processing. For example, consider the flow control system illustrated in Figure 11.14(a). The controller's job is to maintain the flow of a liquid through a pipe at 6 gal/min. This system consists of (1) an electrically operated flow valve, (2) a flow sensor, and (3) the analog controller. The flow valve is operated with a signal of 0–5 V, where 0 V corresponds to completely closed and 5 V is all the way open. The flow sensor provides an output signal of 0–5 V, which correspond to 0–10 gal/min. The system is designed so that a sensor voltage swing of 2.5 V (50% of its range) will cause the flow valve to swing from full off to full on. Therefore, this system has what is called a 50% **proportional band**. This percent-type specification can be translated into a gain factor (K_P) by rearranging Equation 11.1:

Figure 11.14
An analog controller for a proportional system.

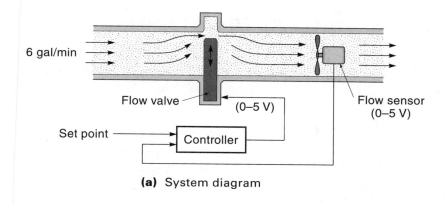

(a) System diagram

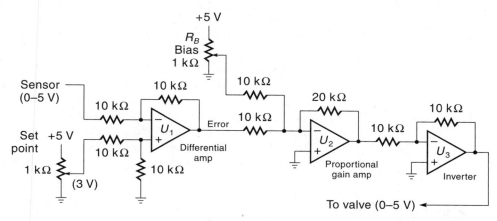

(b) Circuit diagram of an analog controller

$$\text{Output}_P = K_P E$$

$$K_P = \frac{\text{output}_P}{E} = \frac{100\%}{50\%} = 2$$

which is the proportional gain factor.

The analog controller circuit, shown in Figure 11.14(b), consists of three op-amps. The first op-amp (U_1) is acting as a differential amplifier with a gain of 1, subtracting the sensor feedback signal from the set point to create the error voltage. To maintain a flow rate of 6 gal/min, the set point must be 3 Vdc as calculated below using the flow-sensor transfer function:

$$\text{Set point} = 6 \text{ gal/min} \times \underbrace{\frac{5 \text{ V}}{10 \text{ gal/min}}}_{\text{Flow-sensor}} = 3 \text{ Vdc}$$

The output of U_1 (error signal) is fed into op-amp U_2, a simple (inverting) summing-type amplifier whose purpose is to provide the proportional gain (K_P). To make the required gain of 2, the ratio of R_f/R_i (20 kΩ/10 kΩ) is set to 2. Notice that the pot R_B can add a bias voltage to the error signal if necessary. The output of U_2 must be inverted to make the output positive; this is done with U_3, which is a simple inverting amplifier with unity gain.

11.4   INTEGRAL CONTROL

The introduction of **integral control** in a control system can reduce the steady-state error to zero. Integral control creates a restoring force that is proportional to the sum of all past errors multiplied by time, as expressed in Equation 11.4:

$$\text{Output}_I = K_I \Sigma(E\Delta t) \qquad\qquad (11.4)$$

where

Output_I = controller output due to integral control

K_I = integral gain constant (sometimes expressed as $1/T_I$)

$\Sigma(E\Delta t)$ = sum of all past errors (multiplied by time)

For a constant value of error, the value of $\Sigma(E\Delta t)$ will increase with time, causing the restoring force to get larger and larger. Eventually, the restoring force will get large enough to overcome friction and move the controlled variable in a direction to eliminate the error.

An analogy showing the power of integral control is a person who sits down in a comfortable chair to read a book. After a short time, the reader notices the dripping sound of a leaky faucet (steady-state error). The first response of the reader is to do nothing, but as time goes on the sink starts to fill up and spill over, which gets the reader's attention and he or she gets up and turns it off. The point is that the dripping (error) was not increasing, but the *effect* of the steady error was increasing with time until finally the reader (system) was motivated to do something about it.

♦ **EXAMPLE 11.4**

Consider the case where a robot arm has a steady-state error position of 2° due to friction; this error is shown on the graph in Figure 11.15(a). As time elapses, the error remains at 2°. Figure 11.15(b) shows how the restoring torque due to integral control increases with time. The magnitude of the restoring torque at any point in time is equal to the area under the error curve. For example, after 2 s the area under the error curve is 4 deg · s (2 deg × 2 s), as illustrated in Figure 11.15(a). Assuming a gain constant (K_I) of 1 in. · oz/deg · s, the restoring torque is then 4 in. · oz [Figure 11.15(b)]. After about 5 s, the restoring torque gets high enough to overcome friction and nudge the arm the last 2° to remove the error. Once the error goes to zero, the area under the curve stops growing, so the torque stops increasing; *however, it remains at the elevated level of 10 in. · oz.* This last point is important because it allows the gravity problem to be overcome as shown in Example 11.5.

♦ **EXAMPLE 11.5**

The proportional feedback system of Example 11.3 (K_P = 10 in. · oz/deg) has been modified to include integral feedback. The arm has been at rest (at the 30° position) when a weight is placed on the end of the arm, causing a downward torque of 40 in. · oz [Figure 11.16(a)]. Describe how the control system responds to the weight.

Figure 11.15

Graph showing integral control eliminating a steady-state error ($K_I = 1$).

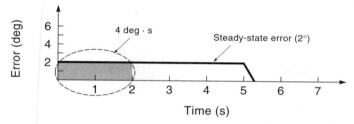

(a) Steady-state error is being reduced to zero.

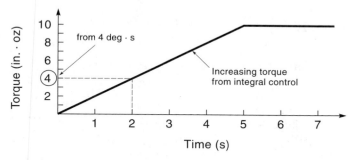

(b) Output of integral controller

Solution

The proportional gain of this system was determined to be 10 in. · oz/deg. As the following calculation shows, 40 in. · oz of torque from the weight would cause the arm to sag 4°:

$$Output_P = K_P E$$

$$E = \frac{output_P}{K_P} = \frac{40 \text{ in.} \cdot \text{oz}}{10 \text{ in.} \cdot \text{oz/deg}} = 4° \text{ sag}$$

which is shown at time A in Figure 11.16 (b).

The proportional control initially provides the entire 40 in. · oz of restoring force to support the arm, resulting in 4° of error. Because there *is* an error, however, integral control starts contributing to the restoring force. As time passes, the area under the error curve increases, and the integral control provides more and more of the 40 in. · oz needed to support the weight, while the proportional control provides less and less [shown as time B in Figure 11.16(b)]. Finally, at time C, the integral control is providing the entire 40 in. · oz, and the error is zero. A new steady state has been reached.   ◆

Integral control response can easily be observed on industrial robots. When a weight is placed on the arm, it will visibly sag and then restore itself to the original position.

Although the addition of integral feedback eliminates the steady-state-error problem, it reduces the overall stability of the system. The problem occurs because integral

Figure 11.16

The system response of proportional plus integral control (Example 11.5).

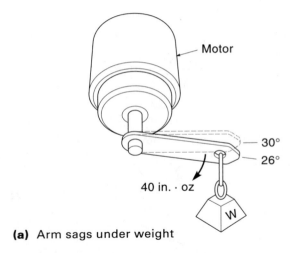

(a) Arm sags under weight

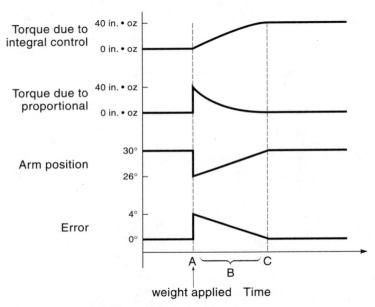

(b) Proportional and integral contributions

feedback tends to make the system overshoot, which may lead to oscillations. One example of this problem is shown in the graphs of Figure 11.17. All mechanical systems have friction, and friction is nonlinear—that is, it takes more force to overcome friction when the object is at rest than it does to keep an object moving (a phenomenon often referred to as *sticktion*). At time = 0, the system in Figure 11.17 has just moved to a new position and stopped, leaving a steady-state error. The restoring force is equal to the contribution from proportional control plus the increasing force from the integral control. For a while the object doesn't move, but finally the combined restoring force overcomes friction and the object "breaks loose." Once moving, the friction force immediately drops so some force is "left over," which goes to accelerating the object. This may cause it to overshoot, and the whole process starts again from the other side.

The problem is that the proportional–integral system has no way (other than friction) to slow the object *before* it gets to the new set point. The system must overshoot before any active braking will be applied. So, unfortunately, the addition of integral feedback solves one problem, steady-state error, but it creates others: overshoot and decreased stability. Also, the response of integral feedback is relatively slow because it takes a while for the error · time area to build up.

Figure 11.17

Integral control may cause overshoot and oscillations.

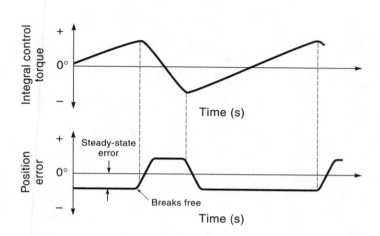

11.5   DERIVATIVE CONTROL

One solution to the overshoot problem is to include derivative control. **Derivative control** "applies the brakes," slowing the controlled variable just before it reaches its destination. Mathematically, the contribution from derivative control is expressed in the following equation:

$$\text{Output}_D = K_D \frac{\Delta E}{\Delta t} \tag{11.5}$$

where

Output_D = controller output due to derivative control

K_D = derivative gain constant

$\dfrac{\Delta E}{\Delta t}$ = error rate of change (slope of error curve)

Figure 11.18

Derivative control
contribution showing
boosting.

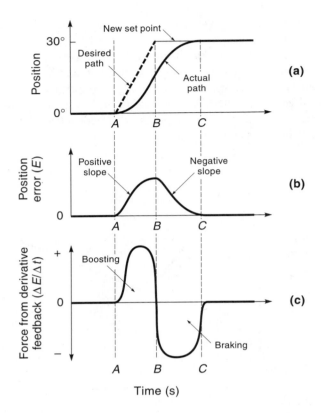

Figure 11.18 shows how a position control system with derivative feedback responds to a set-point change. Specifically, Figure 11.18(a) shows the actual and desired position of the controlled variable, Figure 11.18(b) shows the position error (E), and Figure 11.18(c) shows the derivative control output. Assume the controlled variable is initially at $0°$. Then at time A the set point moves rapidly to $30°$. Because of mechanical inertia, it takes time for the object to get up to speed. Notice that the position error (E) is increasing (positive slope) during this time period (A to B). Therefore, derivative control, which is proportional to error slope, will have a positive output, which gives the object a boost, to help get it moving. As the controlled variable closes in on the set-point value (B to C), the position error is decreasing (negative slope), so derivative feedback applies a negative force that acts like a brake, helping to slow the object.

For process control systems, where the set point is usually a fixed value, derivative control helps the system respond more quickly to load changes. For example, consider a controller that maintains a constant liquid level in a tank. If there is a rapid drop in liquid level (positive-slope error curve), derivative control responds by opening the inlet valve wider than proportional control alone would. Then, when the level is almost restored to the set point and the error is decreasing (negative-slope error curve), derivative control helps shut off the valve.

From the discussions so far, you can see that derivative control improves system performance in two ways. First, it provides an extra boost of force at the beginning of a change to promote faster action; second, it provides for braking when the object is closing in on the new set point. This braking action not only helps reduce overshoot but also tends to reduce steady-state error.

Naturally, the influence of derivative control on the system is proportional to K_D (in Equation 11.5); because derivative feedback improves overall system response, you might think that K_D should be as high as possible. However, like so many things in life, too much of a good thing sometimes brings problems. In this case, too much derivative feedback will slow the system response and magnify any noise that may be present; what is needed is a balance between too much and too little. In the next section, we will describe methods to arrive at a numerical value for K_D. It is important to note that *derivative control has no influence on the accuracy of the system, just the response time, so it is never used by itself.*

11.6   PROPORTIONAL + INTEGRAL + DERIVATIVE CONTROL

Most control systems use a combination of the three types of feedback already discussed: Proportional + Integral + Derivative (PID) Control. The foundation of the system is proportional control. Adding integral control provides a means to eliminate steady-state error but increases overshoot. Derivative control increases stability by reducing the tendency to overshoot. The response of the PID system can be described by Equation 11.6, which simply adds together the three components required:

$$\text{Output}_{PID} = K_P E + K_I \Sigma(E\Delta t) + K_D \frac{\Delta E}{\Delta t} \qquad (11.6)*$$

where

$$\text{Output}_{PID} = \text{output of the PID controller}$$
$$K_P = \text{proportional control gain}$$
$$K_I = \text{integral control gain (often seen as } 1/T_I)$$
$$K_D = \text{derivative control gain (often seen as } T_D)$$
$$E = \text{error (deviation from the set point)}$$
$$\Sigma(E\Delta t) = \text{sum of all past errors (area under the error} \cdot \text{time curve)}$$
$$\Delta E/\Delta t = \text{rate of change of error (slope of the error curve)}$$

---

* The traditional form of the PID equation is

$$\text{output}_{PID} = K_P E + \frac{1}{T_I} \int E dt + T_D \frac{dE}{dt}$$

Comparing this with Equation 11.6, you can see that the integral symbol ($\int$) is another way of saying "the sum of" and the derivitive dE/dt means "change in E divided by change in t."

Analog PID Controllers

The controller that implements the PID equation (11.6) can be either analog or digital, but the trend is toward digital. Figure 11.19 shows a straightforward version of an analog PID controller that uses five differential amplifiers. The first differential amplifier (U_1) subtracts the feedback from the set point to produce the error signal. Op-amps U_2, U_3, and U_4 are configured to be the proportional, integrator, and differentiator amplifiers, respectively. They produce the values for the three terms in the PID equation. The final amplifier (U_5) sums the three terms to produce the output. The capacitor C_I on the integrator accumulates the error in the form of charge, and the capacitor C_D of the differentiator passes only the *change* in error. The constants K_P, K_I, and K_D are selected by adjusting R_1, R_2, and R_3, respectively.

Figure 11.19

An analog PID controller.

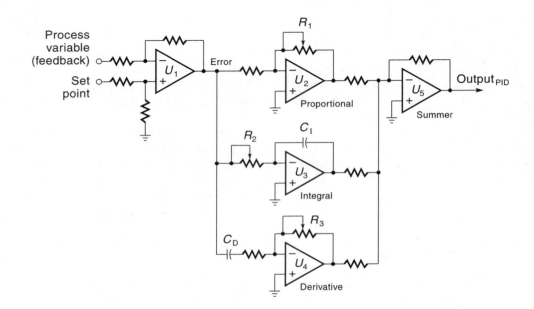

The circuit in Figure 11.19 can exactly implement the PID equation, but practical considerations usually complicate the picture. For example, all real amplifiers have an upper and lower limit at which points they become nonlinear. Large or rapidly changing error signals may cause either the integrator or differentiator amplifier to saturate. Should that happen, its output will temporarily dominate the output signal and may cause the system to go unstable.

One problem associated with the integrator is **windup**. This occurs when a system is subjected to a large disturbance, and the proportional controller (or actuator) in its attempt to correct the problem saturates "full on." Because the system can't provide as much output as is really needed, the error condition lasts longer than it theoretically should, but all this time the integrator is accumulating. Consequently, when the error is finally reduced, the large accumulated integral factor may cause the controlled variable to overshoot. A solution to this problem is to have the integral control section disconnected when the system is saturated.

Another problem with integral control is that a true integrator will sum *all* the error · time area (since the beginning of time). It has been demonstrated that a better system results if the integrator slowly "forgets" the effects of errors from the distant past. This can be done by allowing the charge to slowly leak off capacitor C_I in Figure 11.19.

A problem that can occur with derivative control is as follows: In a real control system, the set point is usually stepped up or down in discrete steps. A step change has an infinitely positive slope, which will saturate the derivative function. A solution to this problem is to base the derivative control on the feedback signal alone (*PV*) instead of the error because the controlled variable (be it temperature, position, or the like) can never actually change instantaneously, even if the set point does. The PID equation for this modified system is

$$\text{Output}_{\text{PID}} = K_P E + K_I \, \Sigma(E\Delta t) + K_D \frac{\Delta PV}{\Delta t} \qquad (11.7)*$$

where

$$\text{Output}_{\text{PID}} = \text{controller output}$$
$$K_P, K_I, K_D = \text{gains for proportional, integral, and derivative}$$
$$E = \text{error } (SP - PV)$$
$$PV = \text{process variable (feedback from sensor)}$$

A practical controller must account for various real-world problems such as those just discussed, but typically the nature and extent of design modifications required are not known until the actual system hardware is tested. Changing the characteristics of the analog controller may require component changes and/or rewiring. A distinct advantage of the digital controller, which we discuss next, is that its response is governed by software, which is relatively easy to change, even at the last minute.

Digital PID Controllers

A PID digital controller is essentially a computer, most likely microprocessor-based. The controller executes a program that performs the same series of operations over and over again. First, the computer inputs values of set point (*SP*) and process variable (*PV*). Then it uses these data to solve the system equation (11.6 or 11.7). Finally, it outputs the result to the actuator. One distinct advantage of the digital system is that control strategies and parameter constants can be changed or fine-tuned by simply modifying the software.

The following is a description of how the PID control equation (11.6) could be implemented with a microprocessor-based controller. The first term of the PID equation ($K_P E$) is simply a multiplication. Most newer microprocessors have a multiply instruction; if not, a multiplication subroutine would be used. Next comes the integral term $K_I \Sigma(E\Delta t)$. As has been discussed, this term represents the area under the error curve. Figure 11.20 shows a sample error versus time curve. The controller samples the error

---

*The traditional form of Equation 11.7 is

$$\text{Output}_{\text{PID}} = K_P E + \frac{1}{T_I} \int E \, dt + T_D \frac{dPV}{dt}$$

where $1/T_I = K_I$ and $T_D = K_D$.

Figure 11.20

A computer can approximate the area under the curve with rectangles.

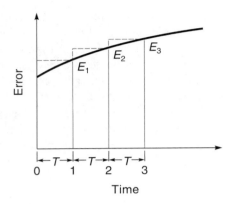

at regular intervals of time (T). A reasonable approximation of the total area under the curve is the sum of the rectangles, where the area of each rectangle is the product of T times the error E. This can be expressed in equation form as follows:

$$K_I \Sigma (E\Delta t) = K_I E_1 T + K_I E_2 T + K_I E_3 T \tag{11.8}$$

where

K_I = integral gain
E_1 = error at time 1
E_2 = error at time 2, and so on
T = time between the samples

Equation 11.8 is called a *difference equation*, and it is very easy for a computer to evaluate. The program simply records the error value (E) at fixed intervals of T and keeps a running total of the $K_I ET$ values. At any time, the integral term $K_I \Sigma (E\Delta t)$ equals this total. Figure 11.21 shows the flowchart for this segment of the program.

The third term of the PID equation is the derivative component. The derivative, or rate of change, is actually the slope of the error curve. As shown in Figure 11.22, the slope is a function of the two most recent error values and the sample time (T):

$$\text{Slope} = \frac{\Delta E}{\Delta t} = \frac{(E_3 - E_2)}{T}$$

Figure 11.21

Flowchart to compute the integral term [$K_I \Sigma (E\Delta t)$].

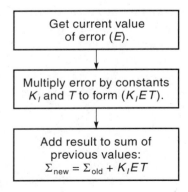

Figure 11.22

A computer can approximate $\Delta E/\Delta t$ by finding the slope.

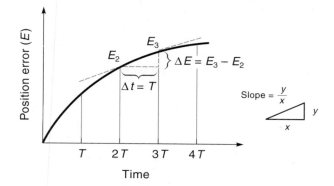

The derivative term in the PID equation (11.6) is formed by multiplying the slope by the constant K_D:

$$K_D \frac{\Delta E}{\Delta T} = \frac{(K_D E_3 - K_D E_2)}{T} \tag{11.9}$$

Equation 11.9 is another difference equation that is easily evaluated by the computer. The program would subtract the value of the older error term ($K_D E_2$) from the newer error term ($K_D E_3$) and then divide the difference by T. Figure 11.23 shows the flowchart for this segment of the program.

Figure 11.23

Flowchart to compute the derivative term $\left(K_D \frac{\Delta E}{\Delta t}\right)$.

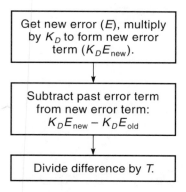

Having shown how the individual terms of the PID equation can be computed, we can now put them together and present Figure 11.24, a flowchart of the entire program. As you can see, the general format of the controller program is an endless loop. For each pass through the loop (called an **iteration** or **scan**), three basic operations are performed:

1. Read in the set point (*SP*) and sensor data (*PV*).

2. Calculate the individual contributions of proportional, integral, and derivative control and then sum these to form the PID output.

3. Send the PID output to the actuator.

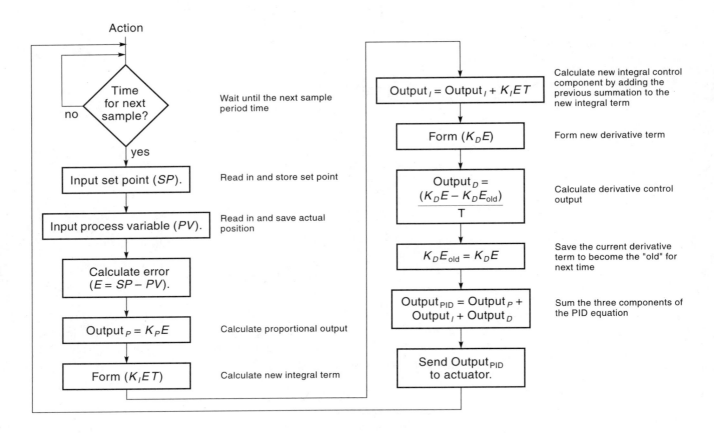

Figure 11.24   Flowchart for computing the PID equation.

Stability

A stable system is one where the controlled variable will always settle out at or near the set point. An unstable system is one where, under some conditions, the controlled variable drifts away from the set point or breaks into oscillations that get larger and larger until the system saturates on each side. Figure 11.25 shows the response of stable and unstable systems. An unstable system is clearly not under control and may in fact be dangerous if large machines are involved.

There are many examples of the "oscillating" kind of instability in everyday life. Imagine holding a large wooden tray in your hands and trying to keep a golf ball (which is rolling about on the tray) centered at a certain spot (Figure 11.26). If the ball is on the right, you would lower the left side of the tray, but that causes the ball to roll past the center; now lower the right side of the tray to bring it back, and the ball rolls past center to the right, and so on. The ball ends up oscillating back-and-forth across the tray.

Figure 11.25

Comparing a stable system and an unstable system.

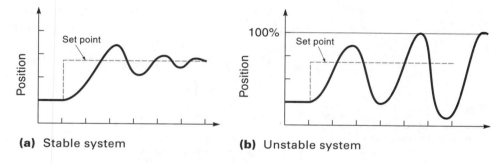

(a) Stable system   **(b)** Unstable system

Figure 11.26

A ball on a tray that shows how control systems tend to oscillate.

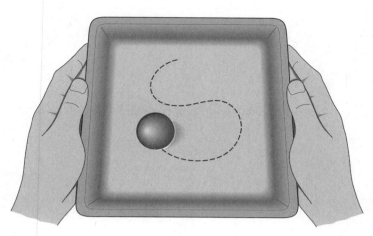

A primary reason control systems go into oscillation is because of phase lag caused by **dead time** or backlash, the interval between when the correcting signal is sent and when the system responds. Consider the case of driving at a moderate speed down a narrow lane. You would be making continuous, small back-and-forth adjustments to the steering wheel in order to stay in the middle of the lane. Now imagine yourself driving down the same road but in a car with a lot of backlash in the steering wheel—in other words, there is half a turn of free play in the steering wheel before the car actually starts turning. If the car starts to drift toward the right, you turn the steering wheel toward the left; with all that free play, however, nothing happens initially, so you turn faster. Eventually, the slack is taken up, but by this time you are turning the steering wheel so fast that the car veers to the left, so you start frantically turning the wheel to the right, and so on.

Closed-loop control systems employ negative feedback, which means that the controller is always pushing the system in the exact opposite direction of the error displacement. If the controlled variable happens to be oscillating, then the controlling force *should* be lagging by 180° because 180° is the opposite direction (for a rotational system). If there is any dead time in the system, the controller's response will lag even more than 180°. In other words, in an oscillating system, *lag time causes phase lag*, but the amount of phase lag depends on the frequency. For example, consider the case of a control system that is cycling at a frequency of 1 cycle per minute. If there is a 1-s lag between the controller output and the system response, then the delay is not very significant (as shown in Figure 11.27(a)). However, if the entire cycle time of the system is only 2 s, then a lag time of 1 s is very significant because the delay adds another 180° of phase lag, making a total lag of 360°. As Figure 11.27(b) shows, 360° lag is the same as being in phase, which means the system has positive feedback.

Figure 11.27

How system lag can make the position of the controlled variable be in phase with the controller output.

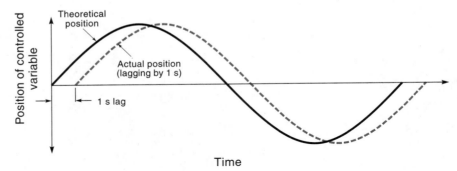

(a) System with long cycle time; so 1-s lag is not significant.

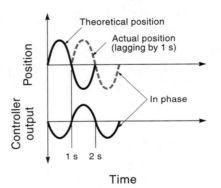

(b) System with 2-s cycle time; 1-s lag causes 180° lag.

Positive feedback occurs when the controller output is in phase with the motion of the controlled variable so that, instead of applying a correcting force, the controller is simply helping the controlled variable go in any direction it wants. It's only a matter of time before some disturbance will cause the controlled variable to be pushed off center, and when it does, positive feedback will take over (if the system gain is at least 1) and push the controlled object to the edge of its range. Therefore, to guarantee that a system will be stable, *the gain must be less than 1 for any oscillating frequency where the lag causes an additional 180° of phase shift.*

A **Bode plot** is a graph that can help determine if a system is stable or not. As shown in Figure 11.28, the Bode plot is actually two curves sharing the same horizontal axis, which is frequency (in radians/second). The top graph shows how the open-loop gain varies with frequency, and the lower graph shows how the phase lag varies with frequency. The curves shown in Figure 11.28 are fairly typical in shape. We can see that the system represented by the Bode plot of Figure 11.28 is stable because, when the phase lag is 180°, the gain is less than 1 (0.3 to be exact). Two terms quantify the stability of a system: the gain margin and the phase margin. The **gain margin** is the gain safety margin (the difference between the actual gain and unity) taken at the frequency that causes 180° of phase shift. For the system shown in Figure 11.28, the gain margin is 0.7 $(1 - 0.3 = 0.7)$. The **phase margin** is the phase safety margin taken at the frequency that causes a gain of 1. For the system shown in Figure 11.28, the phase margin is 70° $(180° - 110° = 70°)$.

Figure 11.28

A Bode plot showing stability considerations.

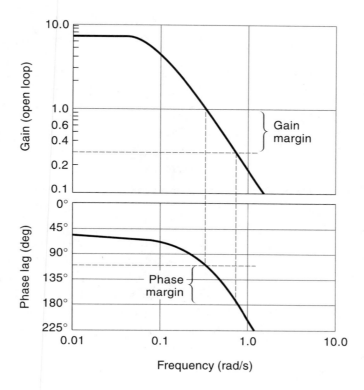

Tuning the PID Controller

The method of arriving at numerical values for the constants K_P, K_I, and K_D depends on the application. Traditionally, PID control was applied to process control systems. However, with the advent of small, fast, off-the-shelf PID modules, PID control is being applied to position control systems (such as robots) as well. In either case, a practical step-by-step procedure can be used to arrive at the PID constants. First, the constants K_P, K_I, and K_D are set to initial values, and the controller is connected to the system. The system could consist of the actual hardware or a computer simulation of same. Then the system is operated, and the response is observed. Based on the response, adjustments are made to K_P, K_I, and K_D, and the system is operated again. This iterative process of adjusting each constant in an orderly manner until the desired system response is achieved is called **tuning**. To make the system stable under all conditions, certain modifications may be necessary to the basic PID equation. Although many methods of tuning PID controllers exist, two of the most common were developed by Zieler and Nichols and are called the continuous-cycle method and the reaction-curve method.

The **continuous-cycle method** (closed-loop method) can be used when harm isn't done if the system goes into oscillation. This method will yield a system with a quick response, which means a step-function input will cause a slight overshoot that settles out very quickly [Figure 11.29(b)]. The tuning procedure is as follows:

Figure 11.29
Waveforms for the continuous-cycle method.

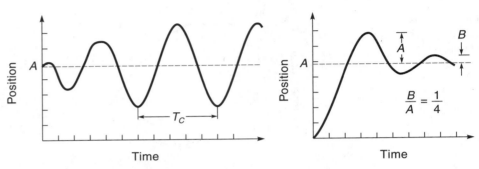

(a) System as forced into oscillation   (b) Resulting response after tuning

1. Set $K_P = 1$, $K_I = 0$, and $K_D = 0$ and connect the controller to the system.
2. Using manual control, adjust the system until it is operating in the middle of its range. Then increase the proportional gain (K_P') while forcing small disturbances to the set point (or the process) until the system oscillates with a constant amplitude, as shown in Figure 11.29(a). Record the K_P' and T_C for this condition.
3. Based on the values of K_P' and T_C from step 2, calculate the initial settings of K_P, K_I,* and K_D* as follows:

$$K_P = 0.6\ K_P' \tag{11.10}$$

$$K_I = \frac{2}{T_C} \tag{11.11}$$

$$K_D = \frac{T_C}{8} \tag{11.12}$$

4. Using the settings from step 3, operate the system, note the response, and make adjustments as called for. Increasing K_P will produce a stiffer and quicker response, increasing K_I will reduce the time it takes to settle out to zero error, and increasing K_D will decrease overshoot. Of course, K_P, K_I, and K_D do not act independently, so changing one constant will have an effect across the board on system response. Tuning the system is an iterative process of making smaller and smaller adjustments until the desired response is achieved [see Figure 11.29(b)].

---

*K_I is sometimes expressed as $1/T_I$, and K_D is sometimes expressed as I_D.

♦ **EXAMPLE 11.6**

A control system is to be tuned using the continuous-cycle method. Initial settings were $K_P = 1$, $K_I = 0$, and $K_D = 0$. By experiment it was found that the system first went into constant amplitude oscillations when $K_P' = 4$. Figure 11.30 shows the system response during the experiment. Determine a first-cut set of values for K_P, K_I, and K_D.

Figure 11.30

Output of a test system (Example 11.6).

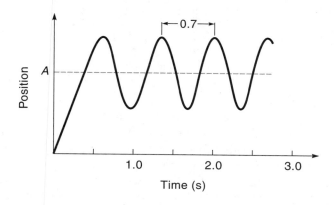

Solution

From the response graph, we see that the period of oscillation is about 0.7 s. Therefore, $T_C = 0.7$ s, and we can calculate the parameters K_P, K_I, and K_D using Equations 11.10, 11.11, and 11.12, respectively:

$$K_P = 0.6\, K_P' = 0.6 \times 4 = 2.4$$

$$K_I = \frac{2}{T_C} = \frac{2}{0.7\,\text{s}} - 2.9/\text{s}$$

$$K_D = \frac{T_C}{8} = \frac{0.7\,\text{s}}{8} = 0.09\,\text{s}$$

♦

The **reaction-curve method** (open-loop method) is another way of determining initial settings of the PID parameters. This method does not require driving the system to oscillation. Instead, the feedback loop is opened, and the controller is manually directed to output a small step function to the actuator. The system response, as reported by the sensor, is used to calculate K_P, K_I, and K_D. Note that the actuator, the process itself, and the sensor are operational in this test, so their individual characteristics are accounted for. Because the loop is open, this procedure will work only for systems that are inherently stable.

Figure 11.31

Test setup and waveform for the reaction-curve method.

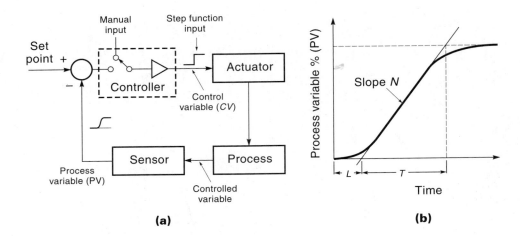

(a)   **(b)**

One test possibility is shown in Figure 11.31(a). Here the loop was opened by placing the controller in the manual mode, then a small step function was manually introduced. This signal caused the controlled variable to move slightly, and the resulting position response was recorded. A typical response curve is shown in Figure 11.31(b). Note that the vertical axis corresponds to the range of the process variable (in percentage). The system constants are calculated based on the response curve, as outlined below:

1. Draw a line tangent to the rising part of the response curve. This line defines the lag time (L) and rise time (T) values. **Lag time** is the time delay between the controller output and the controlled variable's response.

2. Calculate the slope of the curve:

$$N = \frac{\Delta PV}{T} \tag{11.13}$$

where

N = slope of the system-response curve
ΔPV = change in the process variable, as reported by the sensor (in percentage)
T = rise time, from response curve

3. Calculate the PID constants :

$$K_P = \frac{1.2\Delta CV}{NL} \tag{11.14}$$

where

ΔCV = percent step change the in control variable (output of controller)
N = slope, as determined by Equation 11.13
L = lag time [see Figure 11.31(b)]

$$K_I = \frac{1}{2L} \tag{11.15}$$

$$K_D = 0.5\,L \tag{11.16}$$

♦ **EXAMPLE 11.7**

The reaction-curve method is used to tune a PID control system. The system was turned on and allowed to reach a steady-state middle position. Then the controller was placed in the manual mode and directed to output a small step-function signal. The system response (from the sensor) was recorded. Figure 11.32 shows the step signal from the controller (CV) and the resulting response (PV).

Figure 11.32
Waveform for Example 11.7.

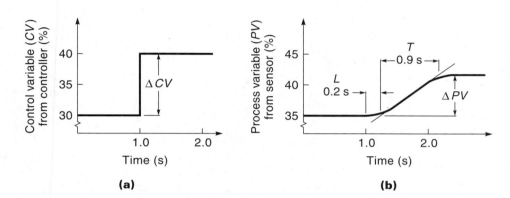

(a)   (b)

Solution   From Figure 11.32(a), we see that the signal from the controller (CV) was a step of 10% (40% − 30% = 10%). Figure 11.32(b) shows the system response (PV). The PV went from 35 to 42% for a change of 7%. After constructing the tangent line on Figure 11.32(b), we read from the graph the values of L and T:

$$L = 0.2 \text{ s}$$

$$T = 0.9 \text{ s}$$

Next we calculate the slope of the response from Equation 11.13:

$$N = \frac{\Delta PV}{T} = \frac{7\%}{0.9 \text{ s}} = 7.8\%/\text{s}$$

Now we can calculate the PID constants by applying Equations 11.14–11.16, respectively:

$$K_P = \frac{1.2\Delta CV}{NL} = \frac{1.2 \times 10\%}{7.8\%/\text{s} \times 0.2 \text{ s}} = 7.7$$

$$K_I = \frac{1}{2L} = \frac{1}{2 \times 0.2 \text{ s}} = 2.5/\text{s}$$

$$K_D = 0.5L = 0.5 \times 2 \text{ s} = 0.1 \text{ s}$$

These values would be used as initial settings; the final settings would come from fine-tuning, as discussed earlier.   ♦

Sampling Rate

In a digital control system, **sample rate** is the number of times per second a controller reads in sensor data and produces a new output value. Generally speaking, the slower the sample rate, the less responsive the system is going to be because the controller would be always working with "old" data—data that were current when the last sample was taken. Another problem with a slow sample rate is that the controller may simply not be able to keep up with what's going on. Shannon's sampling theorem states that the sampling rate must be at least twice the highest frequency being monitored or else **aliasing** can occur, when the collection of sampled data is not sufficient to re-create the original signal. Figure 11.33 shows this condition where the sample rate was slightly less than twice the frequency. Notice that the sample values would suggest a frequency one-third of what the actual frequency was. Of course, in a control system, the controlled variable is not usually tracing out a sine wave; it is more likely to be some random nonlinear motion. How would Shannon's sampling theorem be applied in such a case? To answer this question, we must invoke Fourier's theorem: *Any* function, sinusoidal or not, can be expressed as the sum of sine waves. It turns out that the more nonsinusoidal a signal becomes (say, approaching a square wave), the more significant higher frequencies it contains. This means that, to accurately follow a random path, a control system would have to sample at least twice as fast as the highest component frequency. In practice, a sampling rate of at least ten times the highest (visually apparent) frequency in the system is usually sufficient.

Figure 11.33
Waveforms showing aliasing.

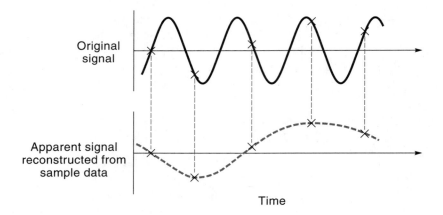

Original signal

Apparent signal reconstructed from sample data

Time

In most systems, sampling is done once at the beginning of each pass through the program loop (that is, one sample for each iteration). Therefore, the sampling rate is determined by how long it takes for the loop to be executed. Also, certain hardware functions take time, like an analog-to-digital conversion (typically, 100 µs or less for 8 bits). Figure 11.34(a) illustrates the sampling cycle; Figure 11.34(b) shows one approach to speed up the sampling rate. In this case, the analog-to-digital conversion occurs at the same time as processing occurs. Although this increases the sample rate, it does not reduce the *throughput time,* the time for any individual sample to pass through the system.

Figure 11.34
Data-sampling cycles.

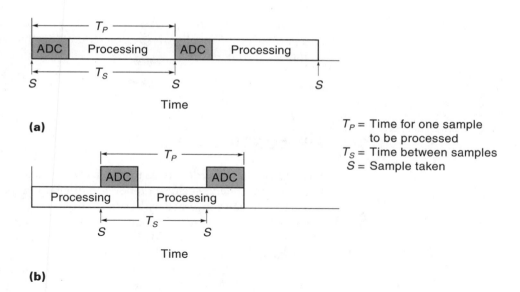

T_P = Time for one sample
to be processed
T_S = Time between samples
S = Sample taken

(a)

(b)

◆ EXAMPLE 11.8

A microprocessor-based control system runs at a clock speed of 1 MHz. The system uses an 8-bit ADC with a 100-μs conversion time. The program loop that processes the analog-to-digital input requires 55 instructions with an average execution time of 4 clocks/instruction. If analog-to-digital conversion is not overlapped with processing, what is the maximum sample rate? What is the highest frequency that this system can monitor?

Solution   Analog-to-digital conversion time = 100 μs

$$\text{Processing time} = (55 \text{ instructions}) \times \frac{(4 \text{ clocks})(1 \text{ μs})}{\text{instruction}}$$

$$= 220 \text{ μs}$$

$$\text{Sample time} = 100 \text{ μs} + 220 \text{ μs} = 320 \text{ μs}$$

$$\text{Maximum sample rate} = \frac{1}{320 \text{ μs}} = 3.125 \text{ kHz}$$

With a sample rate of 3.125 kHz, this system should be able to track a controlled variable with a periodic motion of 312.5 kHz (3.125 kHz/10 = 312.5 Hz).   ◆

Autotuning

Autotuning is the capability of some digital controllers to monitor their own output and make minor changes in the gain constants (K_P, K_I, and K_D). For example, a temperature-control system might decrease its proportional gain slightly if the overshoot is beyond a certain threshold and increase the gain if the response is too slow. Like manual tuning, autotuning is an iterative process, but because it is ongoing, the system can adapt to changes in the process. For this reason, controllers that use autotuning are known as **adaptive controllers**.

11.7   PIP CONTROLLERS

The set point has been defined as the place where you want the controlled variable to be. In a dynamic system, such as a robot arm, the desired position is a moving target, in which case we are concerned with path control. Further, the desired path between two points may not be a straight line. For example, a welding robot needs to follow the path of the seam. There are two ways to implement path control: the "carrot-and-horse" approach and the **feedforward**, or PIP approach.

The servomechanisms discussed so far have all used the carrot-and-horse idea: The controlled variable (horse) is always trying to catch up to the moving set point (carrot). The controller has information about only the past and the present, not where it is going, which is a severe handicap to place on the system. It is like a race-car driver, speeding around the track, who can only see one foot in front of the car. Normally, the driver can see in advance if the road changes direction and make the necessary adjustments to cause a smooth turn. Without the ability to look ahead, the driver would have to go very slow or risk driving off the road (overshoot) on a sharp turn.

A Proportional + Integral + Preview (**PIP**) controller is a system that incorporates information of the future path in its current output. Many systems have this information available—either the entire path is stored in memory or the system is equipped with a preview sensor as illustrated in Figure 11.35 for a welding robot. The equation for a simple PIP system follows:

$$\text{Output} = K_P E + K_I \Sigma (E \Delta t) + K_F (P_{T+1} - P_T) \qquad (11.17)$$

where

$$K_P = \text{proportional gain constant}$$
$$K_I = \text{integral gain constant}$$
$$K_F = \text{feedforward gain constant}$$
$$E = \text{error } (SP - PV)$$
$$P_T = \text{position it should be in } now$$
$$P_{T+1} = \text{position it should be in, in the future (at } T + 1)$$

Notice that the feedforward term, $K_F (P_{T+1} - P_T)$, is proportional to the difference between where the controlled object is and where it must be in the future. If this number is large, the system has a long way to go and should speed up. If the number is small or zero, the system will be stopping and so should begin slowing down. In each case, the value of the feedforward term is added to the controller output so that the controlled object is pushed with more or less force, depending on where it has to be in the near future. Figure 11.36 shows the path of two control systems, one with feedforward and one without. By anticipating the change in direction, the PIP system can begin slowing ahead of time and minimize overshoot.

Figure 11.35

A welder using a preview sensor for PIP control.

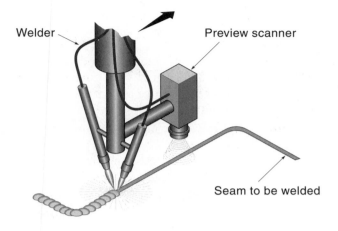

Welder

Preview scanner

Seam to be welded

Figure 11.36

Improved path control with feedforward.

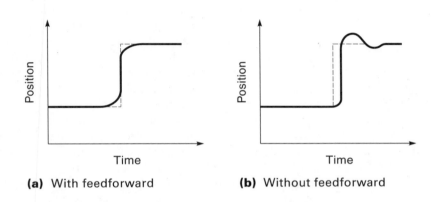

(a) With feedforward   (b) Without feedforward

11.8   FUZZY LOGIC CONTROLLERS

Introduction

Fuzzy logic, a relatively new concept in control theory, is simply the acceptance of principles that have existed since the beginning of time: *Real-world quantities are not usually "all or nothing" or "black and white" but something in-between*. For example, if you are eating an apple bite-by-bite, at what point does it stop being an apple? If you are driving with the windows open at dusk, at what point does "refreshingly cool" turn into "a little chilly"? After the sun sets, at what point would you turn on the headlights? Traditional control systems can handle variations in input, but they handle it in a very one-dimensional way—that is, with a single mathematical model. For example, a proportional controller can respond to any error value between 0 and 100 %, but it handles all cases the same way, that is, by multiplying the error by K_P. Fuzzy logic controllers are modeled after the natural way *people* arrive at solutions:

❏ We apply different solution methodologies (rules), depending on the value of the stimulus. In other words, we might have two or three different types of response to the same general situation, but the specific response we choose will depend on the current stimulus.

❏ We frequently apply more than one of our "rules" at the same time to a single problem, so the actual course of action is the result of a combination of rules, each weighted differently according to the stimulus.

❏ We accept a certain amount of imprecision, which allows us to arrive at workable solutions to problems that are not completely defined and with much less processing time than it would take to arrive at an exact solution.

Let's look at each of these in more detail. The first point: *We apply different solution methodologies to the same problem.* Another way to put this is that an individual has different ways of looking at the same general situation, and his or her response will be different depending on how it is "seen" at the time. For example, we can experience cooler temperatures as being in the "refreshing and brisk" category or "chilly" category, and we have a different range of possibly contradictory responses for each category—for example, in the first case, we may consider opening windows and breathing deeply; in the second case, we might consider putting on a sweater and turning up the heat.

The second point: *We often apply more than one of our rules to the same problem at the same time.* The course of action we end up taking is really a combination of two or more different response strategies, weighted according to the stimulus. For example, if the temperature is 68°F, you might process it as being *both* "refreshing" and "somewhat cool," and your response might be to breathe deeply while trying to remember where your sweater might be. When the temperature drops to 65°F, you might process it as being "too refreshing" (that is, less than 100% refreshing) and "definitely cool"; your response might be to find and put on the sweater.

The third point: *We accept a certain amount of imprecision, which is very important at helping us arrive at workable solutions.* Consider the task of parking a car. Most people can park a car in a minute or two and do a "good enough" job—the car is somewhere between the cars to the front and back, and it is reasonably close to the curb. If the requirement was that you had to be exactly between the cars (to the nearest 0.1 in.) *and* exactly 12 in. from the curb, none of us could do it unless we installed precise sensors; even then, it might take months to maneuver the car to the *exact* place. Control systems (particularly servomechanisms) have traditionally been based on the most precise (and therefore complicated) mathematical model of the system one can make, which requires the controller to do much "number crunching." However, the reality is that *an actual system is never an exact representation of its model* (real systems are riddled with nonlinearities that can never be completely accounted for); therefore, the controller's time spent finding an exact solution to an inaccurate hypothetical model is time not well spent and does not necessarily lead to the best answer.

A fuzzy logic controller mimics the way a knowledgeable human operator would control something. That is, it applies a set of control rules appropriate to the situation, which may overlap and even contradict each other. The final course of action is what we

Figure 11.37

Traditional and fuzzy logic
control block diagrams.

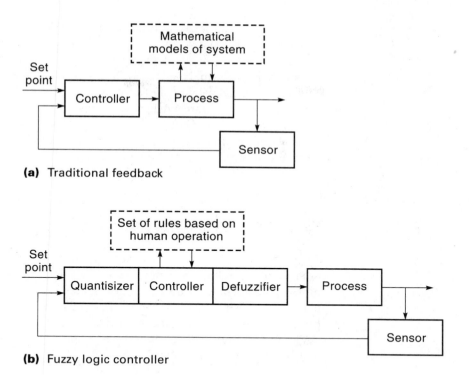

(a) Traditional feedback

(b) Fuzzy logic controller

might call a *judgment,* which is some appropriate combination of all relevant factors. Figure 11.37(a) shows a block diagram of a traditional feedback control system. This controller makes its decisions about what to do based on either a mathematical model of the process or, in the case of a PID controller, a fixed set of mathematical relationships. Figure 11.37(b) shows a block diagram of a fuzzy logic control system. The fuzzy logic controller uses as its guide a set of response rules established by the knowledgeable operator or system engineer. The block marked "Quantisizer" takes the data from the sensor(s) and converts it into a form the fuzzy logic controller can use; for example, the data from a temperature sensor might be converted from degrees into **fuzzy predicates** such as *brisk, cool, cold,* and *very cold.** The output of the fuzzy logic controller is actually a set or range of responses. (Think back at how often you have had "mixed feelings" about a response to a problem.) However, it is clear that an actuator needs a specific signal, such as a voltage for a motor, so it is the job of the "defuzzifier" block [Figure 11.37(b)] to distill the controller's multilevel response into a specific, "crisp," actuator control signal.

Fuzzy logic was first proposed by L. A. Zadeh working at Berkeley in 1965. However, Japanese industry really embraced the idea and developed applications in the area of fuzzy logic control. One of the first major applications of fuzzy logic control was the subway system in the city of Sendai. Performance of this system turned out to be much

---

*Other possibilities are *fuzzy truth values* ("true," "quite true," "not very true"), *fuzzy quantifiers* ("most," "few," "almost"), and *fuzzy probabilities* ("likely," "very likely," "not very likely").

improved over traditional controllers. For example, the cars stop with the doors almost exactly where they should be, and the acceleration is so smooth that patrons hardly have to use the overhead handrail. Other examples are the shifting capabilities in automatic transmissions (Nissan, Honda, and Saturn to name a few). The Nissan system claims to cut fuel consumption by 12–17%. A traditional automatic transmission is built to always shift at certain speeds and engine conditions, which can lead to a lot of shifting. Human drivers not only don't shift as often but also don't always shift at the same speed, for instance, when climbing a hill. Fuzzy logic–controlled washing machines adjust the amount of water, amount of detergent, and cycle time to how dirty and how many clothes are in the load.

Example of a One-Input System

To understand how the fuzzy logic controller actually works, let's look at a simple example of controlling the temperature in a room. The system consists of a room, a gas furnace, a temperature sensor, and the fuzzy logic controller (Figure 11.38). The gas flow to the furnace is regulated by a knob that turns from off to full on (in ten notches). The fuzzy controller will sample the temperature and, then based on a set of rules, adjust the gas knob up or down. The controller is trying to maintain a comfortable temperature (defined as 70°F) in the room.

Figure 11.38

Using a fuzzy logic controller to control the temperature in a room.

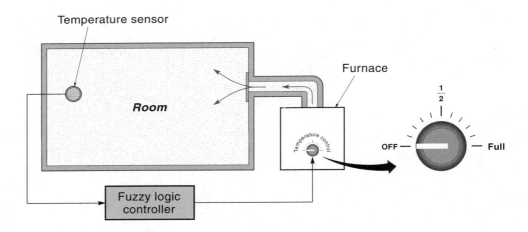

For simplicity, we will define only three temperature conditions: *warm, medium,* and *cool.* Each condition, represented on the graph of Figure 11.39(a), looks like a triangle and is called a **membership function** because it represents a range of temperatures that are all members of the same category. For example, we're willing to say that (for humans) *medium* can be anything between 60 and 80°, but it is most true at 70°. At the same time, we are saying that any temperature below 66° is *cool*; however, using the term *cool* is *more true at 55° than it is at 66°*. The terms *medium* and *cool* are called **fuzzy sets** because they

Figure 11.39

Input and output sets for a
one-input fuzzy logic
controller.

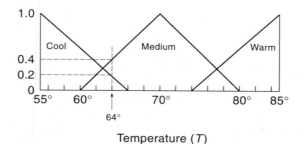

Temperature (*T*)

(a) Input sets for three temperature conditions

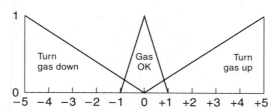

Number of notches on gas knob to turn up or down

(b) Output sets for three kinds of action

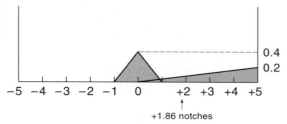

+1.86 notches

(c) Output weighted by input conditions, showing
defuzzified output to be 1.86 notches

express a range of values. Notice that the fuzzy sets overlap (for example, between 60°
and 66°). This simply means that a specific temperature can be a member of more than
one set; for example, 64° is a "little bit cool" and "somewhat medium" at the same time.

The fuzzy controller operates on a set of if–then rules, which are stated in very
linguistic terms. In our example, we might have three rules:

❏ Rule 1: If the temperature is *cool,* then turn up the gas.
❏ Rule 2: If the temperature is *medium* , then the gas is OK.
❏ Rule 3: If the temperature is *warm,* then turn down the gas.

Next we have to define "turning up" and "turning down" the gas. Figure 11.39(b)
shows this in another graph of fuzzy sets. Here you can see that the triangle for "turn gas
down" corresponds to turning the gas knob toward off, anywhere from one to five
notches. "Gas OK" corresponds to not moving the knob, or at most moving it one notch

in either direction, and so on. With these two graphs and the set of rules, the fuzzy logic controller can operate. For example, if the temperature sensor reported 64°, the quantisizer would determine that 64° is 20% *cool* and 40% *medium* [see Figure 11.39(a)]. This means that Rule 1 applies at 20%, Rule 2 applies at 40%, and Rule 3 does not apply at all. We now redraw the output graph with the magnitudes adjusted per the rules [Figure 11.39(c)]. That is, Rule 1 says, "If the temperature is *cool*, then turn up the gas"; thus, because Rule 1 is 20% true, we set the maximum value of "turn up the gas" at 20%. Similarly, Rule 2 is 40% true, so we adjust the "Gas OK" triangle to a 40% maximum value. Technically, the output of the fuzzy logic controller is the union of these two sets [Figure 11.39(c)]. However, the gas knob needs a specific command, so we must "defuzzify" the output set, that is, find some average value. In this case, we find the point on the horizontal axis where the area under the curve is the same on either side. This point is 1.86, meaning we should turn up the gas knob 1.86 notch. Does this make sense? Yes—a temperature of 64° is a little cool, so you would expect the controller to turn up the gas a little.

Example of a Two-Input System

Let's look at a more sophisticated fuzzy logic controller that uses two inputs. Using again the example of a room-temperature controller, let's add a second input: "rate-of-temperature change." So now, the controller knows what the temperature (T) is and how fast it is changing (ΔT). As before, the temperature input will be evaluated as being either *warm, medium,* or *cool* [Figure 11.40(a)]. We also define ΔT in three simple categories: *lowering, steady,* and *raising* [Figure 11.40(b)]. For example, *lowering* is defined as the temperature falling at a rate of 0.2° per minute to 1° per minute. On the graph in Figure 11.40(b), the categories are represented as triangles, where the height represents the extent to which that term is true, going from a probability of 0 to a probability of 1 (or from 0 to 100%).

Now we address the important step of establishing the rules for the controller. There are now nine possible sets of inputs T and ΔT, as given below. For each set of fuzzy inputs, a fuzzy output is specified. Notice that these rules are just simple common sense.

- ❑ Rule 1: If T is *cool* and *lowering,* then increase the gas sharply.
- ❑ Rule 2: If T is *cool* and *steady,* then increase the gas.
- ❑ Rule 3: If T is *cool* and *raising,* then the gas is OK.

- ❑ Rule 4: If T is *medium* and *lowering,* then increase the gas.
- ❑ Rule 5: If T is *medium* and *steady,* then the gas is OK.
- ❑ Rule 6: If T is *medium* and *raising,* then decrease the gas.

- ❑ Rule 7: If T is *warm* and *lowering,* then the gas is OK.
- ❑ Rule 8: If T is *warm* and *steady,* then decrease the gas.
- ❑ Rule 9: If T is *warm* and *raising,* then decrease the gas sharply.

Next, we must assign values to the output conditions of "increase gas," "gas is OK," and "decrease gas." Each output condition could be a fuzzy set (as they were in the

Figure 11.40

Input sets and an output table for a two-input fuzzy logic controller.

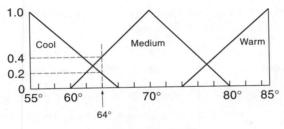

(a) Input sets for three temperature conditions

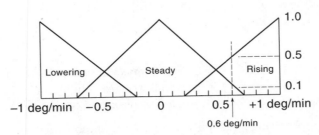

(b) Input sets for three kinds of action

		ΔT	
	Lower	**Steady**	**Raise**
Cool	+5 Rule 1	+2 Rule 2	0 Rule 3
Medium	+2 Rule 4	0 Rule 5	−2 Rule 6
Warm	0 Rule 7	−2 Rule 8	−5 Rule 9

T labels the rows (Cool, Medium, Warm).

(c) Output table

previous example), but this time we will use a table of discrete values [Figure 11.40(c)]. Each position in the table corresponds to a rule and contains a number from 0 to ±5, which corresponds to the number of notches to turn the gas knob up or down. For example, if the temperature is *cool* and *lowering* (Rule 1), then turn the gas way up (five notches). If the temperature is *cool* and *steady*, then turn the gas up a little (two notches), and so on. With the system thus defined, we can now use an example to show how it works.

Assume that the temperature sensor reports 64° (same as previous example) and that the temperature is rising somewhat rapidly at 0.6° per minute. From the graph of temperature-input sets [Figure 11.40(a)], you can see that 64° corresponds to 20% *cool* and 40% *medium*. Then, from the ΔT graph [Figure 11.40(b)] you can see that 0.6° per minute corresponds to 10% *steady* and 50% *raising*. Looking at the list of nine rules, we see that only four rules apply in this case:

❏ Rule 2: If *T* is *cool* and *steady*, then increase the gas.

❏ Rule 3: If *T* is *cool* and *raising*, then the gas is OK.

❏ Rule 5: If *T* is *medium* and *steady*, then the gas is OK.

❏ Rule 6: If *T* is *medium* and *raising*, then decrease the gas.

Because the variables are fuzzy, each rule will contribute to a different degree. The first step toward arriving at a single output value is to compute the compatibility for each rule. The *compatibility* is simply the product of the two probabilities in the rule, as expressed in Equation 11.18:

$$\text{Compatibility} = w_i = A_{i1} \times A_{i2}, \qquad i = 1, 2 \dots \qquad (11.18)$$

where

w_i = compatibility (a measure of the influence of Rule *i*)
A_{i1} = probability of first condition (*T*)
A_{i2} = probability of second condition (ΔT)

Applying Equation 11.18 to the applicable rules of this example,

$$w_2 = 0.2 \times 0.1 = 0.02*$$

$$w_3 = 0.2 \times 0.5 = 0.1$$

$$w_5 = 0.4 \times 0.1 = 0.04$$

$$w_6 = 0.4 \times 0.5 = 0.2$$

Now, using Figure 11.40(c), we can determine the output component for each rule:

$$y_i = w_i \times B_i, \qquad i = 1, 2, 3 \dots \qquad (11.19)$$

where

y_i = output from a single rule
w_i = compatibility (a measure of the influence of Rule *i*)
B_i = appropriate value from the output table

Applying Equation 11.19 to Rules 2, 3, 5, and 6,

$$y_2 = w_2 \times B_2 = 0.02 \times 2 = 0.04$$

$$y_3 = w_3 \times B_3 = 0.1 \times 0 = 0$$

$$y_5 = w_5 \times B_5 = 0.04 \times 0 = 0$$

$$y_6 = w_6 \times B_6 = 0.2 \times -2 = -0.4$$

We now have four outputs from the four rules, but we need to defuzzify this data into a single command for the gas knob. A simple average would not work because some conditions are evoked more strongly than others. One approach is to use a *weighted mean*, as specified in Equation 11.20:

---

*That is, for Rule 2, *cool* = 0.2 and *steady* = 0.1.

$$y_{\text{tot}} = \frac{w_1 y_1 + w_2 y_2 + \ldots}{w_1 + w_2 + \ldots} \tag{11.20}$$

Applying Equation 11.20 to this problem,

$$
\begin{aligned}
y_{\text{tot}} &= \frac{w_2 y_2 + w_3 y_3 + w_5 y_5 + w_6 y_6}{w_2 + w_3 + w_5 + w_6} \\[2mm]
&= \frac{(0.02 \times 0.04) + (0.1 \times 0) + (0.04 \times 0) + (0.2 \times -0.4)}{0.02 + 0.1 + 0.04 + 0.2} \\[2mm]
&= \frac{-0.079}{0.36} = -0.22 \text{ notch}
\end{aligned}
$$

Therefore, the final output of the fuzzy logic controller specifies that the gas should be turned down slightly (0.22 notch). However, the temperature in the room is actually somewhat cool (64°); why does the controller send a signal to turn down the gas? It is mimicking our intelligence, and *we* know that with the temperature rising so fast (0.6° per minute), we had better ease the heat back now or it will surely overshoot.

Closing Thoughts

It seems certain that fuzzy logic controllers will be used in more and more products and systems. On the consumer side, they are already being used in automobiles, camera focusing, air conditioners, and microwave ovens, to name a few. For industrial systems, they are being used in process control for temperature and flow control, and in servo-mechanisms for speed control and helicopter autopilots. Also, fuzzy logic controllers are being used in conjunction with traditional PID controllers, where the job of the fuzzy controller is to adapt the PID parameters to changing conditions.

In many cases, the fuzzy controller is a traditional microcontroller, such as the Intel 8051 or Motorola 68HC11, which has been programmed to implement fuzzy logic. Also, new fuzzy logic control ICs that are designed specifically for this application are becoming available.

A final word: Although fuzzy logic controllers have repeatedly performed better than their traditional control system counterparts, these improvements do not come without effort. Finding the right rule set and specifying the nature and range of the fuzzy variables can be very time-consuming. Arriving at the right set of inputs for the (now famous) subway train in Sendai took engineers months of tuning. However, like the traditional digital controller, the fuzzy logic control algorithm is implemented in software, not hardware, so fine-tuning can be done with minimum system downtime.

SUMMARY

The controller, a part of the control system, directs the actuator to move some parameter, which is called the controlled variable. A feedback controller has two inputs, the set point (SP) and the process variable (PV), and one output. The set point specifies the desired position of the controlled variable. The process variable is the actual position of the controlled variable, as reported by a sensor. The error value (E) is the difference between the set point and the process variable ($E = SP - PV$). Based on the error, the controller creates a control signal for the actuator.

There are two major classifications of feedback control systems: servomechanisms and process control. Servomechanisms are usually mechanical motion systems. Analysis of some servomechanisms may require a highly mathematical approach (not presented in this text). Process control refers to a system that attempts to keep some process at a constant value. Process control systems are usually slower than servomechanisms, and empirical data may need to be used in the design.

There are different levels of control strategy. The simplest are the on–off controllers; with these, the actuator is either on full force or off, as in a thermostatically controlled heating system. Proportional control is more sophisticated; with this system, the controller provides a restoring force that is proportional to the error. One problem with proportional control is that it cannot reduce the error to zero because it provides no force at zero error.

Integral control is sometimes added to proportional control and provides a restoring force based on the sum of past errors. Integral control can reduce the steady-state error to zero, but it may introduce stability problems.

Derivative control, which can also be added to proportional control, provides a restoring force that is based on the rate that the error is increasing (or decreasing) and tends to quicken the response and reduce overshoot.

PID control combines proportional, integral, and derivative controls into one controller and is a very common control strategy, especially for process control systems. The PID controller is tuned to perform in any particular application by adjusting the gain constants for proportional, integral, and derivative. Analog PID controllers use differential amplifiers to create and then sum the three control components. Digital PID controllers work under program control; the program is in the form of an endless loop. With each pass through the loop, the following functions are performed: Read in the process variable and set point, calculate the controller output, and send output to the actuator.

Fuzzy logic controllers are relatively new and use a completely different approach than traditional controllers. Fuzzy logic controllers are not based on a mathematical model of the system but instead implement the same control "rules" that a skilled human operator would. Fundamental in the concept of fuzzy logic is the recognition that rules and conditions come in degrees, as specified in linguistic terms—for example, *warm, warmer,* and *very warm*.

GLOSSARY

adaptive controller   A controller, such as a PID, with the ability to monitor and improve its own performance.

aliasing   The event that occurs when the wrong data are reconstructed from the sampled sensor data because the sample rate is too slow.

autotuning   The ability of some digital controllers to continuously fine-tune themselves by making small changes in their gain constants, based on past performance.

bias   A value that is added to the output of the controller to compensate for a constant disturbing force, such as gravity.

Bode plot   A graph that plots system gain and phase shift versus frequency; it can be used to determine system stability.

closed-loop system   A control system that uses feedback—that is, a sensor tells the controller the actual state of the controlled variable.

continuous-cycle method   A method of tuning a PID controller that involves driving the controller into oscillations and measuring the response; the PID constants are computed based on these data.

controlled variable   The parameter that is being controlled, such as temperature in a heating system.

dead band   The small range, on either side of the set point (of the controlled variable), where the error cannot be corrected by the controller because of friction, mechanical backlash, and the like.

dead time   The interval of time between the correcting signal sent by the controller and the response of the system.

dead zone   *See* **dead band.**

derivative control   A feedback control strategy where the restoring force is proportional to the rate that the error is changing (increasing or decreasing); derivative control tends to quicken the response and reduce overshoot.

error (E)   The difference between where the controlled variable should be and where it actually is ($E = SP - PV$).

feedforward   The concept of letting the controller know in advance that a change is coming.

fuzzy logic   A new control strategy, modeled originally on human thought processes, where decisions are made based on a number of factors and each factor can have such qualities as *maybe* and *almost,* to name only two.

fuzzy predicates   Linguistic terms of degree, used in describing actual variables—for example, *warm, very warm, not so warm,* and so on.

fuzzy set   The range of one fuzzy predicate in a system—that is, all values of temperature that correspond to *very warm.*

gain   A multiplication factor; a steady-state transfer function. The term *system gain* refers to the proportional gain factor (K_P).

gain margin   A form of a safety margin. When the phase lag is 180°, the gain margin is the amount the gain could be increased before the system becomes unstable.

integral control   A feedback control strategy where the restoring force is proportional to the sum of all the past errors multiplied by time. This type of control is capable of eliminating steady-state error but increases overshoot.

iteration   *See* **scan.**

lag time   The time between the change of the set point and the actual movement of the controlled variable.

membership function   Used in conjunction with fuzzy logic control, the distribution of values that have been grouped into one category.

on–off control   *See* **two-point control.**

open-loop system   A type of control system where there is no feedback; consequently, the controller does not know for sure the exact condition of the controlled variable.

overshoot   The event that occurs when the path of the controlled variable reaches and then goes past the set point, before turning around or stopping.

phase margin   A form of a safety margin. When the system gain is 1, the phase margin is the amount the phase lag could be increased before the system becomes unstable.

PID   Stands for Proportional + Integral + Derivative; a control strategy that uses proportional, integral, and derivative feedback.

PIP   Stands for Proportional + Integral + Preview; a control strategy that uses feedforward so that it knows in advance if a change is coming.

positive feedback   The event that occurs when the feedback signal from the sensor is added to the set point (instead of subtracted); positive feedback will likely cause the system to oscillate.

process control system   The type of system where the controller is attempting to maintain the output of some process at a constant value.

process variable (PV)   The actual value of a variable, as reported by the sensor.

proportional band   A term used to describe the gain of a proportional control system. It is the change in error that causes the controller output to change its full swing.

proportional control   A feedback control strategy where the controller provides a restoring force that is proportional to the amount of error.

PV   *See* **process variable.**

reaction-curve method   A method of tuning a PID controller that involves opening the feedback loop, manually injecting a step function, and measuring the system response. The PID constants are computed based on these data.

rise time   The time it takes for the controlled variable to rise from 10 to 90% (of its final value).

sample rate   The times per second (or per minute) that a sensor is read; for a digital controller, the sample rate usually corresponds to the scan rate.

scan   One pass through the program of a digital-type controller.

servomechanism   A mechanical-motion system that uses feedback control, such as a robot.

set point (SP)   An input into the controller, it represents the desired value of the controlled variable.

settling time   The time it takes for the system response to settle down to some steady-state value.

SP   *See* **set point.**

steady-state error The error that remains after the transient response has died away.

step change The event that occurs when a discrete change is made to the set point.

three-position control Similar to the two-point system except the actuator can exert force in two directions, so the system is either off-up-down, off-right-left, and so on.

transient response The actual path that the controlled variable takes when it goes from one position to another.

two-point control A simple type of control system where the actuator is either on full force or off; also called *on–off control*.

tuning When applied to a control system, this term refers to the process of making adjustments to the gain constants K_P, K_I, K_D until system performance is satisfactory.

windup Associated with integral feedback, the problem that occurs when too much error accumulates as a result of the system saturating before it can achieve the state that it theoretically should.

EXERCISES

Section 11.1

1. Assume that Figure 11.25(a) is the response curve of a control system to a step function. Determine the following parameters: rise time, overshoot, and settling time.

2. A robot arm was commanded to go to a new position. Its response was recorded and is shown in Figure 11.41. Determine the rise time, overshoot, settling time, and steady-state error of the response.

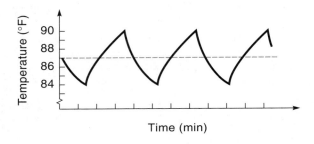

Figure 11.42

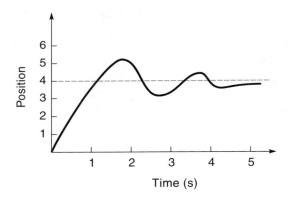

Figure 11.41

Section 11.2

3. Give an example of a two-point control system (other than a heating system) and explain how it works.

4. A thermostat is used to maintain a temperature of 87° F. An accurate recording of the room temperature is shown in Figure 11.42. The respective cut-in and cut-off points are currently 84° and 90°. What would you predict the cycle time would be if the cut-in and cut-off points were respectively changed to 86° and 88°?

5. Describe how three-position control might be used to cause a windmill to always face into the wind. The base of the windmill is rotated with a bidirectional DC motor.

Section 11.3

6. In your own words, explain the principle of proportional control.

7. The system gain for the position system shown in Figure 11.6 is 2 ft · lb/deg. If the set point is 50°, find the torque on the arm when the arm is at 15° and 45°.

8. A proportional controller for a rotating antenna has a gain K_P of 5 in. · oz/deg. The antenna was initially pointing due south but was then commanded to point southeast.
   a. Find the initial torque supplied to the antenna.

b. What is the torque when the antenna gets to within 5° of its set point?

9. Explain why a proportional control system lifting a weight can never reduce the steady-state error to zero.

10. A proportional control system with a K_P of 5 ft · lb/deg controls a robot arm that is 2 ft long. What steady-state error would you expect if the robot lifted a 10-lb object?

11. A linear-motion proportional control system has a gain of $K_P = 5$ lb/in. A friction force acts on the controlled variable with a constant force of 0.25 lb. Find the length of the dead band.

12. A proportional control system is to be used to control the wing flaps of a jet airplane. The wind load is expected to cause as much as 600 in. · lb of torque. The system should be able to keep the flaps to within 5° of the set point. Find the gain K_P necessary to meet the requirements.

Section 11.4

13. Explain how the addition of integral feedback in a proportional control system eliminates steady-state error.

14. A control system with integral control has a 2° steady-state error similar to the graphs in Figure 11.15. The steady-state error is purely the result of a 7 in. · oz friction force. How many seconds will it take the integral control to correct this error? (Assume $K_I = 1$)

15. A proportional control system is working against 20 in. · lb of friction that causes a steady-state error of 5°. If integral feedback is added with a K_I of 2 in. · lb/deg s, how long would it take for the error to be eliminated?

16. What problem is solved and what new problem is created with the addition of integral feedback?

Section 11.5

17. Derivative feedback produces a force that is proportional to what?

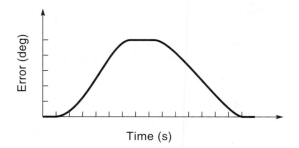

Time (s)

Figure 11.43

18. A plot of error versus time (for a control system) is shown in Figure 11.43. The derivative gain is 2 in. · oz/deg/s. Find the maximum positive and negative values of derivative output. Make a sketch of the derivative output versus time.

19. Explain how derivative feedback makes a control system more responsive to rapid change and how it reduces overshoot.

Section 11.6

20. Explain how the three elements of the PID control system work together to create a practical control system.

21. Draw a block diagram for an analog PID controller, indicating the function that each block performs.

22. How does a digital controller function? List the general steps that a digital controller program follows.

23. How does a digital controller integrate the error signal curve?

24. How does the digital controller find the derivative of the error signal?

25. Write a program in BASIC (or any other language) to implement a PID controller. The sample time is 0.5 s. Assume that the set point and error data are constantly available from READ or INPUT statements. Assume $K_P = 2$, $K_I = 1.5$, $K_D = 1.8$, and all units are compatible. Send out the controller drive signal with an OUT or a PRINT statement.

26. The Bode plot for a control system is shown in Figure 11.44. Is this system stable? Find the gain margin and phase margin.

27. A system oscillates with a period of 5 s. There is a 1-s lag time between the controller output and the controlled-variable movement. How many degrees of phase lag does this system have under these conditions?

28. A control system is to be tuned using the continuous-cycle method. It was found that the system went into oscillation when K_P' was 0.3 V/deg. Figure 11.45 shows the system response. Find K_P, K_I, and K_D.

29. A control system is to be tuned using the reaction-curve method. The test input and resulting output are shown in Figure 11.46. Find K_P, K_I, and K_D.

30. A control system samples a position 30 times per second. If the position signal is cycling back-and-forth, what is the highest frequency the controller can follow? Give both theoretical and practical values.

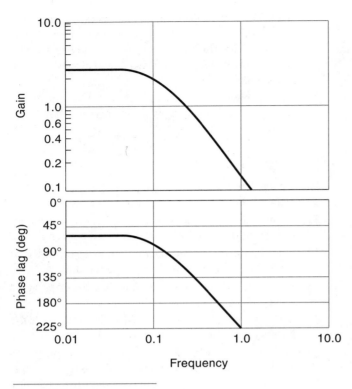

Figure 11.44

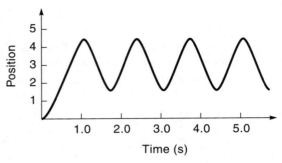

Figure 11.45

31. Is a high sample rate good or bad for a control system? Explain.

Section 11.7

32. What is the necessary condition that allows PIP control to be used?

33. What is *feedforward*, and why does using this concept improve path control over the PID system?

Section 11.8

34. For the single-input fuzzy logic temperature controller specified in Figures 11.38 and 11.39, determine the defuzzified output for a temperature of 76°. Does your answer make sense?

35. For the single-input fuzzy logic temperature controller specified in Figures 11.38 and 11.39, determine the defuzzified output for a temperature of 62°. Does your answer make sense?

36. For the two-input fuzzy logic temperature controller specified in Figure 11.40, determine the defuzzified output for a temperature of 76° and ΔT of -0.3 deg/min. Does your answer make sense?

37. For the two-input fuzzy logic temperature controller specified in Figure 11.40, determine the defuzzified output for a temperature of 62° and ΔT of $+0.7$ deg/min. Does your answer make sense?

Figure 11.46

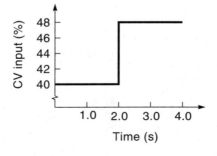

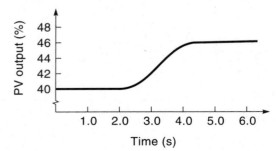

12 Relay Logic and Programmable Logic Controllers

After studying this chapter, you should be able to:

- ❑ Explain the operation of electromechanical relays, time-delay relays, counters, and sequencers.
- ❑ Explain the purpose and operation of a ladder diagram.
- ❑ Explain the operation of a relay-based controller.
- ❑ Understand the concept and purpose of a programmable logic controller (PLC).

- ❑ Understand the hardware and wiring required in a PLC-based system.
- ❑ List the steps that must be taken to make a PLC control system operational.
- ❑ Understand the basic instructions used in a PLC program.
- ❑ Differentiate the ways that a PLC can be programmed.

INTRODUCTION

The automatic control of repetitious mechanical or physical processes is a common control problem, which abounds in major appliances, office machines, the manufacturing industry, and industry in general. Traditionally, this kind of process control was done with electromechanical devices such as relays, timers, and sequencers. With this approach, however, the circuit must be rewired if the control logic changes, which was a particular problem to the automotive industry because of the annual model changeovers. In response to this problem, in the late 1960s, General Motors developed the specifications for a programmable electronic controller that could replace the hard-wired relay circuits. Based on those specifications, Gould Modicon Company developed the first programmable logic controller (PLC). The PLC is a small, microprocessor-based process-control computer that can be connected directly to such devices as switches, small motors, relays, and solenoids, and it is built to withstand the industrial environment. This chapter will introduce the principles of relay logic control and then describe the general operation and programming of the PLC. For those who want a detailed working knowledge of a PLC, a tutorial (published by Allen–Bradley) for connecting, programming, and operating the Allen–Bradley SLC 500 PLC, is included in the appendix.

12.1   RELAY LOGIC CONTROL

Relay Logic

Relays as a device are covered in Chapter 4; however, a quick review of the basics is appropriate here. An **electromechanical relay** (EMR) is a device that uses electromagnetic force to close (or open) switch contacts—in other words, an electrically powered switch. Figure 12.1 shows a diagram of a relay. Relay contacts come in two basic

Figure 12.1

A relay: showing normally closed (NC) and normally open (NO) contacts.

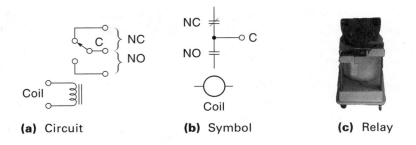

(a) Circuit   **(b)** Symbol   **(c)** Relay

configurations—**normally open contacts** (NO), which are open in the unenergized state (and close when the relay coil is energized), and **normally closed contacts** (NC), which are closed in the unenergized state (and open when the relay is energized). By convention, *the relay symbol always shows the contacts in the unenergized state.* Relays are available with a variety of multiple-contact configurations, two of which are shown in Figure 12.2.

No doubt, relays were first developed to satisfy two electrical control needs: remote control (the ability to turn devices on and off from a remote location) and power amplification. An example of power amplification is the starting relay in a car (a low-current ignition switch energizes a relay, which in turn passes a high current to the starting motor). Electrical designers soon recognized that relays could be used to implement control logic. AND, OR, and NOT gates, as well as flip-flops, can be wired from relays (Figure 12.3). For the AND gate in Figure 12.3(a), relays A *AND* B must be energized for a voltage to be at X. Figure 12.3(b) shows an OR gate. Here, if either relay A *OR* B is energized, the output X receives a voltage. Figure 12.3(c) shows a relay NOT gate. If relay A is energized, the output is 0 V; if it is *NOT* energized, the output receives a voltage.

A flip-flop can be made from a relay using a principle called **sealing**. As seen in Figure 12.3(d), once the relay is energized, an electrical path through the contacts takes over the job of providing power to the coil, and the original A voltage can be removed. (This process is frequently called **latching**. However, a relay catalog uses the term *latching relay* to mean a relay that has a built-in mechanical-latching mechanism.)

Figure 12.2
Multiple-contact relays.

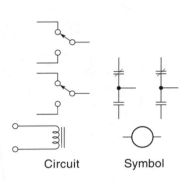

Circuit   Symbol

(a) Double-pole/double-throw (DPDT)

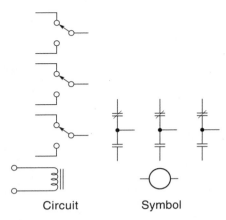

Circuit   Symbol

(b) Triple-pole/double-throw (3PDT)

Figure 12.3
Logic functions from relays (for a 12-V system).

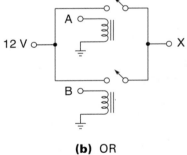

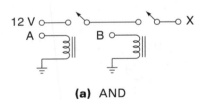

(a) AND

(b) OR

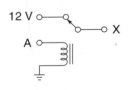

(c) NOT

(d) Sealing or latching relay (flip-flop)

Ladder Diagrams

Switches and relays became widely used in industry for controlling motors, machines, and processes. A switch can turn a single machine on and off, but a relay logic network can control an entire process—turning on one machine, waiting until that operation is done, then turning on the next operation. Decisions can be made by the logic—for example, if a part is too tall, it goes in one bin; otherwise, it goes in the other bin.

Eventually, the **ladder diagram**, a special type of wiring diagram, was developed for relay-and-switching control circuits. Figure 12.4 shows a ladder diagram (notice it resembles a ladder). The ladder diagram consists of two **power rails**, which are placed vertically on each side of the diagram, and *rungs*, which are placed horizontally between the power rails. The power rails are the source of power in the circuit (AC or DC), where the left rail is the "hot" side (voltage) and the right rail is neutral (AC) or ground (DC). Therefore, each rung is connected across the voltage source and is an independent circuit. A rung typically contains at least one set of switch or relay contacts and at least one load such as a relay coil or motor. When the contacts in a particular rung close to make a continuous path, then that rung becomes active, and its load is energized. For example, the top rung in Figure 12.4 contains two switches and a pilot light in series. For the rung to become active, both switches must close, which will then apply voltage to the light. The middle rung has two switches in parallel, so only one of these switches must be closed to make the rung active. The load in this rung is a relay coil (RELAY A). The bottom rung contains a set of NO contacts from RELAY A, and the load is a motor. Consequently, when either SW 3 or SW 4 of the middle rung closes, RELAY A coil is energized, and the motor (in the bottom rung) starts.

Figure 12.4

A relay logic ladder diagram.

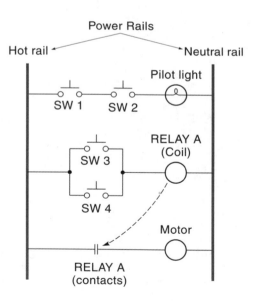

◆ **EXAMPLE 12.1**

In a certain bank, each of three bank officers has a unique key to the vault. The bank rules require that two out of the three officers be present when the vault is opened. Draw the ladder diagram for a relay logic circuit that will unlatch the door and turn on the light when two of the three keys are inserted.

Solution

Three combinations of keys will open the vault: A and B, A and C, and B and C. Each of the three keys—A, B, and C—fits in its own key switch that has two sets of NO contacts. Figure 12.5 shows the completed ladder diagram. The top rung has three branches of switch contacts, one for each acceptable possibility. At least one branch must have continuity so that the relay coil is energized. The bottom rung, activated when the relay contacts close, provides power to the door-latch solenoid and the vault light.

Figure 12.5

A ladder diagram (Example 12.1).

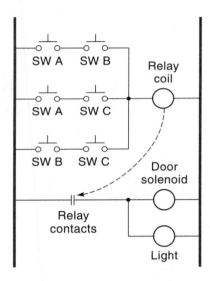

◆ **EXAMPLE 12.2**

A simple pick-and-place robot picks up parts from one conveyer belt and places them on another belt, as shown in Figure 12.6(a). When a part moving along the lower conveyer belt activates Switch 1, a solenoid-powered gripper clamps on the part and carries it toward the upper conveyer belt. When the gripper reaches Switch 2, it releases the part and moves back (empty) to receive the next part. When the gripper reaches Switch 3, it halts and waits for the next part to start the cycle all over again. Draw the relay logic ladder diagram to control this operation.

Figure 12.6

Pick-and-place robot control
(Example 12.2).

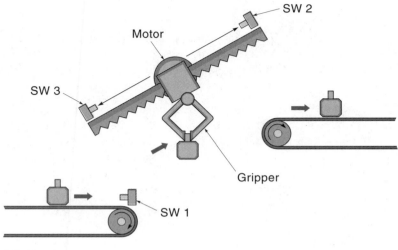

(a) Hardware

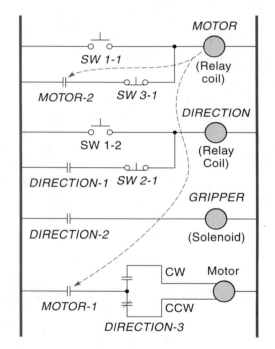

(b) Ladder diagram

Solution   Figure 12.6(b) shows the completed ladder diagram. The cycle starts when a part on the lower conveyer belt activates Switch 1—momentarily closing contacts *SW 1-1* (shown in the top rung). This action energizes the *MOTOR* relay, and the motor starts receiving current through the *MOTOR-1* contacts (bottom rung). The *MOTOR* relay is sealed (latched) via relay contacts *MOTOR-2* (now even when *SW 1-1* opens, the *MOTOR* relay will stay energized through the *MOTOR-2* contacts).

While the motor is getting started, the second set of Switch 1 contacts (*SW 1-2*) in rung 2 energizes the *DIRECTION* relay causing the motor to drive the gripper toward the upper conveyer belt (the *DIRECTION* relay determines the motor direction). Using the same technique as in rung 1, the *DIRECTION* relay is sealed on through one of its own contacts (*DIRECTION-1*).

When the gripper arrives at the upper conveyer belt, it activates Switch 2, which opens the normally closed *SW 2-1* contacts. This action breaks the *DIRECTION* relay seal, allowing it to become de-energized, and the motor switches direction—now going back toward the bottom.

Rung 3 controls the *GRIPPER* solenoid. Controlling the gripper is simple because the *GRIPPER* needs to be activated for the same time period as the *DIRECTION* relay is energized—that is, from the time when the motor starts its upward journey, until just before it descends.

The final action in the cycle is when the gripper descends (empty) and activates Switch 3. This action opens the normally closed contacts (*SW 3-1*), breaking the seal in rung 1 (which had held the *MOTOR* relay energized), and the motor stops.   ◆

Timers, Counters, and Sequencers

Many control situations require that a time delay be inserted at some point in the process. For example, a mixing operation might take 90 s, or a conveyer belt might be allowed 10 s to reach speed before parts are placed on it. A relay control system would use a **time-delay relay** to create the time delay. Once activated, the time-delay relay will wait a predetermined period of time before the contacts open (or close)—the delay may range from less than a second to minutes. Time-delay relays are categorized as being either *on-delay* or *off-delay* and are best explained by example. A 5-s **on-delay relay** with NO contacts will wait 5 s (after being energized) before closing its contacts. When the relay is de-energized, the contacts will open immediately. An application of the on-delay function is a security system that delays activation for a period of time after being turned on, to allow the occupants time to leave. The **off-delay relay** provides a time delay when the coil is de-energized. For example, a 5-s off-delay relay with NO contacts would close its contacts immediately when energized; when it is deenergized, however, the contacts would remain closed for 5 more seconds before opening. An application of the off-delay function is car lights that remain on for a period of time after being turned off (to provide light for the occupants as they are leaving the vehicle). Figure 12.7(a) shows the symbols used for time-delay relays.

Figure 12.7

Time-delay relays.

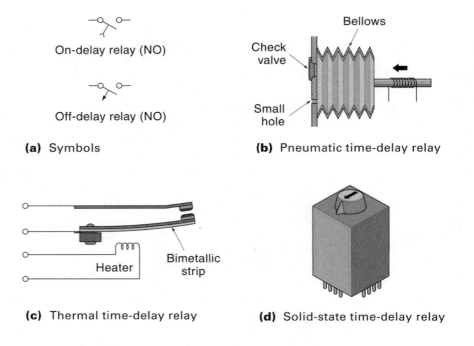

On-delay relay (NO)

Off-delay relay (NO)

(a) Symbols

(b) Pneumatic time-delay relay

(c) Thermal time-delay relay

(d) Solid-state time-delay relay

Several different designs of delay timers, which use different operating principles, have evolved over the years. The **pneumatic time-delay relay** [Figure 12.7(b)] works as follows: When the relay is activated, a small spring-loaded bellows is squeezed closed, causing the air to escape through a check valve. The bellows then is allowed to slowly expand (the air being admitted through a small hole). When the bellows reaches its normal size again, the contacts close. These relays are available with a fixed or an adjustable delay.

A **thermal time-delay relay** [Figure 12.7(c)] uses a temperature-sensitive bimetallic strip. When the relay is energized, a small resistance heater warms the bimetallic strip. As the strip heats, it bends and eventually closes the contacts. The flasher unit in a car, which controls the flashing directional signals, works on this principle.

Solid-state time-delay relays are used in most newer systems that require delay relays. An example is shown in Figure 12.7(d); the knob is for setting the delay. These units are based on the delay involved when (1) charging a capacitor or (2) counting high-speed clock pulses with a digital counter: the higher the count, the longer the delay.

♦ **EXAMPLE 12.3**

A small electric furnace has two heating elements that are energized in stages 3 min apart. That is, when the furnace is turned on, the first heating element comes on right away, and the second element comes on 3 min later. A temperature sensor will shut down the furnace if it gets too hot. Draw the ladder diagram for the control circuit.

Solution On the top rung of the ladder diagram shown in Figure 12.8, a push-button switch energizes the *HEATER* relay, and the relay is sealed by the *HEATER-1* contacts. This

Figure 12.8

Furnace heater control
(Example 12.3).

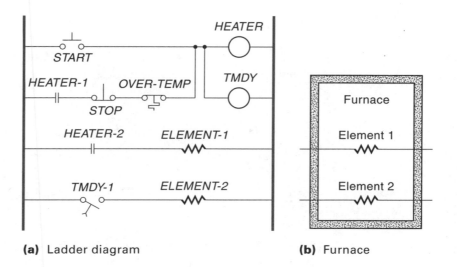

(a) Ladder diagram   **(b)** Furnace

action supplies power to the first heating element (*ELEMENT-1*), via the relay contacts *HEATER-2*, and energizes the *TMDY* delay relay (on-delay type). After 3 min, the time-delay relay activates the second heating element via the *TMDY-1* contacts (bottom rung). At any time if the stop button is pushed or the *OVER-TEMP* sensor contacts open, the seal is broken, and the *HEATER* relay deenergizes, causing both heating elements to shut down.   ◆

Electromechanical counters are devices that count events. The counter, which has a readable output similar to a car's odometer (Figure 12.9), increments once for each electric pulse received. Counters usually simply keep track of the number of times an operation is performed. For example, it could count the number of products coming down the assembly line. Some models will close a set of switch contacts when the count gets to a preset value.

Electromechanical sequencers control the process that has a sequence of timed operations. An example of this is a dishwasher controller. The water is admitted for so many seconds, then the wash cycle is activated for so many seconds, and then the water is pumped out for so many seconds, and so on. As shown in Figure 12.10, the

Figure 12.9

Electromechanical counter.

Figure 12.10
An electromechanical sequencer.

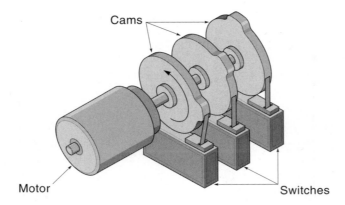

Cams

Motor

Switches

sequencer (known as a *drum controller*) consists of a small timing motor that slowly rotates a cluster of cams, and the cams activate switches. Each switch controls one of the timed operations, and the whole sequence repeats itself with each revolution of the drum. Some sequencers use a solenoid ratchet mechanism to rotate the cams in small discrete steps—one step for each input pulse.

12.2 PROGRAMMABLE LOGIC CONTROLLERS

Introduction

A **programmable logic controller** (PLC) is a small, self-contained, rugged computer designed to control processes and events in an industrial environment—that is, to take over the job previously done with relay logic controllers. Figure 12.11 shows a number of different types of PLCs. Physically, they range in size from half a shoe box to an oscilloscope (and sometimes larger). Wires from switches, sensors, and other input devices are attached directly to the PLC; wires driving motors, lights, and other output devices are also connected directly to the PLC (Figure 12.12). Each PLC contains a microprocessor that has been programmed to drive the output terminals in a specified manner, based on the signals from the input terminals. The PLC program is usually developed on a separate computer such as a personal computer (PC), using special software provided by the PLC manufacturer. Once the program has been written, it is transferred, or downloaded, into the PLC. From this point on, the PLC will operate on its own, as a completely independent controller.

PLC Hardware

Figure 12.13 shows the block diagram for a PLC and includes the fundamental parts found in any microprocessor system (as discussed in Chapter 2). These major functional blocks are explained next.

Figure 12.11
A selection of PLCs.
(Courtesy of Allen–Bradley)

Power Supply

PLCs are usually powered directly from 120 or 240 Vac. The power supply converts the AC into DC voltages for the internal microprocessor components. It may also provide the user with a source of reduced voltage to drive switches, small relays, indicator lamps, and the like.

Processor

The **processor** is a microprocessor-based CPU, and is the part of the PLC that is capable of reading and executing the program instructions, one-by-one.

Figure 12.12
A PLC and its related components.

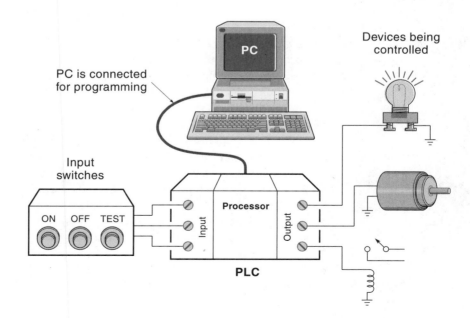

Figure 12.13
Block diagram of a PLC.

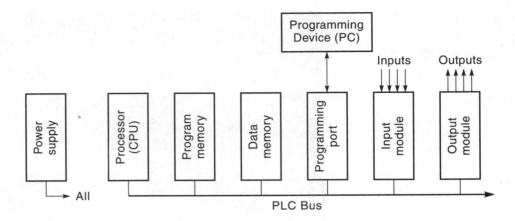

Program Memory

The **program memory** receives and holds the downloaded program instructions from the programming device. If this memory is standard RAM, the program will be lost every time the power is turned off, requiring it to be reloaded. To avoid this bother, the program memory may use an EEPROM (electrically erasable programmable ROM) or a battery-backup RAM, both of which are capable of retaining data even when the power is off. An EPROM (UV erasable PROM) can also store the program, but this device requires a special programming unit that is not part of the PLC system.

Data Memory

Data memory is RAM memory used as a "scratch pad" by the processor to temporarily store internal and external program-generated data. For example, it would store the present status of all switches connected to the input terminals and the value of internal counters and timers.

Programming Port

The **programming port**, an input/output (I/O) port, receives the downloaded program from the programming device (usually a PC). Remember that the PLC does not have an elaborate front panel or a built-in monitor; thus, to "see" what the PLC is doing inside (for debugging or troubleshooting), you must connect it to a PC, as illustrated in Figure 12.12.

Input and Output Modules

The **I/O modules** are interfaces to the outside world. These data ports may be built into the PLC unit or, more typically, are packaged as separate **plug-in modules**, where each module contains a set of ports (see Figures 12.11 and 12.12). The most common type of I/O is called **discrete I/O** and deals with on–off devices. *Discrete input modules* connect real-world switches to the PLC and are available for either AC or DC voltages (typically, 240 Vac, 120 Vac, 24 Vdc, and 5 Vdc). As shown in Figure 12.14, circuitry within the module converts the switched voltage into a logic voltage for the processor. Notice that an AC switch voltage must first be converted into a DC voltage with a rectifier.

Figure 12.14
PLC input module circuits.

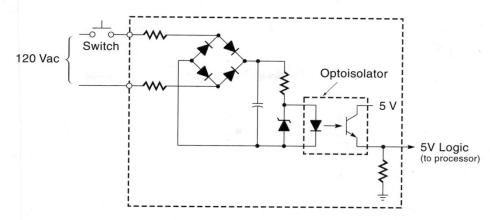

(a) AC input module circuit

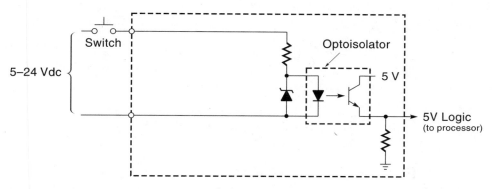

(b) DC input module circuit

Also, an input module will usually include an opto-isolator to prevent possible high-voltage spikes from entering the PLC. A typical discrete input module would have 4, 8, or 16 inputs.

Discrete output modules provide on–off signals to drive lamps, relays, motors, and other devices. As shown in Figure 12.15, several types of output ports are available: Triac outputs control AC devices, transistor switches control DC devices, and relays control AC or DC devices (and provide isolation as well). A typical discrete output module would have 4, 8, or 16 outputs.

Analog I/O modules allow the PLC to handle analog signals. An *analog input module* [Figure 12.16(a)] has one or more ADCs (analog-to-digital converters), allowing analog sensors, such as temperature, to be connected directly to the PLC. Depending on the module, the analog voltage or current is converted into an 8-, 12-, or 16-bit digital word. An *analog output module* [Figure 12.16(b)] contains one or more DACs (digital-to-analog converters), allowing the PLC to provide an analog output—for example, to drive a DC motor at various voltage levels.

Figure 12.15
PLC output module circuits.

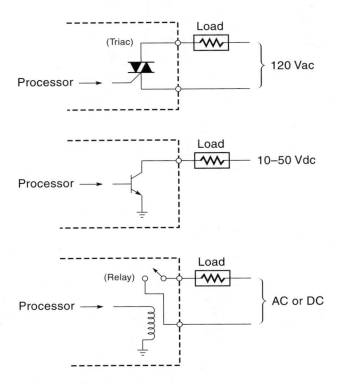

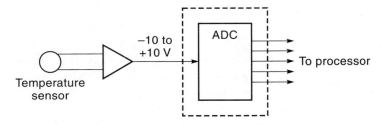

Figure 12.16
Analog I/O modules.

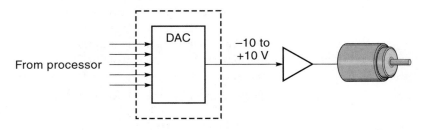

(a) Analog input module

(b) Analog output module

PLC Bus

The **PLC bus** are the wires in the *backplane* of a PLC modular system, which contains the data bus, address bus, and control signals. The processor uses the bus to communicate with the modules.

PLC Setup Procedure

The PLC evolved as a replacement for hard-wired relay logic and was designed to be installed, programmed, used, and maintained by technicians who were familiar with industrial relay and switching circuits. For this reason, manufacturers of PLCs developed a programming language that allows the user to program the desired control logic in the form of a ladder diagram. The ladder diagram may be entered into the system by either drawing it on the PC screen or using a specially adapted switch box. If the PC is used, special software must be installed before the ladder diagram can be entered.

Assuming that the ladder diagram has been entered into a PC (or PC-like terminal device), the next step is to connect the PC to the PLC. In most cases, simply connect the two units with a cable or an interface circuit. A more sophisticated system may have a number of PLCs connected to a single PC on a LAN (local area network). With this system, the PC can communicate with any PLC on the network. Once connection has been established, the program is downloaded from the PC into the PLC.

With the program loaded, the PLC is ready to operate, and in any real application, the next step is testing and debugging. Testing can be done one of two ways: the total software approach, where the inputs are simulated from the programming terminal (PC), or a more real-world approach, where the inputs and outputs are connected to their actual destinations. In either case, the PC screen becomes a "window" into the workings of the program; if a problem becomes apparent, the progran can be corrected immediately and testing continued until all bugs are worked out. When the testing is complete, the PC can be disconnected, and the PLC will continue to execute its stored program as an independent unit.

PLC Operation

The PLC accomplishes its job of implementing the ladder logic control program in typical computer fashion—one step at a time. To see how it does this, let's examine a simple PLC application of turning lamps on and off [Figure 12.17(a)]. Notice that the PLC has one input module and one output module. Two external switches (SW-0 and SW-1) are connected to the PLC via terminals *IN-0* and *IN-1* of the input module. Two terminals of the output module (*OUT-0* and *OUT-1*) drive two indicator lamps (LAMP-0 and LAMP-1). The ladder diagram for this application has only two rungs [Figure 12.17(b)]. The top rung will light LAMP-0 if *both* SW-0 and SW-1 are closed. The bottom rung will light LAMP-1 if *either* SW-0 or *OUT-0* are closed. (The *OUT-0* contacts can be thought of as NO relay contacts of coil *OUT-0* in rung 1.)

Assume that the program has been loaded into the PLC and is residing in the program section of memory. When the PLC is set to **run mode**, program execution

Figure 12.17

A simple PLC application and ladder diagram.

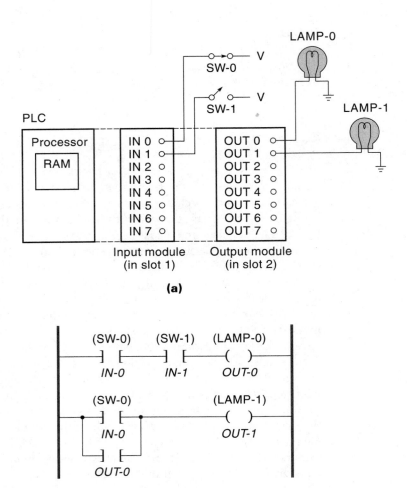

(a)

(b)

begins. Execution is accomplished as a series of steps, which are repeated over and over. One complete cycle of all these steps is a **scan**. The steps are as follows:

1. The PLC reads in data from the input module and stores the values of all eight inputs, as a word, in the Input Data section of memory (Figure 12.18). Each bit in the word represents the present status of one external switch.

2. The PLC turns its attention to the ladder diagram [Figure 12.17(b)], starting with the first rung. The first two symbols in the rung are input switches, so the processor reads their status *from RAM*. It then evaluates the logic indicated by the rung. In this case, SW-0 is on, and SW-1 is off, so *OUT-0* will *not* be energized. This status of *OUT-0* is stored as a 0 bit in the Output Data section of RAM (Figure 12.18).

3. After completely dealing with the first rung, the PLC turns its attention to the second rung. The status of *IN-0* and *OUT-0* are read from RAM. Using these data, the PLC performs the rung logic, which is an OR operation between *IN-0* and *OUT-0* (recall that *OUT-0* is the output of the top rung and is presently off). In this

Figure 12.18

Data in memory, showing the relationship between memory position and the I/O terminals.

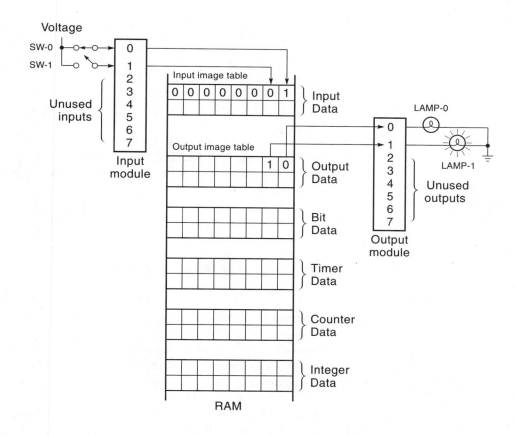

case, because SW-0 is on, *OUT-1 will* be energized, and this result is stored as a 1 bit in the Output Data section of RAM. This completes action on the second rung. If there were more rungs, each would be evaluated in sequence until all had been executed.

4. As a final action of the scan, the updated output data in RAM are sent to the output module, causing all eight output terminals to be updated at once. In this case, LAMP-0 would be off, and LAMP-1 would be on. Now the PLC will go into a rest state for a few seconds or less until the designated scan time has elapsed, and then the whole process starts over again with step 1.

Looking over the process just described, some observations are in order. First, the PLC accomplishes its control mission in the same way that all digital controllers work—by executing a loop (scan) over and over again. For each scan, inputs are read in, outputs are calculated based on the inputs, and then the outputs are sent out to the real world. This means that as much as one scan time's delay may be between an input and an output. Such a delay is not present in a true hard-wired relay circuit where the only delay would be the propagation time of the relays. This leads to the second observation: The order of the rungs in the ladder diagram may be important because the rungs are executed sequentially from top to bottom.

A final observation of the PLC hardware operation is that, for all practical purposes, it can do many independent control operations at the same time. We may be conditioned to think of a computer as being able to do only one thing at a time, albeit very rapidly. And, as we have seen, the PLC does execute its program one rung at a time. However, each rung is a separate circuit, and each control application (as far as the PLC is concerned) is only a set of rungs. Therefore, you could have one PLC running three different control applications at the same time. Each application would have its own assigned I/O terminal and its own set of rungs in the program. Within one scan cycle of the program, the PLC would update the three sets of outputs. This is similar to the chess master who plays three opponents simultaneously by going down the line, making one move on each board.

12.3   PROGRAMMING THE PLC

Ladder Diagram Programming

The most common way to program the PLC is to design the desired control circuit in the form of a relay logic ladder diagram and then enter this ladder diagram into a programming terminal (which could be either a PC, a dedicated PLC programming terminal, or a special handheld switch box). The programming terminal is capable of converting the ladder diagram into digital codes and then sending this program to the PLC where it is stored in memory. There are other ways to program a PLC besides ladder diagrams, and these will be introduced later in this chapter.

There are currently many manufacturers of PLCs, and though they all have basic similarities, the details of the hardware and software will, of course, be different. The following discussion on programming will be kept as general as possible but is patterned after the Allen–Bradley line of PLCs. This seems a logical choice because Allen–Bradley is one of the pioneers in the PLC field and its products are respected and well used by industry.

The PLC deals with two kinds of data that are maintained in memory: a **program file** in RAM or EEPROM, which contains the ladder logic program, and **data files** in RAM, which hold the changing data from the control application, such as switch settings. Table 12.1 lists some of the more common types of data files. Each data file occupies a portion of RAM memory and consists of any number of 8- or 16-bit words. This data arrangement is illustrated in Figure 12.18.

The Input and Output Data files are handled in a special way. The PLC automatically transfers the on–off condition of all input switches directly to the Input Data file in RAM as part of the scan. This is called making an *input image table* (as indicated in Figure 12.18). Notice that a closed switch becomes a 1 in RAM and an open switch becomes a 0. Similarly, data in the Output Data file (called the *output image table*) are automatically transferred to the output-module terminals.

Thus, the PLC program actually deals with only the image data in their various data files. That is why each device on the ladder diagram—be it switch, relay, timer, or

TABLE 12.1   Types of Data Files

	Data file	Description
I	Input Data	The current status of externally connected switches
O	Output Data	The status of those devices specified as outputs
B	Bit Data	The "scratch pad" to store individual bits
T	Timer Data	The data associated with timers
C	Counter Data	The data associated with counters
R	Control Data	The data associated with sequencers and other devices
N	Integer Data	Numbers, such as temperatures (in binary form)

whatever—is identified by its address in RAM. Even though the programmer may think of a certain switch as being, say, an overflow limit switch, the software thinks of it as being a particular bit in a particular data file.

There is an important fundamental difference between a relay logic ladder diagram and a PLC ladder diagram. *The relay ladder diagram is basically a wiring diagram*, and a rung becomes active when current can flow from one power rail to the other; that is, if there is a path for the current through the contacts, then the load will be energized. On the other hand, *the PLC ladder diagram is basically a logic diagram* consisting of symbols (called input and output instructions), and a rung becomes continuous when there is a logical TRUE path from rail to rail. Referring to Figure 12.17(b), if there is a continuous path of TRUE Input instructions on the left side of a rung, then the Output instruction (on the right side) will become TRUE. Therefore, writing a PLC program involves placing instructions in rungs so as to obtain the desired result. Naturally, you can only use instructions that are supported by the language being used, but most PLCs support the basic components such as switches, relays, timers, counters, and sequencers. We will discuss each type of instruction next.

Bit Instructions

Switches and relays are referred to as **bit instructions** because it takes only 1 bit to describe if a switch is open or closed. There are two kinds of switch instructions: one represents a NO switch and the other a NC switch. The following is the symbol and instruction for the NO switch:

NO Instruction: —] [—   (EXAMINE IF ON, Allen–Bradley)

When this symbol is placed in the ladder diagram, it behaves as a NO switch, push button, or set of relay contacts. It can be activated either by a real-world switch or another logical switch on the ladder diagram. In either case, when activated, the NO instruction in the ladder diagram goes TRUE. The state of the NO instruction is dictated by a particular bit in memory, where a 0 means off (FALSE) and a 1 means on (TRUE). The address of this bit is placed on the ladder diagram next to the instruction symbol.

The following is the symbol and instruction for the NC switch:

NC Instruction: —]/[—   (EXAMINE IF OFF, Allen–Bradley)

When this symbol is placed in the ladder diagram, it behaves logically in the same way any NC switch does—that is, it is normally TRUE and when activated goes FALSE (it behaves as a logical inverter). Care must be taken when using this instruction in conjunction with a real-world NC switch.*

The following symbol represents a relay coil or simply an output signal:

OUTPUT   —( )—   (OUTPUT ENERGIZE, Allen–Bradley)

If there is a continuous TRUE path through the rung, then the OUTPUT will go TRUE; thus, if this symbol represents a relay coil, the relay will energize. What really happens inside the PLC is that the bit in memory corresponding to this instruction goes to a 1. This bit may be used to drive an actual hardware relay through an output module or to control an instruction in another rung, or both. In fact, there is no limit to the number of times this bit can be used by other instructions in the program (which is another advantage PLCs have over hard-wired relay logic).

The operation of the three bit instructions are summarized in Table 12.2.

TABLE 12.2 **Summary of Bit Instructions**

If the Data File Bit Is	NO Instruction —] [—	NC Instruction —]/[—	OUTPUT —( )—
Logic 0	FALSE	TRUE	FALSE
Logic 1	TRUE	FALSE	TRUE

In a PLC ladder diagram, each symbol is accompanied by its RAM address (Figure 12.19) because this address tells the PLC where to find the present logical state of the symbol. Recall that the logical state of each symbol is always maintained in RAM. Titles of real-world components that the symbols represent, such as *Overload Sw*, are frequently added to improve clarity, but the PLC itself deals strictly with addresses. The address of an individual bit may be specified by three quantities: the file type, word number, and position of the bit within the word.† As previously stated, I/O signals usually have a direct relationship between their address and the physical connection point on an I/O module. For example, in the top rung of Figure 12.19, the address of *Flow sw* is given as I : 1/0. The I stands for Input Data file, the 1 means that the input module is plugged into slot 1 of the PLC chassis, and the 0 means that the switch is wired to terminal 0 of the module. The following is a more general interpretation of the *Flow sw* instruction address:

---

*This instruction can be confusing when people lose sight of the fact that a real-world NC switch and the NC instruction are *not* the same entity but that *one drives the other*; that is, the physical switch controls the state of the NC instruction. Suppose you actually wired a real NC switch to the PLC and used it to control an NC instruction. Then, if the real switch was activated, it would send a 0 to the PLC, and the logical NC instruction would invert this FALSE to a TRUE, which may be backward from what you expected. To avoid this pitfall, remember that the real-world switch may drive the logical switch, but they are *not* the same thing.

†The address system used in this chapter is a simplified version of the Allen–Bradley system.

Figure 12.19
A simple PLC program.

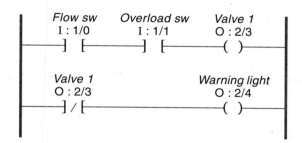

Type of data file (Input in this case)
  Address of word within the data file (or I/O module slot #)
    Bit position within the word, if appropriate (or terminal # of I/O module)
I:1/0

————] [————

The second NO instruction on the first rung (Figure 12.19) represents a switch wired to terminal 1 of the same input module used by *Flow sw*. The OUTPUT instruction activates terminal 3 of the output module plugged into slot 2.

The operation of the top rung in Figure 12.19 is as follows: The *Valve 1* output will go TRUE when there is continuity through the rung. Continuity will be established when both NO instructions are TRUE—in other words, when the contacts of both *Flow sw* and *Overload sw* are closed.

Operation of the second rung illustrates two important concepts. First, notice that the address of the NC instruction is the same as the OUTPUT instruction in the top rung, meaning that the NC instruction will be controlled by the *Valve 1* bit (located at address 0:2/3). Second, because it is an NC instruction, it will go TRUE only when *Valve 1* is FALSE. So *Warning light* will come on when the valve goes off.

Timers

The Timer instruction provides a time delay, performing the function of a time-delay relay. Examples of using time delays include controlling the time for a mixing operation or the duration of a warning beep. The length of time delay is determined by specifying a preset value. The timer is enabled when the rung conditions become TRUE. Once enabled, it automatically counts up until it reaches the Preset value and then goes TRUE (and stays TRUE). The symbol for the Timer instruction is a box or a circle with the pertinent information placed in or about it, as shown and explained in Figure 12.20. Notice that the Timer instruction is placed in the rung. When *Switch* I : 1/3 goes TRUE, the timer starts counting. Five seconds later, when the Accumulator value reaches the Preset value, the DN (done) bit goes TRUE. Notice that the NO instruction in the second rung has the DN bit from Timer T : 1 as its address. Therefore, it will go true after 5 s, causing the OUTPUT instruction at 0 : 2/4 to turn on a light.

In Figure 12.20, the timer is acting as an on-delay relay because it starts the delay when the rung goes TRUE. Off-delay timer instructions are also available.

Figure 12.20
The timer instruction.

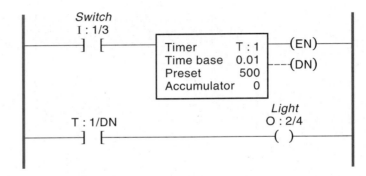

T : 1	The timer address (actually the first of three addresses in RAM) where the first address holds the status bits EN, TT, and DN; the second address holds the preset value; and the third address holds the accumulator value.
Time base	The value 0.01 means that each count corresponds to 0.01 seconds.
Preset	The value 500 means that the delay will last 500 counts, which in this case is 5 s (0.01s × 500 = 5 s).
Accumulator	Holds the value of the current count.
EN	A bit that is TRUE as long as the timer is enabled.
TT	A bit that is TRUE as long as the timer is counting that is, is TRUE for the time delay.
DN	A bit that goes TRUE when the timer is done—in other words, when the count gets to the preset value.

◆ **EXAMPLE 12.4**

A batch process—which involves filling a vat with a liquid, mixing the liquid, and draining the vat—is automated with a PLC. Figure 12.21(a) shows the hardware. The specific sequence of events is as follows: When a switch near the process is switched on,

1. A fill valve opens and lets a liquid into a vat until it's full.

2. The liquid in the vat is mixed for 3 min.

3. A drain valve opens and drains the tank.

Draw the ladder diagram for the PLC program.

Solution   Figure 12.21(b) shows the completed ladder diagram, which is explained rung-by-rung:

1. Rung 1 activates the fill valve and will go TRUE when the local-control on–off switch is on and the tank is less than full. A float-activated switch (Full switch) closes when the vat is full. Therefore, the *Full sw* NC instruction in rung 1 will

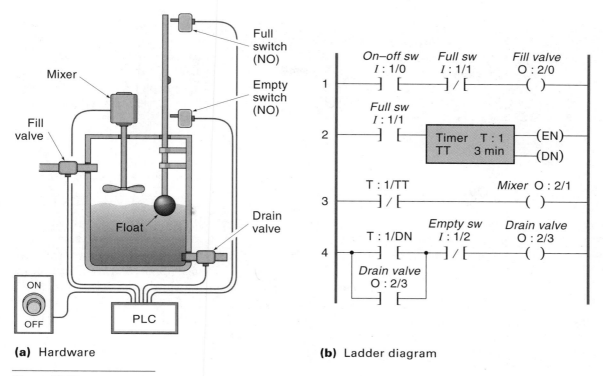

(a) Hardware   **(b)** Ladder diagram

Figure 12.21   Using a PLC timer instruction to control a batch-mixing operation.

be TRUE as long as the vat is not full (it changes to FALSE when the vat is full). As long as both input switches are TRUE, the OUTPUT instruction (*Fill valve*) will be TRUE, which turns on the actual fill valve. (During the fill period, none of the other rungs are TRUE.)

2. Rung 2 activates the 3-min mixing timer. When the vat is full, the physical full switch closes (goes to on), so the NO instruction (*Full sw*) goes TRUE, which starts Timer T : 1 (which has been programmed to provide the 3-min mixing time).

3. Rung 3 controls the mixer motor. As long as the TT bit of the timer is TRUE, the OUTPUT instruction *Mixer* will be TRUE, which activates the mixer motor.

4. Rung 4 activates the drain valve. After 3 min, the timer "times out," and the DN bit (T : 1/DN) goes TRUE, making the first instruction in rung 4 go TRUE. The NC instruction (*Empty sw*) will also be TRUE because the vat is still full (that is, the Empty switch remains unactivated). Therefore, rung 4 will be continuous, causing the OUTPUT instruction *Drain valve* to go TRUE, which opens the drain valve, allowing the liquid to leave. Here's where it gets interesting: As soon as the liquid starts to drain, the full switch will deactivate,

making rung 2 go FALSE, which causes the timer's DN bit to go FALSE. This latter action will cause the first instruction in rung 4 to go FALSE. So, to keep rung 4 TRUE until the vat has completely drained, the T : 1/DN instruction has a parallel branch instruction, which is activated by *Drain valve* itself. In effect, rung 4 is latched on until *Empty sw* activates and breaks the seal. ◆

Counters

A Counter instruction keeps track of the number of times some event occurs. The count could represent the number of parts to be loaded into a box or the number of times some operation is done in a day. Counters may be either count-up or count-down types. The Counter instruction is placed in a rung and will increment (or decrement) every time the rung makes a FALSE-to-TRUE transition. The count is retained until a RESET instruction (with the same address as the Counter) is enabled. The Counter has a Preset value associated with it. When the count gets up to the Preset value, the output goes TRUE. This allows the program to initiate some action based on a certain count; for example, after 50 items are loaded in a box, a new box is moved into place. The symbol for the Counter may be a box or a circle, with all pertinent data placed in or about it, as shown and explained in Figure 12.22.

The ladder diagram in Figure 12.22 is a simple circuit that counts the number of times that *Switch* (address I : 1/2) is opened and closed. The Counter can be reset at any time by closing *Reset sw* because this action activates the RESET instruction. The third rung uses the DN bit from the Counter to turn on a light when the count gets to 50 or higher.

Sequencers

The Sequencer instruction is used when a repeating sequence of outputs is required (recall the example of a dishwasher that cycles through predetermined steps). Traditionally, electromechanical sequencers (Figure 12.10) were used in this type of application (where a drum rotates slowly, and cams on the drum activate switches). The Sequencer instruction allows the PLC to implement this common control strategy.

Figure 12.23 shows the PLC Sequencer instruction. Operation is as follows: The desired output-bit patterns for each step are stored (sequentially) as words in memory—as usual, each bit in the word corresponds to a specific terminal of the output module. Every time the Sequencer instruction steps, it connects the next output pattern in memory to the designated output module. The address of the first pattern (word) in the sequence is given in the Sequencer instruction as the Sequencer File Adr (which is address B : 1 in Figure 12.23). The Sequencer shown in Figure 12.23 steps when the rung makes a FALSE-to-TRUE transition. Other versions of the Sequencer instruction allow the programmer to insert a time value for each step.

Figure 12.22
The Counter instruction.

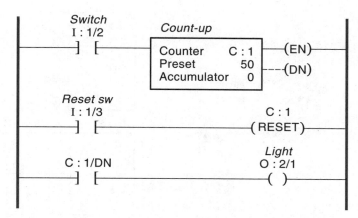

C : 1	The counter address (actually the first of three addresses) where the first address holds such bits as CU and DN, the second address holds the preset value, and the third address holds the accumulated value.
Preset	When the counter gets to the Preset value, it makes the DN bit go TRUE and keeps on counting.
Accumulator	Holds the value of the current count.
EN	Goes TRUE when the counter rung is TRUE.
DN	Stands for *done*—goes TRUE when the count meets or exceeds the PRESET value.
RESET	A separate instruction with the same counter address; when this bit goes TRUE, the Accumulator resets to zero (resets the counter).

Figure 12.23
The Sequencer instruction.

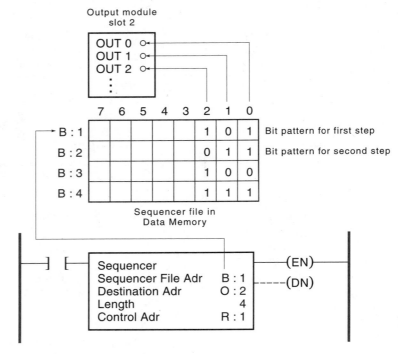

Sequencer Adr	
Sequencer File Adr	The first address of the Sequence file, which stores the sets of outputs.
Destination Adr	The output address (or slot) to which the Sequence file words are transferred with each step.
Length	The length of the Sequence file, that is, the number of steps in the sequence.
EN	Goes TRUE when the Sequencer rung is TRUE.
DN	Stands for *done*—goes TRUE when it has operated on the last word in the Sequence file.
Control Adr	Address that stores the control bits (EN, DN) and words (length) of the sequencer.

♦ **EXAMPLE 12.5**

A process for washing parts requires the following sequence:

1. Spray water and detergent for 2 min (*Wash cycle*).
2. Rinse with water spray only for 1 min (*Rinse cycle*).
3. Water off, air blow dry for 3 min (*Drying cycle*).

The sequence is started with a toggle switch. Draw the ladder diagram for this process (using the Sequencer instruction).

Solution   The completed ladder diagram (Figure 12.24) consists of six rungs. Rungs 1 and 2 generate the 2-min time delay for *Wash cycle*. Rungs 3 and 4 generate the 1-min time delay for *Rinse cycle*, and rung 5 generates the 3-min time delay for *Drying cycle*. Rung 6 has the Sequencer instruction. The Sequencer instruction specifies that the Sequence file starts at address B : 1, that the file is three words long, and that the output module is in slot 2. Looking at the Sequence file, we see that bit 0 controls the water valve, bit 1 controls the detergent valve, and bit 2 controls the blow-dry fan.

Figure 12.24   A Sequencer program (Example 12.5).

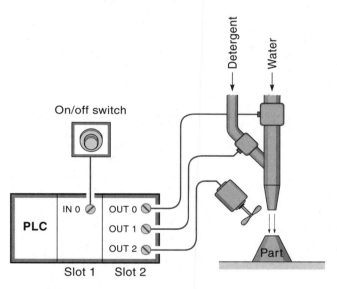

(a) Hardware

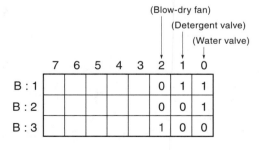

(b) Sequencer File

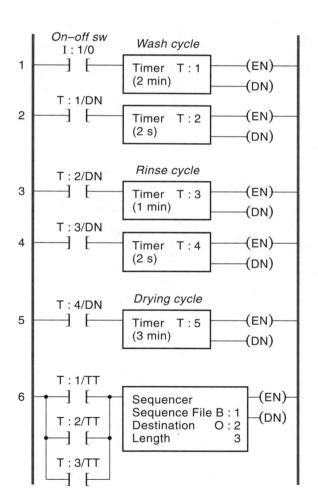

(c) Ladder diagram

Operation Overview

The program consists of five timer rungs and a sequencer rung. The timers are activated, one after the other, down the program—that is, each timer, when finished, starts the timer next in line. The outputs of Timers T : 1, T : 3, and T : 5 are used to step the Sequencer instruction. The Sequencer instruction presents three output-bit patterns to module 2.

Detailed Operation

Operation of the ladder program is as follows: The action starts on rung 1 when the operator switches the on–off toggle switch to on. This provides a FALSE-to-TRUE transition that starts Timer T : 1 (*Wash cycle*), causing its timing bit (TT) to go TRUE for 2 min. This same timing bit (T : 1/TT), makes rung 6 go TRUE, stepping the sequencer to its first position (connecting the word at address B : 1 to output module 0 : 2). Notice that bit 0 and bit 1 of word B : 1 are 1s, which cause terminals OUT-0 and OUT-1 to go on, turning on the water and detergent valves.

After 2 min, the DN bit of Timer T : 1 (T : 1/DN) goes TRUE, which starts the next Timer (T : 2 in rung 2). This timer is only 2 s long, and its purpose is to create a short gap between the steps (to allow the sequencer rung to reset).

When Timer T : 2 is done, its DN bit (T : 2/DN) starts the 1-min Timer T : 3 (*Rinse cycle*). Timer T : 3 timing bit (T : 3/TT) causes the Sequencer instruction to advance to the second step (B : 2). Notice that in word B : 2, only bit 0 is a 1, causing OUT-0 to remain on (leaving the water on, but turning off the detergent).

Following in the same manner, Timer T : 4 provides a short gap, and then Timer T : 5 advances the sequencer to the third and last position (B : 3) for the 3-min *Drying cycle*. Notice that bit 2 of word B : 3 becomes OUT-2, and turns on the blow-dry fan.

There are many possible ways to program this sequence. The method chosen (Figure 12.24) is not necessarily the shortest, but it is a clear, uncomplicated solution that in real life is more valuable than a clever short program (that no one can understand except the programmer!). Also, remember that extra rungs on a PLC ladder program do not mean more hardware (it does mean a slightly longer scan time and memory usage, which in most cases is not an issue).   ◆

Advanced Instructions

So far, we have just examined the basic instructions supported by virtually any PLC. PLCs are getting more sophisticated, and as they do, their instruction sets expand. In fact, many newer PLCs more closely resemble a real-time process-control computer than simply a substitute for relay logic. Many of the advanced instructions operate on a digital word instead of a single bit. Some of these words may be digitized analog data (such as temperature) used with an analog I/O module, which is an ADC or a DAC. Counter and timer accumulation values are stored as words. In many cases, the purpose of processing these number quantities is to arrive at a yes–no decision, where the actual output of the program is still in terms of individual bits (that is, turning a motor on or off). Then too, it is also possible to output a whole digital word to be converted to an analog voltage by an I/O module. Such a voltage could be used to control a motor's speed.

The higher level instructions come in various categories:

❏ **Comparison instructions** compare two words and yield a single bit as a result. For example, a word may be tested to see if it is larger than another word or to see if it is equal to another word. These instructions are demonstrated in Example 12.6.

❏ **Math instructions** perform mathematical operations on words in memory. These operations may include addition, subtraction, multiplication, division, and others. A math instruction might be used to multiple a sensor input word by a constant, for scaling purposes.

❏ **Logical and shift instructions** perform logical operations such as AND, OR, and NOT and Right and Left Shifts on words stored in memory. For example, the AND instruction could be used to mask out certain bits in a word.

❏ **Control instructions** allow for such things as jumps and subroutines. For example, a Jump instruction, when TRUE, will cause the execution to jump ahead (or back) to any designated rung.

❏ **PID control** is available on some PLCs. In some cases, it is built into an I/O module; in other cases, it is programmed as an instruction in the ladder diagram (in the same way that a timer is).

❏ **I/O message and communications instructions** allow for PLCs to send and receive messages and data to other PLCs or a terminal (PC), for those applications where PLCs are networked together.

Using a PLC as a Two-Point Controller

Example 12.6 shows how a PLC can be used to implement a control strategy that requires decisions based on analog values. Although it is a simple example, it demonstrates the use of analog inputs and instructions that deal with analog values.

♦ **EXAMPLE 12.6**

The temperature in an electric oven is to be maintained by a 16-bit PLC at approximately 100°C, using two-point control (actual range: 98–102°). Figure 12.25(a) shows the system hardware: an oven with an electric heating element driven by a high-current relay, an LM35 temperature sensor (produces 10 mV/°C—see Chapter 6), an operator on–off switch, and the PLC. The PLC has a processor and three I/O

Figure 12.25

A two-point controller program (Example 12.6).

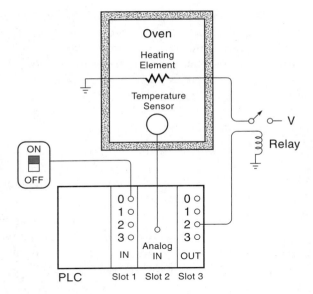

(a) Hardware

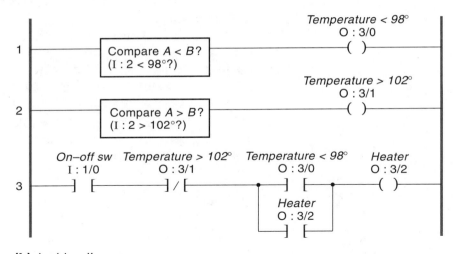

(b) Ladder diagram

modules: a discrete input module (slot 1), a 16-bit analog input module (slot 2), and a discrete output module (slot 3).

Draw the ladder diagram for this system.

Solution   We are asking the PLC to perform the job of a thermostat: When the oven temperature falls below 98°, the heating element will turn on and stay on until the temperature reaches 102°. Thus, the PLC will be giving a discrete output (to the heater), based on an analog input (temperature).

The ladder diagram [Figure 12.25(b)] consists of three rungs that implement the control logic; however, recall that *before* the PLC executes the rungs, it reads in all input module data. In this case, it will read in the condition of the on–off switch from the input module in slot 1 (storing it to memory at address I : 1/0), and it will read in the value of the temperature through the analog input module in slot 2. The analog module contains an ADC that converts the analog voltage to a 16-bit word. The resolution of the module is given as

$$\text{ADC resolution} = 0.305176 \text{ mV} = 1 \text{ LSB*}$$

Using this resolution, we can find the digital equivalent of the lower- and upper-limit temperatures, 98° and 102°:

$$\text{Analog input voltage for } 98° = 98° \times \frac{10 \text{ mV}}{1°C} = \underbrace{980 \text{ mV}}_{\text{Output of LM35}}$$

$$\text{ADC output for } 98° = 980 \text{ mV} \times \frac{\text{LSB}}{0.305176 \text{ mV}} = 3211 \text{ LSB} = 110010001011$$

$$\text{Analog input voltage for } 102° = 102° \times \frac{10 \text{ mV}}{1°C} = \underbrace{1020 \text{ mV}}_{\text{Output of LM35}}$$

$$\text{ADC output for } 102° = 1020 \text{ mV} \times \frac{\text{LSB}}{0.305176 \text{ mV}} = 3342 \text{ LSB} = 110100001110$$

We now know the digital values of the upper- and lower-limit temperatures. We will need these numbers in the program. The operation of the ladder program follows [refer to Figure 12.25(b)]:

1. Rung 1 determines if the temperature is below 98°C and will go TRUE when the conditions of the Compare instruction are met. This particular Compare instruction compares two 16-bit words and goes TRUE if the value of word A is less than word B. In this case, word A is the actual temperature (as converted by the ADC and stored at address I : 2), and word B is the binary equivalent of the lower-limit temperature of 98°. Therefore, the rung will go TRUE if the oven temperature is less than 98°. The value of this rung is stored in bit 0 : 3/0.

2. Rung 2 determines if the temperature is above 102°C. This rung has another Compare instruction, but this one goes TRUE if word A is greater than word B. Notice that word A is again the actual temperature (from the ADC), and word B is the upper-limit temperature of 102°. Therefore, the rung will go TRUE if the oven temperature is greater than 102°. The value of this rung is stored in bit 0 : 3/1.

3. Rung 3 activates the heating element via the control logic that turns the heating element on and off—that is, the OUTPUT instruction *Heater* (0 : 3/2) directly controls the heating element. The rung will go TRUE if the on–off switch is on, *and* the temperature is not over 102° (notice this is an NC instruction), *and* the temperature is less than 98°. Once *Heater* is on, it is sealed on with the parallel

*This ratio would be found on the specification sheet and comes from the following analysis: Input voltage range $= \pm 10 \text{ V} = 20 \text{ V}$; output is 16 bits $= 65,535$ states (2s complement); LSB $= 20 \text{ V}/65,535 = 0.305176 \text{ mV}$.

branch 0 : 3/2 so that, even when the oven temperature rises above 98°, the rung will stay TRUE. With the heating element on, the oven temperature will eventually rise to above 102°, and so the NC instruction (*Temperature* > *102°*) will go FALSE, breaking the seal and turning *Heater* off. The rung will stay FALSE until the temperature drops below 98°, and then the cycle starts over. ♦

Other PLC Programming Languages

Ladder diagram programs are highly symbolic and are the result of years of evolution of industrial control circuit diagrams. The PLC is a relative latecomer, and it was adapted to use traditional ladder logic. However, as we have noted, a PLC program is not actually a wiring diagram but a way to describe the logical relationship between inputs and outputs. Other, more concise ways can convey control logic than ladder diagrams. These programs look more like conventional computer programs, using words and mathematical symbols instead of pictures.

Generally speaking, there are two other types of PLC programming languages (besides ladder diagrams). One type of PLC language used by Allen–Bradley is called simply *ASCII programming*, or "creating an ASCII archive file." Basically, this is a method of conveying a ladder diagram but using words instead of pictures. Each instruction is given a three-letter designation called a **mnemonic**. Table 12.3 lists the more common instructions. In Table 12.3, the first three mnemonics represent such devices as switches and relays, and the last three mnemonics specify how the devices are interconnected logically. In an ASCII program, the ladder diagram is conveyed by placing the mnemonics and addresses in a certain order. These concepts are best shown by Figure 12.26, a simple ladder diagram followed by the corresponding ASCII program.

Another type of PLC programming language uses Boolean logic directly. This involves listing commands such as AND, OR, and NOT in such a way as to describe the program logic. Again, an example is the best way to convey this approach; the following is a Boolean program for the ladder diagram shown in Figure 12.26:

TABLE 12.3 **ASCII Programming Instructions**

Symbol	Mnemonic	Explanation
—] [—	XIC (examine if closed)	Goes TRUE when the contacts close.
—] / [—	XIO (examine if open)	Goes TRUE when the contacts open.
—( )—	OTE (output energize)	Energizes the relay when TRUE.
—┬—	BST (branch start)	Starts a parallel branch.
└ ┤	NXB (next branch)	Ends previous branch, and starts new branch.
___┘	BND (branch ends)	Ends the parallel branches.

Figure 12.26

A ladder diagram and the corresponding ASCII program.

Rung 1

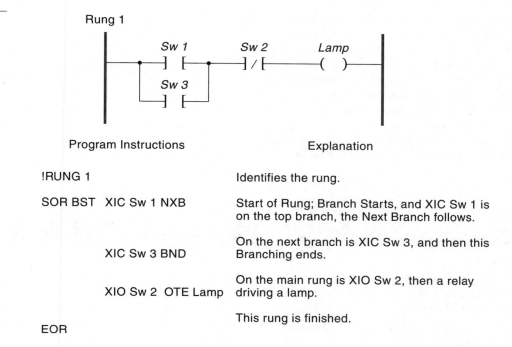

Program Instructions	Explanation
!RUNG 1	Identifies the rung.
SOR BST XIC Sw 1 NXB	Start of Rung; Branch Starts, and XIC Sw 1 is on the top branch, the Next Branch follows.
XIC Sw 3 BND	On the next branch is XIC Sw 3, and then this Branching ends.
XIO Sw 2 OTE Lamp	On the main rung is XIO Sw 2, then a relay driving a lamp.
	This rung is finished.
EOR	

Program Instruction		Explanation
START	Sw 1	Switch 1 is first logical variable.
OR	Sw 3	Switch 3 is in parallel with Switch 1 (which is the same as an OR operation).
AND		The previous result will be ANDed with
NOT	Sw 2	the inverse of the state of Switch 2.
OUT	Lamp	The result of the logic equation is used to activate the Lamp output.

SUMMARY

One common type of control system governs the repetitious sequence of events. Traditionally, this kind of control was implemented with electromechanical relays, timers, and sequencers. Relays can be interconnected to provide some control logic—for example, two relays in series become an AND function. Time-delay relays will delay activating their contacts for some period of time (after being energized). Sequencers use a rotating cam to open and close contacts in some sequence.

A ladder diagram, a special kind of wiring diagram, was developed to document electromechanical control circuits.

This type of diagram (which resembles a ladder) has two vertical wires (rails) on either side of the drawing; these wires supply the power. Each rung of the ladder diagram connects from one rail to the other and is a separate circuit, which typically consists of some combination of switches, relay contacts, relay coils, and motors. It is common for the coil of a relay to be in one rung and the contacts to be in another.

In the late 1960s, the programmable logic controller (PLC) was developed to replace electromechanical controllers. The PLC is a small, microprocessor-based, process-control computer that can be connected directly to switches, relays, small motors, and so on, and is built to withstand the industrial environment. The PLC executes a program that must be written for each specific application. The program is always in the form of a loop, which has the following pattern: The PLC reads and stores the input switch and sensor data; it determines what the output(s) should be; it sends the output control signals to the devices to be controlled.

To use a PLC requires setting up the hardware and software. The hardware installation consists of wiring the PLC to all switches and sensors of the system and to such output devices as relay coils, indicator lamps, or small motors. The control program is usually developed on a PC, using software provided by the PLC manufacturer. This software allows the user to develop the control program in the form of a ladder diagram on the monitor screen. Once the program is complete, it is automatically converted into instructions for the PLC processor. The completed program is then downloaded into the PLC. Once the program is in the PLC's memory, the programming terminal can be disconnected, and the PLC will continue to function on its own.

GLOSSARY

analog I/O module   A module that plugs into the PLC and converts real-world analog signals to digital signals for the PLC and vice versa.

bit instruction   The basic PLC instruction that logically represents a simple switch or relay coil; the state of each switch or coil occupies 1 bit in memory.

comparison instruction   Compares two words and yields a single bit as a result.

control instruction   Allows for such things as jumps and subroutines.

data file   The collection of data in a PLC that represents the present condition of switches, counters, relays, and the like.

data memory   The section of RAM memory in a PLC where the data file is stored.

discrete I/O   I/O that deals with on–off devices. Discrete input modules are used to connect real-world switches to the PLC, and discrete output modules are used to turn on lamps, relays, motors, and so on.

electromechanical counter   A device that counts events. Each time it receives a pulse, it increments a mechanical counter, which can then be read by an operator or used to activate some other process.

electromechanical relay (EMR)   A device that uses an electromagnet to close (or open) switch contacts—in other words, an electrically powered switch.

electromechanical sequencer   A device that provides sequential activation signals; it works by rotating a drum with cams, which activates switches.

EMR   *See* **electromechanical relay**.

I/O message and communications instructions   Allow PLCs to send and receive messages and data to other PLCs or a terminal (PC), for those applications where PLCs are networked.

Input/Output modules   The connection points where real-world switches, relays, and the like are connected to the PLC.

ladder diagram   A type of wiring diagram (which resembles a ladder) used for control circuits; ladder diagrams typically include switches, relays, and the devices they control.

latching   *See* **sealing**.

logical and shift instructions   Perform logical operations such as AND, OR, and NOT and Right and Left Shifts on words stored in memory.

math instructions   Performs mathematical operations on words in memory; these operations may include addition, subtraction, multiplication, division, and others.

mnemonic   An English-sounding abbreviation for PLC logical instructions.

normally closed (NC) contacts   Relay contacts that are closed in the de-energized state and open when the relay is energized.

normally open (NO) contacts   Relay contacts that are open in the de-energized state and closed when the relay is energized.

off-delay relay   A time-delay relay that, after being de-energized, waits a specified period of time before activating.

on-delay relay   A time-delay relay that, after being energized, waits a specified period of time before activating.

PID control   A common control strategy. PID instructions are available on some PLCs.

PLC   *See* **programmable logic controller**.

PLC bus   The wires in the backplane of a modular PLC system that allows the microprocessor (within the PLC) to communicate with plugged-in modules.

plug-in module   An interface circuit to the PLC, packaged as a separate module, which plugs into the PLC. Different modules support different I/O devices, from simple switch and relay output modules to ADC and DAC modules.

pneumatic time-delay relay   A time-delay relay that generates the delay by allowing air to bleed out of a spring-loaded bellows.

power rails   Part of a ladder diagram, the two vertical lines on each side of the diagram, which represent power lines.

processor   Part of the PLC that executes the instructions; it is a small, microprocessor-based computer.

program file   Contains the program that the PLC executes. The program file is part of the PLC memory (the other part being data files).

programmable logic controller (PLC)   A small, self-contained, rugged computer, designed to control processes and events in an industrial environment—to take over the job previously done with relay logic controllers.

program memory   When applied to PLCs, the section of memory (RAM or EEPROM) where the PLC program is stored.

programming port   A connector on a PLC through which the program is downloaded.

run mode   The operational mode of the PLC when it is actually executing the control program and hence performing some control operation.

scan   The event when the processor makes one pass through all instructions in the control program loop.

sealing   A relay wired such that once energized its own contacts take over the job of providing power to the coil; it will stay energized, even if the original energizing signal goes away.

solid-state time-delay relay   A time-delay relay that generates the delay with an electronic circuit.

thermal time-delay relay   A time-delay relay that generates the delay by allowing a bimetallic strip to heat and bend, which activates a switch.

time-delay relay   A relay with an intentional delay action; that is, the contacts close (or open) in a specified period of time *after* the relay is energized (or deenergized).

EXERCISES

Section 12.1

1. Draw a relay wiring diagram for a circuit that implements the simple logic diagram shown in Figure 12.27. The circuit should be in the style of Figure 12.1(a).

2. Repeat Exercise 1 except draw the circuit as a ladder diagram.

3. A small house has three windows and two doors. Each window and door has a switch attached such that the contacts close when a door or window opens. Draw a ladder logic diagram that will turn on a light if one or more windows are open or if both doors are open.

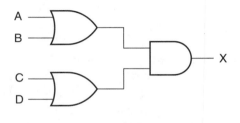

Figure 12.27

4. In a processing plant, jars on a conveyer belt are cleaned out with a high-pressure air jet just prior to being filled (Figure 12.28). When a jar approaches the cleaning station, it activates a switch (with both NO and NC contacts). The conveyer belt stops for 10 s while the air jet is on; then the conveyer belt starts, and the jar moves along. Draw a ladder logic diagram to control this process.

Section 12.2

5. You just bought a PLC, and it's still in the box. List the general steps you will have to take to get the PLC operational in a specific task.

6. A PLC has eight inputs and eight outputs as shown in Figure 12.17(a). Draw a wiring diagram for this PLC that will be used as the controller for the situation of Exercise 3.

7. A PLC has eight inputs and eight outputs as shown in Figure 12.17(a). Draw a wiring diagram for this PLC that will be used as the controller for the situation of Exercise 4.

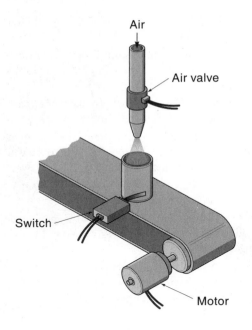

Figure 12.28

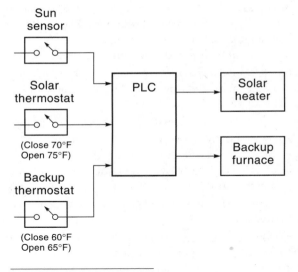

Figure 12.29

8. List the steps the PLC takes to execute the ladder diagram program.

Section 12.3

9. Draw a PLC ladder diagram for the situation described in Exercise 3.

10. A motor is controlled by a NO start button, a NC stop button, and an overload device. The overload device is normally closed and opens if the motor overheats. Draw the PLC ladder diagram for the circuit. Include a light that comes on when the motor overheats.

11. A PLC is to control the solar heating system shown in Figure 12.29. The system has two interrelated parts: (a) A solar thermostat turns on and off the solar heater *if* the sun sensor says the sun is shining, and (b) a backup thermostat turns on and off a conventional furnace *if* the solar energy is insufficient. Both heating systems share the same ductwork, so if the backup thermostat closes, the PLC *must* turn on the backup furnace (and turn off the solar heater if it's on). Draw a PLC ladder diagram for this system.

12. A mixing vat has an inlet valve, an outlet valve, a mixing motor, and a single level-detecting switch (Figure 12.30). Both valves are opened by solenoids. The level switch closes when the vat is full and stays closed until

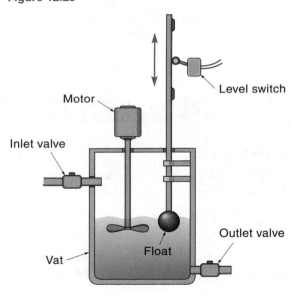

Figure 12.30

the vat is empty. Draw a PLC ladder logic diagram to do the following:
   a. When the start button is pushed, the inlet valve opens until the vat is full.
   b. The mixer motor comes on for 5 min.
   c. The outlet valve opens until the vat is empty.

13. Modify the two-point control program of Example 12.6 so that the oven temperature limits are 72–76°C. The system should be started and stopped with NO push-button switches (instead of toggle switches).

APPENDICES

Appendix A   The Getting Started Guide for APS
(A tutorial for setting up and using the Allen–Bradley
SLC 500 Programmable Logic Controller)   449

Appendix B   Glossary   511

Appendix C   Answers to Odd-Numbered Exercises   527

The Getting Started Guide for APS

Allen-Bradley

The
Getting Started
Guide for APS
(Cat. No. 1747-PA2E)

User
Manual

Important User Information

Solid state equipment has operational characteristics differing from those of electromechanical equipment. "Safety Guidelines for the Application, Installation and Maintenance of Solid State Controls" (Publication SGI-1.1) describes some important differences between solid state equipment and hard–wired electromechanical devices. Because of this difference, and also because of the wide variety of uses for solid state equipment, all persons responsible for applying this equipment must satisfy themselves that each intended application of this equipment is acceptable.

In no event will the Allen-Bradley Company be responsible or liable for indirect or consequential damages resulting from the use or application of this equipment.

The examples and diagrams in this manual are included solely for illustrative purposes. Because of the many variables and requirements associated with any particular installation, the Allen-Bradley Company cannot assume responsibility or liability for actual use based on the examples and diagrams.

No patent liability is assumed by Allen-Bradley Company with respect to use of information, circuits, equipment, or software described in this manual.

Reproduction of the contents of this manual, in whole or in part, without written permission of the Allen-Bradley Company is prohibited.

Throughout this manual we use notes to make you aware of safety considerations.

 ATTENTION: Identifies information about practices or circumstances that can lead to personal injury or death, property damage, or economic loss.

Attentions help you:
- identify a hazard
- avoid the hazard
- recognize the consequences

Important: Identifies information that is especially important for successful application and understanding of the product.

SLC 500, SLC 5/01, SLC 5/02, SLC 5/03, and SLC 5/04 are trademarks of Allen-Bradley Company, Inc.
Gateway 2000 is a trademark of Gateway 2000, Inc.
Windows is a registered trademark of Microsoft, Inc.
IBM is a registered trademark of International Business Machines, Inc.
VERSA is a trademark of Nippon Electric Co. Information Systems Inc.

Summary of Changes

The information below summarizes the changes to this manual since the last printing as 1747–NM001 Series A.

To help you find new information and updated information in this release of the manual, we have included change bars as shown to the right of this paragraph.

Updated Information

Changes from the previous release that require you to perform a procedure differently or that require different equipment are listed below:

The demo unit is now assumed to contain an SLC 5/02 ™ processor rather than an SLC 5/01 ™ processor. See chapter 3, Configuration of SLC 500 Controllers.

Table of Contents

Preface

Who Should Use this Manual P–1
Purpose of this Manual P–1
    Contents of this Manual P–2
    Related Documentation P–2
Common Techniques Used in this Manual P–3
Allen–Bradley Support P–3
    Local Product Support P–3
    Technical Product Assistance P–3
    Your Questions or Comments on this Manual P–3

Setting Up Your Equipment   **Chapter 1**

Hardware Requirements 1–1
Controller Styles 1–2
Setting Up a Demo Unit 1–2
Setting Up a Field–Wired Controller 1–4
Connecting the Controller to a Personal Computer 1–4
Personal Computer Requirements 1–6
Using Extended Memory 1–7
Using Expanded Memory 1–7
Using APS with DOS 1–8
Installing the Software 1–8
Running APS 1–10
APS Display Format 1–11

Control Basics   **Chapter 2**

SLC 500 File Concepts 2–1
    Processor Files 2–1
    Program Files 2–2
    Data Files 2–2
How External I/O Devices Communicate with the Processor 2–3
Addressing External I/O 2–4
    APS Display of Instructions/Addresses 2–4
Ladder Logic Concepts 2–5
    True/False Status 2–5
    Logical Continuity 2–6
    The Processor Operating Cycle 2–7

Creating a Processor File

Chapter 3

Configuration of SLC 500 Controllers 3–1
    Controller Styles 3–2
        Slot Numbers 3–2
    Catalog Numbers 3–2
        Make a Record of Controller Components 3–2
    Arbitrary Controller Used in this Guide 3–4
Creating a Processor File 3–4
    Name the Processor File and Configure the Controller 3–5
    Enter the Ladder Program 3–8
    Add a Rung Comment 3–9
    Save the Processor File 3–10

Online Operations, Quick Edit

Chapter 4

Restoring (Downloading) a Processor File 4–1
    Check the Online Configuration Parameters 4–1
    Go Online and Restore (Download) Processor File GETSTART 4–2
Testing the Program 4–4
Editing the Program with Quick Edit 4–5
Monitoring Data Files 4–7

Creating and Printing Reports

Chapter 5

Creating Reports 5–2
Printing Reports 5–3

Additional Ladder Program Exercises

Appendix A

Entering an Input and Output Branch A–1
    Exercise 1: Entering an Input and Output Branch A–1
    Save the Processor File A–3
    Test the Ladder Program A–3
Entering a Timer Instruction A–4
    Exercise 2: Entering a Timer Instruction A–4
    Save the Processor File A–6
    Test Your Ladder Program A–6

Troubleshooting

Appendix B

APS Error Messages B–1
System LED Status B–3
Processor Error Codes B–4

Glossary

Preface

Read this preface to familiarize yourself with the rest of the manual. This preface covers the following topics:

- who should use this manual
- the purpose of this manual
- terms and abbreviations
- conventions used in this manual
- Allen–Bradley support

Who Should Use this Manual

The Getting Started Guide is intended as an introduction of APS software to first–time users. The simple tasks and practice exercises in this guide do not include important user information for actual control applications.

Purpose of this Manual

This manual is an introductory document, designed to allow you to install APS and begin programming in the shortest time possible. It does this by focusing on a simple controller and a simple program. Basic concepts are presented, but only with enough detail to get you started and let you know that there is more to be learned. Read chapter 1 first. It will acquaint you with the rest of the guide.

Contents of this Manual

Chapter	Title	Purpose
	Preface	Describes the purpose, background, and scope of this manual. Also specifies the audience for whom this manual is intended.
1	Setting up Your Equipment	Lists hardware requirements and shows you how to set up a controller, connect your PC to the controller, and install APS software on your PC.
2	Control Basics	Presents basic information you will need to know before you can begin programming with APS.
3	Creating a Processor File	Shows you how to create a processor file, enter a ladder program and add a rung comment.
4	Online Operations, Quick Edit	Shows you how to restore (download) your processor file to the controller, monitor and test the program, and use Quick Edit.
5	Creating and Printing Reports	Guides you through creating and printing reports. These include program listing, cross reference, processor configuration, and data tables.
Appendix A	Additional Ladder Program Exercises	Introduces you to branching of instructions and the timer instruction.
Appendix B	Troubleshooting Errors	Provides a listing of error messages that you may encounter while working through the guide. Also, offers possible solutions for these errors.
	Glossary	Provides a listing of terms used throughout this guide.

Related Documentation

The table below is a partial list of publications that contain information about installation, programming, and operation of SLC 500 controllers. To obtain a copy, contact your local Allen–Bradley office or distributor.

For	Read this Document	Document Number
A description on how to install and use your *Modular* SLC 500 programmable controller	Installation & Operation Manual for Modular Hardware Style Programmable Controllers	1747–6.2
A description on how to install and use your *Fixed* SLC 500 programmable controller	Installation & Operation Manual for Fixed Hardware Style Programmable Controllers	1747–NI001
A procedural manual for technical personnel who use APS to develop control applications	Allen–Bradley Advanced Programming Software (APS) User Manual	1747–6.4
A reference manual that contains status file data, instruction set, and troubleshooting information about APS	Allen–Bradley Advanced Programming Software (APS) Reference Manual	1747–6.11
A training and quick reference guide to APS	SLC 500 Software Programmer's Quick Reference Guide—available on PASSPORT at a list price of $50.00	ABT–1747–TSG001
A complete listing of current Allen–Bradley documentation, including ordering instructions. Also indicates whether the documents are available on CD–ROM or in multi-languages.	Allen–Bradley Publication Index	SD499
A glossary of industrial automation terms and abbreviations	Allen–Bradley Industrial Automation Glossary	AG–7.1

Common Techniques Used in this Manual

The following conventions are used throughout this manual:

- Bulleted lists such as this one provide information, not procedural steps.
- Numbered lists provide sequential steps or hierarchical information.
- *Italic* type is used for emphasis.
- Text in `this font` indicates words or phrases you should type.
- Key names match the names shown and appear in bold, capital letters within brackets (for example, [ENTER]). A function key icon matches the name of the function key you should press, such as

  SAVE & EXIT
  F8

Allen–Bradley Support

Allen–Bradley offers support services worldwide, with over 75 Sales/Support Offices, 512 authorized Distributors and 260 authorized Systems Integrators located throughout the United States alone, plus Allen–Bradley representatives in every major country in the world.

Local Product Support

Contact your local Allen–Bradley representative for:

- sales and order support
- product technical training
- warranty support
- support service agreements

Technical Product Assistance

If you need to contact Allen–Bradley for technical assistance, please call your local Allen–Bradley representative.

Your Questions or Comments on this Manual

If you find a problem with this manual, please notify us of it on the enclosed Publication Problem Report.

If you have any suggestions for how this manual could be made more useful to you, please contact us at the address below:

Allen–Bradley Company, Inc.
Automation Group
Technical Communication, Dept. J602V, T121
P.O. Box 2086
Milwaukee, WI 53201–2086

Chapter 1

Setting up Your Equipment

This chapter briefly describes hardware requirements and SLC 500 ™ controller styles, then shows you how to set up your equipment in preparation for the exercises in later chapters. Topics include:

- Hardware Requirements
- Controller Styles
- Setting up a Demo Unit
- Setting up a Field–Wired Controller
- Connecting the Controller to a Personal Computer
- Personal Computer Requirements
- Using Extended Memory
- Using Expanded Memory
- Using APS with DOS
- Installing the Software
- Running APS
- APS Display Format

Hardware Requirements

To perform the tasks provided in this manual we recommend the following hardware:

- An SLC 500 modular or fixed controller with external inputs and outputs. An SLC 500 demo unit would be ideal. The programs and examples used in this guide are based on using a modular controller demo unit (catalog no. 1747–DEMO 3 or 1747–DEMO 4).
- A compatible personal computer (PC). A list appears on page 1–6.
- An RS–232/DH–485 Interface Converter (catalog no. 1747–PIC).
- A communications cable for connecting the Interface Converter to the controller (catalog number. 1747–C10). This cable is supplied with the Interface Converter.
- A compatible printer, if you choose to use the "Print Reports" capability described in chapter 5.

Controller Styles

The SLC 500 comes in two different styles: modular and fixed. These styles are illustrated below. The modular controller consists of a chassis, power supply, processor (CPU), and Input/Output (I/O) modules. The fixed controller consists of a power supply, processor (CPU), and a fixed number of I/O contained in a single unit. An expansion chassis can be added to the fixed controller.

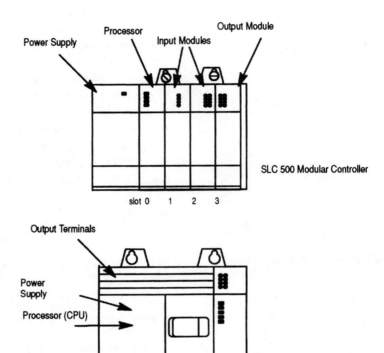

Further information on hardware is found in the Installation and Operation Manuals, catalog number 1747–NI001 (fixed controllers) and publication number 1747–6.2 (modular controllers).

Setting Up a Demo Unit

SLC 500 demo units are available with either a fixed controller or modular controller. This guide assumes you are using a modular controller demo unit for all the programming exercises. If you use a fixed controller demo unit, you will need to use different configuration information and I/O addresses in the exercises. This is explained later.

The figure that follows shows an SLC 500 modular controller demo unit. It is completely wired, with 12 external inputs (6 push buttons and 6 selector switches) and 8 external outputs (pilot lights).

In setting up your system, place the demo unit near your personal computer. Note the On/Off Power Switch and the Power Supply Receptacle on the demo. Make certain that the power switch is Off, then insert one end of the power cord into the power supply receptacle and the other end into an electrical socket.

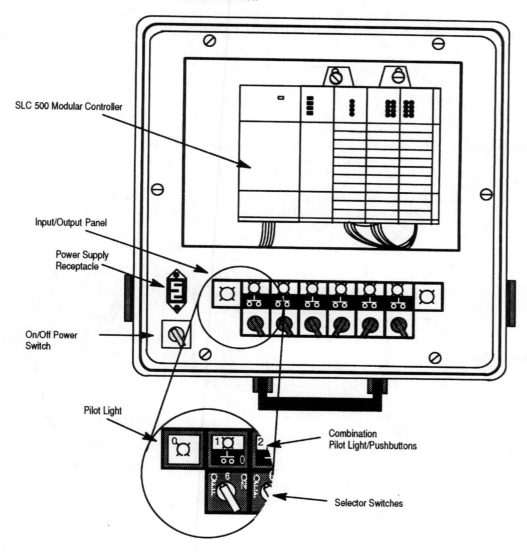

SLC 500 Modular Controller

Input/Output Panel

Power Supply Receptacle

On/Off Power Switch

Pilot Light

Combination Pilot Light/Pushbuttons

Selector Switches

Setting Up a Field–Wired Controller

The details of installing and wiring the controller and external input/output devices are beyond the scope of this guide.

If you are using a field–wired fixed or modular controller, refer to the Installation and Operation Manuals, catalog number 1747–NI001 (fixed controllers) and publication number 1747–6.2 (modular controllers), for information on installation and wiring of the controller and external input/output devices.

We recommend that your controller have at least two external input devices and two external output devices connected to complete the exercises in this guide.

Connecting the Controller to a Personal Computer

To connect the controller to a personal computer, you will require a communications cable, catalog number 1747–C10, and an RS–232/DH–485 interface converter, catalog number 1747–PIC.

1. Locate the communications channel of the controller. The figure below shows where it is located on modular and fixed controllers.

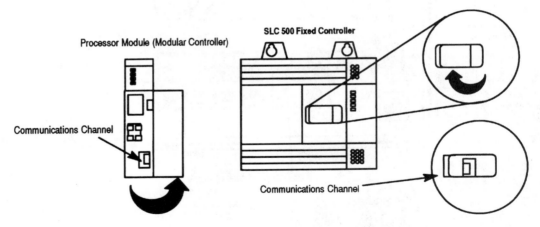

2. Insert one end of the 1747–C10 cable into the communications port of your controller.

3. Insert the other end of the 1747–C10 cable into the DH–485 connector of the 1747–PIC interface converter. The DH–485 connector is shown below.

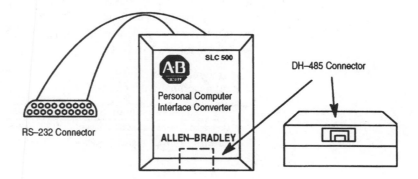

4. Insert the RS–232 connector (see figure above) of the interface converter into the serial communication port of your computer.

If your computer has a 9–pin serial port, use the 9–25 pin adapter provided with the interface converter.

The figure below shows a modular controller connected to a personal computer.

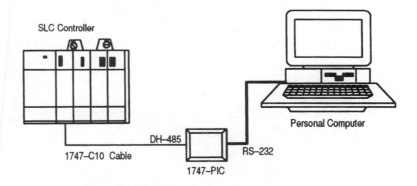

Personal Computer Requirements

The Advanced Programming Software (APS) can be used with an Allen-Bradley T47 or T70 terminal, 386/SX, NEC VERSA™ E Series Notebook, or GATEWAY 2000 models 386DX/25, 386DX/33, 486DX/33, 486DX2/50, and 486DX2/66 personal computers. Your computer must have:

- 640 Kbytes of RAM (2 meg. of extended memory is required)
- 10 Mbyte fixed-disk drive (APS requires 3.5 Mbytes of free disk space.)
- DOS version 3.3 or higher

The amount of free conventional RAM that APS requires depends on what communication drivers you want to load:

If you want to load:	You need:
only the standalone communications drivers	250 Kbytes
Windows drivers (INTERCHANGE™ software)	369 Kbytes

Platform Restrictions

If Using a 386/DX33 or Higher Platform

These platforms operate with no restrictions.

If Using a Platform Lower Than 386/DX33

Some restrictions may apply when operating at these platforms depending on the communication driver you use and its associated baud rate. See the table below.

Important: Communication drivers not listed in the table operate with no restrictions.

	Communication Driver					
	1747–PIC, KF3/KE, DF1 Full–Duplex, DF1 Half–Duplex, and DF1 Micro				Windows 485 (PIC Only)	
Baud Rate / Platform	19200	9600	4800	2400	19200	9600
386/DX25	▲	•	•	•	■	▲
386/SX16	■	▲	•	•	■	■

- • Communication between the driver and APS is supported with or without an expanded memory manager.
- ▲ Communication between the driver and APS is only supported without an expanded memory manager.
- ■ Communication between the driver and APS is not supported.

Important: Use of a platform lower than a 386/SX16 is *not* supported on or offline.

1–6

Using Extended Memory

APS requires a minimum of 2 meg. of extended memory (XMS). To use extended (XMS) memory, add the following command to your CONFIG.SYS file:

Type: `DEVICE=HIMEM.SYS`

If the file HIMEM.SYS is not located in your root directory (C:\), you must specify the directory path to the file so DOS can locate the file (for example, DEVICE = C:\DOS\HIMEM.SYS).

If you use HIMEM.SYS, we strongly recommend that you increase the number of available XMS handles (NUMHANDLES) to 128. To do this add the following line to your CONFIG.SYS file:

`DEVICE=C:\DOS\HIMEM.SYS /NUMHANDLES=128`

For more information about installing the extended memory manager, refer to the README file included with this software. This file is located in the directory in which the APS executable resides.

After you enter this command and save the modified CONFIG.SYS file, restart your computer for the changes to take effect.

Using Expanded Memory

APS can also run with expanded memory manager loaded. Follow the EMS 4.0 and VCPI specifications so that APS can use the expanded memory successfully (for example, DEVICE = C:\DOS\EMM386.EXE). Determine if you can use expanded memory by referring to the chart on page 1–6. If you use a memory manager that is not 100% compatible, your computer may lock up when you attempt to run APS. Contact your local computer software vendor to determine what requirements your memory manager meets.

Important: If an EMS handler (such as EMM386) is installed, it must follow the EMS 4.0 and VCPI specifications. If it does not, contact the manufacturer and obtain the latest release. If you are using an EMS handler (or you want to use other switches than those specified here) refer to the README file included with this software. An exception is the IBM PC-DOS 4.01, which has an EMS manager that is not VCPI compatible.

If you are using one of the following EMS memory managers, make certain you use the switch shown (or an equivalent) in the example below:

```
DEVICE=<path>EMM386 FRAME=NONE
DEVICE=<path>QEMM FRAME=NONE
DEVICE=<path>386MAX NOFRAME
DEVICE=<path>386MAX EMS=0
```

Using APS with DOS

This section shows examples of configuring your personal computer for extended or expanded memory. If you change the CONFIG.SYS file, you must re-boot the system to initialize the file.

Recommended CONFIG.SYS for 386/486 PCs with 4 meg. RAM (uses a combination of extended and expanded memory):

```
DEVICE = C:\DOS\HIMEM.SYS  /NUMHANDLES=128
DOS = HIGH,UMB
DEVICE = C:\DOS\EMM386.EXE 1024 FRAME=NONE
DEVICEHIGH = C:\DOS\ANSI.SYS
FILES=40
BUFFERS=40
DEVICE=C:\DOS\SMARTDRV.SYS 2048
```

Important: The executable **EMM386.EXE** does not automatically exclude the memory used by PC expansion cards. You must exclude the memory used by the all of the PC expansion cards by adding an exclusion option to the memory manager invoke line.

For example, the memory manager invoke line would appear as follows when using a 1784–KT Communication Interface module set up to use address range D400 to DFFF:

```
DEVICE=C:\DOS\EMM386.EXE 1024 x=D400-DFFF FRAME=NONE
```

See the user manual provided with each expansion card for more information.

Example CONFIG.SYS file for *expanded* memory only:

```
FILES=40
BUFFERS=40
DEVICE=C:\DOS\HIMEM.SYS
DEVICE=C:\DOS\EMM386.EXE FRAME=NONE
```

Example CONFIG.SYS file for *extended* (XMS) memory only:

```
FILES=40
BUFFERS=40
DEVICE=C:\DOS\HIMEM.SYS
```

Installing the Software

Before you actually install the software, complete the prepaid postage *Software Updates* registration card and return it. This is important, since it confirms your registration.

We assume that you have installed DOS in your computer. If you have not, do this now, following the instructions supplied with your computer.

To determine if your computer has enough memory for the software, at the DOS prompt type: **CHKDSK**, then press [**ENTER**]. The screen displays the memory configuration of your computer. Check the last line of the display XXXXXX bytes free. APS requires 2 meg. of extended (XMS) or expanded memory (EMM).

Important: Make sure your CONFIG.SYS file contains the following
statements:

```
FILES=40
BUFFERS=40
C:\DOS\SHARE.EXE
```

If running in a windows environment, set the `FILES` parameter
to 46 or higher

These are minimum values. If your CONFIG.SYS file contains
FILES and BUFFERS statements with greater values, there is
no need to change the file. Be aware that these statements may
conflict with the CONFIG.SYS requirements for other software
packages you have installed on your programming terminal.

To change this file, follow the instructions supplied with your computer. If
you change the CONFIG.SYS file, you must re-boot the system to initialize
the file.

Inside the software envelope you will find the APS software on 3.5 inch
disks. If you require 5.25 inch disks and are a registered user, contact the
media exchange service at 1–800–289–2279. Have your software serial
number available when you call.

To install the software, do the following:

1. Insert the diskette labeled Disk 1 into the appropriate disk drive (either
   drive A or drive B). For this example, we are using drive A.

2. Type: `A:INSTALL`, then press [`ENTER`].

   During the installation process, instructions appear on the screen to
   prompt you through the procedure. Follow the instructions and type in
   the information requested.

   If this is the first time you have installed the software, the system prompts
   you for your name, the name of your company, and the serial number of
   your software. This is also the case for any software updates you may
   install.

Important: During the installation process you are asked for the serial
number of your software. *The serial number is not found on the
disks.* The serial number is located in several places:

- the software registration card
- the registration change card
- the outside of the shipping carton

If you enter the serial number incorrectly or enter the wrong
serial number and accept the entry, you will be unable to correct
this situation later. Verify your entry carefully, before
committing your work. The serial number you enter is used to
personalize the software.

Important: You can install APS to run in a windows environment, however, that procedure is beyond the scope of this manual. See the *Advanced Programming Software User Manual*, publication 1747–6.4, for this information.

Running APS

To run APS, follow these steps:

1. If necessary, change the drive specifier to the drive where the software is installed (typically C). To do this, type:

   c: and press [**ENTER**].

2. If you are using the default directory, at the DOS prompt, type:

   CD \IPDS\ATTACH\SLC500 and press [**ENTER**].

   If you specified a different directory path, change to that directory and press [**ENTER**].

3. Type: **AP** and press [**ENTER**]. The main APS menu appears.

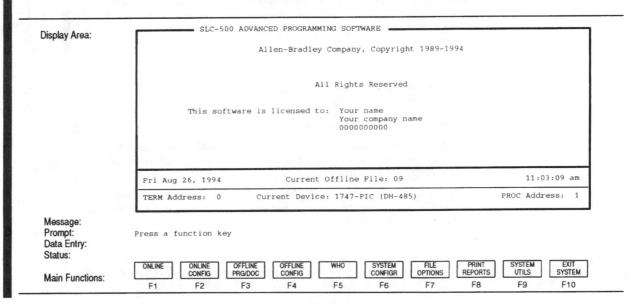

Display Area:

```
┌─── SLC-500 ADVANCED PROGRAMMING SOFTWARE ───────────┐
│                                                     │
│        Allen-Bradley Company, Copyright 1989-1994   │
│                                                     │
│                                                     │
│                   All Rights Reserved               │
│                                                     │
│     This software is licensed to:  Your name        │
│                                    Your company name│
│                                    0000000000        │
│                                                     │
│                                                     │
├─────────────────────────────────────────────────────┤
│ Fri Aug 26, 1994       Current Offline File: 09    11:03:09 am │
├─────────────────────────────────────────────────────┤
│ TERM Address:  0    Current Device: 1747-PIC (DH-485)    PROC Address:  1 │
└─────────────────────────────────────────────────────┘
```

Message:
Prompt:
Data Entry:
Status:

Press a function key

Main Functions:

ONLINE	ONLINE CONFIG	OFFLINE PRG/DOC	OFFLINE CONFIG	WHO	SYSTEM CONFIGR	FILE OPTIONS	PRINT REPORTS	SYSTEM UTILS	EXIT SYSTEM
F1	F2	F3	F4	F5	F6	F7	F8	F9	F10

Exiting the System: You can exit APS software and return to DOS by accessing the APS menu, shown above, and pressing [EXIT SYSTEM] .
F10

APS Display Format

The APS screen is divided into three areas:

- display area
- message, prompt, data entry, and status lines
- main functions

The figure below indicates what appears in these areas.

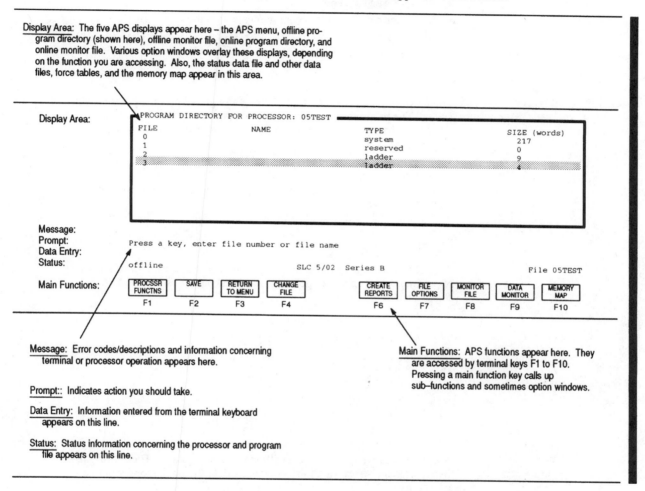

Display Area: The five APS displays appear here – the APS menu, offline program directory (shown here), offline monitor file, online program directory, and online monitor file. Various option windows overlay these displays, depending on the function you are accessing. Also, the status data file and other data files, force tables, and the memory map appear in this area.

Display Area:

```
PROGRAM DIRECTORY FOR PROCESSOR: 05TEST
FILE                    NAME              TYPE                    SIZE (words)
 0                                        system                      217
 1                                        reserved                      0
 2                                        ladder                        9
 3                                        ladder                        4
```

Message:
Prompt:
Data Entry: Press a key, enter file number or file name
Status: offline   SLC 5/02 Series B   File 05TEST

Main Functions:

PROCSSR FUNCTNS	SAVE	RETURN TO MENU	CHANGE FILE		CREATE REPORTS	FILE OPTIONS	MONITOR FILE	DATA MONITOR	MEMORY MAP
F1	F2	F3	F4		F6	F7	F8	F9	F10

Message: Error codes/descriptions and information concerning terminal or processor operation appears here.

Prompt:: Indicates action you should take.

Data Entry: Information entered from the terminal keyboard appears on this line.

Status: Status information concerning the processor and program file appears on this line.

Main Functions: APS functions appear here. They are accessed by terminal keys F1 to F10. Pressing a main function key calls up sub–functions and sometimes option windows.

Chapter 2

Control Basics

This chapter introduces you to basic concepts essential for understanding how the SLC 500 controller operates. It covers:

- SLC 500 file concepts
- How external I/O devices communicate with the processor
- Addressing external I/O
- Ladder logic concepts

SLC 500 File Concepts

The CPU, or processor, provides control through the use of a program you create. The program you create is called a processor file. This file contains other files that break your program down into more manageable sections. These sections are:

- Program Files – provide storage and control of the main program and subroutines.
- Data Files – contains the status of inputs, outputs, the processor, timers, counters, and so on.

Processor Files

Each CPU can hold 1 processor file at a time. The processor file is made up of program files (up to 256 per controller) and data files (up to 256 per controller).

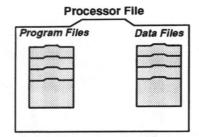

Processor files are created in the offline mode using APS. These files are then restored, also referred to as downloaded, to the processor for online operation.

Program Files

Program files contain controller information, the main control program, and any subroutine programs. The first three program files are required for each processor file. These are:

- **File 0 – System Program**
  This file stores the controller configuration and other system information.
- **File 1**
  This file is reserved for internal controller use.
- **File 2 – Main Ladder Program**
  This file stores the main control program.
- **Files 3 – 255 – Subroutine Ladder Program**
  These files are optional and used for subroutine programs.

Most of your work with program files will be in file 2, the main program file. This file contains your ladder logic program which you create to control your application.

Data Files

Data files contain the data associated with the program files. Each processor file can contain up to 256 data files. These files are organized by the type of data they contain. Each piece of data in each of these files has an address associated with it that identifies it for use in the program file. For example, an input point has an address that represents its location in the input data file. Likewise, a timer in the timer data file has an address associated with it that allows you to represent it in the program file.

The first 9 data files (0 – 8) have default types. You designate the remainder of the files (9 – 255). The default types are:

- **File 0 – Output Data**
  This file stores the state of the output terminals for the controller.
- **File 1 – Input Data**
  This file stores the status of the input terminals for the controller.
- **File 2 – Status Data**
  This file stores controller operation information.
- **Files 3 – 7**
  These files are pre–defined as Bit, Timers, Counters, Control, and Integer data storage, respectively.
- **File 8 – Float Data**
  This file is used by SLC 5/03 ™ with OS301 and SLC 5/04 ™ with OS400 processors for Float data storage.
- **Files 9 – 255**
  These files are user–defined as Bit, Timer, Counters, Control, Integer, and Float data storage.

Most of your work with data files will be in files 0 and 1, the output and input files. Refer to appendix A for an example of the Timer data file.

How External I/O Devices Communicate with the Processor

The figure below applies to a modular controller demo unit having an input module in slot 1 and an output module in slot 3. See page 1–2 for a diagram of the slot location. To simplify the illustration, only pushbutton 0 and pilot light 0 of the external I/O are shown.

Each of the external input circuits is represented by a status bit in the input data file of the processor file. Each of the external output circuits is represented by a status bit in the output data file of the processor file. During controller operation, the processor applies the input data to the program, solves the program based on the instruction you enter, and energizes and de–energizes external outputs.

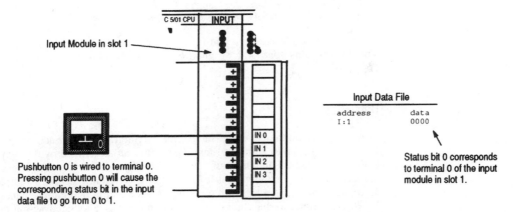

Input Module in slot 1

Pushbutton 0 is wired to terminal 0. Pressing pushbutton 0 will cause the corresponding status bit in the input data file to go from 0 to 1.

Input Data File

```
address          data
  I:1            0000
```

Status bit 0 corresponds to terminal 0 of the input module in slot 1.

Closing an external input circuit changes the corresponding status bit from 0 to 1.

Opening an external input circuit changes the corresponding status bit from 1 to 0.

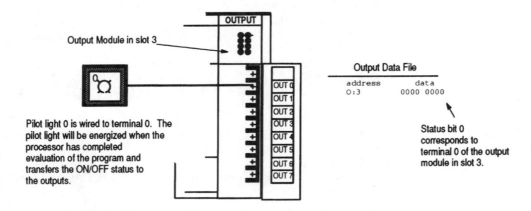

Output Module in slot 3

Pilot light 0 is wired to terminal 0. The pilot light will be energized when the processor has completed evaluation of the program and transfers the ON/OFF status to the outputs.

Output Data File

```
address          data
  O:3          0000 0000
```

Status bit 0 corresponds to terminal 0 of the output module in slot 3.

When an output data file status bit has been solved as a 1, the corresponding external output circuit will be energized (ON).

When an output data file status bit has been solved as a 0, the corresponding external output circuit is de–energized (OFF).

Addressing External I/O

As pointed out in the last section, external inputs and outputs are linked to the input data file and output data file of the processor file. Each status bit in these files has an address. You specify the appropriate address when you enter an instruction in your ladder program.

For our purposes, input addresses have the form **I:e/b**
where
I = Input data file
: = Element or slot delimiter
e = Slot number of the input module
/ = Bit or terminal delimiter
b = Terminal number used with input device

Similarly, output addresses have the form **O:e/b**
where
O = Output data file
: = Element or slot delimiter
e = Slot number of the output module
/ = Bit or terminal delimiter
b = Terminal number used with output device

Examples:
I:1/0 = Input, slot 1, terminal 0
I:2/0 = Input, slot 2, terminal 0
O:3/0 = Output, slot 3, terminal 0
O:3/7 = Output, slot 3, terminal 7
O:0/7 = Output, slot 0, terminal 7 (fixed controllers only because of slot 0)
I:0/4 = Input, slot 0, terminal 4 (fixed controllers only because of slot 0)

Eventually, you will be addressing other data files, such as Status, Bit, Timer, Counter, Control, Integer, String, ASCII, and Float. Addressing of these files is discussed in the APS programming manual.

APS Display of Instructions/Addresses

APS displays I/O addresses as shown below.

When you enter an XIC instruction (defined later) and the address I:1/0, APS will display the address with the instruction as follows:

```
          I:1.0
        ——] [——
            0
```

Explanation:

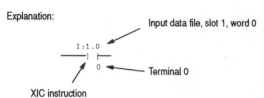

Ladder Logic Concepts

As we mentioned earlier, the program files you create contain the program used for your controlling application. The programs are written in a programming language called Ladder Logic. This name is derived from its ladder–like appearance.

A ladder logic program consists of a number of rungs, on which you place instructions. Instructions each have a data address associated with them and based on the status of these instructions the rung is solved.

The figure below shows a simple 1–rung ladder program. The rung includes two input instructions and an output instruction. Note, in the example below each instruction has a name (Examine if Closed), a mnemonic (XIC), and an address (I:1/0).

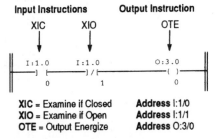

XIC = Examine if Closed   **Address** I:1/0
XIO = Examine if Open   **Address** I:1/1
OTE = Output Energize   **Address** O:3/0

A simple rung, using bit instructions.

True/False Status

The data file bits that these instructions are addressed to will be either a logic 0 (OFF) or a logic 1 (ON). This determines whether the instruction is regarded as "true" or "false":

If the data file bit is	The status of the instruction is		
	XIC Examine if Closed —] [—	XIO Examine if Open —]/[—	OTE Output Energize —( )—
Logic 0	False	True	False
Logic 1	True	False	True

Logical Continuity

During controller operation, the processor evaluates each rung, changing the status of instructions according to the logical continuity of rungs. More specifically, input instructions set up the conditions under which the processor will make an output instruction true or false. These conditions are:

- When the processor finds a continuous path of true input instructions in a rung, the OTE output instruction will become (or remain) true. We then say that "rung conditions are true".

- When the processor does *not* find a continuous path of true input instructions in a rung, the OTE output instruction will become (or remain) false. We then say that "rung conditions are false".

The figure below indicates the data file conditions under which the rung is true:

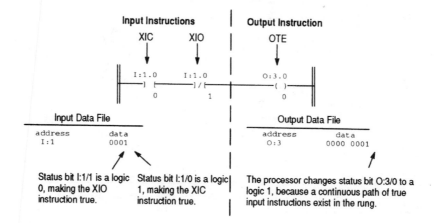

In the above example, if the input data file was 0000, then the rung would be false and the output data file would read as 0000 0000.

The Processor Operating Cycle

The diagram below indicates the events that occur during the processor operating cycle. This sequence is repeated many times each second.

Event	Description	
Input Scan	Input Data File address data I:1 0001	The status of external input circuits is read. The input data file is updated with this information.
Program Scan	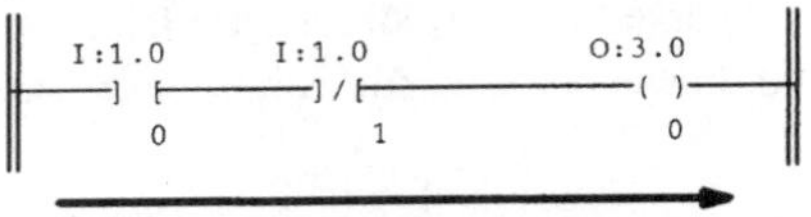	The ladder program is executed. The input data file is evaluated, the ladder rung is solved, and the output data file is updated.
Output Scan	Output Data File address data O:3 0000 0001 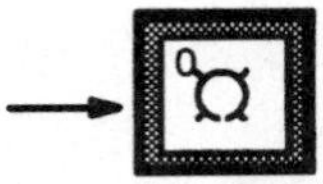 Illuminated	The output data file information is transferred to the external output circuit, thus energizing or de–energizing it.
Communications	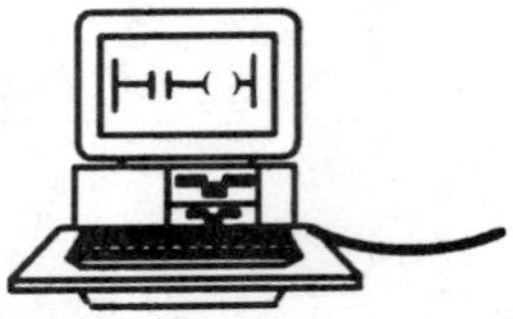	Communications with the programming terminal and other network devices takes place.
Housekeeping	Processor internal housekeeping takes place.	

Chapter 3

Creating a Processor File

In this chapter you create a processor file. The tasks you will perform:

- For modular controllers: Make a record of the processor module catalog number, the chassis catalog number(s), the I/O module catalog numbers, and the slot locations of I/O modules.

- For fixed controllers: Make a record of the controller catalog number (and I/O module catalog numbers and slot locations if you are using the 1746–A2 expansion chassis).

- Run APS software and initiate the creation of a processor file.

- Name the processor file GETSTART.

- Enter the controller configuration.

- Enter a 1–rung ladder program.

- Add a rung comment.

- Save the processor file to disk.

Configuration of SLC 500 Controllers

The following paragraphs briefly describe SLC 500 controllers and indicate the location of catalog numbers on the devices. This information will help you when you create a processor file and enter the specific controller configuration that will run the file.

To make the best use of this guide, you should have access to an SLC 500 Demonstration Unit, which includes completely wired external inputs and outputs. For the exercises in this guide, we arbitrarily assumed that you are using a Demo unit using a modular controller with the components listed on page 3–4.

Controller Styles

As previously mentioned, SLC 500 controllers are available in two styles –the fixed controller and the modular controller. Examples are shown in the figure below.

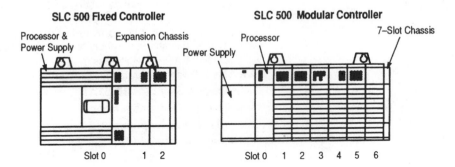

The fixed controller combines a power supply, processor (CPU), and a fixed number of I/O points in a single unit. You have the option of adding a 2–slot expansion chassis if you want to add I/O points.

The modular controller consists of a power supply, 1–3 I/O chassis, a processor module which you insert in slot 0 of the first chassis, and various I/O modules which you insert in the remaining slots of the chassis.

Slot Numbers

Note that slot numbers are indicated in the figure above. In fixed controllers, slot 0 applies to the processor and fixed I/O points; slots 1 and 2 apply to I/O modules located in the expansion chassis. In modular controllers, slot 0 is always reserved for your processor module; the remaining slots apply to the various I/O modules you have inserted.

Catalog Numbers

When you configure your controller, you must specify the processor catalog number, chassis catalog numbers, and I/O module catalog numbers as required. The location of the catalog number on the various components is shown in the following figures.

Make a Record of Controller Components

We recommend that you make a list of the processor, chassis, and I/O catalog numbers, and also the chassis numbers assigned to the chassis and the slot locations of all I/O modules. You can then refer to this list as you configure your controller.

Catalog Number Location – SLC 500 Fixed Controllers

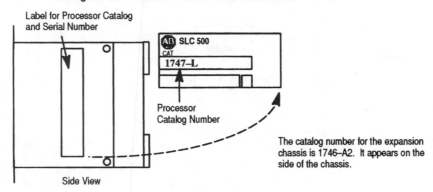

Label for Processor Catalog
and Serial Number

Side View

AB SLC 500
CAT
1747–L

Processor
Catalog Number

The catalog number for the expansion
chassis is 1746–A2. It appears on the
side of the chassis.

Catalog Number Location – SLC 500 Modular Controllers

Processor (CPU) Modules

I/O Modules

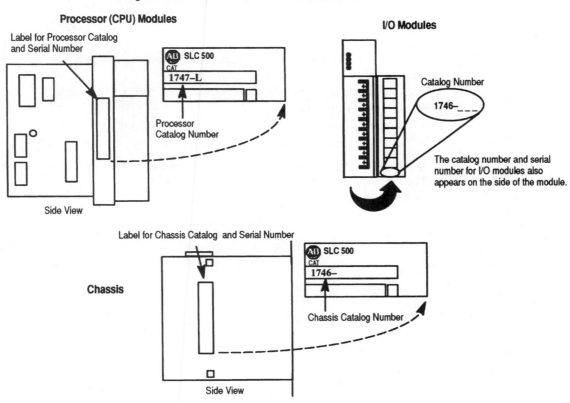

Label for Processor Catalog
and Serial Number

Side View

AB SLC 500
CAT
1747–L

Processor
Catalog Number

Catalog Number

1746–___

The catalog number and serial
number for I/O modules also
appears on the side of the module.

Label for Chassis Catalog and Serial Number

Chassis

AB SLC 500
CAT
1746–

Chassis Catalog Number

Side View

3–3

Arbitrary Controller Used in this Guide

In the following procedures, we have arbitrarily assumed that the controller you are configuring in your processor file is a modular demo unit including the following components:

- Chassis 1746–A4, 4–slot chassis
- Processor 1747–L524 in slot 0
- Input module 1746–IA4 in slot 1
- Input module 1746–IA8 in slot 2
- Output module 1746–OA8 in slot 3

The ladder program shown on page 4–8 contains I/O addresses that are consistent with the configuration indicated above. If you are using some other controller configuration, keep in mind that these addresses may not be valid for your controller.

Creating a Processor File

A processor file is always created offline, in the terminal workspace. In creating the processor file, you will:

- Name the file and configure the controller.
- Enter a ladder program.
- Add a rung comment.
- Save the processor file to disk.

If you are not already running APS, refer to "Running APS," page 1–10. The following procedure begins at the APS menu display.

Name the Processor File and Configure the Controller

Complete the following steps:

1. Access the create processor file window.

   Press 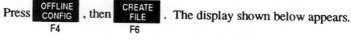 . The display shown below appears.

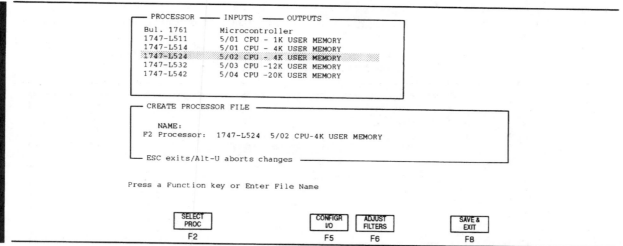

2. Enter the name GETSTART.

   The prompt line asks you to enter a file name. Type GETSTART, then press [ENTER]. GETSTART appears in the Create Processor File window.

3. Enter the appropriate processor catalog number.

   The Create Processor File window lists the default processor, 1747–L524. This is correct for our controller. If you are using a different processor, use the cursor keys to locate the appropriate processor in the upper option

   window, then press 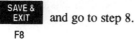 .

4. If you have selected a fixed controller and are not using an expansion chassis, the controller configuration is complete at this point. Press

   [SAVE & EXIT / F8]   and go to step 8.

5. Configure the chassis of your controller.

Press **CONFIGR IO** **F5** . The following option window appears. Note that chassis 1 is specified as 1746–A4, the default selection. This is correct for our controller. If you are using a different chassis, press **MODIFY RACKS** **F4** , then

RACK 1 **F1** . Select the appropriate chassis, using the cursor keys, and press **[ENTER]**. If you are using more than one chassis, follow the same procedure for chassis 2 and 3.

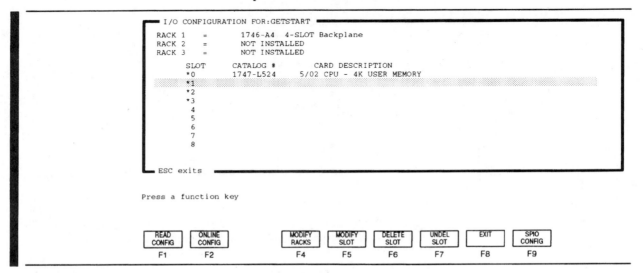

```
 ┌─ I/O CONFIGURATION FOR:GETSTART ─────────────────────────┐
 │ RACK 1   =      1746-A4  4-SLOT Backplane                 │
 │ RACK 2   =      NOT INSTALLED                             │
 │ RACK 3   =      NOT INSTALLED                             │
 │                                                          │
 │          SLOT    CATALOG #       CARD DESCRIPTION        │
 │          *0      1747-L524       5/02 CPU - 4K USER MEMORY│
 │          *1                                              │
 │          *2                                              │
 │          *3                                              │
 │           4                                              │
 │           5                                              │
 │           6                                              │
 │           7                                              │
 │           8                                              │
 │                                                          │
 └─ ESC exits ──────────────────────────────────────────────┘

Press a function key
```

READ CONFIG	ONLINE CONFIG		MODIFY RACKS	MODIFY SLOT	DELETE SLOT	UNDEL SLOT	EXIT	SPIO CONFIG
F1	F2		F4	F5	F6	F7	F8	F9

Note the asterisks next to slots 0 thru 3. This indicates that we have configured these slots and can now configure I/O modules. Slot 0 is already configured with our processor.

6. Configure the I/O modules.

The cursor is located on slot 1. To configure it, press **MODIFY SLOT** **F5** . The following option window appears:

```
┌─ I/O MODULE SELECTION FOR SLOT: 1 ─────────────────────┐
│ ┌──────────────────────────────────────────────────┐  │
│ │ CATALOG        CARD DESCRIPTION                    │  │
│ │ 1746-I*8      Any 8pt. Discrete Input Module       │  │
│ │ 1746-I*16     Any 16pt. Discrete Input Module      │  │
│ │ 1746-I*32     Any 32pt. Discrete Input Module      │  │
│ │ 1746-O*8      Any 8pt. Discrete Output Module      │  │
│ │ 1746-O*16     Any 16pt. Discrete Output Module     │  │
│ │ 1746-O*32     Any 32pt. Discrete Output Module     │  │
│ │ 1746-IA4       4 - Input 100/120 VAC               │  │
│ │ 1746-IA8       8 - Input 100/120 VAC               │  │
│ │ 1746-IA16     16 - Input 100/120 VAC               │  │
│ │ 1746-IB8       8 - Input (SINK) 24 VDC             │  │
│ │ 1746-IB16     16 - Input (SINK) 24 VDC             │  │
│ │ 1746-IB32     32 - Input (SINK) 24 VDC             │  │
│ └──────────────────────────────────────────────────┘  │
└─ ESC exits ────────────────────────────────────────────┘


Press ENTER to select I/O Module
Enter Module ID Code ▓
```

This window allows you to select a module for slot 1. Use the up/down
cursor keys to place the cursor on the appropriate module catalog number,
then press **SELECT MODULE** / F2 . This returns the display to the I/O configuration
window with the selected module indicated. Cursor down to the next
open slot and repeat the configuration steps. For our controller, the
completed option window appears as follows:

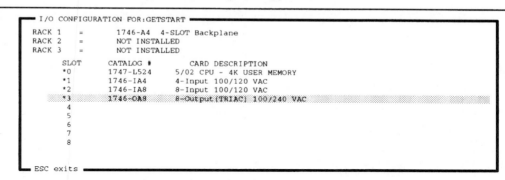

```
┌─ I/O CONFIGURATION FOR:GETSTART ──────────────────────────────────────┐
│ RACK 1   =      1746-A4   4-SLOT Backplane                             │
│ RACK 2   =      NOT INSTALLED                                          │
│ RACK 3   =      NOT INSTALLED                                          │
│                                                                        │
│      SLOT      CATALOG #        CARD DESCRIPTION                       │
│      *0        1747-L524        5/02 CPU - 4K USER MEMORY              │
│      *1        1746-IA4         4-Input 100/120 VAC                    │
│      *2        1746-IA8         8-Input 100/120 VAC                    │
│      *3        1746-OA8         8-Output (TRIAC) 100/240 VAC           │
│       4                                                                │
│       5                                                                │
│       6                                                                │
│       7                                                                │
│       8                                                                │
└─ ESC exits ────────────────────────────────────────────────────────────┘


Press a function key
```

READ CONFIG	ONLINE CONFIG		MODIFY RACKS	MODIFY SLOT	DELETE SLOT	UNDEL SLOT	EXIT	SPIO CONFIG
F1	F2		F4	F5	F6	F7	F8	F9

7. Create the archive file GETSTART.

Press ![EXIT] , then ![SAVE & EXIT] . Archive file GETSTART is created on
   F8   F8
your computer hard disk and placed in the Offline Processor File window.

8. Press ![SAVE TO FILE] to save Getstart as the new defualt fle.
         F9

The file GETSTART is in the terminal workspace, and you have returned
to the APS menu.

Enter the Ladder Program

The following rung consists of an XIC input instruction and an OTE output
instruction. The addresses conform to the controller configuration indicated
on page 4–4. *If you have entered a different controller configuration, make
certain that the addresses are consistent with your configuration.* It is also
important that you have an external input, such as a pushbutton, and an
external output, such as a pilot light, at the terminal addresses used. You will
be using these external devices in later chapters of this guide.

The rung can be entered by completing the following steps:

1. Access the Program Directory of file GETSTART.

Press ![OFFLINE PRG DOC] .
       F1

2. Monitor Program File 2.

Press ![MONITOR FILE] .
       F8

3. Insert a rung.

Press ![EDIT] , then ![INSERT RUNG] .
       F10   F4

4. Enter the Input instruction on the rung.

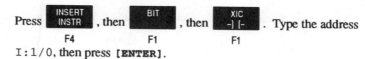

Press [INSERT INSTR], then [BIT], then [XIC -] [-]. Type the address
     F4   F1   F1
I : 1 / 0, then press [ENTER].

5. Enter the Output instruction on the rung.

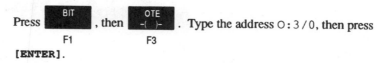

Press [BIT], then [OTE -( )-]. Type the address O : 3 / 0, then press
     F1   F3
[ENTER].

6. Accept the rung.

Press [ESC], then [ACCEPT RUNG], then [ESC].
                   F10

Add a Rung Comment

Complete the following steps to add a rung comment:

1. Configure the display so that rung comments will be visible.

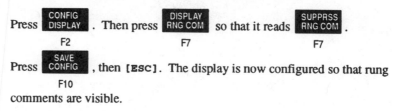

Press [CONFIG DISPLAY]. Then press [DISPLAY RNG COM] so that it reads [SUPPRSS RNG COM].
     F2   F7   F7

Press [SAVE CONFIG], then [ESC]. The display is now configured so that rung
     F10
comments are visible.

2. Add the rung comment.

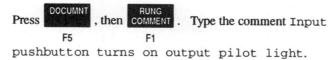

Press [DOCUMNT], then [RUNG COMMENT]. Type the comment Input
     F5   F1
pushbutton turns on output pilot light.

3. Accept and save the comment.

Press [ACCEPT EXIT], then [SAVE DOCUMNT], then [ESC].
     F8   F10

Your completed ladder program and rung comment should look like this:

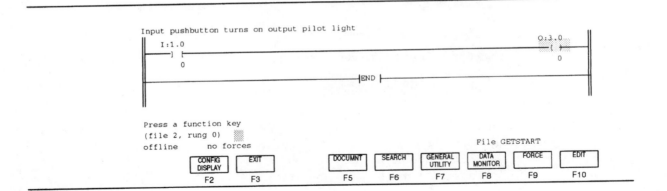

Save the Processor File

Complete the following steps to save the processor file to disk:

1. Return to the Program Directory.

Press .

2. Save the file to disk.

Press . Accept the default Save options by pressing  .

3. Return to the APS menu.

Press RETURN TO MENU .

F3

Chapter 4

Online Operations, Quick Edit

In this chapter wou will complete the following tasks:

- Download (restore) processor file GETSTART.
- Monitor the ladder program in the run mode.
- Test the program.
- Edit the program using Quick Edit.
- Test the edited program.
- Monitor the input and output data files.

Restoring (Downloading) a Processor File

There are two tasks to complete in restoring processor file GETSTART to the processor:

- Check the Online Configuration parameters.
- Go online and download (restore) processor file GETSTART.

The procedures begin at the APS menu display.

Check the Online Configuration Parameters

Complete these steps:

1. Access the Online Configuration window.

Press . From the menu, choose a 1747–PIC (DH–485) by

cursoring to it and pressing
```
DRIVER
CONFIG
```
F2
 . This display appears:

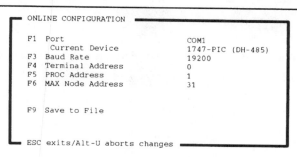

```
┌─ ONLINE CONFIGURATION ──────────────
│
│   F1  Port                      COM1
│          Current Device         1747-PIC (DH-485)
│   F3  Baud Rate                 19200
│   F4  Terminal Address          0
│   F5  PROC Address              1
│   F6  MAX Node Address          31
│
│
│   F9  Save to File
│
│
└─ ESC exits/Alt-U aborts changes ───────
```

Press a function key

PORT	SELECT DEVICE	BAUD RATE	TERM ADDRESS	PROC ADDRESS	MAX ADDRESS		SAVE TO FILE
F1	F2	F3	F4	F5	F6		F9

2. Verify the parameters.

The default values are shown for items F1 to F6. If you used the COM1 port of your computer and used a catalog 1747–PIC Interface Converter when connecting your computer to the controller, chances are that you will be able to establish processor–computer communications. If any of these default parameters are incorrect, change them with the function keys, then:

Press **SAVE TO FILE** , then [ESC]. This returns the display to the APS menu.
F9

Go Online and Restore (Download) Processor File GETSTART

Complete these steps:

1. Access the Restore File window.

Press **ONLINE** .
F1

If the message MESSAGE TIMEOUTS – LOSS OF COMMUNICATIONS appears, one or more of the Online Configuration parameters is incorrect and/or there is an improper connection between the computer and the processor. Refer to appendix B.

Once you establish communications with the processor, the program directory display will appear. Do one of these three things:

A. If the default program directory appears, (the directory is named DEFAULT, and only the system file is listed) press **RESTORE** .
F2

B. If a file exists in the processor, and no matching disk file is found on the computer harddisk, you will be asked "Read Processor Program?".
Press **NO** , then press **RESTORE** .
F10   F2

C. If a file exists in the processor, and a matching disk file is found on the computer harddisk, press **SAVE RESTORE** , then **RESTORE PROGRAM** .
F2   F4

After you have done a), b), or c), the following display appears:

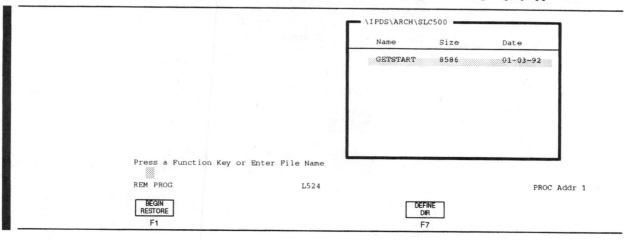

```
┌─ \IPDS\ARCH\SLC500 ──────────────────────────┐
│                                               │
│     Name          Size          Date          │
│                                               │
│     GETSTART      8586          01-03-92       │
│                                               │
│                                               │
│                                               │
│                                               │
│                                               │
│                                               │
└───────────────────────────────────────────────┘

Press a Function Key or Enter File Name
▓▓

REM PROG                    L524                         PROC Addr 1

  ┌─────────┐                        ┌─────────┐
  │ BEGIN   │                        │ DEFINE  │
  │ RESTORE │                        │ DIR     │
  └─────────┘                        └─────────┘
     F1                                  F7
```

2. Select and accept file GETSTART.

The cursor is located in the righthand window, which lists all of the processor files saved on disk. Move the cursor to the file GETSTART if it is not already there. Press [BEGIN RESTORE F1]. If the processor is in the program mode, the file is restored (downloaded). If the processor is in the run mode, you are asked "Change Processor Mode to Program?". Press [YES F8]. File GETSTART is restored (downloaded).

When the restoring (downloading) process is complete, you are asked to "Press Any Key to Continue". After you press any key, the program directory for file GETSTART appears:

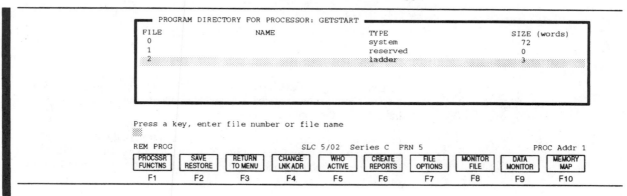

```
┌─ PROGRAM DIRECTORY FOR PROCESSOR: GETSTART ──────────────────────────┐
│  FILE              NAME              TYPE              SIZE (words)    │
│   0                                  system            72             │
│   1                                  reserved          0              │
│   2                                  ladder            3              │
│                                                                       │
│                                                                       │
│                                                                       │
└───────────────────────────────────────────────────────────────────────┘

Press a key, enter file number or file name
▓▓

REM PROG              SLC 5/02  Series C  FRN 5               PROC Addr 1
┌────────┬────────┬────────┬────────┬────────┬────────┬────────┬────────┬────────┬────────┐
│PROCSSR │ SAVE   │RETURN  │CHANGE  │ WHO    │CREATE  │ FILE   │MONITOR │ DATA   │MEMORY  │
│FUNCTNS │RESTORE │TO MENU │LNK ADR │ ACTIVE │REPORTS │OPTIONS │ FILE   │MONITOR │ MAP    │
└────────┴────────┴────────┴────────┴────────┴────────┴────────┴────────┴────────┴────────┘
   F1       F2       F3       F4       F5       F6       F7       F8       F9       F10
```

Testing the Program

To test the ladder program you entered in chapter 4, we will now monitor program file 2, and change the processor mode from program to run. Then activate the external input having address I:0/0 and observe the effect on the external output at address O:3/0.

Begin at the program directory display for processor file GETSTART.

1. Monitor program file 2 and enter the Run mode.

Press 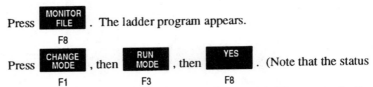 . The ladder program appears.

Press ████ , then ████ , then ████ . (Note that the status line now indicates REM RUN instead of program.) If you get a fault code on the status line, refer to appendix B to clear the fault.

2. Test the program.

The following diagram shows the rung you entered if you are using the modular controller demo unit discussed on page 3–4. If you are using some other controller configuration, make certain that your external input device and output device are wired to the controller input and output that you addressed in your ladder program.

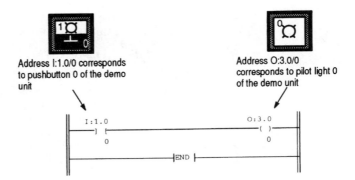

Address I:1.0/0 corresponds to pushbutton 0 of the demo unit

Address O:3.0/0 corresponds to pilot light 0 of the demo unit

To test the program, press pushbutton 0. Pilot light 0 should go on. The display should show both the XIC and OTE instructions highlighted to indicate that they are true.

Processor operation: When you pressed pushbutton 0, the input instruction went from false to true. This resulted in a path of true input instructions in the rung, causing the output instruction to go from false to true.

Now release the pushbutton. Pilot light 0 should go off. Neither instruction in the rung should be highlighted. When you released pushbutton 0, the input instruction went from true to false; this broke the path of true input instructions, causing the output instruction to go from true to false.

Editing the Program with Quick Edit

The Quick Edit feature of APS software allows you to move quickly from online monitoring to offline editing, then back to online monitoring. To give you experience at doing this, we will edit the program by adding an input instruction on the rung. The effect of the edit: Selector switch 6 must be on (closed) to allow pushbutton 0 to turn on pilot light 0.

We will place an XIC instruction in series with (to the right of) the XIC instruction already entered. It will have address I:2/2, corresponding to selector switch 6 of the demo unit. See the figure below.

Complete the following six steps to edit and test the edited program. The starting point for this procedure is the online monitor file display, with the processor in the run mode:

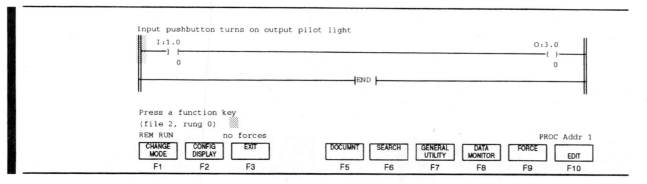

1. Go offline and edit the disk version of the file.

   Press  . Note that the status line of the display

   F10   F3

   now indicates that you are offline, at file GETSTART.

2. Select Modify Rung and position the cursor.

   Press **MODIFY RUNG** . We want to append an instruction to the XIC

   F5

   instruction, so use the cursor key to position the cursor on the XIC instruction.

3. Enter an XIC instruction, address I:2/2.

   Press **APPEND INSTR** , then **BIT** , then **XIC -] [-** .

   F3   F1   F1

   Type the address "I:2/2", then press **[ENTER]**, then **[ESC]**.

4. Accept the rung.

Press .

5. Save the edit and go back online.

Press ![SAVE GO ONLINE F1] . Accept the default Save options by pressing ![YES F8] .

Before the software restores the program it asks "Change Processor Mode to Program?". Press ![YES F8] . When the program is successfully restored, the software asks "Change Processor Mode to Run?". Press ![YES F8] .

You are now back online with the edited program, in the run mode.

6. Test the edited program.

The following diagram shows the rung you have modified if you are using the modular controller demo unit discussed on page 4–4. If you are using some other controller configuration, make certain that your external input devices and output device are wired to the controller inputs and output that you addressed in your ladder program.

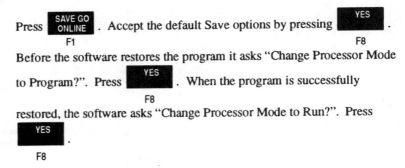

Address I:1.0/0 corresponds to pushbutton 0 of the demo unit

Address I:2.0/2 corresponds to selector switch 6 of the demo unit

Address O:3.0/0 corresponds to pilot light 0 of the demo unit

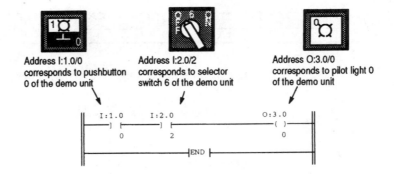

To test the program, first turn selector switch 6 to the on position. Note that the corresponding input instruction in the rung is highlighted, indicating that it is true. Now press pushbutton 0. Pilot light 0 should go on. The rung should show all instructions highlighted to indicate that they are true.

Processor operation: When you pressed pushbutton 0, the corresponding input instruction went from false to true. This resulted in a path of true input instructions in the rung, causing the output instruction to go from false to true.

Now turn selector switch 6 to the off position. Note that the corresponding input instruction in the rung is no longer highlighted, indicating that it is false. Press pushbutton 0. Note that the corresponding input instruction is highlighted, but the output instruction does not go from false to true. This is because a continuous path of true input instructions does not exist in the rung.

Monitoring Data Files

In this procedure, you will monitor the input data file and the output data file. These files include a status bit for each of the configured I/O terminals of the controller. You will monitor data file changes as you operate pushbutton 0 and selector switch 6. To end the exercise, you will go offline to the APS menu.

The starting point is the online monitor file display with the processor in the run mode:

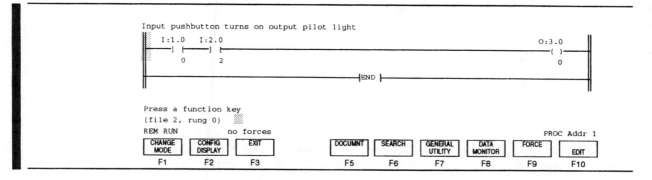

1. Position the ladder cursor and access the input data file.

   Use the cursor key to position the cursor on the XIC instruction having address I:1.0/0, then press **DATA MONITOR** (F8). The input data file appears, with the cursor located on status bit I:1.0/0. This is shown below.

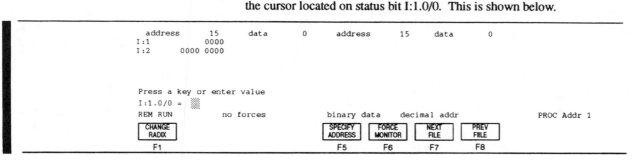

4-7

2. Monitor input data changes resulting from input device operaton.

    Press pushbutton 0. Note that the status bit goes from 0 to 1, as the instruction goes from false to true. Now turn selector switch 6 to the on position. Note that status bit I:2.0/2 goes from 0 to 1, as the instruction goes from false to true.

3. Access the output data file.

    The output data file precedes the input data file in the data table. Press

    PREV FILE . The output data file appears. Since we didn't specify a

    F8

    particular bit address, the cursor is located on the status bit having the lowest address, O:3.0/0. This is also the status bit for pilot light 0 in our program. This is shown below.

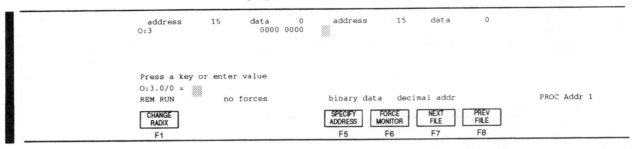

```
          address      15    data      0    address     15    data      0
          O:3                  0000 0000          ▒▒

          Press a key or enter value
          O:3.0/0 = ▒▒
          REM RUN           no forces          binary data   decimal addr          PROC Addr 1
          ┌──────────┐                    ┌──────────┐┌──────────┐┌──────┐┌──────┐
          │ CHANGE   │                    │ SPECIFY  ││ FORCE    ││ NEXT ││ PREV │
          │ RADIX    │                    │ ADDRESS  ││ MONITOR  ││ FILE ││ FILE │
          └──────────┘                    └──────────┘└──────────┘└──────┘└──────┘
             F1                                F5          F6         F7      F8
```

4. Monitor output data changes resulting from input device operation.

    Press pushbutton 0 with selector switch 6 in the on position. Note that status bit O:3.0/0 goes from 0 to 1, as the output instruction of our program goes from false to true.

    Continue to press pushbutton 0, as you turn selector switch 6 to the off position. Note that bit O:3.0/0 goes from 1 to 0. This is because there is no longer a path of true input instructions in the rung, causing the output instruction to go false.

5. Return to the APS menu.

    Press [ESC]. This returns you to the online monitor file display.

    Press **EXIT** . This returns you to the online program directory

    F3

    display.

    Press **RETURN TO MENU** . This takes you offline, returning you to the APS menu

    F3

    display.

Chapter 5

Creating and Printing Reports

This chapter shows you how to create and print reports. The following four hard copy reports can be created and printed:

- Program Listing – Can include a) the main program file and all subroutine files, b) a single file, c) a range of files, or d) a range of rungs.
- Cross Reference – Provides an alphabetical list of addresses and their rungs, in either address or symbol order.
- Processor Configuration – Details the configuration of the processor and associated hardware in the system.
- Data Tables – Details the contents of the offline or online data files.

If you do not have a printer set up in your system, we suggest that you go through these procedures anyway, to familiarize yourself with report capabilities.

Creating Reports

A report can be created at the program directory display, either offline or online. In the following procedure, reports are created offline. The starting point is the APS menu.

Complete the following steps:

1. Access the documentation (reports) and options windows.

Press OFFLINE PRG DOC F3 . This accesses the program directory display.

Press CREATE REPORTS F6 . The following windows appear in the display area:

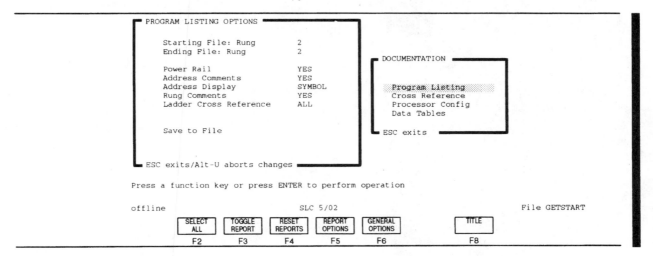

```
┌─ PROGRAM LISTING OPTIONS ──────────────────┐
│                                            
│     Starting File: Rung      2             
│     Ending File: Rung        2             
│                                            
│     Power Rail               YES      ┌─ DOCUMENTATION ──────────┐
│     Address Comments         YES      │                          
│     Address Display          SYMBOL   │    Program Listing        
│     Rung Comments            YES      │    Cross Reference        
│     Ladder Cross Reference   ALL      │    Processor Config       
│                                       │    Data Tables            
│     Save to File                      │                          
│                                       └─ ESC exits ──────────────┘
│                                            
└─ ESC exits/Alt-U aborts changes ──────────┘

Press a function key or press ENTER to perform operation

offline                          SLC 5/02                      File GETSTART
      ┌─────────┬─────────┬─────────┬─────────┬─────────┐   ┌───────┐
      │ SELECT  │ TOGGLE  │ RESET   │ REPORT  │ GENERAL │   │ TITLE │
      │ ALL     │ REPORT  │ REPORTS │ OPTIONS │ OPTIONS │   │       │
      └─────────┴─────────┴─────────┴─────────┴─────────┘   └───────┘
         F2        F3        F4        F5        F6             F8
```

2. Specify documentation.

The "Documentation" window lists the four reports you can create. The cursor is located on the "Program Listing" report. Options for the Program Listing are shown in the window at the left. Function key F5 allows you to change items in the options window.

Move the cursor to "Cross Reference", then "Processor Config", then "Data Tables". Note that as you do this, the option window changes to match the report the cursor is located on.

An explanation of the various options is beyond the scope of this guide. For our purposes, the default options are suitable.

3. Press ▐ SELECT ALL ▌ . In doing this, you have selected all four reports.

    F2

This is verified by the appearance of asterisks at the left of each report in the Documentation window.

4. Specify a title and create the reports.

Press ▐ TITLE ▌ . Type GETPRINT in the window that appears.

    F8

Press [ENTER] to accept the title. Then press [ENTER] to perform the create reports operation. When the reports have been created, DOCUMENTATION COMPLETE appears in the display area, and PRESS A KEY TO CONTINUE appears on the prompt line.

Press any key. The program directory appears.

5. Return to the APS menu.

Press ▐ RETURN TO MENU ▌ . You will be asked to "Save Current Work?" Since we

    F3

have not changed the ladder program in any way, a Save is not required.

Press ▐ NO ▌ . The APS menu appears.

    F10

Printing Reports

Printing reports is done from the APS menu. Complete the following steps:

1. Access file GETSTART in the Report Directory.

Press 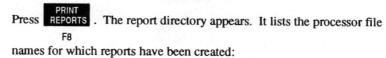 . The report directory appears. It lists the processor file

names for which reports have been created:

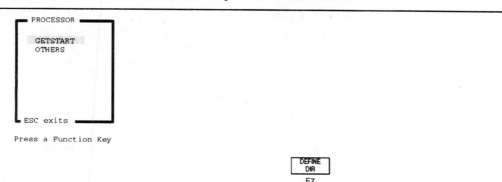

```
┌─ PROCESSOR ─┐
│             │
│  GETSTART   │
│   OTHERS    │
│             │
│             │
│             │
│             │
└─ ESC exits ─┘

Press a Function Key
```

```
┌──────┐
│DEFINE│
│ DIR  │
└──────┘
  F7
```

Use the up/down cursor key to move the cursor to GETSTART, then press
[ENTER]. The display shows the reports you have created for file
GETSTART:

```
┌─ PRINT GETSTART ──────────────────────────────┐
│  Report                    Size        Date    │
│  ───────────────────────────────────────────  │
│  Program Listing           2660     01-04-92   │
│  Cross Reference           2064     01-04-92   │
│  Data Table                3657     01-04-92   │
│  Processor Config          1739     01-04-92   │
│                                                 │
└─ ESC exits ─────────────────────────────────── ┘

Press a function key or press ENTER to perform operation
```

```
┌────────┐ ┌────────┐ ┌───────┐ ┌─────────┐ ┌─────────┐ ┌───────┐
│ SELECT │ │ TOGGLE │ │ CLEAR │ │ PRINTER │ │ SELECT  │ │ PRINT │
│  ALL   │ │ SELECT │ │  ALL  │ │ CONFIG  │ │ PROCESS │ │ FILES │
└────────┘ └────────┘ └───────┘ └─────────┘ └─────────┘ └───────┘
    F2         F3         F4         F5          F6          F7
```

2. Configure the printer and prepare it for operation.

Press  . Change configuration parameters if necessary, then

press **[ENTER]**. Prepare the printer for operation.

5–3

3. Select the reports to be printed and initiate printing.

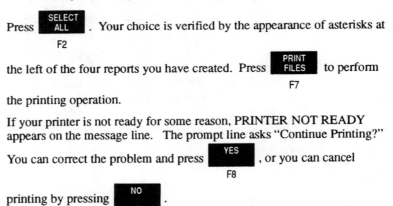

Press [SELECT ALL] (F2). Your choice is verified by the appearance of asterisks at the left of the four reports you have created. Press [PRINT FILES] (F7) to perform the printing operation.

If your printer is not ready for some reason, **PRINTER NOT READY** appears on the message line. The prompt line asks "Continue Printing?" You can correct the problem and press [YES] (F8), or you can cancel printing by pressing [NO] (F10).

After the printing operation, you can return to the APS menu by pressing [ESC].

Appendix A

Additional Ladder Program Exercises

This appendix lets you apply what you have learned in the previous chapters. It covers:

- Entering a program with a I/O branches
- Entering a program with a timer instruction

Entering an Input and Output Branch

The important feature of this program is the output and input branch. The input branch is based on what is called OR or parallel logic. This means that if either input #0 OR input #1 is true, then output #0 and #1 turn on.

Exercise 1: Entering an Input and Output Branch

We are assuming you have created a new file, configured it and you are now ready to begin entering an instruction. See chapter 4 for help with the above. Begin offline at the edit screen.

1. Enter the rung and XIC instruction.

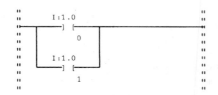

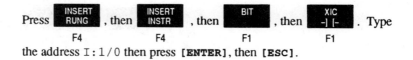

Press [INSERT RUNG] (F4), then [INSERT INSTR] (F4), then [BIT] (F1), then [XIC -] [-] (F1). Type the address I:1/0 then press **[ENTER]**, then **[ESC]**.

2. Enter a branch and another XIC instruction.

Cursor left once, so your cursor is on the XIC instruction. Press

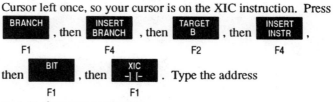

[BRANCH] (F1), then [INSERT BRANCH] (F4), then [TARGET B] (F2), then [INSERT INSTR] (F4),

then [BIT] (F1), then [XIC -] [-] (F1). Type the address

I:1/1 then press **[ENTER]**.

3. Enter an OTE instruction.

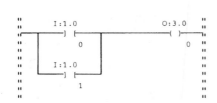

Cursor up, then cursor right so your cursor is at the far right power rail.

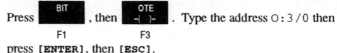

Press [BIT] (F1), then [OTE -( )-] (F3). Type the address O:3/0 then

press **[ENTER]**, then **[ESC]**.

4. Enter a branch and another OTE instrucion.

Cursor left once so you are on the OTE instruction. Press 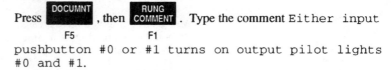 , then

INSERT BRANCH (F4) , then TARGET C (F3) , then INSERT INSTR (F4) , then BIT (F1) , then

OTE -( )- (F3) . Type the address O:3/1 then press [ENTER], then [ESC].

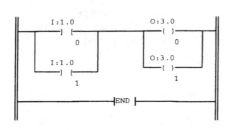

5. Accept the rung.

Press ACCEPT RUNG (F10) , then [ESC].

6. Enter the rung comment.

Press DOCUMNT (F5) , then RUNG COMMENT (F1) . Type the comment Either input pushbutton #0 or #1 turns on output pilot lights #0 and #1.

7. Accept and save the comment.

Press ACCEPT EXIT (F8) , then SAVE DOCUMNT (F10) , then [ESC].

Your completed ladder program and rung comment should look like this:

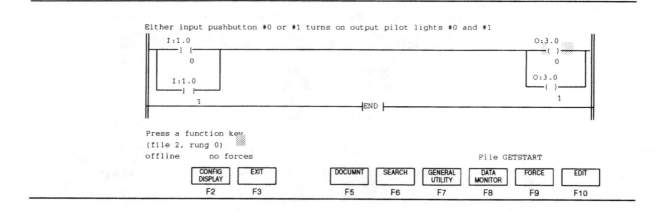

Save the Processor File

Complete the following steps to save the processor file to disk:

1. Return to the Program Directory.

   Press .

   F3

2. Save the file to disk.

   Press . Accept the default Save options by pressing  .

   F2   F8

3. Return to the APS menu.

   Press RETURN TO MENU .

   F3

Test the Ladder Program

Complete the following steps to test the processor file:

1. Go online with your processor and restore the new file. Refer to chapter 4 for help.

2. Monitor the file. Refer to chapter 4 for help.

3. Place the processor in the RUN mode. See chapter 4 for help.

4. Press pushbutton #0. Outputs #0 and #1 turn ON.

5. Release pushbutton #0. Outputs #0 and #1 turn OFF.

6. Press pushbutton #1. Outputs #0 and #1 turn ON.

7. Release pushbutton #1. Outputs #0 and #1 turn OFF.

Entering a Timer Instruction

In exercise 2, you enter a timer instruction with a time delay of 10 seconds. Two different types of timer status bits activate output pilot lights #0 and #1. The first type, called a "timer timing" status bit turns on output #0 for 10 seconds. The second type, called a "done" status bit, turns on output #1 *after* 10 seconds.

Exercise 2: Entering a Timer Instruction

We are assuming you have created a new file, configured it and you are now ready to begin entering an instruction. See chapter 3 for help with the above. Begin offline at the edit screen.

1. Enter a rung and an XIC instruction.

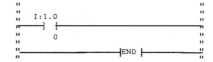

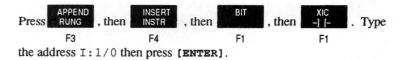

the address I : 1 / 0 then press **[ENTER]**.

2. Enter the timer instruction.

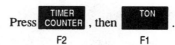

Type the address T4 : 0 then press **[ENTER]**. This is the Timer Address.

Type the timebase . 01 then press **[ENTER]**. This is the timebase in seconds.

Type 1000 then press **[ENTER]**. This is the Timer Preset Value in hundredths of a second.

Type 0 then press **[ENTER]**. This is the Timer Accumulated Value.

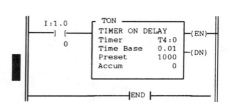

3. Accept the rung.

Press **[ESC]**, then
ACCEPT RUNG .
F10

4. Enter a second rung and an XIC instruction.

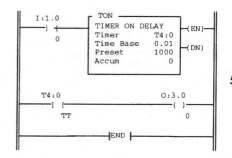

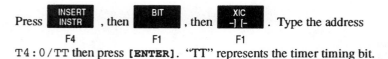

T4 : 0 / TT then press **[ENTER]**. "TT" represents the timer timing bit.

5. Enter an OTE instruction.

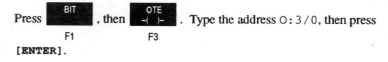

[ENTER].

6. Accept the rung.

Press [ESC], then 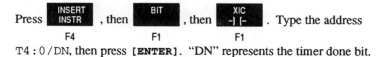 .

7. Enter a third rung and an XIC instruction.

Press [INSERT INSTR / F4] , then [BIT / F1] , then [XIC -] [- / F1] . Type the address T4 : 0 / DN, then press [ENTER]. "DN" represents the timer done bit.

8. Enter an OTE instruction.

Press [BIT / F1] , then [OTE -( )- / F3] . Type the address O : 3 / 1, then press [ENTER].

9. Accept the rung.

Press [ESC], then [ACCEPT RUNG / F10] , then [ESC].

10. Exit the edit mode.

Press [ESC]. Your completed ladder program should look like this:

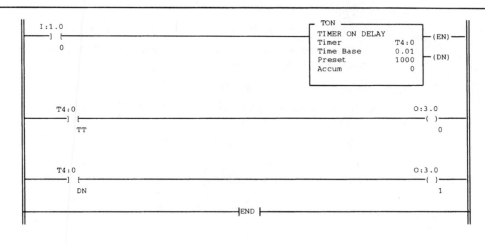

Save the Processor File

Complete the following steps to save the processor file to disk:

1. Return to the Program Directory.

   Press .

   F3

2. Save the file to disk.

   Press . Accept the default Save options by pressing  .

   F2   F8

3. Return to the APS menu.

   Press RETURN TO MENU .

   F3

Test Your Ladder Program

Complete the following steps to test the timer instruction file:

1. Go online with your processor and restore the new file. Refer to chapter 4 for help.

2. Monitor the file. Refer to chapter 4 for help.

3. Place the processor in the RUN mode. See chapter 4 for help.

4. Press pushbutton #0 for at least 10 seconds. During the first 10 seconds, output #0 turns ON and #1 stays OFF.

5. After 10 seconds, output #0 turns OFF and output #1 turns ON.

6. Release pushbutton #0; the timer resets and both outputs #0 and #1 turn OFF.

Appendix B

Troubleshooting

This appendix shows you how to identify and correct errors that you may encounter while working through this guide. They include:

- APS error messages
- System LEDs status
- Processor error codes

APS Error Messages

Table B.A details APS error messages:

Table B.A
APS Error Messages

Error Message	Cause	Corrective Action
APS Timeout – Loss of Communications	Wrong baud rate	Select different baud rate in F2 "Online Config"; Processor default is 19200.
	Wrong processor node address	Select different processor address in F2 "Online Config"; Processor default is 1.
	Wrong device type	Device type in F2 "Online Config" should be 1747–PIC.
	Incompatible or wrong computer Serial Port	Select different COM port in F2 "Online Config"; verify PC COM port works.
	Bad cable	Check continuity in 1747–C10 cable; contact local Allen–Bradley distributor for replacement.
	Bad 1747–PIC	Contact your local Allen–Bradley distributor for replacement.
	Incompatible 9–25 Pin Adaptor	Consult PC manual for Serial Port type (DCE or DTE); 9–25 Pin Adaptor supplied with 1747–PIC is for a DTE Serial port. If serial port is DCE, you may need a null–modem adaptor.
	Not enough power to 1747–PIC	Check line power to SLC power supply; check position of power supply jumper for modular systems.
Database Read Error	Files and buffers are not set up correctly	Use a word processor or DOS Edline to verify/change your CONFIG.SYS file to contain minimum values of Files = 40 and Buffers = 40. (Minimum values of 46 are required if running APS in a Windows™ environment.) If the file is modified, re–boot PC.

Table B.A
APS Error Messages (continued)

Error Message	Cause	Corrective Action
Fatal Communication Hardware Error	Incompatible or non–existent Serial COM port on PC	Select different COM port by pressing F2 "Online Config"; verify COM port works.
Illegal Data or Parameter Value	Maximum node address of the processor exceeds 31	Reduce the maximum node address of the processor to 31 by pressing F5 "WHO"; F5 "Who Active"; F7 "Max Address".
I/O Address Not Configured	Processor/system configuration does not match entered addresses	Verify correct address format (I:slot/terminal or O:slot/terminal); verify system configuration by pressing F3 "Offline Prg/Doc"; F1 "Procssr Functns"; F1 "Change Procssr"; F5 "Configr I/O".
No Matching Disk File Found	Processor program does not exist on hard disk	To read the processor program (upload), press F8 "Yes"; otherwise press F10 "No" to continue with other online activities.
No Memory Left or Not Enough Memory to Load Communication Driver	PC does not have enough free RAM memory to continue	Verify your PC has >250K of free RAM to execute APS. (>369K free RAM is required if you are running the INTERCHANGE™ software). Exit APS and type "CHKDSK" at DOS prompt. The last line should read ">250K bytes free" (or >369K). If not, disable TSR, drivers, menus, shells, etc. loaded in AUTOEXEC.BAT or CONFIG.SYS that may be running in background. Re–boot PC.
Incompatible Processor Type	The processor configuration of the program you are restoring does not match your hardware	Verify that processor configuration of your program matches your hardware by pressing F3 "Offline Prg/Doc"; F1 "Procssr Functns; F1 "Change Procssr".

System LED Status

The system LEDs are located at different places on the modular system and the SLC fixed controller. Refer to the installation and operation manual for more information on system LED status. See Figure B.1.

Figure B.1
System LEDs

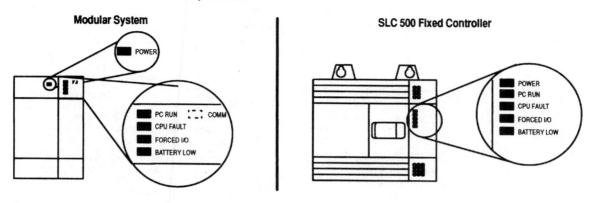

Table B.B
LED Status – Error Conditions

Processor LED	When It Is	Indicates that
RUN (Color: red)	On (steadily)	The processor is in the Run mode.
	Off	The processor is in a mode other than Run.
CPU FAULT (Color: red)	Flashing (at power up)	The processor has not been configured.
	Flashing (during operation)	The processor detects a major error either in the processor, expansion chassis or memory.
	On (steadily)	A fatal error is present (no communication).
	Off	There are no errors.
FORCED I/O (Color: red)	Flashing	One or more input or output addresses have been forced to an On or Off state but the forces have not been enabled.
	On (steadily)	The forces have been enabled.
	Off	No forces are present or enabled.
BATTERY LOW (Color: red)	On (steadily)	The battery voltage has fallen below a threshold level or the battery and the battery jumper are missing.
	Off	The battery is functional, or the battery jumper is present.
COMM (Color: red)	On (steadily)	The SLC 5/02 is receiving data.
	Off	The SLC 5/02 is not receiving data.

Processor Error Codes

Table B.C details some of the processor error codes. Refer to the Advanced Programing Software Reference Manual, publication 1747–6.11, for a complete list of error codes and troubleshooting information.

Table B.C
Processor Error Codes

Error Code	Cause	Corrective Action
0001	RAM program is corrupt due to noise, lightning, improper grounding or loss of capacitor or battery back–up.	Check wiring, layout, grounding. If using a 4K CPU, verify that a battery is installed to retain RAM memory when power is removed. See CPU FAULT–Flashing under system LED status. Restore the program using APS or an HHT.
0012	RAM program is corrupt or RAM itself is bad due to noise, lightning, improper grounding, or loss of capacitor or battery back–up.	Check wiring, layout, grounding. If using a 4K CPU, verify that a battery is installed to retain RAM memory when power is removed. See CPU FAULT–Flashing under system LED status. Restore the program using APS or an HHT.
XX50, XX51, XX52 XX53, XX54, XX55 (xx = slot #)	I/O module configuration/conflict or runtime problem.	Verify that processor configuration matches your hardware by pressing F3 "Offline Prg/Doc"; F1 "Procssr Functns"; F1 "Change Procssr"; F5 "Configr I/O"; See CPU FAULT–Flashing under the system LED status.
0056	Rack configuration error.	Verify that rack configuration in your program matches your hardware by pressing F3 "Offline Prg/Doc"; F1 "Procssr Functns"; F1 "Change Procssr"; F5 "Configr I/O". If multiple rack system, verify proper installation of rack interconnect cable. See CPU FAULT – Flashing under system LED status.

Glossary

The following terms are used throughout this manual. Refer to the Allen–Bradley Industrial Automation Glossary, publication number AG–7.1, for a complete guide to Allen–Bradley technical terms.

address: A character string that uniquely identifies a memory location. For example, I:1/0 is the memory address for the data located in the Input file location 1/0.

APS (Advanced Programming Software): Software used to monitor and develop SLC 500 ladder logic programs.

bit: Binary digit. The smallest unit of information in the binary numbering system. Represented by the digits 0 and 1. The smallest unit of memory.

branch: A parallel logic path within a ladder logic rung.

chassis: A hardware assembly that houses devices such as I/O modules, adapter modules, processor modules, and power supplies.

comment: Text included within a program to explain what the program is doing. Comments do not affect the operation of the program in any way.

communication scan: A part of the SLC CPU's operating cycle in which communication takes place with other devices, such as APS on a personal computer.

controller: A unit, such as a programmable controller or relay panel, that controls machine or process elements.

CPU (central processing unit): The decision–making and data storage section of a programmable controller.

cross reference: A report listing addresses, instructions, and their rung numbers where used.

data file: An area within a processor file that contains the status of inputs, outputs, the processor, timers, counters, and so on.

data table report: A report documenting the contents of the data files.

DOS: The operating system used to operate a personal computer.

edit: To create or modify a ladder program.

expansion rack: A 2–slot rack used only with fixed controllers.

false: The status of an instruction that does not provide a continuous logical path on a ladder rung.

file: A collection of information organized into one group.

fixed controller: A controller with a power supply, CPU, and I/O integrated into a single package.

function keys: Keys on a personal computer keyboard labeled F1, F2, F3, and so on. The operation of each of these keys is defined by software and a key may have a different function for each menu display.

hard disk: A disk storage device for storing relatively large amounts of data.

hardware: Mechanical, electrical, and electronic components and assemblies.

I/O (Inputs and Outputs): Consists of input and output devices that provide and/or receive data from the programmable controller.

input device: A digital or analog device, such as a limit switch, push–button switch, pressure sensor, or temperature sensor, that supplies input data through an input circuit to a programmable controller.

input scan: A part of the SLC's operating cycle. Status of the input modules are loaded into the Input data file.

instruction: A mnemonic and data address defining an operation to be performed by the processor. A rung in a program consists of a set of input and output instructions. The input instructions are evaluated by the controller as being true or false. In turn, the controller sets the output instructions to true or false.

interface converter: An Allen–Bradley device, Catalog Number 1747–PIC, used to establish communication between the personal computer and an SLC 500 programmable controller.

ladder logic: A program written in a format resembling a ladder–like diagram. The program is used by a programmable controller to control devices.

mnemonic: A simple and easy to remember term that is used to represent a complex or lengthy set of information.

modular controller: SLC 500 system consisting of a power supply, chassis, CPU, and input and output modules.

module: An interchangeable plug–in device that may inserted into a rack.

network: A series of stations (nodes) connected by some type of communication medium. A network may be made up of a single link or multiple links.

off line: Describes devices not under direct communication. For example, when programming the software.

online: Describes devices under direct communication. For example, when the software is monitoring the program file in a controller.

operating cycle: The sequential order of operations performed by a programmable controller when in the run mode.

OTE (OuTput Energize): An instruction that energizes when a rung is true and de–energizes when a rung is false.

output device: A device, such as a pilot light or a motor starter coil, that receives data from the programmable controller.

output scan: A part of the SLC's operating cycle. During this scan the output data file information is transferred to the output modules.

processor: A Central Processing Unit. (See CPU.)

processor configuration: A report detailing the configuration of the processor.

processor file: The set of program and data files used by the SLC to control output devices. Only one processor file may be stored in the SLC at a time.

processor overhead: The part of the operating cycle used for housekeeping and setup purposes.

program file: The area within a processor file that contains the ladder logic program.

program listing: A report containing a range of program files or a range of rungs.

program mode: When the SLC is not executing the processor file and all outputs are de–energized.

program scan: A part of the SLC's operating cycle. During the scan the ladder program is executed and the Output data file is updated based on the program and the Input data file.

rack: See chassis.

read: To acquire data from a storage place. For example, the processor READs information from the input data file to solve the ladder program.

report: A printable document containing information about a processor file. For example, a ladder listing, a cross reference, the data tables, and the processor configuration.

restore: To download (transfer) a program from a personal computer to a SLC.

run mode: When the processor file in the SLC is being executed, inputs are read, the program is scanned, and outputs are energized and de–energized.

rung: Ladder logic is comprised of a set of rungs. A rung contains input and output instructions. During Run mode, the inputs on a rung are evaluated to be true or false. If a path of true logic exists, the outputs are made true. If all paths are false, the outputs are made false.

save: To upload (transfer) a program stored in memory from a SLC to a personal computer; OR to save a program to a computer harddisk.

SLC (Small Logic Controller): A controller that comes in 1 of 2 styles: fixed or modular.

slot: The area in a rack that a module plugs into.

software: Executable programming package used to develop SLC ladder diagrams.

status: The condition of a circuit or system, represented as logic 0 (OFF) or 1 (ON).

terminal: A point on an I/O module that external I/O devices, such as a pushbutton or pilot light, are wired to.

true: The status of an instruction that provides a continuous logical path on a ladder rung.

write: To copy data to a storage device. For example, the processor WRITEs the information from the output data file to the output modules.

eXamine If Closed (XIC): An input instruction that is logically true when the status of the bit located at its address is a 1; false when it's a 0.

eXamine If Open (XIO): An input instruction that is logically true when the status of the bit located at its address is a 0; false when it's a 1.

absolute optical encoder   An optical rotary encoder that outputs a binary word representing the angular position.

absolute pressure   The pressure difference between the measured value and an absolute vacuum.

AC servomotor   A two-phase motor used in control applications; designed to have a near-linear torque–speed curve.

acceleration   The process of increasing velocity; uniform acceleration is when the velocity is increasing at a constant rate.

accumulator   A temporary, digital data-storage register in the microprocessor used in many math, logic, and data-moving operations. Used in a hydraulic system, a spring-loaded tank connected into the high-pressure side of the system; the accumulator can store some fluid to be used at peak-demand periods.

active filter   A circuit that incorporates an op-amp.

actuator   The source of energy or motion in a system; typical actuators would be electric motors or hydraulic cylinders.

adaptive controller   A controller, such as a PID, with the ability to monitor and improve its own performance.

ADC   *See* **analog-to-digital converter**.

address   A number that represents the location of 1 byte of data in memory or a specific input/output port.

address bus   A group of signals coming from the microprocessor to memory and I/O ports, specifying the address.

aftercooler   Used in pneumatic systems, a device that removes moisture from the air supply by chilling and thereby condensing any water vapor.

air compressor   Used in pneumatic systems, a device that creates the supply of high-pressure air.

aliasing   The event that occurs when the wrong data are reconstructed from the sampled sensor data because the sample rate is too slow.

ALU   *See* **arithmetic logic unit**.

analog drive   A method of controlling an electric motor's speed by varying the DC supply voltage.

analog I/O module   A module that plugs into the PLC and converts real-world analog signals to digital signals for the PLC and vice versa.

analog switch   A solid-state device that performs the same function as a low-power mechanical switch.

analog-to-digital converter (ADC)   A device (usually an IC) that can convert an analog voltage into its digital binary equivalent.

anode   One of three terminals of an SCR and PUT.

arithmetic logic unit (ALU)   The part of the CPU that performs arithmetic and logical operations.

armature   The part of a motor that responds to the magnetic field; typically, the armature is the rotating assembly of an electric motor.

assembly language   A computer program written in mnemonics, which are English-like abbreviations for machine-code instructions.

autotuning   The ability of some digital controllers to continuously fine-tune themselves by making small changes in their gain constants, based on past performance.

axial piston pump   A type of hydraulic pump that uses a number of small pistons to pump the fluid, this is a variable-displacement pump because the stroke of the pistons is determined by the adjustable angle of the swash plate.

backlash   In a gear pass, the small amount of free play between gears so that the teeth don't bind on each other.

ballast resistor   A resistor placed in series with the motor coils to improve the torque at higher stepping rates; it works by reducing the motor-time constant.

ball bearing screw   A ball bearing type of a leadscrew, recirculating balls roll between the threaded shaft and the nut to remove almost all friction.

band-pass filter   A circuit that allows only a specified range of frequencies to pass and attenuates all others above and below the pass band.

base   One of three terminals of a transistor.

baud   The rate at which the signal states are changing; frequently used to mean "bits per second."

BDCM   *See* **brushless DC motor**.

bias   A value that is added to the output of the controller to compensate for a constant disturbing force, such as gravity.

biasing circuit   The part of a transistor circuit that generates the forward bias voltage.

bidirectional control valve   Used in hydraulic and pneumatic systems, it can direct the fluid into either end of the actuator cylinder, allowing the piston to be moved in either direction.

bilevel drive   A technique that uses two voltages to improve torque at higher stepping rates. A higher voltage is applied to the motor at the beginning of the step, and then a lower voltage is switched in.

bimetallic temperature sensor   A strip made from two metals with different coefficients of thermal expansion; as the strip heats up, it bends.

bipolar junction transistor (BJT)   Known as a *transistor*, a solid-state device that can be used as a switch or an amplifier.

bipolar motor   A motor that requires polarity reversals for some of the steps; a two-phase motor is bipolar.

bit   The smallest unit of digital data, which has a value of 1 or 0.

bit instruction   The basic PLC instruction that logically represents a simple switch or relay coil; the state of each switch or coil occupies 1 bit in memory.

BJT   *See* **bipolar junction transistor**.

Bode plot   A graph that plots system gain and phase shift versus frequency; it can be used to determine system stability.

bonded-wire strain gauge   A force sensor consisting of a small pattern of thin wires attached to some structural member; when the member is stressed, the stretched wires slightly increase the resistance.

Bourdon tube   A type of pressure sensor consisting of a short bent tube closed at one end; pressure inside the tube tends to straighten it.

brush   A stationary conductive rod that rubs on the rotating commutator and conducts current into the armature windings.

brushless DC motor (BDCM)   The newest type of DC motor; it does not use a commutator or brushes; instead, the field windings are turned on and off in sequence, and the permanent magnet rotor (armature) is pulled around by the apparently rotating magnetic field.

byte   An 8-bit digital word.

CAD   *See* **computer-aided design**.

CAM   *See* **computer-aided manufacturing**.

capacitor start motor   A single-phase AC induction motor that uses a capacitor to phase-shift the AC for the start windings.

cathode   One of three terminals of an SCR and PUT.

CEMF   *See* **counter-EMF**.

central processing unit (CPU)   The central part of a computer, the CPU performs all calculations and handles the control functions of the computer.

chatter   the condition that occurs when the output of a comparator oscillates when the input voltage is near the threshold voltage.

CIM   *See* **computer-integrated manufacturing**.

circular pitch   The distance along the pitch circle of a gear of one tooth and valley.

class A operation   The event when a transistor is biased so the collector current is approximately half of its maximum value.

class B operation   The event when a transistor is biased so that the collector current is just turning on.

class C operation   The event when a transistor is biased below cutoff and used as an on–off switch.

closed-loop control system   A control system that uses feedback. A sensor continually monitors the output of the system and sends a signal to the controller, which makes adjustments to keep the output within specification.

closed-loop gain   The gain of an amplifier when feedback is being used. The value of closed-loop gain is less than open-loop gain (for negative feedback).

closed-loop system   A control system that uses feedback—that is, a sensor tells the controller the actual state of the controlled variable.

cold junction One of two temperature-sensitive junctions of the thermocouple temperature sensor. The cold junction is usually kept at a reference temperature.

collector One of the three terminals of a transistor.

commutator The part of an armature that makes contact with the brushes.

comparator A type of op-amp that is used open-loop to determine if one voltage is higher or lower than another voltage. Part of the control system that subtracts the feedback signal (as reported by the sensor) from the set point, to determine the error.

comparison instruction Compares two words and yields a single bit as a result.

compensating gauge A nonactive strain gauge that is used solely for the purpose of canceling out temperature effects.

compound motor A motor that has both shunt and series field windings.

computer-aided design A computer system that makes engineering drawings.

computer-aided manufacturing A computer system that allows CAD drawings to be converted for use by a numerical control (NC) machine tool.

computer-integrated manufacturing A computer system that oversees every step in the manufacturing process, from customer order to delivery of finished parts.

constant current chopper drive A drive circuit for stepper motors that uses PWM techniques to maintain a constant average current at all speeds.

contactor A heavy-duty relay that switches power directly to motors and machinery.

continuous-cycle method A method of tuning a PID controller that involves driving the controller into oscillations and measuring the response; the PID constants are computed based on these data.

continuous-level detector A sensor that can determine the fluid level in a container.

control bus A group of timing and control signals coming from the microprocessor to memory and I/O ports.

control instruction Allows for such things as jumps and subroutines.

control strategy The set of rules that the controller follows to determine its output to the actuator.

control system A system that may include electronic and mechanical components, where some type of machine intelligence controls a physical process.

control valve Used to control the flow of fluids. A valve that has a built-in valve-operating mechanism, allowing it to be controlled remotely by a signal from the controller.

control winding One of two windings in an AC servomotor.

controlled variable The ultimate output of the process; the actual parameter of the process that is being controlled, such as temperature in a heating system.

controller The machine intelligence of the control system.

counter-EMF (CEMF) A voltage that is generated inside an electric motor when running under its own power; the polarity of the CEMF is always opposite to the applied voltage.

CPU *See* **central processing unit.**

creep The property that explains why two pulleys connected with a belt will not stay exactly synchronized.

critically damped A system that is damped just enough to prevent overshoot.

cumulative compound motor A compound motor where the magnetic fields of the shunt and series fields aid each other.

cumulative error Error that accumulates; for example, a cumulative error of 1° per revolution means that the measurement error would be 5° after 5 revolutions.

current loop In a signal transmission system, a single loop of wire that goes from the transmitter to the receiver and back to the transmitter. The signal intelligence is conveyed by the current level instead of the voltage. This system is immune to voltage drops caused by wire resistance.

current-to-voltage converter An op-amp based circuit used as a transmitter for a current-loop system.

cutoff frequency The frequency at which a filter circuit starts to attenuate.

DAC *See* **digital-to-analog converter.**

damping A drag force on a system from sliding or viscous friction, which acts to slow movement (make it sluggish).

data bus A group of signals going to and from the microprocessor, memory, and I/O ports. The data bus carries the actual data that are being processed.

data communication equipment (DCE) One of two units specified by the RS-232 standard (for serial data transfer); the DCE is usually a modem.

data file The collection of data in a PLC that represents the present condition of switches, counters, relays, and the like.

data memory The section of RAM memory in a PLCS where the data file is stored.

data terminal equipment (DTE)   One of two units specified by the RS-232 standard (for serial data transfer). The DTE is usually the computer.

DCC   *See* **distributed computer control**.

DCE   *See* **data communication equipment**.

DC link converter   A circuit that converts line voltage at 60 Hz to AC power at different voltage and frequencies, for the purpose of controlling the speed of AC motors.

DC offset voltage   The small voltage that may occur on the output of an op-amp, even when the inputs are equal; DC offset can be eliminated with a resistor adjustment.

DDC   *See* **direct digital control**.

dead band   The small range, on either side of the set point (of the controlled variable), where the error cannot be corrected by the controller because of friction, mechanical backlash, and the like.

dead time   The interval of time between the correcting signal sent by the controller and the response of the system.

dead zone   *See* **dead band**.

delta-connection   One of two ways in which to connect three-phase motor or generator coils. With the delta-connection, each of the three coils is connected between two phase wires; there is no neutral.

derivative control   A feedback control strategy where the restoring force is proportional to the rate that the error is changing (increasing or decreasing); derivative control tends to quicken the response and reduce overshoot.

desiccant dryer   Used in hydraulic systems, a device that dries the air supply by passing the air through a water-absorbing chemical.

detent torque   A magnetic tug that keeps the rotor from turning even when the power is off; also called *residual torque*.

diac   A bistable, two-terminal, solid-state device that is used to trigger the triac.

diametral pitch   The ratio of the number of gear teeth per inch of pitch diameter; in practice, the number that describes the tooth size when specifying gears.

differential amplifier   A circuit that produces an output voltage that is proportional to the instantaneous voltage difference between two input signals. Op-amps by themselves are difference amplifiers; however, a practical differential amplifier incorporates additional components.

differential compound motor   A compound motor where the magnetic fields of the shunt and series windings oppose each other.

differential pressure   The difference between two pressures where neither may be ambient.

differential voltage   A signal voltage carried on two wires, where neither wire is at ground potential.

differentiator   An op-amp circuit that has an output voltage proportional to the instantaneous rate of change of the input voltage.

digital control system   A control system where the controller is a digital circuit, typically a computer.

digital-to-analog converter (DAC)   A circuit that translates digital data into an analog voltage.

DIP switch   A set of SPST switches built into the shape of an IC (DIP stands for dual in-line package).

direct current tachometer   Essentially, a DC generator that gives an output voltage proportional to angular velocity.

direct digital control   An approach to process control where all controllers in a large process are simulated by a single computer.

discrete I/O   I/O that deals with on–off devices. Discrete input modules are used to connect real-world switches to the PLC, and discrete output modules are used to turn on lamps, relays, motors, and so on.

discrete-level detector   A sensor that can determine if the fluid in a container has reached a certain level.

distributed computer control   An approach to process control where each process has its own local controller, but all individual controllers are connected to a single computer for programming and monitoring.

double-acting cylinder   A hydraulic or pneumatic cylinder and piston assembly in which the piston can be driven in either direction.

double-pole/double-throw switch (DPDT)   A switch contact configuration.

download   To transfer a computer program or data into a computer (from another computer).

DPDT   *See* **double-pole/double-throw switch**.

drain   One of three terminals in an FET.

dryer   Used in pneumatic systems, a dryer removes the moisture from the air supply.

DTE   *See* **data terminal equipment**.

duty cycle   The percentage of cycle time that the PWM pulse is high—that is, a true square wave has a duty cycle of 50%.

dv/dt effect   The event when the SCR turns on if the anode–cathode voltage rises too quickly.

dynamic torque   The motor torque available to rotate the load under normal conditions.

earth ground   The voltage at (or connection to) the surface of the earth at some particular place.

efficiency   When describing energy conversions, the percentage of input energy that is converted to useful output energy.

eight-step drive   A two- or four-phase motor being driven in half-steps; the sequencing pattern has eight steps.

electric field   A condition that exists in the space between two objects that are at a different voltage potential. A wire in this space will assume a "noise" voltage proportional to the field strength.

electromechanical counter   A device that counts events. Each time it receives a pulse, it increments a mechanical counter, which can then be read by an operator or used to activate some other process.

electromechanical relay (EMR)   A device that uses an electromagnet to close (or open) switch contacts—in other words, an electrically powered switch.

electromechanical sequencer   A device that provides sequential activation signals; it works by rotating a drum with cams, which activates switches.

embedded controller   A small microprocessor-based controller that is permanently installed within the machine it is controlling.

emitter   One of three terminals of a transistor.

EMR   *See* **electromechanical relay.**

energy   The amount of work it takes to do a job; energy has different units for chemical, thermal, mechanical, and electrical systems.

enhancement mode   Property of some MOSFETs where the gate voltage can always be positive (for N-channel).

error (E)   In a control system, the difference between where the controlled variable should be (set point) and where it actually is ($E = SP - PV$).

event control system   A control system that cycles through a predetermined series of steps.

event-driven operation   In a sequentially controlled system, an action that is allowed to start or continue based on some parameter changes. This is an example of closed-loop control.

expansion card/slot   An expansion card is a printed circuit card that plugs into an expansion slot on the motherboard of a personal computer (PC). The expansion card usually interfaces the PC to the outside world.

extension rate   The rate at which a linear actuator can extend or retract; the term is usually applied to electrical linear actuators.

feedback   The signal from the sensor, which is fed back to the controller.

feedforward   The concept of letting the controller know in advance that a change is coming.

fetch–execute cycle   A computer cycle where the CPU fetches an instruction and then executes it.

FET   *See* **field effect transistor.**

field   The part of an electric motor that provides a magnetic field; typically, it is the stationary part.

field effect transistor (FET)   A three-terminal solid-state amplifying device that uses voltage as its input signal.

field winding   A stationary electromagnet used to provide the magnetic field needed by the armature.

flow sensor   A sensor that measures the quantity of fluid flowing in a pipe or channel.

follow-up system   A control system where the output follows a specified path.

forced commutation   The process of turning off an SCR by momentarily forcing the anode–cathode voltage to 0 V (or below).

forward bias voltage   The DC offset base voltage required to start the transistor conducting.

forward breakover voltage   The voltage across a thyristor that causes it to switch into its conduction state.

forward conduction region   The operating range of a thyristor when it is conducting.

forward current gain (h_{FE})   The gain of a transistor; the collector current divided by the base current.

forward path   The signal-flow direction of the controller to the actuator.

four-phase stepper motor   A motor with four separate field circuits; this motor does not require polarity reversals to operate and hence is unipolar.

four-step drive   The standard operating mode for two- or four-phase stepper motors taking full steps; the sequencing pattern has four states.

fractional horsepower (hp) motor   A motor with less than 1 hp.

frame   A basic part of an AC motor, an overall case that supports the stator and the rotor bearings.

fuzzy logic   A new control strategy, modeled originally on human thought processes, where decisions are made based on a number of factors and each factor can have such qualities as *maybe* and *almost,* to name only two.

fuzzy predicates Linguistic terms of degree, used in describing actual variables—for example, *warm, very warm, not so warm,* and so on.

fuzzy set The range of one fuzzy predicate in a system—that is, all values of temperature that correspond to *very warm.*

gain A multiplication factor; the steady-state relationship between input and output of a component. (In this text, *gain* and *transfer function* are used interchangeably, although this is a simplification.) The term *system gain* refers to the proportional gain factor (K_P).

gain-bandwidth product (f_T) A constant for an amplifier that is the product of the open-loop gain and the frequency at that gain; can be read as the frequency when a transistor's gain has been reduced to 1.

gain margin A form of a safety margin. When the phase lag is 180°, the gain margin is the amount the gain could be increased before the system becomes unstable.

gate One of three terminals in an SCR, PUT, FET, or triac.

gauge pressure The difference between measured and ambient pressure (ambient is 14.7 psi or 101.3 kPa).

gearhead In smaller motors, a gear train attached or built-in to the motor assembly, which effectively gives the motor more torque at less rpm.

gear motor A simple hydraulic motor in which fluid pushes on the teeth of a set of constantly meshed gears (in a housing), causing them to rotate.

gear pass Two gears in mesh.

gear pump A simple positive-displacement hydraulic fluid pump that uses two constantly meshed gears (in a housing).

gear ratio The ratio of the number of teeth of two gears in mesh. (Also the ratio of pitch diameters.)

gear train A gear system consisting of more than one gear pass.

GFI *See* **ground-fault interrupter.**

Grey code A sequence of digital states that has been designed so that only 1 bit changes between any two adjacent states.

ground-fault interrupter (GFI) A saftey device that will shut off the power if it senses a current leakage to the safety ground.

half-steps By alternating the standard mode with the dual excitation mode, the angle of step will be half of what it normally is.

Hall effect The phenomena of a semiconductor material generating a voltage when in the presence of a magnetic field; used primarily as a proximity sensor.

harmonic drive A unique device using a flexible gear that can provide a large gear ratio with virtually no backlash.

head The fluid pressure in a tank, which is caused by the weight of the fluid above and is therefore proportional to the level.

heat sink A piece of metal, possibly with cooling fins, used to dissipate heat from a power device to the air.

high-pass filter A circuit that allows higher frequencies to pass but attenuates lower-frequency signals.

holding current (I_H) Once the SCR has been turned on, the small current (to the SCR) necessary to keep the SCR in the conduction state.

holding torque The motor torque available to keep the shaft from rotating when the motor is stopped but with the last field coil still energized.

Hooke's law The "spring law" that states the amount a spring deflects is proportional to applied force. The deformation in a spring is directly proportional to the force on the spring (spring force = constant × spring deformation).

hot junction One of two temperature-sensitive junctions of the thermocouple temperature sensor. The hot junction is used on the probe.

hybrid solid-state relay A device that uses a reed relay to activate a triac.

hybrid stepper motor A motor that combines the features of the PM and VR stepper motors—that is, it can take small steps and has a detent torque.

hydraulic cylinder A linear actuator powered by fluid under pressure, it consists of a cylinder, enclosed at both ends with a piston inside. The piston is moved by admitting the pressurized fluid into either end of the cylinder.

hydraulic system A system that uses pressurized fluid (oil) to power actuators such as hydraulic cylinders.

hydrostatic pressure The pressure that exists inside a closed hydraulic system; the significance is that the pressure is the same everywhere within the system (assuming no velocity effects).

I See **moment of inertia**.

I_{DSS} The highest possible drain current for a particular JFET; occurs when the gate voltage is 0 V.

ice-water bath A traditional way to create a known reference temperature for the cold junction of a thermocouple.

impact pressure   The pressure in an open (pitot) tube pointed "upstream."

incremental optical encoder   An optical rotary encoder that has one track of equally spaced slots; position is determined by counting the number of slots that pass by a photo sensor.

inertia   The property that explains why an object in motion will tend to stay in motion. Inertia is directly related to mass; the more mass an object has, the more energy it takes to get it moving or to stop it.

input/output (I/O)   Data from the real world moving in and out of a computer.

Input/Output modules   The connection points where real-world switches, relays, and the like are connected to the PLC.

instruction set   The set of program commands that a particular microprocessor is designed to recognize and execute.

instrument gears   Smaller gears with pitch in the 48–28 range, found in office machines and smaller industrial machines.

instrumentation amplifier   A practical differential amplifier, usually packaged in an IC, with features such as high input resistance, low output resistance, and selectable gain.

integral control   A feedback control strategy where the restoring force is proportional to the sum of all the past errors multiplied by time. This type of control is capable of eliminating steady-state error, but increases overshoot.

integral horsepower (hp) motor   A motor with 1 hp or more; that is, a large motor.

integrator   An op-amp based circuit that has an output voltage proportional to the area under the curve traced out by the input voltage.

interfacing   The interconnection between system components.

inverting amplifier   A simple op-amp voltage amplifier circuit with one input, where the output is out of phase with the input; one of the most common op-amp circuits.

inverting input   The minus $(-)$ input of an op-amp; the output will be out of phase with this input.

I/O message and communications instructions   Allow PLCs to send and receive messages and data to other PLCs or a terminal (PC), for those applications where PLCs are networked.

I/O   *See* **input/output**.

isolation circuit   A circuit that can transfer a signal voltage without a physical electrical connection.

iteration   One pass through the computer program being executed by the digital controller; each iteration "reads" the set-point and sensor data and calculates the output to the actuator. *See* **scan**.

JFET   *See* **junction FET**.

jogging   The practice of "inching" a motor into position by repeated short bursts of power.

junction FET   One type of an FET.

Kelvin scale   The Kelvin temperature scale starts at absolute zero, but a Kelvin degree has the same temperature increment as a Celsius degree ($0°C = 273°$ K).

ladder diagram   A type of wiring diagram (which resembles a ladder) used for motor-control circuits; ladder diagrams typically include switches, relays, and the devices they control.

lag time   The time between the change of the set point and the actual movement of the controlled variable.

latching   *See* **sealing**.

leadscrew linear actuator   A type of linear actuator that uses an electric motor to rotate a threaded shaft; the linear motion is created by a nut advancing or retreating along the threads.

least significant bit (LSB)   The rightmost bit in a binary number, it represents the smallest quantity that can be changed—that is, the difference between two successive states.

limit switch   A switch used as a proximity sensor—that is, a switch mounted so that it is activated by some moving part.

linearity error   A measurement error induced into the system by the sensor itself, it is the difference between the actual quantity and what the sensor reports it to be.

linear variable differential transformer (LVDT)   A type of linear-motion position sensor. The motion of a magnetic core, which is allowed to slide inside a transformer, is proportional to the phase and magnitude of the output voltage.

line voltage   A single number that represents the voltage of a power system; for three-phase power, the line voltage is the vector sum of two of the three-phase voltages.

loading error   An error that may occur in an analog voltage signal when the circuit being driven draws too much current, thus "loading down" the voltage.

load torque   The torque required of the motor to rotate the load.

lockup property   The property of high-ratio worm gears that cannot be driven backward.

logical and shift instructions Perform logical operations such as AND, OR, and NOT and Right and Left Shifts on words stored in memory.

logical variable A single data bit in those cases where a single bit is used to represent an on–off switch, motor on–off control, and so on.

low-pass filter A circuit that allows lower-frequency signals to pass but attenuates higher-frequency signals.

L/R drive A stepper motor driver circuit that uses ballast resistors in series with the motor coils to increase torque at higher stepping rates.

LSB *See* **least significant bit.**

LVDT *See* **linear variable differential transformer.**

machine language The set of operation codes that a CPU can execute.

magnetic field noise An unwanted current induced in a wire because the wire is in a time-varying magnetic field.

main winding One of two windings in an AC servomotor.

mass The amount of material in an object; mass is related to weight in that the more mass it has, the more it will weigh.

math instructions Performs mathematical operations on words in memory; these operations may include addition, subtraction, multiplication, division, and others.

membership function Used in conjunction with fuzzy logic control, the distribution of values that have been grouped into one category.

membrane switch Usually a keypad with a flexible membrane over the top.

memory The part of the computer that stores digital data. Memory data is stored as bytes, where each byte is given an address.

memory-mapped input/output A system where I/O ports are treated exactly like memory locations.

metal-oxide semiconductor FET (MOSFET) One type of an FET.

microcontroller An integrated circuit that includes a microprocessor, memory, and input/output; in essence, a "computer on a chip."

microprocessor A digital-integrated circuit that performs the basic operations of a computer but requires some support-integrated circuits to be functional.

microstepping A technique that allows a regular stepper motor to take fractional steps; it works by energizing two adjacent poles at different voltages and by balancing the rotor between.

microswitch A small push-button switch with a very short throw distance.

minimum holding voltage Minimum voltage needed to keep a relay activated.

mnemonic An English-like abbreviation of an operation code for PLC logical instructions.

modem A circuit that converts serial data from digital form into tones that can be sent through the telephone system.

momentary-contact switch A switch position that is spring-loaded.

moment of inertia (I) A mechanical property of an object that is based on its shape and mass and the axis of rotation; the larger the moment of inertia, the more torque it takes to spin the object about the designated axis.

momentum The property of a moving object that tends to keep it moving in the same direction.

MOSFET *See* **metal-oxide semiconductor FET.**

most significant bit (MSB) The leftmost bit in a binary number.

MSB *See* **most significant bit.**

MT$_1$ One of three terminals of a triac.

MT$_2$ One of three terminals of a traic.

multiplexing The concept of switching input signals (one at a time) through to an output; is typically used so that multiple sensors can use a single ADC (analog-to-digital converter).

natural resonant frequency In mechanical systems, the frequency at which a part or parts will vibrate. The resonant frequency is a function of the mass and spring constant.

NC *See* **numerical control.**

negative feedback A circuit design where a portion of the output signal is fed back and subtracted from the input signal. This results in a lower but predictable gain and other desirable properties.

neutral The return wire or cold side in an AC power-distribution system; the neutral wire is at ground potential.

Newton Unit of force in the SI system (1 N = 0.224 lb).

no-load speed The speed of a motor when there is no external load on it; it will always be the maximum speed (for a particular voltage).

noninverting amplifier A simple op-amp voltage amplifier circuit with one input, where the output is in phase with the input.

noninverting input   The positive (+) input of an op-amp; the output will be in phase with this output.

nonvolatile memory   Computer memory such as ROM that will not lose its data when the power is turned off.

normal force   In friction calculations, the force pushing the sliding surfaces together.

normally closed (NC) contacts   Relay contacts that are closed in the deenergized state and open when the relay is energized.

normally open (NO) contacts   Relay contacts that are open in the deenergized state and closed when the relay is energized.

normally open (NO), normally closed (NC)   The state of momentary switch or relay contacts when the device is not activated.

notch filter   A circuit that attenuates a very narrow range of frequencies.

NPN, PNP   The two basic types of transistors, the difference being the direction of current and voltage within the device.

null modem   A cable that allows two DTE units to communicate with each other (see RS-232).

numerical control   A digital control system that directs machine tools, such as a lathe, to automatically machine a part.

off-delay relay   A time-delay relay that, after being deenergized, waits a specified period of time before activating.

on–off control   See two-point control.

open-loop control system   A control system that does not use feedback. The controller sends a measured signal to the actuator, which specifies the desired action. This type of system is not self-correcting.

open-loop gain   The gain of an amplifier when no feedback is being used; it is usually the maximum gain possible.

open-loop system   A type of control system where there is no feedback; consequently, the controller does not know for sure the exact condition of the controlled variable.

operational amplifier (op-amp)   A high-gain linear amplifier packaged in an integrated circuit; the basis of many special-purpose amplifier designs.

operation code (op-code)   A digital code word used by the microprocessor to identify a particular instruction.

optical coupler   An isolation circuit that uses a light-emitting diode (LED) and a photocell to transfer the signal.

optical rotary encoder   A rotary position sensor that works by rotating a slotted disk past a photo sensor.

optical tachometer   Mounted next to a rotating shaft, a photo sensor that gives a pulse for each revolution (a stripe is painted or fixed on the shaft).

orifice plate   A type of flow sensor whereby a restriction is placed in a pipe causing a pressure difference (between either side of the restriction) that is proportional to flow.

overdamped   A system that has so much drag (from static or viscous friction) that its response is sluggish.

overload device   A device that can stop the motor if an overload condition is detected. Overload is detected by sensing the temperature of the motor or the current that the motor is drawing.

overshoot   The event that occurs when the path of the controlled variable (mechanical or electrical) approaches its destination too fast and goes beyond the set point before turning around or stopping. Underdamped systems tend to overshoot.

parallel interface   A type of data interface where 8 bits enter or leave a unit at the same time on eight wires.

PC   See personal computer.

permanent magnet (PM) motors   A motor that uses permanent magnets to provide the magnetic field. The PM motor has a linear torque–speed relationship, making it desirable for control applications.

permanent magnet (PM) stepper motor   A motor that uses one or more permanent magnets for the rotor; this motor has a detent torque.

permanent-split capacitor motor   A capacitor start motor where the start winding remains engaged during operation—that is, the motor does not have a centrifugal switch.

personal computer (PC)   A microprocessor-based, self-contained, general-purpose computer (usually refers to an IBM or compatible computer).

phase   The number of separate field winding circuits.

phase-control circuit   Circuit that can delay conduction for part of the AC cycle for the purpose of reducing average output voltage.

phase margin   A form of a safety margin. When the system gain is 1, the phase margin is the amount the phase lag could be increased before the system becomes unstable.

phase sequence   The sequence of phase voltages possible with three-phase power: ABC and ACB.

phase voltage   The individual voltage in a multiphase power system.

photodiode   A type of optical sensor, it increases its reverse-leakage current when exposed to light.

photo resistor   A type of optical sensor; its resistance decreases when exposed to light.

photo transistor   A type of optical sensor; light acts as the base current and turns on the transistor.

photovoltaic cell   A device that converts light into electrical energy; used as an optical sensor, or as a solar cell.

pick-and-place robot   A simple robot that does a repetitive task of picking up and placing an object somewhere else.

PID   Stands for Proportional + Integral + Derivative; a control strategy that uses proportional, integral, and derivative feedback.

PID control   A common control strategy, PID instructions that are available on some PLCs.

piezoresistive effect   A property of semiconductors in which the resistance changes when subjected to a force.

pinion   The small driver gear in a gear pass.

PIP   Stands for Proportional + Integral + Preview; a control strategy that uses feedforward so that it knows in advance if a change is coming.

pitch circle   If gears were solid disks, the pitch circle would be the theoretical circle that meshed gears roll on.

pitch diameter   The diameter of the pitch circle.

pitot tube   A velocity sensor for fluids whereby a small open tube is placed directly into the flow; the pressure in the tube is proportional to fluid velocity.

PLC   See **programmable logic controller**.

PLC bus   The wires in the backplane of a modular PLC system that allows the microprocessor (within the PLC) to communicate with plugged-in modules.

plug-in module   An interface circuit to the PLC, packaged as a separate module, which plugs into the PLC. Different modules support different I/O devices, from simple switch and relay output modules to ADC and DAC modules.

PM motor   See **permanent magnet motor**.

pneumatic control valve   Used in pneumatic systems, a valve that controls air flow to the actuators.

pneumatic cylinder   A linear actuator powered by compressed air, it consists of a cylinder, enclosed at both ends with a piston inside. The piston is moved by admitting the compressed air into either end of the cylinder.

pneumatic system   A system that uses compressed air to power actuators such as pneumatic cylinders.

pneumatic time-delay relay   A time-delay relay that generates the delay by allowing air to bleed out of a spring-loaded bellows.

pony motor   A separate motor used to start a synchronous motor.

port   The part of a computer where I/O data lines are connected; each port has an address.

positive-displacement pump   A hydraulic pump that expels a fixed volume of fluid for each revolution of the pump shaft.

positive feedback   The event that occurs when the feedback signal from the sensor is added to the set point (instead of subtracted). Positive feedback will likely cause the system to oscillate.

potentiometer   A variable resistor that can be used as a position sensor.

power   A property that describes how fast energy is being used; in other words, power is energy per unit time.

power factor   The cosine of the angle between the current and voltage. A power factor of 1 means the current and voltage are in phase, which is desirable. A lagging power factor means the current is lagging the voltage, probably due to the inductive load of motors.

power rails   Part of a ladder diagram, the two vertical lines on each side of the diagram, which represent power lines.

power transistor   A transistor designed to carry a large current and dissipate a large amount of heat.

pressure-control valve   Used in a hydraulic system, an adjustable spring-loaded valve that can maintain a set pressure in a system by bleeding off excess fluid.

pressure regulator   Used in pneumatic systems, a regulator that maintains a specified pressure.

pressure tank   Used in pneumatic systems, a reservoir for compressed air.

prime mover   A component that creates mechanical motion from some other energy source; the source of the first motion in the system.

process   The physical process that is being controlled.

process control system   The type of system where the controller is attempting to maintain the output of some process at a constant value.

process variable (PV)   The actual value of a variable, as reported by the sensor.

processor   Part of the PLC that executes the instructions; it is a small, microprocessor-based computer.

program counter   A special address-holding register in a computer that holds the address of the next instruction to be executed.

programmable gain instrumentation amplifier An instrumentation amplifier with fixed gains that can be selected with digital inputs.

programmable logic controller (PLC) A small, self-contained microprocessor-based controller used primarily to replace relay logic controllers.

programmable unijunction transistor (PUT) A solid-state device that performs the same function as a UJT except that the trigger voltage is adjustable.

program file Contains the program that the PLC executes. The program file is part of the PLC memory (the other part being data files).

program memory When applied to PLCs, the section of memory (RAM or EEPROM) where the PLC program is stored.

programming port A connector on a PLC through which the program is down loaded.

proportional band A term used to describe the gain of a proportional control system. It is the change in error that causes the controller output to change its full swing.

proportional control A feedback control strategy where the controller provides a restoring force that is proportional to the amount of error.

proximity sensor A sensor that detects the physical presence of an object.

psi Pounds per square inch, a unit of pressure.

pull-in current Minimum current needed to activate a relay.

pull-in voltage Minimum voltage needed to activate a relay.

pull-out torque The maximum torque that an AC motor can provide just before it stalls. However, unlike the DC motor, this maximum torque condition occurs at about 75% of the unloaded speed.

pulse-width modulation (PWM) A method of controlling an electric motor's speed by providing pulses that are at a constant DC voltage. The width of the pulses is varied to control the speed.

push-button switch A momentary switch activated by pushing.

PUT *See* **programmable unijunction transistor.**

PV *See* **process variable.**

PWM *See* **pulse-width modulation.**

RAM *See* **random-access memory.**

random-access memory (RAM) Sometimes called read/write memory, a memory arrangement using addresses where data can be written in or read out; RAM loses its contents when the power is turned off.

Rankine scale The Rankine temperature scale starts at absolute zero, but a Rankine degree has the same temperature increment as a Fahrenheit degree ($0°F = 460°R$).

rated speed The speed of a motor when producing its rated horsepower.

reaction-curve method A method of tuning a PID controller that involves opening the feedback loop, manually injecting a step function, and measuring the system response. The PID constants are computed based on these data.

read-only memory (ROM) Similar to RAM in that it is addressable memory, but it comes preprogrammed and cannot be written into; also, it does not lose its data when the power is turned off.

read/write (R/W) line A control signal that goes from the microprocessor to memory.

real time Refers to a computer that is processing data *at the same time* that the data are generated by the system.

reciprocating piston compressor An air compressor that uses pistons and one-way valves to compress the air.

reduced voltage–starting circuit A special circuit that limits the motor current during start-up.

reed relay A small relay with contacts sealed in a tube and activated by a magnetic field.

regulator system A control system that maintains an output at a constant value.

resistance temperature detector (RTD) A temperature sensor based on the fact that the resistance of a metal wire will increase when the temperature rises.

resolution In digital-to-analog conversion, the error that occurs because digital data can only have certain discrete values. The smallest increment of data that can be detected or reported. For a digital system, the resolution is usually the value of the least significant bit.

rise time The time it takes for the controlled variable to rise from 10 to 90% (of its final value).

robot A servomechanism control system in the form of a machine with a movable arm.

ROM *See* **read-only memory.**

rotary switch A rotating knob that activates different switch contacts.

rotating field The phenomenon of what is apparently happening in the stator coils of an AC motor. Even though the coils themselves are stationary, the magnetic field is transferred from coil to coil sequentially, producing the effect that the field is rotating.

rotor A basic part of an AC motor, the part in the center that actually rotates. If the field poles are stationary (as they usually are), then the rotor is known as the armature.

RS-232 standard A serial data transmission standard that specifies voltage levels and signal protocol between a DTE (computer) and a DCE (modem).

RTD *See* **resistance temperature detector**.

run mode The operational mode of the PLC when it is actually executing the control program and hence performing some control operation.

run windings The main stator windings in a single-phase motor.

R/W *See* **read/write line**.

sample rate The times per second (or per minute) that a sensor is read; for a digital controller, the sample rate usually corresponds to the scan rate.

sample-and-hold circuit A circuit that can temporarily store or remember an analog voltage level.

sampling rate The times per second a digital controller reads the sensor data.

scan The event when the processor makes one pass through all instructions in the control program loop. *See* **iteration**.

SCR *See* **silicon-controlled rectifier**.

sealed current Current required to keep a relay energized.

sealing A relay wired such that once energized its own contacts take over the job of providing power to the coil; it will stay energized, even if the original energizing signal goes away.

Seebeck effect The property used by a thermocouple, a voltage proportional to temperature developed in a circuit consisting of junctions of dissimilar metal wires.

sensor Part of the control system that monitors the system output (such as temperature, pressure, or position), the sensor converts the output movement of the system into an electric signal, which is fed back to the controller.

sequentially controlled system A control system that performs a series of actions in sequence, an example being a washing machine.

serial interface A type of interface where data are transferred 1 bit after the other on a single wire.

series-wound motor A motor that has the field windings connected in series with the armature; the series-wound motor has a high starting torque and a high no-load speed.

servomechanism An electromechanical feedback control system where the output is linear or rotational movement of a mechanical part, such as a robot.

set point (SP) An input into the controller, it represents the desired value of the controlled variable.

settling time The time it takes for the system response to settle down to some steady-state value.

shaded pole A small metal ring around the end of the electromagnetic pole of an AC relay, for the purpose of keeping the relay from "buzzing" at 60 Hz.

shaded-pole motor A single-phase induction motor, usually small, that does not use a start winding. Instead, a modification to the single-phase poles causes the motor to have a small starting torque.

shunt-wound motor A motor that has the field windings connected in parallel with the armature windings; the shunt-wound motor has a measure of natural speed regulation—that is, it tends to maintain a certain speed despite load changes.

signal common The common voltage reference point in a circuit, usually the negative terminal of the power supply (or battery).

signal return *See* **signal common**.

silicon-controlled rectifier (SCR) A semiconductor device that provides speed control to a DC motor from an AC-power source (without the need of a power supply).

single-acting cylinder A spring-loaded pneumatic cylinder; air pressure drives the piston in one direction, and spring pressure returns it.

single-board computer A premade microprocessor-based computer assembled onto a single printed-circuit card.

single-ended voltage A signal voltage that is referenced to the ground.

single-phase AC The type of AC supplied to residences (and industry). It consists of a single alternating waveform that makes 60 complete cycles per second.

single-point ground A single connection point in a system, where all signal commons are connected together and then connected to an earth ground.

single-pole/double-throw switch (SPDT) A switch contact configuration.

single-pole/single-throw switch (SPST) A switch contact configuration.

single-step mode Operating the motor at a slow enough rate so that it can be stopped after any step without overshooting.

slew mode   Stepping the motor at a faster rate than the single-step mode; used to move to a new position quickly. The motor will overshoot if the speed is not ramped up or down slowly.

slide switch   Similar to a toggle switch except the handle slides back-and-forth.

slider (wiper)   The moving contact in a potentiometer, usually the center of three terminals.

sliding friction   The frictional drag force on two dry sliding objects.

slip   The difference between the rotor speed of an induction motor and the synchronous speed of the stator field; the amount of slip is typically less than 10%.

slip rings   Sliding electrical contacts that allow power to be fed to the rotor of the synchronous motor.

slotted coupler   An optical proximity sensor that is activated when an object "cuts" a light beam.

snubber   A circuit that prevents a fast voltage rise across an SCR, for the purpose of keeping it from false firing.

solenoid   An electromechanical device that uses an electromagnet to produce short-stroke linear motion.

solid-state relay (SSR)   A solid-state switching device used as a relay.

solid-state time-delay relay   A time-delay relay that generates the delay with an electronic circuit.

source   One of three terminals in an FET.

SP   *See* **set point.**

SPDT   *See* **single-pole/double-throw switch.**

speed regulation   In general, a motor's ability to maintain its speed under different loads; specifically, a percentage based on no-load speed and full-load speed.

split-phase   A general term applied to an induction motor that has two sets of windings but operates on single-phase AC with some provision to phase-shift the AC for the second set of windings.

split-phase control motor   A two-phase motor, usually small, powered by single-phase AC using a capacitor for the phase shift. The direction is reversible, making these motors useful in controlling back-and-forth motion.

spool   The moving part within the bidirectional control valve.

SPST   *See* **single-pole/single-throw switch.**

spur gear   A type of circular gear with radial teeth (the most common type of gear).

squirrel cage rotor   The type of rotor used in an induction motor—named as such because someone thought it looked like a squirrel cage.

SSR   *See* **solid-state relay.**

stall   When the load on the AC motor is increased (about 75% of unloaded speed), the motor torque drops back abruptly, and the motor stalls and stops.

stalling   A situation wherein the motor cannot rotate because the load torque is too great.

stall torque   The torque of the motor when the shaft is prevented from rotating; it will always be the maximum torque (for a particular voltage).

start windings   A second set of windings (besides the run windings) in the single-phase induction motor; the start windings are usually only engaged during the starting process.

starting torque   The motor torque available when the motor is first started; the starting torque of most AC motors is less than its maximum torque but more than its rated load.

static friction   The friction force that must be overcome to get an object at rest to move; for a particular object, static friction is greater than sliding friction.

static pressure   The pressure measured when the open tube is directed perpendicularly to the flow.

stator   A basic part of an AC or stepper motor, a stationary collection of coils that surrounds the rotor and consists of field poles (electromagnets).

steady-state error   The error that remains after the transient response has died away.

step change   The event that occurs when a discrete change is made to the set point.

stepper motor   A motor that rotates in steps of a fixed number of degrees each time it is activated.

strain   The deformation (per unit length) as a result of stress.

stress   Subjecting an object to tension or compression forces. Stress is the force per unit area within the object.

stroke   The overall linear travel distance of a linear actuator.

summing amplifier   An op-amp circuit that has multiple inputs and one output. The value of the output voltage is the sum of the individual input voltages.

swash plate   Part of the axial piston pump; when the angle of the swash plate is changed, the flow from the pump is changed.

switch wafer   Part of a rotary switch.

synchronous condenser A term applied to a synchronous motor that is being used solely for power factor correction.

synchronous motor An AC motor that runs at the synchronous speed, which means an exact multiple of the line frequency. A synchronous motor has no slip.

synchronous speed The speed of the rotating field in the stator, which is always an exact multiple of the line frequency. A synchronous motor spins at the synchronous speed, whereas an induction motor turns somewhat slower than the synchronous speed.

TF *See* **transfer function**.

thermal time-delay relay A time-delay relay that generates the delay by allowing a bimetallic strip to heat and bend, thus activating a switch.

thermistor A temperature sensor based on the fact the resistance of some semiconductors will decrease as the temperature increases.

thermocouple A temperature-measuring sensor made from the junction of two dissimilar metal wires; when the junction is heated, a small voltage is generated.

three-phase AC The type of AC available to industry. It consists of three AC waveforms (called phases) on three wires, where each phase is delayed 120° from the previous phase and all phases operate at 60 Hz.

three-phase stepper motor A motor with three separate sets of field coils; usually found with VR motors.

three-position control Similar to the two-point system except the actuator can exert force in two directions, so the system is either off-up-down, off-right-left, and so on.

three-position switch Switch with a center position.

threshold detector A circuit that provides a definite off–on signal when an analog voltage rises above a certain level.

thumbwheel switch Switch that rotates a drum to select numeric data.

thyristor A class of four-layer semiconductor devices (such as PNPN) that are inherently bistable; the SCR and triac are thyristors.

time-delay loop A programming technique where the computer is given a "do-nothing" job such as counting to some large number for the purpose of delaying time.

time-driven operation In a sequentially controlled system, an action that is allowed to happen for a specified period of time. This is an example of open-loop control.

timer relay (TR) A relay with a built-in time delay mechanism so that the contacts close or open a specified time after the relay is energized.

toggle switch A manually operated device that connects or disconnects power.

torque In general, the measure of the motor's strength in providing a twisting force; specifically, the product of a tangential force times the radius. Torque is used in rotational systems just as force is used in linear systems.

torque–speed curve A graph of a motor's torque versus speed; can be used to predict the motor's speed under various load conditions.

TR *See* **timer relay**.

transconductance (g_m) The gain of an FET, which is the change in drain current divided by the change in gate voltage.

transducer A term used interchangeably with *sensor*. Literally means that energy is converted, which is what a sensor does.

transfer function A mathematical relationship between the input and output of a control system component: TF = output/input. (In this text, *transfer function* and *gain* are used interchangeably, although this is a simplification.)

transient response The actual path that the controlled variable takes when it goes from one position to another.

triac A bistable, three-terminal, solid-state device that switches power.

trigger circuit A circuit used to generate the turn-on pulse for an SCR or a triac.

tuning When applied to a control system, this term refers to the process of making adjustments to the gain constants K_P, K_I, K_D until system performance is satisfactory.

turbine A flow sensor based on having the fluid rotate a propeller of some kind.

two-capacitor motor A capacitor start motor that uses one value of capacitance for starting and switches to a lower value of capacitance for running.

two-phase stepper motor A motor with two field circuits. This motor requires polarity reversals to operate and hence is bipolar.

two-point control A simple type of control system where the actuator is either on full force or off; also called *on–off control*.

two-position switch A toggle or slide switch with two positions.

UART *See* **universal asynchronous receiver transmitter**.

UJT *See* **unijunction transistor**.

underdamped A system that has relatively little damping so that it responds quickly and tends to overshoot.

unijunction transistor (UJT)   A bistable, three-terminal solid-state device that primarily triggers an SCR.

unipolar motor   A motor that does not require polarity reversals. A four-phase motor is unipolar.

universal asynchronous receiver transmitter (UART)   A special purpose integrated circuit that converts data from parallel to serial format and vice versa.

universal motor   An electric motor that can run on AC or DC power; it is essentially a series DC motor.

vane motor   A device that creates rotary motion from pneumatic or hydraulic pressure. The construction is similar to a vane pump.

vane pump   A hydraulic pump that uses an offset rotor with retractable vanes.

variable-displacement pump   A hydraulic pump that can be adjusted to change the flow while keeping the speed of rotation constant.

variable reluctance (VR) stepper motor   A motor that uses a toothed iron wheel for the rotor and consequently can take smaller steps but has no detent torque.

venturi   A type of flow sensor whereby fluid is forced into a smaller channel, which increases its velocity; the higher velocity fluid has a lower pressure (than the fluid in the main channel), and the pressure difference is proportional to velocity.

$V_{GS(off)}$ The gate voltage necessary to turn the drain current off (for a JFET).

virtual ground   A point in a circuit that will always be practically at ground voltage but that is not physically connected to the ground.

viscous friction   A drag force experienced when a lubricant is used between the objects so that the objects do not actually touch; the drag force, which is proportional to velocity, comes from the layers of lubricant slipping over each other.

volatile memory   Computer memory such as RAM that will lose its data when the power is turned off.

voltage follower   A very simple but useful op-amp circuit with a voltage gain of 1, used to isolate circuit stages and for current gain.

weight   Technically, a downward force exerted by a mass, caused by gravity.

window comparator   A comparator with built-in hysteresis, that is, two thresholds: an upper switch-in point and a lower switch-in point.

windup   Associated with integral feedback, the problem that occurs when too much error accumulates as a result of the system saturating before it can achieve the state that it theoretically should.

wiper   The center and/or moveable contact in a switch. *See* **slider**.

wire-wound potentiometer   A variable resistor that uses a coil of resistance wire for the resistive element; can be used as a position sensor.

word   A unit of digital data that a particular computer uses; common word sizes are 4, 8, 16, and 32 bits.

worm   The spiral-looking gear in a worm gearbox.

worm gear   The circular gear that the worm meshes with in a worm gearbox.

wye-connection   One of two ways in which to connect three-phase motor or generator coils; with the wye-connection, each of the three coils is connected between a phase wire and neutral.

Young's modulus   A constant that relates stress and strain for a particular material (Young's modulus = stress/strain).

zero-voltage switching   As applied to SSRs, the delay in switching until the AC voltage crosses the zero point.

Section 1.1

1.1

a. Set point $\rightarrow$ $\boxed{\text{Controller}}$ $\rightarrow$ $\boxed{\text{Actuator}}$ $\rightarrow$ $\boxed{\text{Process}}$ $\rightarrow$ Controlled variable

b. Set point specifies the desired system output. The controller generates a drive signal to the actuator. The actuator moves the process. There is no feedback in an open-loop control system, so the controller cannot correct any errors that may occur "downstream."

c. The operating characteristics of the components must be well-known and stable or predictable.

d. Simplicity, less hardware

1.3

The error is initially 15° (45°–30°). This error signal causes the controller to drive the robot arm toward the 30° position. When the arm reaches 30°, there is no more error, so the controller directs the arm to stop moving.

1.5

$$V_{pot} = \frac{.1\,V}{deg} \times 45\ deg = 4.5\ V$$

1.7

Two sets of input/output data are given; either can be used.

$$TF_{motor} = \frac{500\ rpm}{6\ V} = 83.3\ rpm/V$$

Section 1.2

1.9

Analog control system	Digital control system
Uses analog circuits such as op-amps, which are hard-wired to do the job.	Uses a microprocessor or microcontroller under program control.
No appreciable delay time from input to output.	Works on the basis of a program loop, which is a 3-step process: 1. Inputs read 2. Outputs calculated 3. Outputs sent out
	Usually requires using ADCs or DACs.

Section 1.3

1.11

A **process control system** is usually trying to keep some parameters constant in a continuous process, such as the thickness of a steel plate from a rolling mill, or it is controlling a sequence of events, such as filling a tank, mixing, and draining.

A **servomechanism** usually is a position-control system, which controls the movement of some device, such as a robot arm.

1.13

a. Time driven—defrost cycle on a "frost-free" refrigerator

b. Event driven—Automatic choke on a car (closes until engine heats)

c. Time and event driven—A batch mixing process whereby (1) the fill valve opens until vat is full; (2) mixes for 2 minutes

Section 2.1

2.1

ALU (Arithmetic Logic Unit)—performs calculations.

Control Unit—manages the data flow within the computer, such as reading and executing program instructions.

CPU (Central Processing Unit)—The "heart" of a computer, includes the ALU and control unit. A microprocessor is a CPU.

Memory—Consists of memory cells organized in groups of 8 bits (byte). Each byte is given a unique address. Data are stored and retrieved on the basis of their address. The memory is used to store the program, as well as the program results.

I/O (Input/Output)—The input port is a channel through which data from the "outside world" enter the computer. The output port is a channel through which data leave the computer.

2.3

Address bus—A group of wires that transports the memory or I/O address from the microprocessor.

Data bus—A group of wires that transports the data to and from the CPU, from memory and I/O devices.

Control bus—A group of wires that transports timing and control signals from the CPU to memory and I/O.

Section 2.2

2.5

Final value in accumulator = 4

Section 2.3

2.7

a. Parallel data port—Digital data is transferred from the computer to the "outside world" in 8- (or 16) bit chunks on 8 (or 16) wires.

b. Serial data port—Digital data are transferred on a single wire bit by bit (one bit at a time).

2.9

$$V_{out} = \frac{204 \times 9\,V}{256} = 7.17\,V$$

2.11

$$Output = \frac{3.7\,V \times 255}{12\,V} = 78.6 \approx 79$$

$$= 1001111_{Binary}$$

Section 2.4

2.13

"Real time" computing means that the computer is connected directly to the "real world" system and is performing calculations on the system data as it comes in. Real time computing is necessary for control systems because the controller response must be calculated immediately upon receipt of sensor data.

2.15

A computer generates a time delay by using a "time-delay loop," which is a section of the program that counts up to a large number: the larger the number, the longer the delay.

Section 2.5

2.17

A **microcontroller** is a "computer on a chip." It contains a **microprocessor**, memory, I/O ports, and perhaps other features. Newer microprocessors are faster and have a more sophisticated set of instructions than a microcontroller, but they require external components, such as memory, to work.

2.19

1. An expansion card(s) for the PC with at least two input ports (for the two position sensors), and two output ports (for the two motors). Each input port requires an ADC and each output port requires a DAC (the ADCs and DACs may be already on the expansion card).

2. Two power amplifiers for the motors, to take the signal from the DACs and to boost the power to drive the motors

3. Software to perform the control operation. The software would have to read in the sensor data from the input expansion card, and send the motor drive signals to the output expansion card.

Section 3.1

3.1

a. $10(2 - 1) = 10$ V
b. $10(-3 - -2.5) = -5$ V
c. $10(2.5 - -3) = 55$ V
d. $10(-2.5 - -4) = 15$ V
e. $10(-1.5 - 1) = -25$ V
f. $10(1.5 - 3) = -15$ V

3.3

The input goes directly into the + input of the op amp; therefore, the output will be in phase with the input, and the input resistance is very high. The output is connected back to the − input, and because the two op amp inputs are virtually the same voltage, the output voltage will be the same as the input voltage, so the gain is 1.

3.5

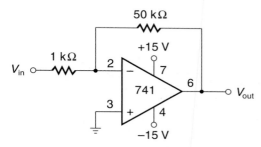

Select $R_i = 1$ kΩ, then $R_f = -A R_f = - -50 \times 1$ kΩ $= 50$ kΩ

3.7

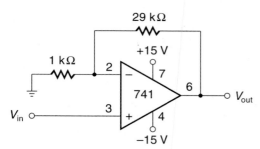

Select $R_i = 1$ kΩ

$$A_V = 30 = \frac{R_f}{1 \text{ k}\Omega} + 1$$

$$R_f = 29 \times 1 \text{ k}\Omega = 29 \text{ k}\Omega$$

3.9

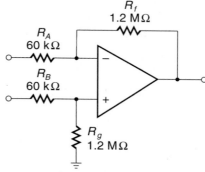

Select R_A and R_B to be ten times higher than source impedance:

$$R_A = R_B = 10\, R_{\text{source}} = 60 \text{ k}\Omega$$
$$R_f = R_g = 20 \times 60 \text{ k}\Omega = 1.2 \text{ M}\Omega$$

3.11

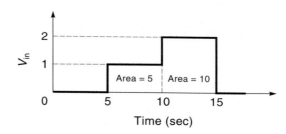

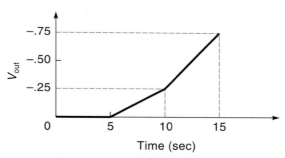

$V_{\text{out}}(5 \text{ s}) = -0.05 \times 0 = 0$ V
$V_{\text{out}}(10 \text{ s}) = -0.05 \times 5 = -0.25$ V
$V_{\text{out}}(15 \text{ s}) = -0.05(5 + 10) = -0.75$ V

3.13

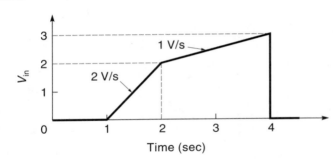

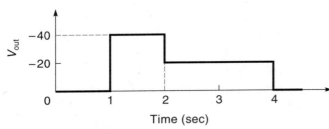

$V_{out(0-1\ s)} = -20\ s \times 0\ V/s = 0\ V$

$V_{out(1-2\ s)} = -20\ s \times 2\ V/s = -40\ V$

$V_{out(2-4\ s)} = -20\ s \times 1\ V/s = -20\ V$

$V_{out(4s+)} = -20\ s \times 0\ V/s = 0\ V$

3.15

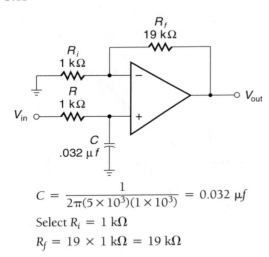

$C = \dfrac{1}{2\pi(5 \times 10^3)(1 \times 10^3)} = 0.032\ \mu f$

Select $R_i = 1\ k\Omega$

$R_f = 19 \times 1\ k\Omega = 19\ k\Omega$

3.17

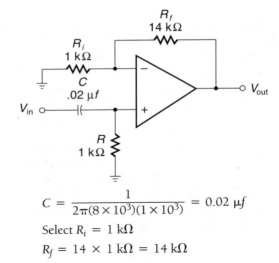

$C = \dfrac{1}{2\pi(8 \times 10^3)(1 \times 10^3)} = 0.02\ \mu f$

Select $R_i = 1\ k\Omega$

$R_f = 14 \times 1\ k\Omega = 14\ k\Omega$

3.19

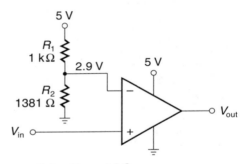

Select $R_1 = 1\ k\Omega$

Solve for R_2. $R_2 = 1381\ \Omega$

Section 3.2

3.21
(See figure at top of next page)

$$R = \frac{V}{I} = \frac{1\ V}{5\ mA} = 200\ \Omega$$

If we select the receiver resistor (R_{rec}) to be 200 Ω, the gain of the differential amp should be 1.

for Gain = 1, $R_f = R_A$ Select 10 kΩ

3.23

$$C = \frac{t}{5\ R_s} = \frac{0.2s}{5 \times 1k\Omega} = 40\ \mu f$$

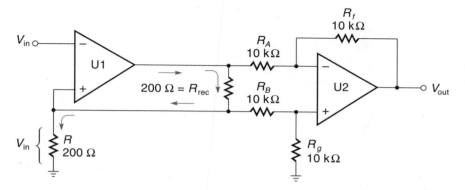

Section 3.3

3.25
A **ground loop** is a current which can exist in a wire that is connected to the earth at two different places. Large currents can result, because even a few volts' difference across a low resistance ground wire develops a large current.

3.27
When V_{in} is pulled to ≈ 0 V by the TTL logic circuit going "low," a voltage is developed across the LED. Within the TIL 112, light from the LED strikes the base of a photo transistor and turns it on. The transistor conducts and pulls V_{out} to ≈ 0 V. A pull-up resistor on V_{out} pulls it up to 5 V when the transistor is off.

3.29
The principle of **magnetic shielding** is that the shield material (like steel) is so conducive to magnetic fields, that the field would rather go in the shield, leaving the adjacent area relatively free of magnetic fields.

3.31
A **single-point ground** is where all the shields and grounds are connected together at one place, and connected to the earth. It is important because it prevents "ground loops."

Section 4.1

4.1
DPDT toggle switch:

Wipers C_1 and C_2 move together: when up, they make contact with B_1 and B_2; when down, they make contact with A_1 and A_2.

4.3
The 2SHX191 switch is a two-pole on–off–on momentary switch. The rest position is off in the middle. It may be activated up or down under spring pressure.

Section 4.5

4.5
DPDT relay:

When the coil is energized, contacts C_1 and C_2 are pulled down.

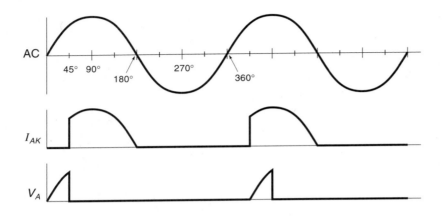

4.7

For the Y4-V52 relay:

coil voltage = 6 Vdc (from table)

$$\text{coil current} = \frac{V}{R} = \frac{6\,\text{V}}{52\,\Omega} = 115\,\text{mA}$$

contact voltage 125 Vac or 28 Vdc
contact current 2A typ (3A max)

4.9

All relays in table meet the 1.5 A requirement. Auto relay should use 12 Vdc coil.

Best choice: Y2 − V185

Section 4.3

4.11

$$I_B = \frac{I_C}{h_{FE}} = \frac{5\,\text{A}}{60} = 83\,\text{mA}$$

4.13

$$I_C = h_{FE}\,I_B = 40 \times 25\,\text{mA} = 1\text{A}$$

Approximate: $I_E = I_C = 1\,\text{A}$

Exact: $I_E = I_C + I_B = 1\,\text{A} + 25\,\text{mA} = 1.025\,\text{A}$

4.15

$$V_B = V_{tot}\frac{R_2}{R_1+R_2} = 10\,\text{V} \times \frac{1\,\text{k}}{13\,\text{k}+1\,\text{k}} = 0.714\,\text{V}$$

Using input curves,

$$I_B \approx 1.1\,\text{mA}.\ I_C = h_{FE}I_B = 70 \times 1.1\,\text{mA} = 77\,\text{mA}$$

4.17

Transistor operating modes:

Class A—Transistor is biased so that it is approximately half on, with no signal applied. This way, the output can swing either up or down (from the midpoint).

Class B—Transistor is biased so that it is at the brink of turning on, but it is off if no signal is applied. This way, any positive-going input voltage will begin to turn the transistor on.

Class C—Transistor is biased so it is completely off, and the input voltage, when applied, will turn the transistor completely on. So the transistor acts like a switch, either completely on, or completely off.

4.19

Select MJ4400 $I_C = 5\,\text{A}$ $h_{FE(min)} = 12$

$P_p = 100\,\text{W s}$

4.21

JFET	MOSFET
1. Gate voltages must be negative.	1. Gate voltages can be positive.
2. Input resistance is very high.	2. Input resistance is almost infinite because gate is capacitively coupled.

4.23

For $V_{GS(th)} = 2\,\text{V}$:

$$V_{GS(active)} = 6\,\text{V} - 2\,\text{V} = 4\,\text{V}$$

$$I_D = V_{GS}\,g_{fs} = 4\,\text{V} \times 2.5\,\text{mho} = 10\,\text{A (max)}$$

For $V_{GS(th)} = 4.5\,\text{V}$:

$$V_{GS(active)} = 6\,\text{V} - 4.5\,\text{V} = 1.5\,\text{V}$$

$$I_D = V_{GS}\,g_{fs} = 1.5\,\text{V} \times 2.5\,\text{mho} = 3.75\,\text{A (min)}$$

Section 4.4

4.25

Turning an SCR off:

1. Open the load circuit with another device, like a switch.
2. Momentarily short the SCR with another device.
3. Use a second SCR, capacitively coupled to the primary SCR (pulls down anode voltage)

4.27
(See figure at top of previous page)
Waveforms for SCR, which are triggered at 45°.

4.29
Select 2N6403. Load current = 16 A, Voltage limit = 400 V

Section 4.5

4.31
Load current (I_m) waveform for Triac triggered at 60°.

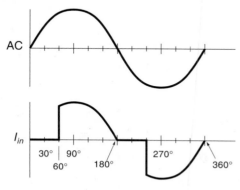

4.33
Delay time = 8.33 ms − 2 ms = 6.33 ms

4.35
Conduction time = 8.33 ms − 3.52 ms = 4.81 ms

Section 4.6

4.37
The purpose of R_1 is to limit the current that is charging capacitor C. By controlling the resistance of R_1, the triggering point of the UJT is controlled.

4.39
When the voltage between R and C get high enough, the diac switches into its conduction mode, which in turn triggers the triac into conduction.

Section 5.1

5.1
Static friction—"Start-up" friction; the force it takes to get something moving from a rest state
Sliding friction—"Running" friction; the continuous drag on an object once it's moving. Sliding friction is less than static friction.

Viscous friction—The drag on an object running with a lubricant. Viscous friction is proportional to velocity.

5.3
$$k = \frac{F}{\text{deflection}} = \frac{1 \text{ oz}}{0.25 \text{ in.}} = 4 \text{ oz/in.}$$

5.5
$$F = ma = \frac{0.313 \text{ lb} \cdot \text{s}^2}{\text{ft}} \times \frac{2 \text{ ft}}{\text{s}^2} = 0.63 \text{ lb}$$

5.7
$$F = ma = 5 \text{ kg} \times \frac{0.5 \text{ m}}{\text{s}^2} = 2.5 \text{ N}$$

5.9
The heavier part will take more power to accelerate, because it has a larger moment of inertia (I).

5.11
$$\alpha = \frac{T}{I} = 1.33 \text{ radians/s}^2$$

$$t = \frac{w}{\alpha} = 7.87 \text{ s}$$

5.13
$$\alpha = \frac{T}{I} = 0.152 \text{ rad/s}^2$$

$$t = \frac{w}{\alpha} = 68.9 \text{ s}$$

Section 5.2

5.15
$$\text{Power} = VI = 220 \text{ V} \times 15 \text{ A} = 3300 \text{ w}$$
$$= 11281 \text{ Btu/hr}$$

5.17
$$I = \frac{P}{V} = \frac{26.6 \text{ w}}{120 \text{ V}} = 0.22 \text{ A}$$

5.19
$$I = \frac{P}{V} = \frac{28.8 \text{ w}}{120 \text{ V}} = 0.24 \text{ A}$$

Section 5.3

5.21
A car that keeps on bouncing would be **underdamped**.
A car body that dipped just once after going over the bump would be **critically damped**.
A car with very stiff shock absorbers would not even dip after going over the bump; this is **overdamped**.

5.23

spring constant $K = 80$ lb/in

mass $= 0.0625$ lb $\cdot$ s^2/ft

resonant frequency $= 19.7$ Hz

Section 5.4

5.25

a. $N_G = 4$
b. $N_G = 2.5$

5.27

$\theta_2 = 180°$

$\theta_2 = 288°$

5.29

$N_{Gtot} = N_{G1} \times N_{G2} \times N_{G3} = 1.75 \times 3 \times 2 = 10.5$

Gear output $= \dfrac{500 \text{ rpm}}{10.5} = 47.6$ rpm

5.31

With a gear ratio of 7.7, the motor speed would be stepped down to:

$w_{roller} = \dfrac{500 \text{ rpm}}{7.7} = 65$ rpm $\therefore$ motor is OK.

5.33

Backlash occurs if there is a space between meshed gear teeth. Then, if one gear is held stationary, the other gear has a small amount of free movement, called "backlash." Backlash increases in long gear trains, proportional to the gear ratio.

5.35

When the gear ratio of a worm gear is high enough, then **lock-up** occurs. This is when the friction is high enough to prevent "back spinning" of the worm gear by the worm.

5.37

$N_h = \dfrac{200}{(200 - 196)} = 50$

Section 5.5

5.39

fan speed $= 1750$ rpm $\times \dfrac{3 \text{ in.}}{12 \text{ in.}} = 438$ rpm

5.41

fan speed $= 1750$ rpm $\times \dfrac{8 \text{ cm}}{30 \text{ cm}} = 467$ rpm

5.43

Similarities   1. connect parallel shafts
            2. all power is transmitted on one side
            3. can be used for speed change

Differences   1. roller chain is lubricated
            2. roller chain is not tight (does not use friction)
            3. roller chain does not creep

Section 6.1

6.1

$V_{pot} = 4.63$ V

6.3

Loading error $= V_{NL} - V_L = 5$ V $- 4.76$ V

$= 0.24$ V

6.5

$\Delta\theta = \dfrac{\text{error} \times \theta_{tot}}{100} = \dfrac{0.25 \times 350°}{100} = 0.875°$

6.7

a. maximum shaft rotation $= \dfrac{350°}{3} = 117°$

b. maximum shaft error $= \dfrac{0.7°}{3} = 0.23°$

6.9

LSB ? 0.0196 V/step (for ADC)

θ_{LSB} ? 1.37°

With 4 : 1 gear ratio: $\theta_{LSB} = 0.34°$; system is OK.

6.11

$\dfrac{360°}{3°/\text{state}} = 120$ states

7 bits 1 128 states $\therefore$ 7 tracks is sufficient

6.13

resolution of encoder $= \dfrac{360°}{500 \text{ slots}} = 0.72°/\text{slot}$

current angle $= \dfrac{0.72°}{\text{slot}} \times 355 \text{ slots} = 255.6°$

Section 6.2

6.15

Velocity is the change in position divided by the corresponding change in time:

$\text{vel} = \dfrac{d_2 - d_1}{t_2 - t_1}$

To get velocity from position sensors, take two position readings (d_2 and d_1) at known time interval ($t_2 - t_1$) and apply the formula.

6.17

$$\text{vel} = \frac{d_2 - d_1}{t_2 - t_1} = \frac{135° - 133°}{0.25 \text{ s}} = \frac{2°}{0.25 \text{ s}} = 8°/s$$

Section 6.3

6.19
1. Switch that detects if car door is closed
2. Switch that detects if refrigerator door is closed

6.21
A slotted coupler contains an LED and photo transistor in one package. Anything (opaque) that moves into the slot will break the light path and be detected. An example would be to detect if a disk is at the correct angle by having the outer edge of the disk pass through the coupler. A hole in the disk would activate the coupler.

Section 6.4

6.23

$$k = \frac{\text{Force}}{\text{extension length}} = \frac{180 \text{ lb}}{1.25 \text{ in.}} = 144 \text{ lb/in.}$$

6.25
The strain gauge consists of an array of fine wires. If the gauge is extended or compressed, the wires change resistance (slightly). The strain gauge is bonded to a material whose spring constant is known, so that by knowing the elongation of the wires, one can calculate the force that caused the elongation.

6.27

$$F = \rho \, A = 2400 \, \frac{\text{lb}}{\text{in}^2} \times 0.785 \text{ in}^2 = 1884 \text{ lb}$$

6.29

$$F = \rho \, A = 1325 \, \frac{\text{N}}{\text{cm}^2} \times 3.14 \text{ cm}^2 = 4161 \text{ N}$$

Section 6.5

6.31
When a **Bourdon tube** is pressurized it tends to straighten out. The amount that the tube straightens is detected with a position sensor.

6.33

$$\text{differential pressure} = 18.7 \, \frac{\text{lb}}{\text{in}^2} - 12 \, \frac{\text{lb}}{\text{in}^2}$$
$$= 6.7 \text{ lb/in}^2$$

6.35

$$\text{differential pressure} = 121.3 \text{ kPa} - 90 \text{ kPa}$$
$$= 31.3 \text{ kPa}$$

Section 6.6

6.37
The **cold junction** of a thermocouple:
1. Is maintained at a constant temperature with a control system.
2. Is compensated for by reading in the ambient temperature and applying a correction factor.
3. Is compensated for by using a temperature-sensitive diode in the thermocouple interface circuit.

6.39

$$V_{net} = 12 \text{ mV} + 2 \text{ mV} = 14 \text{ mV}$$

From the graph, 14 mV 1 Temp ≈ 400°F

6.41

$$\text{Temperature} = 30 \, \Omega \times \frac{°C}{0.39 \, \Omega} = 77°C$$

Section 6.7

6.43
Flow sensors:
   a. **orifice plate**—Fluid is forced through a restriction in pipe. Pressure drop across restriction is proportional to flow.
   b. **venturi**—Pipe diameter is gently reduced, which speeds up flow and reduces pressure. Flow is proportional to amount of reduced pressure.
   c. **Pitot tube**—A small tube faces directly into the flow and detects the impact pressure, which is proportional to flow.
   d. **turbin-type**—A small paddle wheel is placed in the flow. Rotational velocity of the wheel is proportional to flow.
   e. **magnetic**—For fluids which are at least slightly conductive, a magnetic field is placed around a section of pipe. The moving fluid generates a voltage proportional to flow.

Section 6.8

6.45

Locate one detector at the 3-ft level and the other at the 4-ft level. When the level drops to 3 ft, turn on the pump. When the level reaches 4 ft, turn off the pump.

6.47

$$P = dH = 9800 \ \frac{N}{m^3} \times 2 \ m = \frac{19{,}600 \ N}{m^2}$$

$$= 19{,}600 \ Pa$$

6.49

$$H = \frac{P}{d} = 12{,}000 \ \frac{N}{m^2} \times \frac{m^3}{6000 \ N} = 2 \ m$$

Section 7.1

7.1

A motor and a generator are fundamentally the same, so when a motor is running, it is also generating an internal voltage called the CEMF (counter-EMF). This internal voltage is the opposite polarity of the applied voltage, and so it has the effect of reducing the applied voltage. Therefore, as the rpm increases, the CEMF increases, the actual voltage available to the motor decreases, and the torque decreases.

7.3

$$\text{At 0 rpm: } I_A = \frac{V_{In} - CEMF}{R_A} = \frac{10 \ V - 0 \ V}{20 \ \Omega} = 0.5 \ A$$

$$\text{At 1,500 rpm: } I_A = \frac{V_{In} - CEMF}{R_A} = \frac{10 \ V - 6 \ V}{20 \ \Omega}$$

$$= 0.2 \ A$$

Section 7.2

7.5

A series-wound motor has the armature and field windings connected in series. This motor has a large starting torque and tends to "run away" (go faster and faster) when unloaded.

7.7

A shunt wound motor has the armature and field windings connected in parallel. This type of motor has some natural speed regulation, that is, if the load increases, the motor speed tends to stay constant. (Actually, the speed decreases, but not as much as for a series-wound motor.)

7.9

$$\text{Speed regulation} = \frac{S_{NL} - S_{FL}}{S_{FL}}$$

$$= \frac{2000 - 1750}{1750} \times 100 = 14.3\%$$

Section 7.3

7.11

The PM motor is particularly suited to control applications because it has a straight (linear) torque-speed curve. This simplifies the math operations required of the controller.

7.13

Motor voltage $\approx 11.5 \ V$

7.15

No-load speed $= 10{,}100$ rpm
Stall torque $= 2.19$ in. $\cdot$ oz
Voltage $= 9 \ V$

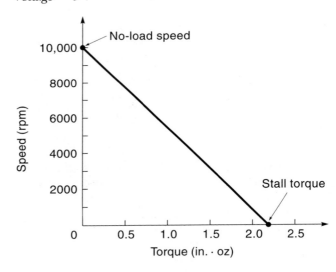

Section 7.4

7.17

A Class A power amplifier can be extremely inefficient, because the transistor "absorbs" the power that does not go into the motor. At lower speeds, the transistor may dissipate *more* power than the motor does.

7.19

$$P_{dis} = VI = 5 \ V \times 1 \ A = 5 \ W$$

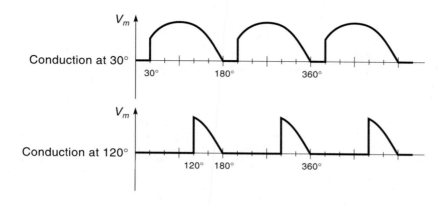

7.21

$$\text{Duty cycle} = \frac{1500 \text{ rpm}}{2000 \text{ rpm}} \times 100\% = 75\%$$

7.23

1. PWM is more power efficient, so it doesn't generate as much heat, allowing the components to be smaller and less costly.
2. PWM is controlled by a digital signal, so no DAC may be required.

7.25

(See figure above)

Section 7.5

7.27

$$\text{Stall torque} = 0.628 \text{ N} \cdot \text{m}$$
$$\text{Motor voltage} \approx 9 \text{ V}$$

Section 7.6

7.29

1. More reliable
2. More efficient
3. Less maintenance
4. Easily controlled with electronics

Section 8.1

8.1

Final position = 2 complete revolutions + 330° cw

8.3

A stepper motor can be **operated open loop** because the controller can reliably know the motor's position by keeping track of the number and direction of steps (providing the motor starts from a known reference position).

8.5

Lifting: Weight = 26.7 oz
Holding (power on): Weight = 46.7 oz
Holding (power off): Weight = 6.7 oz

8.7

Step	Coil voltages
1	$+A \ -B$
2	$+A \ -B$ and $+C \ -D$
3	$+C \ -D$
4	$-A \ +B$ and $+C \ -D$
5	$-A \ +B$
6	$-A \ +B$ and $-C \ +D$
7	$-C \ +D$
8	$+A \ -B$ and $-C \ +D$

Section 8.2

8.9

A **variable reluctance stepper motor** uses a toothed iron wheel for the rotor. When a field pole is energized, the nearest tooth is pulled into alignment. By having more teeth than poles, the step size can be made quite small.

Section 8.3

8.11

The **hybrid stepper motor** combines the features of the PM and VR stepper motors. It has a permanent magnet between two toothed iron rotors. The magnetic field from the permanent magnet gives the motor a detent torque.

Section 8.4

8.13

Position	Pole 1	Pole 2
1	5 V	0 V
2	4.5 V	0.5 V
3	4 V	1 V
4	3.5 V	1.5 V
5	3 V	2 V
6	2.5 V	2.5 V
7	2 V	3 V
8	1.5 V	3.5 V
9	1 V	4 V
10	0.5 V	4.5 V

8.15

Bilevel drive is a technique for improving the torque at higher stepping rates. At the beginning of each step, a higher voltage is applied to the motor coil, which causes a steep current rise. After the current is established, a lower voltage is switched in for the remainder of the step period.

Section 8.5

8.17

a. Steps to advance 1.25 in. = 6000 steps
b. Linear advance per step = 0.000208 in./step

Section 9.1

9.1

a. 270° ? −170 V
b. 90° (for prong B which is neutral) ? 0 V
c. 360° ? 0 V

9.3

A delta-connection generator and load

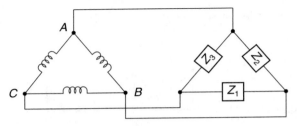

9.5

$$E = \sqrt{3}\, V_p = 1.732 \times 240\ \text{V} = 416\ \text{Vac}$$

Section 9.2

9.7

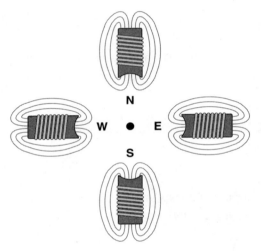

9.9

$$P = \frac{120f}{S_s} = \frac{120 \times 60}{1200} = 6 \text{ pole motor}$$

9.11

$$S_s = \frac{120f}{P} = \frac{120 \times 400}{8} = 6000 \text{ rpm}$$

8 poles per phase, and 1 phase ? 8 individual poles

9.13

$$S_{\text{Rotor}} = S_s - S_s\ (\text{Slip}) = 1200 - 0.05(1200)$$
$$= 1140 \text{ rpm}$$

9.15

The *three phase motor is reversible*, by reversing any two of the three phase voltages.

9.17
Split-phase motor:

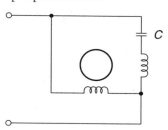

The split-phase motor has two windings: one connected directly across the power lines, and one connected through a capacitor, which phase shifts the power. These two windings provide a true rotating field, and so the motor is self-starting. In essence, it is a two-phase motor running on single-phase power.

9.19
Split-phase control motor:

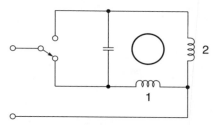

This motor is bidirectional. If the switch is down, winding 1 gets the power directly, and winding 2 is phase-shifted. When the switch is up, winding 2 gets the power directly and winding 1 is phase-shifted.

Section 9.3

9.21
1. The synchronous motor runs at synchronous speed.
2. The synchronous motor has slip rings to get power to the rotor.
3. The synchronous motor is not self-starting (even 3-phase type).

9.23
1. Cause the motor frequency to start slow and then speed up; the motor will follow the rotating field.
2. Use another motor, called a pony motor, to get the synchronous motor up to speed.
3. Insert some "squirrel cage" rotor bars in the rotor. The motor starts as an induction motor.

Section 9.5

9.25
Reversing circuit for 3-phase motor
(See figure at top of next page)

9.27
1. Changing the voltage will change the speed somewhat, because a reduced voltage will increase the slip.
2. Changing the frequency will change the speed, because it changes the speed of the rotating field. The frequency can be changed with a circuit such as a DC link converter.

Section 10.1

10.1
A leadscrew linear actuator works as follows: an electric motor turns a threaded shaft. The output motion is taken from a "nut," which takes a linear motion along the threaded shaft.

Leadscrew linear activators are less messy and don't require additional hardware (pumps, tubing) that hydraulic systems do.

Leadscrew linear-actuators move slower than hydraulic systems, and can't store up energy for short, big, bursts of power.

10.3
$$\text{Time} = 12 \text{ in.} \times \frac{1 \text{ s}}{0.85 \text{ in.}} = 14.1 \text{ s}$$

10.5
Type 885 meets the specifications.

Section 10.2

10.7
$$F = P \times A = 500 \frac{\text{lb}}{\text{in}^2} \times 12.57 \text{ in}^2 = 6283 \text{ lb}$$

10.9
$$F = 27,475 \text{ N}$$

10.11
$$\text{Diameter} = 3.0 \text{ in.}$$

10.13
$$F = PA = \frac{104 \text{ lb}}{\text{in}^2} = 12.56 \text{ in}^2 = 1306 \text{ lb}$$

10.15
See Figure 10.19.

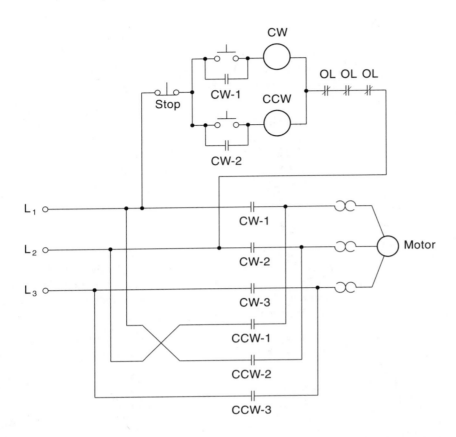

Section 10.3

10.17

$$F = PA = \frac{100\ lb}{in^2} \times 0.785\ in^2 = 78.5\ lb$$

10.19

Piston diameter = 0.92 in.

10.21

Piston force left over (after spring):
35.4 lb − 12 lb = 23.4 lb

Section 11.1

11.1
Rise time: 1.0
Overshoot: 0.7
Settling time: 6

Section 11.2

11.3
The level of liquid in a tank is to be maintained at 30 in. deep. A level detector at 29 in. turns on the pump. Another level detector at 31 in. turns off the pump.

11.5
A wind direction sensor mounted on the windmill indicates if the wind direction is more than say, 20° either side of straight on. If the wind is coming from the right (more than 20°), the motor comes on and rotates the windmill CW. If the wind is coming from the left (more than 20°), the motor comes on and rotates the windmill CCW (until it's within 20°).

Section 11.3

11.7

Torque when arm is 15° = 70 ft · lb

Torque when arm is 45° = 10 ft · lb

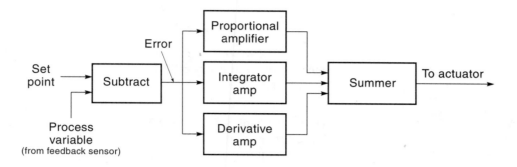

11.9

A simple proportional system can never reduce the position error to zero when lifting a weight. This is because the proportional system *only* provides a force when there *is* an error. Therefore, to provide the lifting force, there would have to be an error (the weight would sag an amount proportional to its weight).

11.11

> Error = ± 0.05 in.
>
> Total dead band = 2E = 0.1 in.

Section 11.4

11.13

Integral feedback will provide increasing restoring force *as long as there is any error*. Therefore, if a system has steady-state error, the restoring force will continue to increase until the error is eliminated.

11.15

> Time = 2 s

Section 11.5

11.17

Derivative feedback produces a restoring force which is proportional to the rate-of-change of the error, that is, if the error is changing rapidly, the derivative control provides a large force, and if the error is changing slowly, a small force.

11.19

Derivative feedback makes a control system more responsive to rapid change, because a quick change will have a large error rate-of-change, which produces a large, immediate restoring force. Derivative feedback reduces overshoot because it "applies the brakes" *before* reaching the set point.

This occurs because, as the controlled object slows, it has a negative error rate-of-change, which results in a negative restoring force (brakes).

Section 11.6

11.21

Block diagram of an analog PID controller.
(See figure above)

11.23

The digital controller *integrates* the error signal by evaluating a "difference equation."

$$K_I \Sigma (E \, \Delta t) = K_I E_1 T + K_I E_2 T + K_I E_3 T + \cdots$$

In other words, the controller accumulates the area under the error curve as rectangles, one new rectangle for each new sample.

11.25

BASIC program to implement a PID controller.

$$K_P = 2 \qquad K_I = 1.5 \qquad K_D = 1.8$$

Assume the following port adrs SP = 100, PV = 120, OUTPID = 140

```
OUTI = 0   ⎫
OLDDV = 0  ⎬ Initialize to 0
           ⎭
INP 100, SP      Read in set point
INP 120, PV      Read in feedback sensor
E = SP − PV      Calculate error
OUTP = 2 * E     Calculate output from proportional control
OUTI = OUTI + (1.5 * E * .5)  Calculate output from integral
                              control
NEWDV = 1.8 * E
OUTD = (NEWDV − OLDDV)/.5  ⎫ Calculate output from derivative
OLDDV = NEWDV             ⎬ control
                          ⎭
OUTPID = OUTP + OUTI + OUTD  Calculate total PID output
OUT 140, OUTPID             Send output to actuator
```

Program would wait ≈ .5 s with a time delay loop, then loop back

11.27

Lag $= 72°$

11.29

$$N = \frac{\Delta PV}{T} = \frac{6\%}{1.25\ s} = \frac{4.8\%}{s}$$

$$K_P = \frac{1.2\ \Delta CV}{NL} = \frac{1.2 \times 8\%\ s}{4.8\% \times 0.7\ s} = 2.9$$

$$K_I = \frac{1}{2L} = \frac{1}{2 \times 0.7\ s} = .7/s$$

$$K_D = 0.5\ L = 0.5 \times 0.7\ s = 0.35\ s$$

11.31

A high sample rate is good because it allows the controller to know more accurately what the controlled variable is doing, in particular, the controller can respond more quickly to any rapid changes in the system.

11.33

"Feedforward" is when the controller knows in advance the path it should take. This allows the controller to make better decisions about when to slow down and when to speed up. A PIP controller uses feedforward.

Section 11.7

11.35

Rule 2 (gas OK) applies 26%.
Rule 1 (turn gas up) applies 33%.
Turn gas up 2.53 notches.

11.37

For temperature $= 62°$: Cool ? 33%, Medium $= 26\%$
For $\Delta T = +.7°$/min: Steady ? 0%, Rising $= 60\%$
Weighted mean output $= -0.137$ notches

Section 12.1

12.1

(See figure below)

12.3

(See figure at top of next page)

Section 12.2

12.5

Installing a PLC
Hardware
1. Decide where the PLC will be located and install cables to I/O devices.
2. Connect input devices (switches, sensors, etc.) to input port connectors on PLC.
3. Connect output devices (motors, relays, etc.) to output port connectors on PLC.

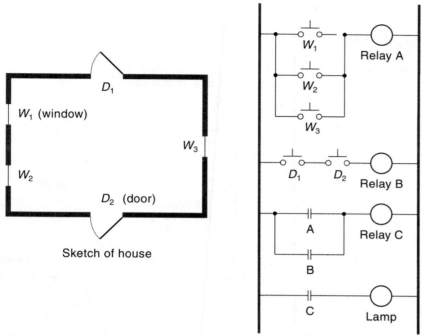

Exercise 12.3

PLC controls a conveyer belt

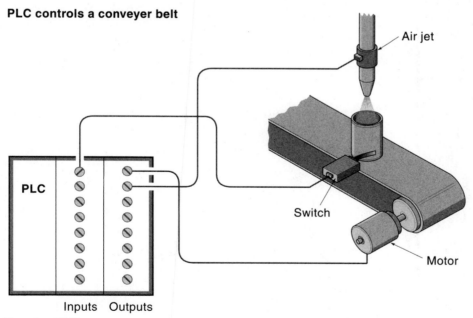

Exercise 12.7

Software
1. Develop a ladder diagram of the program.
2. Using software (from the PLC manufacturer), enter the program into a PC.
3. Connect the PC to the PLC and download the program into the PLC.
4. Test the operation of the program, make modifications as required.
5. When done, disconnect the PC and allow the PLC to operate on its own.

12.7

(See figure at bottom of previous page)

Section 12.3

12.9

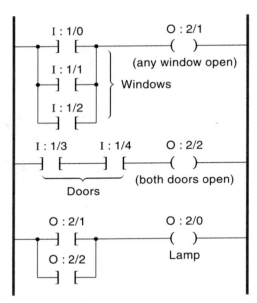

12.11

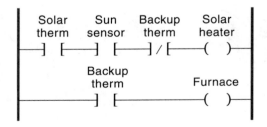

12.13

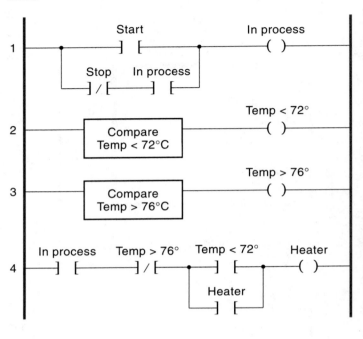

Absolute optical encoders, 183
Absolute pressure, 209
Acceleration, meaning of, 134–135
Accumulators, hydraulic, 348–349
Accumulators, microprocessor, 26
AC motor control, 323–330
  jogging, 325–326
  reduced-voltage starting, 326–328
  start-stop control, 323–325
  variable-speed control, 328–330
AC motors, 302–331
  advantages of, 233, 303
  fractional horsepower motors, 303
  induction motors, 308–319
  integral horsepower motors, 303
  single-phase AC motors, 304–305
  synchronous motors, 319–322
  three-phase AC motors, 306–308
  universal motors, 323
AC power, discovery of, 303–304
AC servomotors, operation of, 319
Active filters, 67–68
Actuators, 3, 223, 362
  control valves, 357–358
  electric linear actuators, 335–339
  function of, 334–335
  hydraulic actuators, 345
  pneumatic actuators, 355–357
Adaptive controllers, 396
Address, of microcomputer, 24
Address bus, of input/output (I/O) system, 24–25
Aftercooler, 352

Air compressors, 352
  reciprocating piston, 352
Alnico magnet, 243
Analog control systems, elements of, 9
Analog drive
  DC motor-control, 251, 252–255
Analog PID controllers, 382–383
Analog proportional controllers, 374–375
Analog switch circuit, 73–74
Analog-to-digital conversion, 29–32
Anode, 109, 120
Armature, 93
Armature winding, 234–236
Assembly language, 26
Asynchronous data, 34
Autotuning
  and adaptive controllers, 396
  digital controllers, 395–396
  meaning of, 395
Axial piston pump, 345

Backlash, 162
Ballast resistors, 295
Ball bearing screw, 336
Band-pass filter, 66, 67, 68
Base, of transistor, 98, 119
Baud rate, 34
Bellows, 209–210
Belts, 164–167
  creep and slip, 166–167
  operation of, 165–167
  types of, 165

Bias, control strategy
  and gravity problem, 374
  proportional control, 374
Biasing, in bipolar junction transistors, 100–102
Bidirectional control valve, 349
Bifilar winding, 287
Bilevel drive, 295
Bimetallic temperature sensors, 212
Bipolar, meaning of, 280
Bipolar junction transistors, 98–104
  biasing in, 100–102
  operation of, 99
  styles/mounting, 102–104
Bit instructions, programmable logic controllers, 429–431
Bode plot, 388–389
Bonded-wire strain gauges, 202–206
Bourdon tubes, 209
Brushes, motor, 234
Bus
  address bus of I/O system, 24–25
  of programmable logic controllers, 425
Byte, 23

Capacitor start motor, 316–317
Cathode, 109, 120
Central processing unit (CPU), 23–25
Chatter, 69
Chemical energy, 141–142
Chopper drive, 296
Circular pitch, 154
Class A operation, 101
Class B operation, 102

Class C operation, 102
Closed-loop control systems, 361–362
  elements of, 5–6, 361–362
Closed-loop gain, 56
Coil springs, 130
Cold junction, 212–213
Collector, of transistor, 98
  common collector configuration, 252
  open collector, 291
Common collector configuration, 252
Common emitter configuration, 252
Common return, 282
Commutators, motor, 234
Comparators, 5, 68–70
  chatter, 69
  function of, 68–69
  window comparator, 69–70
Compensating gauge, 202–203
Compilers, 37
Compound motors, 243
  cumulative compound motors, 243
  differential compound motors, 243
Compression springs, 130, 131
Computer-aided design (CAD), nature of,
  15
Computer-aided manufacturing (CAM),
  nature of, 15
Computer-integrated manufacturing
  (CIM), nature of, 16
Computers
  microprocessor system, 22–35
  personal computers, 41–42
  programmable logic controllers,
    40–41, 420–443
  single-board computers, 39–40
  single-chip microcomputers, 37–39
Constant-power region, 330
Constant-torque region, 330
Contactor, 95
Continuous-cycle method, PID tuning,
  390
Continuous-duty solenoid, 340
Continuous-level detectors, 225–227
Control bus, 25
Controlled variable, meaning of, 3,
  362
Controller programming, 35–37,
  383–387

elements of, 35–36
Controllers, 3
  function of, 362
  stepper motors, 287
Control strategy, simple/complex
  strategies, 5–6
Control systems
  analog, 9, 374–375, 381–383
  basic components of, 3
  classification of, 2
  closed-loop control systems, 5–6,
    361–363
  derivative control, 379–381
  digital control systems, 9–10, 35–37,
    383–386
  direct digital control, 11–13
  fuzzy logic control, 397–405
  history of, 2–3
  integral control, 376–379
  nature of, 1–3
  numerical control, 15–16
  open-loop control systems, 3–4, 361
  performance criteria, 363–364
  PID control, 381–385, 389–393
  PIP control, 396–397
  process control, 10–14, 362
  proportional control, 366–369
  robotics, 16
  sequentially controlled systems,
    13–14
  servomechanisms, 14–18, 362–363
  three position control, 365–366
  transfer functions, 6–9
  two-point control, 364–365
Control valves
  as actuators, 357–358
  hydraulic control valves, 349–351,
    357–358
  pneumatic control valves, 355
  pressure control valves, 347–348
Control winding, servomotor, 319
Conventional current flow, 98
Counter-EMF, 236–237
Counter instruction, programmable logic
  controller programming, 434–435
Counters, electromechanical, 419
Creep, and belts, 166–167
Critically damped, 148, 149

Cumulative compound motors, 243
Cumulative error, 272
Current loop, 47, 71–73
Current-to-voltage converter, 72–73
Cutoff frequency, 66

Damping, 147–149
  meaning of, 147
  types of damping, 148
Data bus, 25
Data communication equipment (DCE),
  35
Data memory, of programmable logic
  controller, 422
Data terminal equipment (DTE), 34
DC link converter, 328–329
DC motor-control, 251–264
  analog drive, 251, 252–255
  control circuits, 259–261
  for larger motors, 261–264
  pulse-width modulation, 251,
    256–261
  reversing rotation direction of motor,
    255–256
DC motors
  advantages of, 233
  brushless DC motor, 267–269
  compound motors, 243
  motor-application problem, 264–267
  permanent magnet motors, 243–247
  series-wound motors, 238–241
  shunt-wound motors, 240–243
  wound-field DC motors, 238–243
DC offset voltage, 53
Dead band, dead zone, 369–370
Dead time, and stability/instability of
  system, 387
Delta-connection, 306
Depletion mode, 106
Derivative control, 379–381
Desiccant dryer, 352
Detent torque, 277–278
Diac, 120–121
Diametral pitch, 154–155
Difference equation, 384
Differential amplifier, 60–62
Differential compound motors, 243
Differential pressure, 209

Differential voltage, 60
Differentiator circuit, 64–65
Digital control systems, 9–10, 35–37, 383–386
Digital PID controllers, 383–385
Digital-to-analog conversion, 27–29
Dimmer switch, 116
DIP switches, 91
Direct current tachometers, 193–196
Direct digital control, pros and cons of, 11–13
Discrete I/O, 422–423
Discrete-level detectors, 224–225
Double-acting cylinder, 345, 346
Double-pole/double-throw switch, 88
Downloading, of microprocessor program, 40
Drain, of FET, 104
Driver amplifiers, 287–288
Drive screw, 335
Drum controller, 420
Dryers, types of, 352
Dummy gauge, of strain gauge, 202–203
Duty cycle, 259
    meaning of, 256–257
Dv/dt effect, 112
Dynamic torque, 278

Earth ground, 77–78
Eddy currents, 340
Edison, Thomas, 303
Eight-step drives, 281
Electric-actuated valve, 357
Electrical energy, 141–142
    AC power, 305–308
Electric linear actuators, 335–339
    leadscrew linear actuators, 335–337
    solenoids, 337–339
Electric shield, noise from, 81
Electromechanical counters, 419
Electromechanical delay, 412
Electromechanical relays, 93–95
    operation of, 93–94
    types of, 95
Electromechanical sequencers, 419–420
Electromotive force, 236
    counter-EMF, 236–237

Electron flow, 98
Embedded controllers, 37
Emitter, of transistor, 98, 119
    common emitter configuration, 252
Energy, 141–146
    conversion factors, 142–144
    definition of, 142
    efficiency of conversion, 144
    forms of, 141–142
    heat transfer, 144–146
Enhancement mode, of FET, 106
Equipment protection, ground-fault interrupters for, 305
Error, and control system, 5, 362
    due to gravity, 371–373
    steady-state error, 364, 369–371
Event control system, functions of, 2
Event-driven systems, 14
Excitation, of synchronous motor, 320
Expansion cards, 41
Expansion slots, 41
Extension rate, of leadscrew actuator, 336
Extension springs, 130

Feedback, meaning of, 5
Feedback control
    analog controllers, 374–375, 382
    closed-loop system, 5, 361–362
    derivative control, 379–381
    digital controllers, 383–386
    fuzzy logic controllers, 397–406
    integral control, 376–379
    on–off controllers, 364–366
    proportional control, 366–375
    Proportional + Integral + Derivative (PID) Control, 381–396
    Proportional + Integral + Preview (PIP) Control, 396
Feedforward approach, 396
Ferrite magnet, 243
Fetch-execute cycles, 26
Field
    of AC motors, 308–310
    of DC motors, 234, 238, 241, 244
    shielding from electric field, 81–82
    shielding from magnetic field, 80–81
Field effect transistors, 104–109
    junction FET (JFET), 104–106

    metal oxide semiconductor FET (MOSFET), 106–109
Filters
    active filters, 67–68
    band-pass filter, 66, 67, 68
    high-pass filter, 66, 67
    low pass filter, 66, 67
    notch filter, 66, 67
Flat springs, 130, 131
Flex spline, 163
Flip-flop, using relay, 412
Flow-control valves, 349–351, 357–358
    bidirectional control valve, 349–351
    electric-actuated valve, 357
    pneumatic-actuated valve, 357, 358
    solenoid-actuated valve, 357
Flowmeters, turbine flow sensors, 222–223
Flow sensors, 220–223
    magnetic flowmeters, 223
    pressure-based flow sensors, 220–222
    turbine flow sensors, 222–223
Flyback diodes, 258
Follow-up system, functions of, 2
Forced commutation, 111
Forward-bias voltage, of transistor, 100
Forward breakover voltage, 110
Forward conduction region, 110
Forward current gain, of transistor, 99
Forward path, 5
Forward-reverse switching, 255–256
Four-phase stepper motors
    control of, 290–293
    operation of, 281–282
Four-pole AC motor, 310
Four-step drives, 280–281
Fractional horsepower motors, 303
Frame, AC motor, 308
Free-running motor, 250
Free-wheeling diodes, 258
Friction, 127–129
    of different materials, 128
    and normal force, 128
    overcoming force of, 369–370
    sliding friction, 127
    steady-state error from, 369–371
    viscous friction, 128
Full-wave phase-control circuit, 112

Fuzzy logic
  development of, 399–400
  principles of, 397–398
Fuzzy logic controllers, 363, 397–406
  basis of design, 398–399
  one-input system, 400–402
  two-input system, 402–405
Fuzzy predicates, 399
Fuzzy sets, 400–401

Gain, meaning of, 7, 366
Gain-bandwidth product, 104
Gain margin, 389
Gas-pressurized accumulator, 349
Gate, of FET, 104, 109, 115, 120
Gauge factor, 203–204
Gauge pressure, 209
Gear motor, 345, 346
Gear pass, 154
Gear pump, 343–344
Gears, 153–164
  harmonic drive, 163–164
  long gear trains, 162
  ratio, 156
  spur gears, 154–155
  to transfer power, 160–162
  used to change speed, 155–160
  worm gears, 163
Gear train, 159–160
Gravity problem
  and bias, 374
  proportional control, 371–373
Grey code, 184
Ground, meaning of, 77
  single-point ground, 82
  virtual ground, 55
Ground-fault interrupters, 305–306
  for equipment protection, 305
  for life protection, 305
Ground loops, 77–78

Half-effect proximity sensors, 198–201
Half-steps, 280–281
Half-wave phase-control circuit, 112
Hall effect, meaning of, 199–200
Hall-effect proximity sensors, 198–201, 222–223
Hall, E. H., 198

Harmonic drive, 163–164
Head, pressure, 226
Heat sink, 103, 144
Heat transfer, process of, 144–146
High-level languages, 36–37
High-pass filter, 66, 67
Holding current, 110, 262
Hooke's law, and springs, 129–130, 201
Hot junction, 212
Hybrid solid-state relay, 98
Hybrid stepper motors, 286–287
  features of, 286
  operation of, 287
Hydraulic actuators, 345
Hydraulic cylinder, 345
Hydraulic pumps, 343–345
  axial piston pump, 345
  gear pump, 343–344
  positive-displacement pump, 344
  vane pump, 344
  variable-displacement pump, 344
Hydraulic systems, 340–351
  accumulators, 348–349
  air compressor, 352
  dryers, 352
  flow-control valves, 349–351
  functions of, 340
  hydraulic actuators, 345
  hydraulic pumps, 343–345
  operation of, 341–343, 351–352
  compared to pneumatic systems, 351, 352
  pressure-control valves, 347–348
  pressure tank, 352–353
Hydrostatic pressure
  meaning of, 341
  principle of, 341–343

Ice-water bath, 213
Impact pressure, 222
Incremental optical encoders, 184–186
  interfacing to computer, 188–189
Induction motors, 308–319
  AC servomotors, 319
  classification of, 313
  components of, 308
  operation of, 308–313
  single-phase motors, 314–317

split-phase control motors, 318
  three-phase motors, 314
Inertia, 133–134
  moment of inertia, 133, 134, 250
  nature of, 133–134
Input/output (I/O) system, 24–25
  analog I/O, 423–424
  components of, 24–25
  discrete I/O, 422–423
  of programmable logic controllers, 422–425
Instruction set
  of microcontrollers, 37–38
  of microprocessor system, 25
Instrumentation amplifier, 62–64
Instrument gears, 155
Integral control, 376–379
Integral horsepower motors, 303
Integrated circuit, temperature sensors, 217, 219
Integrator circuit, operational amplifiers, 64
Interface circuits
  analog switch circuit, 73–74
  current loop, 71–73
  instrumentation amps, 62–64
  op-amps, 53–60
  sample-and-hold circuit, 75–76
Interfacing
  interface problems, 46–47
  meaning of, 46
Intermittent-duty solenoid, 340
Inverting amplifier, 54–56
Inverting input, of op-amp, 48
Isolation circuits, 78–80
Isothermal block, 213
Iteration, 9, 36, 385

Jogging
  AC motor control, 325–326
  meaning of, 325–326
Junction FET (JFET), 104–106

Kelvin scale, 219

Ladder diagrams, 94, 323–324, 414–416

components of, 413–416
functions of, 413
programmable logic controllers,
428–429
relay control logic, 413–416, 429
Lag time, 387–388
Latch, 188
Latching, meaning of, 412
Latching the relay, 325
meaning of, 412
Leadscrew linear actuators, 335–337
Leaf springs, 130, 131
Least significant bit, 23, 179
Life protection, ground-fault interrupters
for, 305
Limit switches, 90, 196–197
Linear actuators, electric linear
actuators, 335–339
Linearity error, potentiometers, 177–178
Linear systems, equations of motion,
134–138
Linear variable differential transformer,
190–191
Line voltage, 307
Liquid-level sensors, 224–227
continuous-level detectors, 225–227
discrete-level detectors, 224–225
Load cells, 206
Loading error, potentiometers, 175–176
Load sensors, 201–208
bonded-wire strain gauges, 202–206
low-force sensors, 207–208
semiconductor force sensors, 206
Load torque, 250
Lockup property, 163
Logic variables, 27
Long gear trains, 162
Low-force sensors, 207–208
Low pass filter, 66, 67
L/R drives, 295
LSB. *See* least significant bit

Magnetic field shielding, 81
Magnetic flowmeters, 223
Magnets, of permanent magnet motors,
243
Main winding, of AC servomotor, 319
Mass, 132–133

nature of, 132
Mechanical energy, 141–142
Mechanical resonance, natural resonant
frequency, 149–151
Mechanical systems
belts, 164–167
damping, 147–149
energy, 141–146
equations of motion, linear systems,
134–138
equations of motion, for rotational
systems, 138–141
friction, 127–129
gears, 153–164
inertia, 133–134
mass, 131–133
mechanical resonance, 149–153
operation of mechanical system,
147–153
roller chain, 168
springs, 129–131
Mechanical torque, 250
Membership function, of fuzzy logic,
400
Memory of computer
nonvolatile memory, 24
random-access memory (RAM), 24
read-only memory (ROM), 24
volatile memory, 24
Memory-mapped input/output, 25
Metal oxide semiconductor FET
(MOSFET), 106–109
characteristics of, 107
power MOSFET for analog drive, 254
Mho, 107
Microcontrollers, 37–38
operation of, 37–38
types of, 38
Microprocessor system, 22–35
central processing unit (CPU), 23–25
components of, 25–26
control bus, 25
controller programming, 35–36
data bus, 25
design in control systems, 22
input/output (I/O) system, 24–25
instruction set, 25
operation of, 25–27

parallel interface, 27–33
personal computers, 41–42
programmable logic controllers, 40–41
serial interface, 34–35
single-board computers, 39–40
single-chip microcomputers, 37–39
steps in program execution, 26–27
Microstepping, 281
stepper motor control, 293
Microswitches, 90
Minimum holding voltage, 94
Mnemonic, 26, 442
Modem
baud rate, 34
interface with PC, 34–35
Modicon, Gould, 411
Modulus of elasticity, 204
Momentary-contact switches, 88–89
Moment of inertia, 133, 134, 250
Momentum, nature of, 132
Most significant bit (MSB), 23
Motherboard, 41
Motion
acceleration, 134–135
equations for linear systems,
134–135
equations for rotational systems,
138–141
Motors
AC motors, 233, 302–331
brushless DC motor, 267–269
classification of, 233
DC motor-control, 251–264
motor-application problem, 264–267
theory of operation, 234–238
permanent magnet (PM) motors,
243–251
speed regulation, 241–242
stepper motors, 272–299
wound-field DC motors, 238–243
Motor springs, 130, 131
MSB. *See* most significant bit
Multiplexing, analog multiplexing, 74

National Electrical Manufacturers
Association, classification of
induction motors, 313
Natural resonant frequency, 149–151

Negative feedback
  causes of, 387–388
  operational amplifiers, 50
Neutral, wire, 305
Newtons, meaning of, 132
Noise
  from electric field, 81
  magnetic field noise, 81
  and optical couplers, 79–80
  and shielding, 80–83
No-load speed, 240
Noninverting amplifier, 56–58
Noninverting input, of op-amp, 48
Nonvolatile memory, 24
Normal force, and friction, 128
Normally closed contacts, 412
Normally closed switches, 90
Normally open contacts, 412
Normally open switches, 90
Notch filter, 66, 67
NPN transistors, 99
Null modem, 35
Numerical control, elements of, 15–16

Off-delay relay, 417–418
Offset null, 53
On-delay relay, 417
On–off controllers, 364–366
  function of, 364
  three-position control, 365–366
  two-point control, 364–365
Op-code, 26
Open collector, 291
Open-loop control systems, 361
  elements of, 3–4, 361
Open-loop gain, 48
Operational amplifiers
  active filters, 67–68
  characteristics of, 47
  comparators, 68–70
  components of, 47–48
  differential amplifiers, 60–62
  differentiator circuit, 64–65
  instrumentation amplifier, 62–64
  integrator circuit, 64
  inverting amplifier, 54–56
  inverting/noninverting input, 48
  negative feedback, 50

noninverting amplifier, 56–58
summing amplifier, 58–60
types of, 51
voltage follower, 53–54
Operation code, 26
Optical couplers, 79–80
Optical proximity sensors, 197–198
  photodetectors used, 197–198
  slotted coupler, 198
Optical rotary encoders, 182–189
  absolute optical encoders, 183
  decoding operation in, 186–187
  incremental optical encoders, 184–186
  interfacing to computer, 188–189
Optical tachometers, 193
Optointerrupter, 198
Orifice plate, 220–221
O rings, 355
Oscillating instability, 386–387
  causes of, 386–387
Overdamped, 148, 149
Overshoot, 148, 149, 364

Parallel interface, 27–33
  analog-to-digital conversion, 29–32
  digital-to-analog conversion, 27–29
  parallel ports in control system,
    32–33
Parallel ports, in control system, 32–33
Performance criteria
  of control systems, 363–364
  steady-state parameters, 363–364
  transient response, 363–364
Permanent magnet (PM) motors,
    243–251
  circuit model of, 247–251
  reversing of, 255–256
  torque-speed curve, 244–246
  types of magnets used, 243
Permanent magnet (PM) stepper motors,
    273–283
  benefits of, 273–274
  four-phase stepper motors, 281–282
  load, effect on, 274–276
  single-step mode, 276
  slew mode, 276–277
  torque-speed curves, 277–278
  two-phase stepper motors, 278–281

Permanent-split capacitor motor, 317
Personal computers, 41–42
  components of, 41–42
Phase, of stepper motor, 279
Phase-control circuit, 112
Phase lag, 387–388
Phase margin, 389
Phase sequence, 306
Phase voltages, 306
Photodetectors
  photodiode, 197
  photoresistor, 197
  photo transistor, 197
  photovoltaic cell, 198
Photodiode, 197
Photoresistor, 197
Photo transistor, 197
Photovoltaic cell, 198
Pick-and-place robots, 16–17
PID controllers. *See* Proportional +
    Integral + Derivative (PID)
    controllers
Piezoresistive effect, of silicon, 206
Pinion, 157
PIP controllers. *See* Proportional +
    Integral + Preview (PIP) controllers
Pitch circle, 154
Pitch diameter, 154
Pitot tube, 222
Plug-in modules, 422
Pneumatic time-delay relay, 418
Pneumatic actuators, 355–357
Pneumatic control valves, 355
Pneumatic systems, 351–357
  compared to hydraulic systems, 351,
    352
  pneumatic actuators, 355–357
  pneumatic control valves, 355
  pressure regulators, 353–354
PNP transistors, 99
Pneumatic-actuated valve, 357, 358
Poles, of toggle switches, 88, 89
Poles per phase, 310
Port, of input/output (I/O) system, 24
Position sensors, 174–191
  linear variable differential transformer,
    190–191
  optical rotary encoders, 182–189

potentiometers, 174–182
velocity from, 191–192
Positive-displacement pump, 344
Positive feedback
causes of, 388
conditions for, 388
Potentiometers, 174–182
linearity error, 177–178
loading error, 175–176
operation of, 174–175
wire-wound potentiometer, 179
Power, AC, 303–308
single-phase, 304–305
three-phase, 306–308
Power, mechanical
definition of, 142
shaft-to-shaft transmission, 164–168
transfer using gears, 160–162
Power-factor correction, synchronous
motors, 321
Power factor, meaning of, 321
Power rails, of ladder diagram, 414–417
Power transistors
bipolar junction transistors, 98–104
field effect transistors, 104–109
uses of, 98
Pressure
absolute pressure, 209
differential pressure, 209
gauge pressure, 209
impact pressure, 222
static pressure, 222
Pressure-based flow sensors, 220–222
Pressure-control valves, hydraulic
systems, 347–348
Pressure regulators, 353–354
Pressure sensors, 208–210
bellows, 209–210
Bourdon tubes, 209
semiconductor pressure sensors, 210
Pressure tank, 352–353
Prime mover. See Actuators
Process, meaning of, 3
Process control, 10–14
elements of, 362–363
Processor, of programmable logic
controller, 421
Process variable, 366

Program counter, 26
Programmable gain instrumentation
amplifier, 63–64
Programmable logic controller
programming
advanced instructions, 438–439
ASCII programming, 442
bit instructions, 429–431
Boolean logic programming, 442–443
comparison instructions, 439
control instructions, 439
counter instruction, 434–435
I/O message/communication
instructions, 439
ladder diagram programming,
428–429
logical and shift instructions, 439
math instructions, 439
PID control, 439
sequence instruction, 435–438
timer instruction, 431–434
Programmable logic controllers, 14,
40–41, 420–443
components of, 40
functions of, 420
hardware for, 420–425
networking of, 40
operation of, 425–428
setup procedure, 425
as two-point controller, 439–442
Programmable unijunction transistors,
120
Program memory, of programmable logic
controller, 422
Programming languages, high-level,
36–37
Programming port, of programmable
logic controller, 422
Proportional band, 374–375
Proportional control, 366–375
analog proportional controllers,
374–375
bias, 374
elements of, 366–369
gravity problem, 371–373
steady-state error problem, 369–371
Proportional + Integral + Derivative
(PID) controllers, 381–396

analog PID controllers, 381–383
autotuning, 395–396
digital PID controllers, 383–385
flowchart, 386
PID equation, 383–385
sampling rate, 394–395
tuning of controller, 389–393
Proportional + Integral + Preview (PIP)
controllers, 396
equation for, 396
Proximity sensors, 196–201
half-effect proximity sensors,
198–201
Hall-effect proximity sensors,
198–201
limit switches, 196–197
optical proximity sensors, 197–198
Pull-in current, 94
Pull-in voltage, 94
Pull-out torque, 313
Pulse sequence generator, 287
Pulse-width modulation
advantages over analog drive,
257–258
chopper drive, 295
control circuits, 259–261
DC motor-control, 251, 256–261
produced with microcontroller,
260–261
Push-button switches, 90

Random-access memory (RAM), 24
Rankine scale, 219
Rare-earth magnet, 243
Rated speed, of motor, 238
Reaction-curve method, PID tuning,
391–393
Read-only memory (ROM), 24
Read/write (R/W) line, 25
Real time computing, 35
Reciprocating piston pump, 352
Reduced-voltage starting circuits, 264
AC motor control, 326–328
AC motors, 326–328
Reed relay, 95
Reference junction, 212
Regulator system, functions of, 2
Relay contacts, configurations for, 412

Relay control logic
  electromechanical counters, 419
  electromechanical delay, 412
  electromechanical sequencers, 419–420
  ladder diagrams, 413–416
  relay logic, elements of, 412
  time-delay relay, 417–418
Relays, 93–98, 412–417
  electromechanical relays, 93–95
  solid-state relays, 95–98
Resistance temperature detectors (RTD),
    215–216
Resolution
  in digital systems, 179
  meaning of, 179
  nature of, 28–29
Resonance, mechanical, 149–153
Rise time, 364
Robotics
  elements of, 16
  integral control in, 377–379
  pick-and-place robots, 16–17
Roller chain, 168
Rotary switches, 91
Rotating field, 309
Rotational systems
  equations of motion, 138–141
  torque, 138–139
Rotors, 267, 272, 308
RS-232 standard, 34
RTD. *See* resistance temperature detector
Run windings, 315

Safety ground, 305
Sample-and-hold circuit, 75–76
Sampling rate, 36
  digital controllers, 394–395
  meaning of, 394
Scan, 9, 36, 385
Sealed current, 94
Sealing, to make flip-flop from relay, 412
Sealing the relay, 325
Semiconductor force sensors, 206
Semiconductor pressure sensors, 210
Sensors
  flow sensors, 220–223
  function of, 173
  functions of, 5

liquid-level sensors, 224–227
load sensors, 201–208
position sensors, 174–191
pressure sensors, 208–210
proximity sensors, 196–201
temperature sensors, 210–220
velocity sensors, 191–196
Sequence instruction, programmable
    logic controller programming,
    434–438
Sequencers, electromechanical
    sequencers, 419–420
Sequentially controlled systems,
    elements of, 13–14
Serial interface
  conversion for data, 34
  and modem, 34–35
  RS-232 standard, 34–35
  UART, 34
Series-wound motor, torque-speed curve
    for, 239–241
Servomechanisms, 14–18
  controller for, 362–363
Set point, meaning of, 3, 362
Settling time, 364
Shaded-pole motors, 317
Shaded poles, 94
Shielding, 80–83
  shield-ground factors, 82–83
  to single-point ground, 82–83
Shunt-wound motor, 240
Signal common, 77
Signal conditioning
  comparators, 68–70
  filters, 66–68
  interface circuits for, 71–76
  voltage follower, 53–54
Signal return, 77
Signal transmission
  attenuation prevention using voltage
    follower, 54
  current loop, 71–73
  earth ground, 77–78
  ground loops, 77–78
  isolation circuits, 78–80
  shielding, 80–83
  single-point ground, 82
Silicon, piezoresistive effect of, 206

Silicon-controlled rectifier, 109–115
  motor control with, 262–263
  operation of, 110–11
  terminals of, 109
  turning on/off, 111–113
Single-acting cylinder, 355
Single-board computers, 39–40
  components of, 40–41
Single-chip microcomputers, 37–39
Single-ended voltage, 60
Single-phase AC motors, operation of,
    304–305
Single-phase induction motors
  operation of, 314–317
  types of, 317
Single-point ground, 82
Single-pole/double-throw switch, 88
Single-pole/single-throw switch, 88
Single-step mode, stepper motors,
    276
Six-pole motor, 310
Slew mode, stepper motors, 276–277
Slider, 174
Slide switches, 89
Sliding friction, 127
Slip
  and belts, 166–167
  AC motors, 312
Slot period, 191, 192
Slotted coupler, 198
Snubber, 113
Software
  to control stepper motor, 296–299
  to implement PID, 383–386
  to program a PLC, 428–443
Solenoid-actuated valve, 357
Solenoids, 338–339
  continuous-duty solenoid, 339
  DC and AC solenoids, 339
  intermittent-duty solenoid, 339
  limitations of, 338
  types of, 340
Solid-state relays, 95–98
  operation of, 95–97
  types of, 98
Solid-state time-delay relay, 418
Source of FET, 104
Speed, changing with gears, 155–160

Speed regulation
  induction motors, 313
  motors, 240–242
Split-phase control motors, operation of, 315, 318
Spool, 349
Springs, 129–131
  and Hooke's law, 129–130
  types of, 130
Spur gears, 154–155
Squirrel cage rotor, 311
Stability/instability of system, 386–389
  Bode plot to determine stability, 388–389
  and dead time, 387
  gain margin, 389
  and negative feedback, 387–388
  oscillating instability, 385, 387
  phase margin, 389
  positive feedback, 388
  stable system, meaning of, 385
  unstable system, meaning of, 385
Stalling, meaning of, 274
Stall torque, 240
Starting torque, 313
Start-stop control, AC motor control, 323–325
Start windings, 315
Static pressure, 222
Stators, 272, 308
Steady-state error, 364, 369–371
  from friction, 369–371
  gravity problem, 371
Step change, 363
Stepper motor control, 287–296
  four-phase stepper motor, 290–293
  to improve torque, 294–296
  microstepping, 293
  with software, 296–299
  two-phase stepper motor, 289–290
Stepper motors, 272–299
  bipolar, 278, 281
  detent torque, 273–277
  effect of load on, 274–276
  elements of, 272–273
  half steps, 280–281
  hybrid stepper motors, 286–287

permanent magnet stepper motors, 273–283
single step mode, 276
slew mode, 276
unipolar, 281–282
variable reluctance stepper motors, 283–286
Strain, meaning of, 204
Stress, meaning of, 203
Stroke, 338
Summing amplifier, 58–60
Swash plate, 345
Switches
  DIP switches, 91
  limit switches, 90
  membrane switches, 92–93
  microswitches, 90
  push-button switches, 90
  rotary switches, 91
  slide switches, 89
  thumbwheel switches, 92–93
  toggle switches, 88–89
Switch wafers, 91
Synchronous data, 34
Synchronous motors, 319–322
  features of, 319–320
  operation of, 320–321
  power-factor correction, 321
  small motors, 321–322
Synchronous speed, 310

Tachometers, 193–196
  direct current tachometers, 193–196
  optical tachometers, 193
Temperature sensors, 210–220
  bimetallic temperature sensors, 212
  integrated circuit temperature sensors, 217, 219
  resistance temperature detectors (RTD), 215–216
  thermistors, 216–217
  thermocouples, 212–215
Tesla, Nikola, 304
Thermal energy, 141–142
Thermal time-delay relay, 418
Thermistors, 216–217
Thermocouples, 212–215

Three-phase AC motors, operation of, 306–308, 314
Three-phase stepper motor
  control of, 285
  operation of, 285–286
  compared to PM type, 285–286
Three-position control, 365–366
Three-position switch, 88
Threshold detector, 200
Thumbwheel switches, 92–93
Thyristors, silicon-controlled rectifier, 109–115
Time-delay loop, 36
Time-delay relay, 417–418
  off-delay relay, 417–418
  on-delay relay, 417
  pneumatic time-delay relay, 418
  solid-state time-delay relay, 418
  thermal time-delay relay, 418
Time-driven systems, 14
Timer instruction, programmable logic controllers, 431–434
Toggle switches, 88–89
  types of, 88
Torque, 138–139
  constant-torque region, 330
  detent torque, 277–278
  dynamic torque, 278
  equation for, 138–139
  load torque, 250
  meaning of, 235
  mechanical torque, 250
  of motor, 235–236
  and power transfer, 160
  pull-out torque, 313
  stall torque, 240
  starting torque, 313
  static torque, 276
  stepper motors, improving at higher stepping rates, 294–296
  torque/speed relationship, 244–246
Torque-speed curve
  for induction motor, 313
  permanent magnet motors, 244–246
  permanent magnet (PM) stepper motors, 277–278
  for series-wound motor, 239–241
  for shunt-wound motor, 241

Torsion springs, 130, 131
Transconductance, 106–107
Transducers, 173
  See also Sensors
Transfer functions, 6–9
  equation for, 6
Transient response, 363–364
Transistors
  bipolar junction transistors, 98–104
  common collector configuration, 252
  common emitter configuration, 252
  field effect transistors, 104–109
  NPN and PNP transistors, 99
  open collector configuration, 291
  See also Bipolar junction transistors
Transmission gate, 73
Triacs, 115–119
  conduction periods, 118–119
  delay, calculation of, 118–119
  operation of, 115–117
  trigger conditions, 115, 118
Trigger circuits, 119
Trigger devices
  diac, 120–121
  programmable unijunction transistors,
    120
  unijunction transistors, 119
Tuning
  continuous-cycle method, 390
  of PID controller, 389–393
  reaction-curve method, 391–393

Turbine flow sensors, 222–223
Two-capacitor motor, 317
Two-phase stepper motors, control of,
  279–281, 289–290
Two-point control, 364–365
Two-point controller, programmable
  logic controller as, 439–442
Two-position switch, 88

UART. *See* Universal asynchronous
  receiver transmitter
Underdamped, 148
Unijunction transistors, 119
Unipolar, meaning of, 281
Universal asynchronous receiver
  transmitter (UART), 34
Universal motors, 323
  operation of, 323

Vane motors, 357
Vane pump, 344
Variable-displacement pump, 344
Variable reluctance stepper motors,
  283–286
  three-phase stepper motor, 285
Variable-speed control, AC motors,
  328–330
Velocity
  meaning of, 191
  from position sensors, 191–192
Velocity sensors, 191–196

tachometers, 193–196
Venturi, 222
Virtual ground, 55
Viscous friction, 128
Volatile memory, 24
Voltage-current conversion, 71–73
Voltage follower, 53–54

Wave drive, 281
Wave generator, 163
Weight, meaning of, 132
Weighted accumulator, 349
Westinghouse, George, 304
Windage, 250
Windings, of single-phase motor, 315
Window comparator, 69–70
Windup, problem of, 382
Wiper, 88, 174
Wire-wound potentiometer, 179
Words, of binary data, 23
Worm gears, 163
Wound-field DC motors, 238–243
  compound motors, 243
  series-wound motors, 238–241
  shunt-wound motors, 241–243
Wye-connection, 306–308

Young's modulus, meaning of, 204

Zadeh, L. A., 399
Zero-voltage switching, 97